THE GLOBAL ECONOMICS OF FORESTRY

This book traces the economic and biological pattern of forest development from initial settlement and harvest activity at the natural forest frontier to modern industrial forest plantations. It builds from diagrams describing three discrete stages of forest development, and then discusses the management and policy implications associated with each, supporting its observations with examples and data from six continents and from both developed and developing countries. It shows that characteristic distinctions between the three stages make forestry unusual in natural resource management and that effective policy requires different, even contrasting, decisions at each stage.

Hyde's comprehensive discussion covers a wide range of issues, including the impacts of both specific forest policies, such as forest taxes and incentive payments, and broader macroeconomic policies, the unique requirements of current issues such as global warming, biodiversity and tourism, and the complexities of the different forest products industries. Concluding chapters review the roles of the newer institutional landowners, of smaller private and farm landowners, and of public agencies. This highly-original volume reaches far beyond forest economics; it explains what forestry can do for regional development and environmental conservation and what policies designed for other sectors and the macro-economy can do for forestry.

William F. Hyde has been a Senior Associate at Resources for the Future and at the Center for International Forestry Research in Indonesia, a professor at Duke University and Virginia Tech, and a visiting professor at Gothenburg and Beijing Forestry Universities. In a career that spans thirty-five years he has been a consultant to more than two dozen organizations in more than thirty countries and has authored or co-authored eight books and numerous other publications.

"This unique book covers the fundamental economic questions of forestry, but does not limit itself to a particular region or (time) frame. It comprehensively complements the contributions the author has made to our understanding of the economics of forestry with the broader research in the field. All of the topics are covered with sufficient clarity to provide a unified perspective on the fundamental questions facing forestry in the future."

—*David Newman, SUNY College of Environmental Science and Forestry*

"Relevant worldwide for many important policy decisions, including pressing climate issues, and equally helpful for training students across disciplines and levels of experience, Hyde's stages framework is clear and easy to grasp. Whether you prefer words, diagrams or equations, a tour of points and places is on offer from somebody who really has been there."

—*Alexander Pfaff, Duke University*

"Lives up to its title ... will be a classic."

—*Runsheng Yin, Michigan State University*

THE GLOBAL ECONOMICS OF FORESTRY

William F. Hyde

First published 2012
by RFF Press
Routledge, 711 Third Avenue, New York, NY 10017

Simultaneously published in the UK
by RFF Press
Routledge, 2 Park Square, Milton Park, Abingdon, Oxon OX14 4RN

RFF Press is an imprint of the Taylor & Francis Group, an informa business

© 2012 Taylor & Francis

The right of William F. Hyde to be identified as author of this work has been asserted by him in accordance with sections 77 and 78 of the Copyright, Designs and Patents Act 1988.

All rights reserved. No part of this book may be reprinted or reproduced or utilised in any form or by any electronic, mechanical, or other means, now known or hereafter invented, including photocopying and recording, or in any information storage or retrieval system, without permission in writing from the publishers.

Trademark Notice: Product or corporate names may be trademarks or registered trademarks, and are used only for identification and explanation without intent to infringe.

Library of Congress Cataloging in Publication Data
Hyde, William F.
  The global economics of forestry / William F. Hyde.
  p. cm.
  Includes bibliographical references and index.
  1. Forests and forestry—Economic aspects. 2. Forest management. I. Title.
  SD393.H94 2012
  577.3—dc23
  2011041692

ISBN13: 978-0-415-51828-4 (hbk)
ISBN13: 978-0-203-12269-3 (ebk)

The findings, interpretations, and conclusions offered in RFF Press publications are those of the authors. They do not necessarily represent the views of Resources for the Future, its directors, or its officers. Similarly, any geographic boundaries and titles depicted in RFF Press publications do not imply any judgment or opinion about the legal status of a territory on the part of Resources for the Future.

Typeset in Bembo and Stone Sans
by EvS Communication Networx, Inc.

Printed and bound in Great Britain by CPI Group (UK) Ltd, Croydon, CR0 4YY

ABOUT RESOURCES FOR THE FUTURE ***AND*** RFF PRESS

Resources for the Future (RFF) improves environmental and natural resource policymaking worldwide through independent social science research of the highest caliber. Founded in 1952, RFF pioneered the application of economics as a tool for developing more effective policy about the use and conservation of natural resources. Its scholars continue to employ social science methods to analyze critical issues concerning pollution control, energy policy, land and water use, hazardous waste, climate change, biodiversity, and the environmental challenges of developing countries.

RFF Press supports the mission of RFF by publishing book-length works that present a broad range of approaches to the study of natural resources and the environment. Its authors and editors include RFF staff, researchers from the larger academic and policy communities, and journalists. Audiences for publications by RFF Press include all of the participants in the policymaking process—scholars, the media, advocacy groups, NGOs, professionals in business and government, and the public.

Resources for the Future

Board of Directors

Board Leadership
W. Bowman Cutter, *Chair*
John M. Deutch, *Vice Chair*
Frank E. Loy, *Vice Chair*
Lawrence A. Hinden, *Treasurer*
Philip R. Sharp, *President*

Board Members
Vicky A. Bailey
Anthony Bernhardt
Trudy Ann Cameron
Red Cavaney
Preston Chiaro
Mohamed T. El-Ashry
Linda J. Fisher
Deborah S. Hechinger
Peter Kagan
Sally Katzen
Rubén Kraiem
Richard G. Newell
Richard Schmalensee
Robert N. Stavins
Lisa A. Stewart
Joseph Stiglitz
Mark R. Tercek

Chair Emeriti
Darius W. Gaskins, Jr.
Robert E. Grady

Editorial Advisers for RFF Press

Walter A. Rosenbaum, *University of Florida*
Jeffrey K. Stine, *Smithsonian Institution*

CONTENTS

List of Figures ix
List of Tables xi
Foreword xiii
Preface xv

1 Introduction 1
   *Appendix 1A: Fundamental Questions within the General Subject
      Matter of Forest Economics, Policy, and Management   9*

2 The Pattern of Forestry in the Course of Economic Development 13
   Appendix 2A: Forestry Data   53

3 Forest Development in the Long Run 64
   Appendix 3A: The Faustmann Model   93
   Appendix 3B: Allowable Cut   100

4 Forest Policy 108
   Appendix 4A: Contemporary Policy Objectives   152

5 Forest Concessions: A Specialized Topic in Forest Policy and
   Management 173
   Appendix 5A: Contracts, Rents, and Royalties   194

6 The Effects of Trade, Macroeconomics, Growth and Development 201
   Appendix 6A: Forestry in the National Accounts   251

7 Industrial Forestry 263
   Appendix 7A: A Note on Factor Costs and Industry Location   304

8	Institutional Investors	307
	Appendix 8A: CAPM 316	
9	Non-Industrial Private Landowners	318
	Appendix 9A: Myths and Fallacies in the Tradition of Non-Industrial Private Forestry 344	
10	Public Landowners	350
	Appendix 10A: Least Cost Plus Loss 385	
11	Forests and Local Human Communities	386
	Appendix 11A: Nonseparable Household Production and Consumption 425	
	Appendix 11B: The Generalized Economy of a Forest-based Community 429	
12	Summary, Conclusions, Policy Implications	432

Index   *463*

LIST OF FIGURES

2.1	A new forest frontier	16
2.1A	A new forest frontier continued, stage I	16
2.2	Shifting cultivation	19
2.3	A developing frontier, stage II	21
2.4	Labor and capital opportunity costs, deforestation and forest degradation	24
2.5	A mature forest frontier, stage III	31
2.6	Multiple market values: Brazil nuts, mahogany, and lower-valued timber	42
2.7	Market-valued opportunity costs	44
2.8	Property rights	46
2.9	An aerial view, with roads and environmental barriers	48
3.1A	Agricultural technologies on new frontiers	74
3.1B	Agricultural technologies on developing frontiers	75
3.1C-I	Land-saving agricultural technologies on mature frontiers	76
3.1C-II	Neutral and land-using agricultural technologies on mature frontiers	77
3.2A	Improved rural infrastructure and logging technologies on new frontiers	79
3.2B	Improved infrastructure, logging, and forest utilization on developing frontiers	81
3.2C	Improved infrastructure, logging, and forest utilization on mature frontiers	83
4.1	Income taxes	112
4.2	Forest incentive programs	117
4.3	Agricultural policies	133

4.4	Policies targeting the wood processing industries	135
4.5A	Improved roads: Direct effects	139
4.5B	Improved roads: Indirect effect	139
4.6	Improved property rights	142
4A.1	Sustainable forestry and the control of deforestation	166
4A.2	Sustainability and deforestation: the effects of development	168
5.1	Contracting for mature timber	175
5.2	Rent under a uniform royalty	179
6.1	The land use impacts of trade between two regions in the mature stage of forest development	206
6.2	Trade and land use—the remaining cases	208
6.3	The environmental Kuznets curve	240
8.1	Asset performance: Risk vs. average return	311
9.1	The three-stage model, repeated	321
9.2	Multiple motives	324
10.1	Recreation demand and supply	366
10.2	The wildfire optimization problem	372
11.1A	Relative importance of bamboo income in Anji County—by household income class, 1989–1990	393
11.1B	Relative importance of bamboo income in Anji County—by household income class, 1994–1995	393
11.2	The relationship between integration into the cash economy (%) and NWFP contribution to household income (%) for 61 communities	409

LIST OF TABLES

2.1	The global forest	35
2.2	Predicting degraded open access forest	36
2A.1	Characteristic thresholds for national definitions of forest stock	57
3.1A	Categories of technological and institutional change and their impacts on land use	85
3.1B	Stages of forest development and causes of marginal land use change	86
4.1	Common regulations of private forestlands	120
6.1	Effects of trade on consumers, employment and production, forest management and on the unmanaged natural forest, by importing and exporting region	209
6.2	Forest sectors as components of national economies	230
6.3	Forest sectors in the United States as components of regional economies	232
6.4	Regression results for the sources of change in managed and natural forest cover in Hainan, 1957–1994	236
6A.1	Summary measures of national income—along with official estimates of these measures for the U.S. accounts for 2002 (in billion US$)	251
6A.2	2002 U.S. gross national product and gross national expenditure (in billion US$)	252
6A.3	Modified Philippine income and product accounts for 1988 (in million pesos)	260
7.1	Characteristics of the forest products industry in three municipalities in Brazil's eastern Amazon	268

7.2A	Measures of industry and firm size for select manufactured wood products industries in the United States, 2002	275
7.2B	Operating characteristics with respect to major input categories for select manufactured wood products industries in the United States, 2002	277
7.2C	Growth in raw material consumption and shipment of products in select manufactured wood products industries in the United States, 1997–2002	282
7.3A	Measures of industry and firm size for select furniture manufacturing industries in the United States, 2002	284
7.3B	Operating characteristics with respect to major input categories for select furniture manufacturing industries in the United States, 2002	285
7.4A	Measures of industry and firm size for select paper manufacturing industries in the United States, 2002	288
7.4B	Operating characteristics with respect to major input categories for select paper manufacturing industries in the United States, 2002	289
7.5A	Measures of industry and firm size for the logging industry in the United States, 1997	296
7.5B	Operating characteristics with respect to major input categories for independent logging contractors in the United States, 1997	296
8.1	CAPM comparisons of regional timberland portfolios with portfolios of various financial assets, 1960–1994	312
10.1	Forest recreation on public lands: Activities and economic instruments for allocation	363
11.1	Sources of difference in per capita farm income in Anji County, 1989–1990 to 1994–1995	394
11.2	Vietnam	399
11.3	Cambodia	402
11.4	Laos	404

FOREWORD

Frank J Convery
UNIVERSITY COLLEGE, DUBLIN IRELAND

The forestry profession is not merely a way of making a living; it's a vocation. This special character is nicely captured in *The Forester*, by Irish poet John Locke (1847–1889):

> Envy him — he works with God
> To raise the prayer of trees to Heav'n
> To serve the needs of fellow men
> From wooden crib to cask of death
> Envy him, you men of towns
> Who tread his paths through woodland ways
> His are the clean-sweet smells of earth
> His is the music of the trees.

But, while nobility guarantees good intentions, it does not guarantee good decisions. It does not help answer questions such as: when to harvest trees, how much to invest, how ownership—public, industrial, nonindustrial, community—influences decisions and why, how to value nonmarket ecosystem and other services, and the relevance of diversity in this context, the role of technology and international trade in shaping choices, etc.

The future is made by those who understand the scientific and social significance of what is happening, and act wisely on this intelligence. This understanding is especially important when it comes to the management of our forests. Jared Diamond has documented how some civilisations wilted and then disappeared when they failed in their stewardship of their forests.[1] What is true of individual civilisations is true also of our planet and its forest endowment. And this is what

---

1 Diamond, Jared. 2005. *Collapse – how societies choose to fail or succeed*. New York. Penguin.

this book is about; how to mobilise the best insights that economics has to offer to help guide how we manage our forests. And understanding the context is critical. For the first time in modern human history, the economic development torch has passed to a new generation. The hundreds of millions in the OECD family who have already made the transition to good nutrition, long life, and relatively high incomes are being joined by the billions in Asia, South America, and Africa. This is a step change phenomenon of enormous import and poses the question: as a global community can we make the transition in ways that meet our material needs, and maintain our life support systems, including keeping the planet's temperature at levels that limit the disruptiveness of climate change? World trade has also grown rapidly, and the digital revolution and remote sensing and Geographic Information Systems have revolutionised how we measure forests, including our understanding of the extent to which they are being augmented or depleted. Foresters have to simultaneously understand the forces that are shaping their world that they do not control, and those over which they do have some influence.

Hyde's book is about enriching our understanding of context and finding our way towards better decisions. He does this by combining the theoretical frameworks offered by economics with a very skilful and convincing use of evidence, addressed to the central questions. Forests are in an important sense the canaries in our planetary mine. If they thrive, we thrive, and vice versa. Leonard Woolf observed that 'history is made, not by wickedness, but by stupidity'. Hyde shows us how to be wise.

PREFACE

This book is the synthesis arising from almost forty years of experience as a natural resource economist specializing in forestry. Or perhaps it was almost fifty years, beginning when I first started thinking about forest management as a member of hotshot crews fighting wildfire in the U.S. Pacific Northwest.

Over that period, I have had the privilege of working with some of the best natural resource and environmental economists, and also some of the best more specialized forest economists of my own and prior vintages. I have also had the privilege of working introductions to the environmental management and policy problems of forests in almost two score of countries on six continents. The former taught me the great merit of sound fundamental economics as an organizing tool. The latter, and particularly the working experience with practicing on-the-ground forest managers, showed me that my economics was only reliable when built together with an understanding of both the biological, physical, and institutional characteristics of local forest organization and the demographic characteristics of the local markets. Those working foresters, when I was wise enough to listen and perceptive enough to understand, were some of the best teachers.

These experiences teach an unavoidable respect for local differences. The local problems of on-the-ground managers are real. While I respect the basic organization of academic research and environmental economics, I believe the most important problems of forest economics, management, and policy can be understood only with reference to the nuances of local data—which, invariably, were not collected for my purposes (or those of any economist) and which, therefore, must be modified to enable meaningful inquiry. I am disturbed by the volume of economic analysis that overlooks this point and I set as a personal objective the minimization of my own errors from overlooking the special characteristics of the available data.

Nevertheless, my teachers and my experiences have also shown that there are underlying economic principles which are global—and these provide the justification for a presentation such as this book.

As many who have had careers in natural resources, I grew up around my resource, hiking and camping, then working on remote ranches and fighting wildfires. Subsequently, in my introduction to college economics, first Francis Gathof, then Charles Wilbur and Vito Tanzi showed me the analytical power of that discipline. Later Bob Gregory, Bill Bentley, Fred Knight, and the Michigan Society of Environmental Fellows introduced the specific economic, biological, and institutional characteristics of forestry and other natural resources. The reading assignments of Bob and Bill, combined with the fire experience, showed the merit of some traditional norms of forestry, but raised my doubts about others. Meanwhile, Frederic Scherer taught me and all of my classmates the importance of both personal motivation and the modifications necessary to understand dynamic processes within neoclassical economics.

In my initial professional experience at Resources for the Future, Marion Clawson challenged me to focus on the greater problems within any broad topic of inquiry and John Krutilla encouraged me—and many others—to identify and trust our own strengths without hesitating to ask for assistance from those with other strengths—and then to share credit where it is due. Both insisted on building from fundamental principles of good microeconomics. Both Tom Stoel at the young Natural Resources Defense Council and Hank deBruin ("Smokey"), then the director of U.S. Forest Service fire management, neither of them economists, reinforced the perspectives of Marion and John and encouraged me to challenge traditional norms within forestry.

My personal inquiry into policy and management problems in forestry began with a question about fire, insect, and disease control and a related interest in population dynamics. This has remained an interest of mine, but its primacy was rapidly replaced with a short-term assignment to examine the tradeoff between timber supply and the various, recreation, and environment and other non-market demands on forests. As I completed this assignment, the Reagan administration raised questions about regulation—and deregulation. Reagan received little response within the forestry community—so I entered upon a second large project (with Roy Boyd) to address the question. Subsequently, Bob Buckman, the U.S. Forest Service Associate Chief for Research, invited me—and others (including Dave Newman and Barry Seldon)—to examine the benefits of forestry research, and to compare the experience of forestry research with that of agricultural research.

About the same time, international opportunities associated with the question of forestry's role in economic development began to present themselves. Over time, these only reinforced a belief in the universal merit of fair applications of microeconomics in forestry. They also showed, as we should all expect, that (a) many regions and countries share similar experiences, if at different moments in time, and (b) even those experiences of one country that are not shared at present are often instructive to another country which may share the experience in the future. A sabbatical in Thailand, numerous conversations with Barin Ganguli, twenty-five years of activity with Chinese scholars and forest managers, and a variety of observations in Malawi, Zambia, Ethiopia, and South Africa, and then in Bolivia, Argentina, Colombia, and Chile have been convincing for me.

Were these experiences simply a fortuitous sequence of events? At first they were. However, before finishing the regulation/deregulation project, it was apparent to me that some greater organization, and not just fortuitous opportunity, must underlie any attempt to comprehend the full breadth of forestry. It seemed to me that the common classification of forestry into first three, then four, categories of landowners, and the separation of the roles of the public category into land management, research, and extension, is universally justifiable. It also became apparent that each category of land managers and each public role contained its own fads and fallacies. In order to understand forestry, one has to understand each of these categories and roles for itself as well as for its relationship to the others. Those first three projects began to rough out my understanding of public and (often regulated) non-industrial private land management, and also the research and extension responsibilities of the public forestry agencies.

The international experience and my personal knowledge from more remote parts of the United States provided insight to forestry's role in economic development—and to my eventual understanding of the three-stage pattern of forestry development discussed in this book. Meanwhile, I began searching for ways to fill the remaining gaps. At various times, George Staebler, Grant Aiscough, Bambang Hortono, Bill Stuart, Luc LeBel, Runsheng Yin, Jintao Xu, Herath Gunatilake, Sjur Baardsen and numerous conversations with millowners and managers helped with these, particularly with the behaviors and responsibilities of industrial landowners, for me the most complex and least understood landowner category. Altogether, this provides the background, as well as the organization, for this book.

Over the course of forty years many others have made their imprints on my comprehension of forestry and, therefore, on this book. Josh Bishop, George Taylor, Javed Mir, Doug Barnes, Tim Brown, Bill Bruner, Neil Byron, Manuel Ruiz-Perez and Brian Belcher, Jeff Sayer, Yves Dube, Wen YaLi and Han Xiao, Bill Magrath, and George Dutrow each arranged the opportunity for a crucial experience and shared their own insights. Ann delos Angeles' own dissertation prompted me to begin thinking about the general merit of the von Thunen model for explaining the association of forestry with its adjacent sectors. Frank Convery arranged my first formal presentation of the three-stage model in 1994. That presentation began an enduring association with Thomas Sterner, Gunnar Kohlin, their outstanding Environmental Economics Unit at Gothenburg University, and its many fine associates around the world. David Pearce enthusiastically encouraged further work on the model and suggested additional applications. Other colleagues each contributed one or more selective insights: Peter Pearse, Clark Binkley, Ken Chomitz, Steve Stone, Henry Peskin, Keshav Kanel, Carlos Young, LeCong Uan, Juan Seve, Greg Amacher, Shashi Kant, Peter Berck and Diane Burton, Steve Daniels, Knox Lovell, Adrian Whiteman, Chona Cruz, Jacek Siry, Arild Angelsen, Randy Bluffstone and Priscilla Cooke, and Ron Johnson will each find his or her imprint on one or more chapters. As will Yaoqi Zhang (who wrote the finest forestry dissertation I have read), Daowei Zhang, Marty Luckert, Sara Scherr, Sam Bwalya, Mohammed Rafig, Tekie Alemu, Resham Dangi, and Natasha Landell-Mills. Roger Sedjo has been a good friend and fair critic for, it seems, forever. This is a long list—but the book has taken a long time to assemble and many have contributed.

I hope this book accurately reflects what these experiences and these colleagues have taught me. (Of course, where it does not, it is my responsibility and not theirs.) I hope it adds a bit to the global understanding of forest management and policy, but I also look forward to learning the corrections and improvements that others make in the time to come.

William F. Hyde

1
INTRODUCTION

The economics of forestry contains a number of discrete and fundamental problems. A list of just those most common to forest management textbooks would include the long-term sufficiency of timber supply, the optimal path to an even flow of annual timber harvests from forests that were mature and natural when initially exploited but which are being replaced with managed stands, the different characteristic behaviors of various classes of forest landowners, forest regulation by government agencies to satisfy public objectives, and the impacts of both public and private forest-based resource activities on ecosystems and local human communities.

Historically, these problems of forestry, and of forest economics, have focused on commercial timber. More recently, all of the contributions of forests to human welfare have become fair topics for economic inquiry. This means that discussions of non-consumptive forest-based ecosystem services like forest recreation and erosion control now find their way into the discussion, just as market-based industrial products like sawlogs and pulpwood have always been a part of the discussion. Discussions of the thinner and less complete markets for subsistence household products like fuelwood and various non-timber forest products also find their way into the modern discussion.

The objective of this book is to examine the global economics of forestry and all of its components, including each of those five classic problems from the first paragraph, all of the welfare enhancing products and forest ecosystem services identified in the second paragraph, and also a series of newer issues attracting critical global attention today, issues such as deforestation, the protection of a diversity of flora and fauna, carbon sequestration to mitigate global warming, and the legal recovery and distribution of revenues created by activities in the forest, as well as the control of illegal forest trespass. (In addition, we will discover that the original five problems are preface for a catalogue of related questions. The appendix to this chapter identifies a few of these.)

Organization

The book begins with a simple but comprehensive organizing construct that is representative of forestry in all lands and cultures whether for timber or non-timber commodities, and whether in the historical past or today. In an important sense, this organizing construct and its universal application are the real contributions of this book. Subsequent discussion later in the book only demonstrates its historical and contemporary, local and global, applications. Empirical examples from around the world, introduced during the development of the construct, and throughout the book thereafter, substantiate its validity.

Forestry is about trees, but it is also about land and the location of human activity in the forest. The latter two features are best described within the framework of economic geography—and that is the basis of our organizing construct. That is, its focus is on marginal shifts in land use—not marginal product per land unit—and within that focus, the attention is on the geographic progression in the use of forest land, and on the progress of the forest sector as local markets and regional economies develop. (In general, shifts in the numbers of hectares of forestland have greater impacts on the production of most forest-related goods and services than shifts in physical productivity per hectare. Therefore, the decision to focus on marginal shifts in land use is a preferred choice.) We will observe that a three-stage development process, conveniently summarized within a graph of only three functions, summarizes the economic progress of forest development. Recognition of the different biological and economic characteristics of forests and local markets in each of the three stages is a reliable basis for decisions pertaining to forest management and policy. Indeed, proper recognition of the differences between the three stages would save us from numerous modern errors of forest policy.

Begin by considering that the consumption of both timber and non-timber forest products, and also the provision of forest ecosystem services such as erosion control and habitat for critical biodiversity, is largely the result of production from natural forests even today, although managed forests and even forest plantations do play an increasing role. Therefore, any comprehensive description of forestry must include both natural and managed forests, and these tend to be distinguished from each other in their locations, natural forests generally occurring in more remote locations and managed forests generally occurring closer to concentrated human activity and its demands for a long and varied list of forest products.

Furthermore, any basic description of forests must accommodate a dynamic element as both the natural and the managed forests are constantly changing. They grow but, just as importantly, the geographic loci of their exploitation shift with time. For example, many industrial forestry operations and many rural households that extracted products from the natural forest forty years ago manage their own trees and forests today. These newly managed resources generally occur in different locations than the harvest activities on either the original natural forest of forty or more years ago or the often diminished natural forest of today. Our organizing construct incorporates these features, natural and managed forests, in progressive snapshots of economic development that permit us to consider their changing character over time.

This description of the forest introduces the idea of at least three different land use margins, the intensive and extensive margins of managed forests and the frontier of the natural forest. In some cases a fourth margin also becomes important for forest production: the margin of production from the somewhat degraded forest that is not actively managed, yet is not on the natural forest frontier either. These four margins for forestry contrast with the two (intensive and extensive) margins that are standard for most other sectors of economic activity. They are the unusual feature that makes forestry different from most other economic activities.

Chapters 2–6: The Pattern of Forest Development

The next five chapters of the book lay out the construct first for its short-run and then its long-run applications, discuss first the local and then the macroeconomic market and policy implications to be taken from them and, finally, examine the special case of contracts for harvests of mature timber.

The economic distinction between the short and long run is the distinction between the decision to use an existing forest resource and the decision to invest and capture that forest's future potential. The short run is the period of current harvest activity—for timber or any extractive forest resource: fruits, nuts, mushrooms, Christmas greenery, resin, latex, fuelwood, forage, fodder, or whatever. We observe many cases around the world where the harvest activity for any of these products is locally unsustainable. That is, repetitions of harvests and growth and then renewed harvests at the same location are uncommon despite official public commitment and a history of professional forestry training to the contrary. An understanding of the geographic pattern of forest development, together with an appreciation of the difficulties involved in establishing secure rights for some forest properties, will demonstrate why local sustainability can be an elusive objective for forestry. Furthermore, the lower incomes and sparser labor opportunities that characterize human activity in many forested rural areas and many developing countries combine to make sustainability at any particular site even more elusive. Our construct will demonstrate this important point, and the reason for it as well.

If the long run is a period sufficient to incorporate returns from all forms of forest investment, then it must be long enough to include everything from improved current harvest opportunities to improvements in existing young forest stands and even to site preparation and regeneration of an entirely new forest. For many timber production cases, the period between the preparation of a site for a new forest and final harvests from the eventual mature forest on that site can be longer than fifty years. Technological change is inevitable in periods this long and even a modest rate of technical change over periods as long as fifty years often has a larger effect on forest production, and also on the forest industry's pattern of consumption from the forest, than either biological growth or the improvements in it due to most commonly known forest management practices. Yet discussions of forest policy and management generally overlook the expectation of technological change. Despite its importance for forestry, we will observe that technological change has generally progressed at a rate that has been slower for most forest activities than the faster rate observed for the rest of the economy. We can consider the source of this difference but, regardless of the source, different rates of technical change for forestry than for the land uses that compete with it suggest different effects on the three—or four—land use margins. In any event, technical change is another reason why the creation of a "permanent forest estate" is unlikely, and why policies that attempt to enforce one are almost certain to fail. That is, the geographic boundaries of forests will continue to change with time, like it or not.

Modern discussions of forest policy range from explicit and direct forest policies to agricultural policies that spillover to affect the forest, from changes in the administration of local property rights to policies that affect wood processing and foreign trade, and even to the effects of the overall macroeconomic policy environment on the forest. The traditional policy discussion in forestry concentrated on the security of long-run timber supply. This discussion continues, but modern discourse adds even greater attention to the general forest environment. We will examine both in chapter 4—and draw the interesting, and potentially disconcerting, observation that some direct forest policies have opposite effects on managed and natural forests and, therefore, contradictory effects on the overall forest environment and also on timber supply.

The procedure for selling forest resources, particularly timber from mature public forests receives special attention these days. Mature natural forests are the source of well over half of the world's industrial timber, and probably an even greater share of its non-timber forest products. They form a special case for policy and management, a case that incorporates the questions of concession management, royalties and rents, enforcement, and illegal activity in the forest. These questions seem to be especially troublesome for developing countries and for tropical forest management today, although the terms of harvest contracts for mature timber and the optimal arrangements for the management of forest concessions occupy a central place in discussions of temperate forest management in developed countries as well. Forest trespass and illegal harvesting also occurs in developed countries, although this fact is not as widely recognized in current discussions of global forest policy. The fifth chapter sorts through these management and policy questions.

In many cases, policies that are not normally considered forest policies contribute substantially to changes in forest cover. For example, agricultural policies are widely perceived to have important affects on the natural forest frontier. We will see that this is not always true, but that it is true in important select cases. It is also true that either select macroeconomic policy or the overall macroeconomic policy environment can overwhelm the effects of direct forest regulations, taxes, and subsidies in some cases. In these cases, the desired effects of even the best-designed forest policies may be of little consequence, and policymakers desirous of modifying the forest environment should be encouraged to turn their attention to the broader issues that can have greater impact. The corollary to this argument is that policymakers with broader responsibilities who also have a concern for the environment in general, and for the forest environment in particular, must recognize that their macroeconomic policy recommendations can have crucial unanticipated effects on the forest.

Finally, trade, since it involves multiple forested regions, multiple markets, and multiple countries, is an important category of macroeconomic activity, and trade policy is an important category of macroeconomic policy. Trade and trade-related issues like the impacts of trade liberalization or the impacts of the certification of forest management receive particular attention today, especially from those with global environmental interests. However, the discussions of these issues often focus on specific forested regions or specific classes of forests without recognizing that their impacts on forests in one region or country almost always export accompanying but opposite impacts to forests in another region or country that trades with the first region. The effects in both regions are important, as is the net effect on the global environment. The final chapter of this first section of the book examines these important questions having to do with the macroeconomic environment, macroeconomic policy, and trade policy on the forest.

Categories of Forest Ownership: Chapters 7–10

A second group of four chapters examines the landowner classes common to forestry. These chapters each review the unique objectives and behavioral characteristics of a single broad class of forest landowners and place each ownership class in its own position within our overall organizing construct.

Three of these landowner classes, industrial or integrated land-owning and wood-processing firms, non-industrial private forest landowners (NIPFs), and public land management agencies, are traditional to forestry. An emerging fourth class is composed of institutional landowners who hold forests as investments but who possess neither logging equipment nor the facilities necessary to process their wood. This emerging class exists primarily in North America, and especially in the U.S. South. Pension funds are an example. Forty-two pension funds invested an estimated $7.25 billion in managing 2.7 million hectares of forest between the mid-1980s and year 2000. Their investment has continued to grow since then until Weyerhaeuser Company, the last large landowner among forest industrial firms, recently converted its landholdings into a real estate investment trust (REIT). These pension funds, REITs, and other institutional investors too, purchase land and timber as an investment and eventually sell their timber according to the income requirements of their broader investment portfolios. We can anticipate that this new landowner class will become more important globally as local timber values continue to increase almost everywhere and as the unique characteristics of investments in timber management become important to a wider range of investors. Identifying these unique characteristics is central to understanding the behavior of institutional landowners. We can also anticipate that some non-industrial landowners will begin to understand and to take advantage of the same investment characteristics. In fact, we already observe that farm foresters in some developing countries share some of the same risk averting strategies associated with institutional forest landowners in the United States.

Some forest landowners, particularly some government agencies who manage the public's forests, establish long-term contractual arrangements for land management with industrial, community, and even individual private managers. These contractual arrangements are a topic of previous discussion in chapter 5. Within the terms of their contracts, the behavior of the long-term contractors is similar to the behavior of either industrial or NIPF landowners. Therefore, we include these contracting land managers within the discussions focusing on those other two landowner classes.

The literature on the economic decisions of industrial landowners is sparse but we can anticipate that the behavior of the most capital-intensive industrial operations (typically pulp and paper mills, rather than sawmills) merits special attention. Their investments in mills and logging equipment are generally much greater than their investments in forestland. For those that still own forestland, their decisions to protect their larger investments in manufactured capital and also their dominant market positions with respect to multiple local timber suppliers explain many apparent variations from the straightforward textbook expectations for independent timber management. Furthermore, differences between firms with regard to their relative costs of manufactured capital and with respect to the degree of the firms' competitive positions for the basic wood resource explain many characteristics of individual firm behavior. These differences anticipate a wide range of contractual arrangements between industrial operations, loggers, and NIPF landowners. They also anticipate the geographic and physical limits of industrial extraction from the natural forest.

The experience of non-industrial private landowners is better-understood, particularly for NIPF landowners in North America and Northern Europe. It is less-well-understood that farm foresters and some community forestry operations in developing countries share many of the same behavioral characteristics and often make similar forest management decisions. In both cases, the landowner's primary financial activity and his or her own comparative advantage is often in some activity other than forestry; e.g., a local farmer or businessman who owns and manages a few hectares of forest but concentrates his or her management effort on the farm or business rather than the small forest property. Irregular financial returns from forest activities may not be a problem and may even be the personal preference of these landowners if the returns can be designed to coincide with some other landowner objective. In fact, wealthier NIPF landowners in developed countries increasingly emphasize non-market objectives for holding forestland—although they may continue to look forward to periodic financial returns as well. Non-market objectives such as erosion control or the production of non-marketed goods for household consumption are also important for farm forest managers in developing countries.

A good case can be made for similarities in behavior around the world within each of these landowner or land manager classes. Nevertheless, some important distinctions remain and, just as we will find that distinctions in the capital intensity of their processing operations are central to the behavior of industrial forest managers, we will also observe that distinctions in relative income or wealth levels are important for understanding the decisions of NIPF land managers.

Public agencies, the fourth and final class of landowners, administer eighty percent of the forests in developed countries and they may administer an even larger share in the developing world.[1] Even in the United States, one of the more market-oriented developed countries, the public owns almost one-third of all forestland (USDA Forest Service 2010). The public values protected by different government forest management agencies distinguish these agencies; e.g., flood control in Bangladesh, more general watershed management in Portland, Oregon, in the United States, and community development in interior Canada, as well as an array of recreational, aesthetic, and other environmental values that varies with the local natural environment and with the relative importance of either global, regional, or local human demands on the forest. The chapter on public forestlands relates these different public responsibilities to the organizing construct of the first group of five chapters and also to the established subject matter of joint production ("multiple use" in forest terminology) and welfare economics.

Community Impacts: Chapter 11

At the close of this group of four chapters on classes of forest landowners, we add a chapter discussing the experience of communities with significant forestry sectors. Three hundred fifty million people live within or adjacent to the world's forests, and as many as one billion may be dependent on forests for a portion of their livelihood (World Bank 2001, 2003). The central question for them, and for this chapter, has to do with their dependence on the forest and forest-based processing enterprises, the characteristics of the most affected households, and the potential role of the forest in regional development for their benefit.

Conclusion

A final chapter summarizes and draws conclusions for management and policy. Perhaps the most important conclusions have to do with the roles for public

---

1   Government agencies administer 34 percent of the forests in the Unites States, Japan, and the EU countries and 92 percent of the forests in the other developed countries (UN-ECE 2000). Summary data are unavailable on the ownership patterns of many developing countries. However, for those for which the data are available, the public share often exceeds eighty percent of all forestland. For tropical forests alone, White and Martin (2002) and Molnar, Scheer, and Khare (2004) estimate that central governments administer 71 percent and local communities administer another 13 percent.

regulatory and land management agencies. Markets take care of many of the problems of forest allocation, including the allocation of some goods like fuelwood and some varieties of recreation that are often perceived to be non-market goods. However, it is insufficient to say "the market will take care of it." Private land managers must understand their own markets, including the effects of market information and the role of market power. They must also understand the activities of the public regulatory and land management agencies—because public regulations generally target the private lands and because the public lands often provide products and services that compete with those from the private lands. Furthermore, the public needs to understand the effectiveness of the public forestry and environmental agencies in their current activities if it expects these agencies to do justice to the public interest. Finally, and more fundamentally, the public needs to understand where markets fail and the most effective means of addressing these failures.

Some of the activities of some public forestry agencies do not serve the public well. Moreover, the policy instruments these agencies prefer are not always those that are best-suited to achieve either their own objectives or the public's. On the other hand, the out-dated objectives and instruments of the agencies often have their roots in a reasonable understanding of economic conditions that prevailed in a previous time and place. The institutional rules for determining the annual allowable cut and sustainable yield of commercial timber and for practicing multiple use management or protecting dependent communities, reducing rural poverty, and encouraging regional development are examples. Understanding the historic context of these issues may help us understand the manner in which modern public forestry institutions address them, and it may help understand the effort needed to bring about change. Therefore, we will introduce the institutional context with anecdotes and empirical evidence within the chapters and in appendices to some chapters. The final chapter will build on these, on the obvious implications of our organizing construct from chapters 2–6, and on the general redistributive and market-failure-correcting responsibilities of public agencies to recommend socially meaningful roles for public forestry agencies and their regulatory, extension, research, and land management offices.

Building a comprehensive model to address the full range of all of these issues is a tall order. It is necessary, however, if we seek to gain perspective on the specialized problems of forest economics, policy, and management; their connections with the institutions of forestry; and their interactions with the economic activities of the rest of the world. This book is an attempt to develop that perspective.

Literature Cited

Molnar, A., S. Scheer, and A. Khare. 2004. *Who conserves the world's forests? A new assessment of conservation and investment trends*. Washington, DC: Forest Trends.

USDA Forest Service. 2010. Forest inventory data online, http://fiatools.fs.fed.us/fido/standardrpt.html (accessed April 3, 2010).

UN Economic Commission for Europe. 2000. Forest resources of Europe, CIS; North America, Australia, Japan, and New Zealand. Geneva: UN Economic Commission for Europe.

White, A., and A. Martin. 2002. *Who owns the world's forests?* Washington, DC: Forest Trends.
World Bank. 2001. *A revised forest strategy for the World Bank Group.* Washington, DC: The World Bank.
World Bank. 2003. *Sustaining forests: A World Bank strategy.* Washington, DC: The World Bank.

Appendix 1A:
Fundamental Questions within the General Subject Matter of Forest Economics, Policy, and Management

Forest economists, managers, and policy analysts confront a long list of largely unsolved questions. Some we acknowledge. Some we seem to disregard in our pursuit of other issues. Some are related. Many we address without complete success because we address them independently of their association with the rest of forest economics or broader regional economics. Moreover, if experienced forest economists and policy analysts have not been more explicit about many of these questions and their broader associations, then surely the likelihood of misplaced assumptions and misguided conclusions on the part of policy analysts and forest advocates with less specific experience should not be surprising.

There is no established order to these questions. However, it is the responsibility of any attempt to establish a comprehensive perspective of forest economics to develop an orderly means for assessing all of the questions related to that topic. That is the objective of the three-stage construct developed in chapter 2.

It should be useful to identify a few of these questions at the outset, before we develop the three-stage construct for their assessment. The questions themselves will help us recognize the uncertainties of our subject matter as we proceed through the chapters that follow. They may also help identify those problems that remain unsolved.

- What is the optimal harvest age or timber rotation? This is the most central of all questions in the forest economics literature—but why do so few forest managers practice the economically optimal behavior described by this literature? (Chapters 3, 7, 9, and 10)
- What is different about investment in forestry? And how can we reconcile the classic description of forestry as a renewable resource with secure property rights, with observations of rising real prices for basic forest products, with widespread harvesting of mature natural forests without subsequent reforestation, and with global observations of forest trespass and illegal extraction of forest products? (Chapters 2, 3, and 5)
    o What is the optimal level of investment in forestry for an industrial ownership? Why do we observe a complete spectrum of choice ranging from firms that own no land or trees to firms that control enough timber to supply the entire demand of their own processing facilities? (Chapter 7)
    o Why do many industrial landowners possess only a limited amount of forestland, yet invest in intensive silvicultural practices on these limited lands? Could their objectives be satisfied with greater investment

in land but less investment in silviculture? Why do some industrial landowners hold their timber beyond the harvest age preferred by local NIPF landowners who operate on land of similar quality and in the very same market as their industrial neighbors? (Chapter 7)
  o How can we explain the willingness of some millowners and some timber procurers to pay less for local stumpage and more for stumpage that is a greater distance from their mills? Does this behavior conflict with the expectations of location and economic rent? (Chapter 7)
- What are the most effective contractual instruments for harvests from the forest, and how can we improve on the current degree of sustainability in forest resource use? In what sense is forest sustainability a reasonable objective and in what sense is sustainability futile? Are secure property rights the answer? (Chapters 2, 4, and 5) And consider the related questions:
  o Why do some industrial forest landowners employ far-sighted managers who reforest subsequent to their harvest activities and otherwise practice "good forestry," while other successful and established industrial firms seem to "rape and pillage" the land without regard for a future supply of timber? Under what different conditions is each class of actions "economically rationale"? (Chapters 2 and 7)
  o How serious is forest depletion (globally, nationally, locally)? What are the primary contributing factors, and what are the most effective responses? (Chapters 2, 4, and 6)
- Do local communities manage better than government forestry agencies? Under what conditions? Does local community management have beneficial implications for women and the poor? Many forestry development projects today are designed with the expectation of affirmative answers to these two questions. But what is the evidence? (Chapters 4 and 11)
  o In fact, do women and the poor really suffer more than men from forest depletion, and do they gain more from forest improvement? (Chapter 11)
  o In general, what are the distributive merits of forest-based activities? Distributive arguments underlie many forest policies and poverty reduction is a major objective of many international donor activities in their association with the forest and environmental agencies of developing countries. Yet very few evaluations of distributive impacts exist in the policy analytic literature of forestry and, therefore, the empirical merit of distributive programs in forestry is largely unknown. (Chapters 4, 9, and 11)
- Who does plant trees, and under what conditions? What are the best interventions for inducing forest management and when are they most likely to be successful? What are the best policy actions for protecting natural forests? How do these compare with interventions designed to induce improvements in forest management? (Chapters 2, 4, and 10)
- Do consumptive uses of the forest decline with improvements in non-forest employment opportunities, or with increases in real income? (Is there an environmental Kuznets' curve for trees?) (Chapters 3 and 10)

- ○ Do older timberstands generally offer better protection for non-market environmental values? The policy positions taken by many forest biologists and many environmentalists seem to suggest that they do. Some widely cited economics literature accepts and develops this position as well. Yet it is reasonable that forests with mixed stands of various ages and forest types support greater biodiversity than a landscape composed solely of mature forest. We do know that some forms of diversity in both stand age and forest type enhance both wildlife production and human recreational experience. (Chapter 4)
- ○ Continuing this theme: If diversity is preferable to a simple formula of "older trees are better," then there must be conflicts between some of the multiple uses of the forest. Can we anticipate the primary conflicts, their locations of greatest importance, and the most effective policy responses? Is multiple use efficient or can we gain from specialization in different forest stands? (Chapters 2, 3, and 4)
- ○ If older trees are not always better, then what about younger trees? Under what conditions are plantations an adequate substitute for natural forests? What are the best ways to preserve the ecosystem services of natural forests? (Chapters 4 and 7)
- How do the policies of other sectors spillover to affect forests and forestry? (Chapters 4 and 6)
  - ○ In particular, how can we reconcile the obvious impacts of some agricultural policies on some forestlands, while observing that similar agricultural activities and policies seem to have no impact on other forests? (Chapter 4)
  - ○ And a related question about agriculture and agricultural policy: Is agriculture the major source of global forest degradation and deforestation? (Chapters 2, 3, and 4)
  - ○ What is the impact of trade? Does trade liberalization lead to increased deforestation or does it encourage investment in forest management? Is the promotion of value-added processing a socially beneficial objective of public policy? Do log export restrictions help? Why do forests decline in some regions as they simultaneously recover in others? (Chapters 4 and 6)
  - ○ And how important is a stable policy environment in general? Economists know that instability undermines long-term investment. We also know that many developing countries suffer extended periods of economic and political instability and that their instability has substantial effects on marginal human populations and natural environments. Finally, we know that investments in trees and in the infrastructure that supports sustainable timber management are long-term investments. Yet instability is almost entirely overlooked as a factor affecting forest management. Just how important is its impact? (Chapters 4 and 6)
  - ○ If forestry is a long-term investment, then technological change in forestry and technological change in the adjacent sectors; agriculture, logging, and wood processing; must have important effects on long-run

timber production. This is another issue that deserves more attention than it has received in the literature of forest economics and policy. What are its impacts on harvest levels? On the forest environment? (Chapter 3)
- What important characteristics of forestry do we overlook as our economics research and our practical policy discussions focus on the forest itself and only less completely inquire into the conduct and performance of the logging and wood products industries? (Chapters 3 and 7)
    - How do the different industrial structures and behaviors of industrial enterprises affect forest management and the forest environment itself? (Chapter 7)
    - We presume that changes in resource availability affect the wood processing industry. Yet we observe different responses to similar declines in resource availability. Some mills close, others modify their operations, still others seem unaffected. Sawmills seem to respond differently than pulp and paper operations, and plywood operations respond differently yet. What explains these differences? (Chapter 7)
- Finally, the basic forest data on which nearly all analysis resides are physical data. Economic distinctions are entirely absent from these data. In fact, these data generally include some forests that are uneconomic for almost any purpose, and they ignore other forests and trees that serve multiple economic objectives. How do these data bias our analyses and how great are the implicit errors? Can we modify our analyses to address the data problem and still address the most important problems of forest policy and management? (Chapter 2)
- After considering all of these questions, would we agree with the classic argument that the unique characteristics of forestry are (a) its long production period and (b) the fact that its input (a growing forest) is also its output (timber)? Does the long growth period of forestry have an economic effect that is different from even longer-term investment activity in some other industries? How is the input and product relationship in forestry different from that for other renewable resources? How about the multiple product characteristic of forestry? Is it distinctive? And do we observe other characteristics that truly do distinguish behavior in forestry from that in other human endeavor? (Chapters 2 and 3)

2

THE PATTERN OF FORESTRY IN THE COURSE OF ECONOMIC DEVELOPMENT

Once more, forestry is about trees, but the economics of forestry is about land and the location of human activity in the forest. It is as much about the human consumption of forest products as it is about the trees themselves. These characteristics are best understood within the framework of economic geography and a fundamental analytical organization first proposed by von Thunen in 1826.[1]

Consumption from forests in almost all regions of the world is still largely a product of unmanaged natural growth, although managed and even planted forests play an increasing role. Therefore, any comprehensive description of forestry must include both natural and managed forestlands.

The practice of forest management is limited by the institutions and the transactions costs associated with property rights—the costs of defining, maintaining, enforcing, and transferring claims over forest resources. Furthermore, the primary products obtained from forests are often geographically dispersed and low-valued relative to the costs of their collection or extraction. Therefore, the importance of these transactions costs is relatively greater for forestry than for many other economic activities. It means that the lessons of the literature on property rights are crucial to any comprehensive description of forests.[2]

In addition, any fundamental description of the economics of forestry must accommodate a dynamic element as both the economic and biological characteristics of both natural and managed forests change in gradual but very important ways over the significant periods of time that are generally involved in forestry. For example, many industrial forestry operations and many rural households that extracted products from the natural forest thirty or forty years ago manage their

---

1 See Samuelson (1983) for a review.
2 See Coase 1937; Demsetz 1967; Alchian and Demsetz 1973; Anderson and Hill 1975; North 1990; and Alston, Libecap, and Mueller 1999.

own trees and forests today. Most of their newer managed resources are in different locations than either their prior harvest activities in the original natural forest or the residual natural forest that remains today. Their newer managed trees and forests are younger and faster growing than the natural forests, but the remaining natural forests tend to be more biologically diverse. (That difference in diversity creates advantages for industrial uses of managed forests but also for the non-consumptive uses of the remaining natural forest. We will revisit this difference in later chapters.) Meanwhile, the same industrial operations and the same households, or their neighbors, continue to extract some products and they continue to obtain some ecosystem services from the depleted and less accessible natural forests that remain today.

Each of these elements—economic geography or location, natural and managed forests, transactions costs, and dynamic change in the forest properties—must be a part of the basic forestry model. This chapter sets out a model that incorporates all of them. It also introduces the effect of relative input costs on forest activities—and it will become apparent that relative labor opportunities in particular have a crucial spatial effect that makes them an important determinant of forest degradation and deforestation, as well as of the illegal extractive activities observed in forests around the world.

The majority of this chapter focuses on a general perspective of the economic value and location of forest activity. Special characteristics may modify this general perspective, but they do not alter its fundamental form. The chapter closes with a discussion of four of those special characteristics of forests that are almost always important in real applications: multiple values originating from the forest, including non-market values; various possible configurations of property rights; and a heterogeneous landscape.

Conceptual models like the one discussed in this chapter beg empirical support. The chapter provides examples and cites supporting evidence. Data are a problem, however. We can anticipate difficulties with data for the many non-market values obtained from forests—but data are a problem even for commercial forest values. For example, timber is the most widely consumed market-valued product of the forest and transactions prices for timber are widely available. Yet the common data on timber prices and timber harvest volumes do not represent uniform points of either harvest and market access or market delivery. The latter is a crucial factor for products like timber with high ratios of transportation cost to primary product value. That is, meaningful comparisons of similar products must refer to uniform points of access, and alternative sources of timber volume are not good substitutes unless their access to the processing facility is similar. Therefore, the most common physical data on forest stocks, with their widely varying differences in accessibility, are imprecise estimates of the forest inventory that is necessary to address modern policy questions about either market-valued products like timber or fuelwood or non-market values like biodiversity or carbon sequestration and global warming. An appendix to the chapter discusses these data problems, speculates on the magnitude of error associated with them, and suggests analytical approaches for addressing them.[3]

The Pattern of Forest Development

A common pattern of forest development emerges from observations taken almost anywhere around the world and from almost any period in time. This is a pattern of new settlement, followed by deforestation around the settlement and gradually increasing scarcity of forest products eventually combined with regional economic development and the rising prices that induce the forest investment that limits further depletion of the remaining natural forest. Figures 2.1–2.5 capture the basic elements and three fundamental stages of this pattern. They also provide the key reference points necessary for further reflection in subsequent chapters of this book, reflections on forest management and investment timing, technological change, institutional constraints, trade, and the policies affecting forestry, as well as on the decisions of different classes of forest landowners.

Begin with von Thunen's starting point, a homogeneous environmental plain containing an unsettled frontier of forests and grassland. Generally, the first permanent settlers in any new frontier are farmers or miners. Even those miners are often supported by a few nearby vegetable farmers. They and some other new settlers convert some forest and grassland to agricultural production. They purchase some family necessities and sell some of their production in the small market of a local town.[4] Therefore, this description of the frontier contains a settlement (a town and a market) and agricultural land, as well as a mature natural forest.

A New Forest Frontier, Stage I

Figures 2.1 and 2.1A describe this new frontier. The vertical axes measure economic value while the horizontal axes measure access to the market. The settlement and its local market occur at point A. The value of agricultural land around this settlement is a function of the net farmgate price of agricultural products—which is greatest when the farm is nearest the local market and transportation costs are minimal. The land's value in agricultural use, described by the function V_a, declines with decreasing access to the market or with increasing transportation costs—which is related to decreasing access and, similarly, to increasing distance from the market center. These characteristics make the manufactured goods

---

[3] This and all future chapters relegate more technical and mathematical material to appendices—in an attempt to make the fundamental material in the body of each chapter more accessible.

[4] Even the most subsistence-oriented of these farmers exchange some goods and services in a local market. The original discussion of dual economies was clear on this point. See Boeke (1948, 1953) and Furnival (1939, 1948), or see Ginsberg (1973) for a review. Boeke and Furnival describe societies that are predominantly subsistence-oriented. Nevertheless, all households in the societies they observe participate in market exchange at least at some minimal level.

16 The Pattern of Forestry in the Course of Economic Development

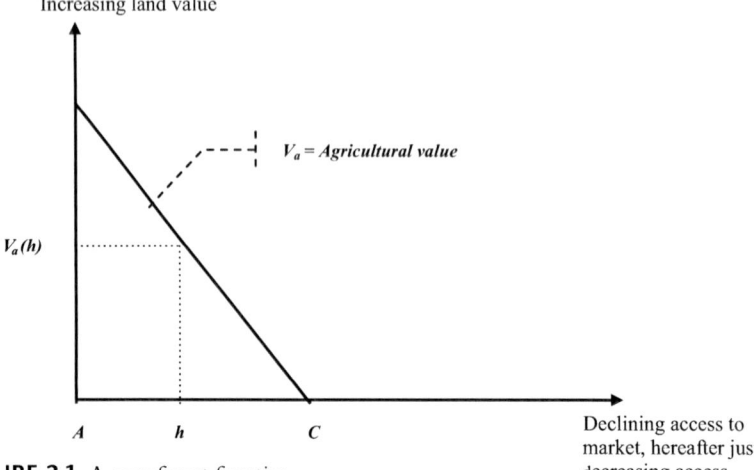

FIGURE 2.1 A new forest frontier

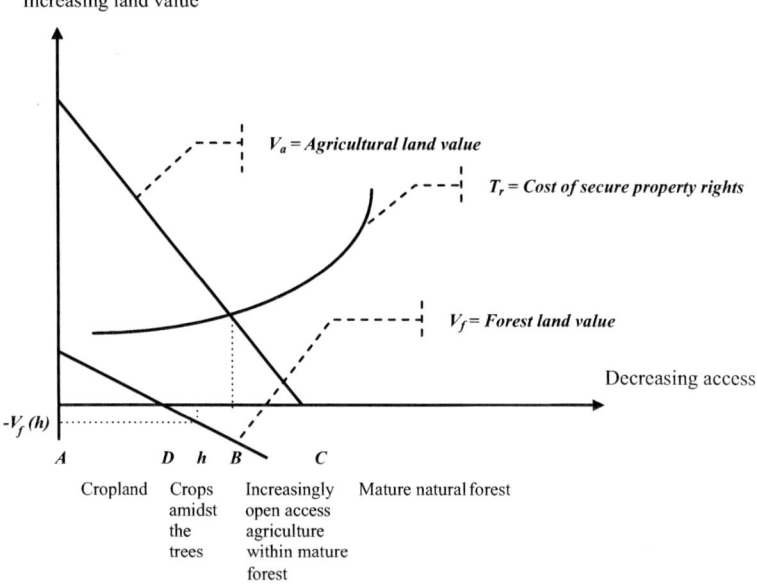

FIGURE 2.1A A new forest frontier continued, stage I

used in agricultural production more expensive at increasing distances from the market. They also reduce the net farmgate value of agricultural products from lands with more limited access, and these are inducements to decrease the intensity of cultivation, decrease the population density, and decrease the ratio of labor to land with increasing distance or decreasing access from the market. The function V_a, declines all the way to zero at some point like C, after which the costs of agricultural use exceed any return potentially available to the household, and agricultural use and development is unrewarding.

The vertical distance between any point on V_a and the corresponding point on the horizontal axis measures the economic rent for the unit of bare land described by that point. For example, the net economic rent accruing to the hectare around point h is the distance from h to V_a, or $V_a(h)$. $V_a(h)$ is the sum of discounted net expected returns to this unit of land obtained over time by the agent that manages it. It is the discounted sum of periodic returns from market revenues and household subsistence values minus the discounted sum of production costs incurred to obtain these returns. These costs include both market costs and the opportunity costs of the agent's own labor and capital.

The agents that manage agricultural land, particularly land at the frontier, tend to be individual households. Each household absorbs certain transactions costs in order to insure the proceeds of its productive activities. These transactions costs are different from the agricultural management costs incorporated in V_a. They include the costs of establishing and maintaining the rights to use the land and its resources or, alternatively, the costs of excluding trespassers and their unwanted activities. They include, for example, the registration of formal rights or the establishment of customary rights and they certainly include the cost of fencing and patrolling boundaries where those are necessary. The function T_p in Figure 2.1A describes the transactions cost level for each unit of land. The cost of maintaining any particular level of these property rights increases as the levels of public infrastructure and effective control decline for properties that are less accessible to (and more distant from) the market center at A.

The functions explaining agricultural land value and transactions costs intersect at a point like B. New farm households manage lands between points A and B for permanent agricultural activities. They use land between points B and C as a resource to be exploited for short-term advantage. That is, they harvest the products that grow naturally in this second region, crops like forage and fodder for their grazing livestock, native fruits and nuts for their own consumption, and fugitive resources like wildlife also for their own consumption. They are unlikely to invest even in the most modest improvements on the lands between B and C because the cost of protecting the improvements would be greater than the expected returns. Therefore, their use of these lands is unsustainable except for periodic extraction from regrown natural vegetation.

Some households and some local communities protect lands beyond point B in declining degrees farther from their homes—for example, by sending children out as shepherds for their grazing livestock. Nevertheless, the costs associated with maintaining their property rights continue to increase for lands farther and farther beyond point B until eventually the costs rise to a level such that no reasonable number of shepherds or other resource guards can fully exclude some competing open access users of remote lands. Therefore, we can say that the lands between B and C have increasingly unenforceable property rights, or that they are lands with an increasing tendency to open access.

What about the trees and forests that grow naturally across this landscape? Initially, these resources interfere with agricultural production and their removal is costly. Settlers remove trees wherever the agricultural value of converted forestland plus the value of the trees in consumption (e.g., for construction timber or fuelwood) exceeds their removal and delivery costs. Otherwise, they leave the trees standing. Indeed, farmers in some frontier settlements only girdle the trees and leave standing dead trees in the middle of their croplands. Girdling kills the trees and allows light to penetrate to the agricultural crops planted below while allowing the farmer to avoid or defer the difficult job of removing the dead trees and stumps.

The imposition on labor and capital associated with removing unwanted trees means that the new function V_f describing forest value probably lies below the agricultural value gradient. Some part of it may even lie below the horizontal axis. Where it does, as for the land identified by point h in Figure 2.1A, the negative value $V_f(h)$ implies a net burden imposed on farmers for removing the forest, or for leaving it to interfere with agricultural production. During this initial stage of frontier settlement, the forest value gradient never extends as far as point C where the agricultural value gradient declines to zero.

The market prices for agricultural and forest products must be sufficient to cover the costs of their production—otherwise no one would produce and sell these products. In the first stage of development, the costs of harvest and delivery are the only costs of forest production. There is no management cost associated with growing forest products because the existing forest is mature. There are no transactions costs for forest products because there are no successful prior claims on the forest. The act of harvesting establishes the first successful claim, but it is only a claim on the harvested product and not an effective claim on the forestland.

If the market price for forest products just equals the cost of harvest and delivery, then the in situ price of the product at the point of harvest, point D in the figure, must be zero and the value of forestland at this point is also zero. A mature natural forest of no value to individual local settlers remains standing at all less accessible points beyond D.

Two different cases fall within this characterization of the first stage of forest development: shifting cultivation and permanent settlement.

Shifting cultivation: Shifting cultivation refers to the subsistence practice of farming a unit of land until the agricultural pests become intolerable or until the soil nutrients are depleted, then abandoning this land and moving on to clear and farm new land until its pests also become intolerable or the nutrient value of its soil is also depleted. Abandoned agricultural land eventually reverts to forest, the concentration of agricultural pests declines, and the soil recovers some of its nutrients before some local households eventually return to clear the recovered forest once more and use the land again. Shifting cultivators are not as mobile as many settlers of new frontiers. They tend to be permanent inhabitants of regions of low population density and plentiful land.[5]

---

5 Nair (1993) describes the basic elements of shifting cultivation. The publications of the ASB (Alternatives to Slash and Burn) project of the International Centre for Research in Agroforestry are a rich reservoir of information about its practice in many tropical countries (e.g., Tomich et al. 1998, 2001). Angelsen (2007) provides a formal economic model.

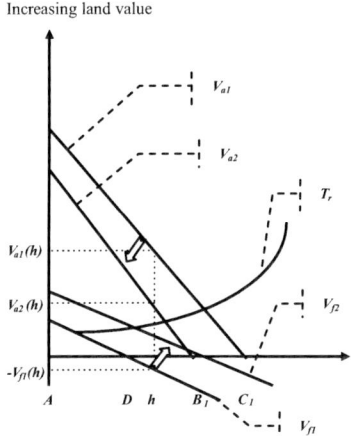
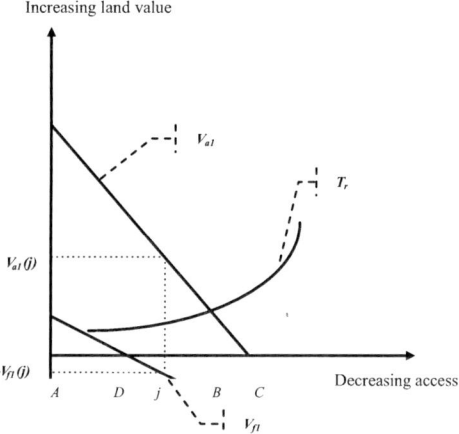

First home territory in the first period (V_{a1}, V_{f1}), and in the second period (V_{a2}, V_{f2}) after some depletion of soil nutrients and some forest removal.

A newer second home territory after further depletion of the soil nutrients at the first site. The shift to a second site provides greater productive agricultural opportunity.

FIGURE 2.2 Shifting cultivation

Shifting cultivation has been practiced in some temperate regions and it remains a common practice among traditional communities in more isolated regions of many tropical countries today.[6] As many as sixty million indigenous people may live in the world's forests and most of them practice some amount of shifting cultivation (World Bank 2003). The estimates of shifting cultivation as a source of all tropical deforestation range from 5.7 percent to as high as 45 percent (FAO 2001a; Lanly 2003).

Figure 2.2 summarizes the shifting cultivator's experience. V_{a1} and V_{f1} are the initial agricultural and forest value gradients. The net value to a shifting cultivator household of a unit of land identified by point h is $V_{a1}(h) - V_{f1}(h) - T_p(h)$.

With time, growth in the pest population, and the depletion of soil nutrients, the agricultural value gradient declines to V_{a2}. This induces the household to clear additional land near the existing homestead. Meanwhile, members of the household continue to remove small amounts of products from the natural forest for domestic consumption. As a result, the household gradually removes the remaining forest farther and farther from the homestead itself, the act of obtaining forest products requires more and more effort, the collection of forest products requires more time and they become more valuable, and the forest value gradient shifts upward and to the right to V_{f2}.

Eventually pest populations and nutrient depletion arising from agricultural use drive the agricultural value gradient to a point where it is in the household's interest to move on, to convert another forested site rather than to continue

---

6 Modern developed countries have shared in the experience, some in a fairly recent past. For example, soil profiles within 100 km. of Helsinki show evidence of shifting cultivation in Finland as recently as the turn of the twentieth century.

20 The Pattern of Forestry in the Course of Economic Development

farming the old site. The household "shifts" its agricultural activities to a new site when the net return from undeveloped land identified by point j, $V_{a1}(j) - V_{f1}(j) - T_p(j)$, is greater than the net return from the depleted soils on the old land, $V_{a2}(h) - T_p(h)$. (Interfering trees would have been cleared from site h by this time.) Similarly, the household uses the new land at point j until its productivity also declines to the point where the use of an undeveloped third site becomes more rewarding.

Shifting cultivators continue their periodic cycle through the land as long as a forest of sufficient quality remains available for conversion; or until the population grows, labor opportunity costs decline, and the quality of available unused land declines to the point where more intensive investment in the original land unit is more rewarding than shifting to new land.[7] With time, new access to previously unavailable agricultural technologies can also improve the returns to more intensive management and improve the incentives for permanent and non-shifting agricultural activity.[8]

Permanent agricultural households on new frontiers: Once settlements of shifting cultivators begin adopting more advanced management practices, they become similar to the stationary and permanent agricultural settlements on the new forest frontiers previously described by Figure 2.1A.

For households in these settlements, local consumer demand justifies the removal of some products from the trees and forests on the new frontier. If the local market retains its vitality and the demand for forest products continues to justify resource extraction at each new moment in time, then the forest frontier and the forest value gradient must extend gradually outward and upward, respectively. The frontier at point D shifts outward because all trees closer to the market were removed in previous periods and the only forest resources remaining to satisfy the local demands for construction and fuel are now farther from the market center. That is, forest resources are now economically scarcer. The value gradient shifts upward because the costs of harvest and delivery from the forest are now greater and no producer can justify supplying the market unless the market prices rise to a level sufficient to compensate the additional production costs. Consumers must either pay the greater price or switch to substitute products.

A Developing Forest Frontier, Stage II

Eventually the demands for agricultural land, construction timber, and fuelwood justify the removal of the natural forest to the margin of sustainable agricultural use at point B. If market and household demands for forest products remain

---

7 Templeton and Scherr (1999) make this case with reference to a broad range of seventy empirical studies.
8 Stevens (1988) provides a clear discussion of this transition for the traditional population of the Annapurna Sanctuary in Nepal over the course of the twentieth century. Muller and Zeller (2002) describe a similar pattern in Vietnam's Central Highlands.

strong, then the removal of trees and forest continues even beyond this time and the forest value gradient eventually extends farther yet, following the arrows in Figure 2.3, such that point D identifying the frontier of forest value shifts outward beyond the frontier of agricultural value at point C. Forest product values,

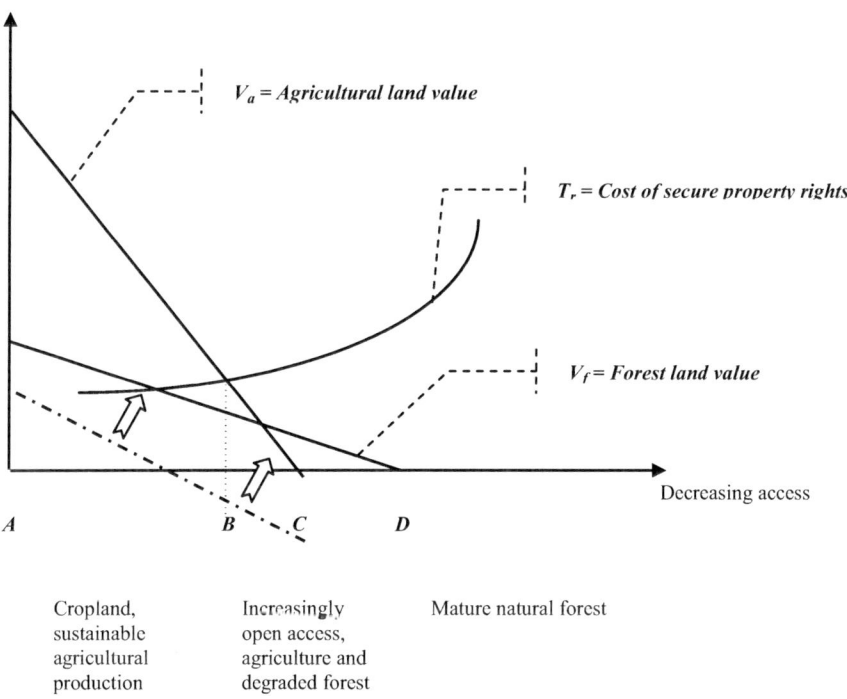

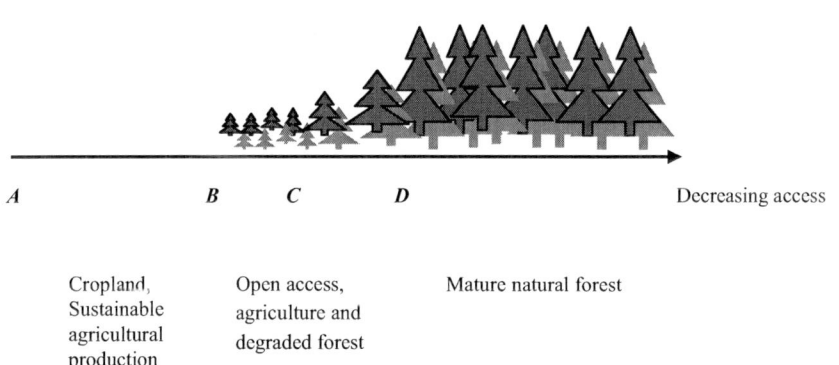

FIGURE 2.3 A developing frontier, stage II

rather than agricultural values, now explain the extent of this new region of increasingly open access between points B and D. This is an important distinction. Agricultural land conversion was the source of deforestation in the initial stage for regional development. Now, in this second stage, commercial forest activities such as logging are the cause of deforestation. Agriculture continues to make use of some open access lands, but the open access lands devoted to agriculture (*BC* in Figure 2.3) are now only a share of all open access lands (*BD*) within the region. Much as logging followed agriculture in stage I, agriculture may now follow logging into the interior of regions in stage II.

Once more, the cost of obtaining and protecting the property rights insures that the land between points *B* and *D* remains an open access resource. Of course, many governments protect some land beyond point *B* and even some land beyond point *D*, but they must absorb the increasing protection cost—and even then some trespass and *de facto* open access exploitation occurs. For example, protected forested parklands around the world suffer encroachment and government timber reserves around the world experience illegal logging.[9]

Forest degradation, a feature in the region of increasingly open access: The region of open access will not be fully deforested. Rather, the forest on these lands will be selectively harvested and degraded to the level where the expected return from remaining lower quality products is less than the opportunity cost of the labor and capital necessary for their extraction. The graphic below Figure 2.3 illustrates the effect on a uniform forest. The standing volume of the degraded resource tends to be larger (degradation is less) at point *D* than at point *B* because the opportunity costs of labor and capital used in resource extraction are greater at the greater distance from the market represented by *D*.

The standing resource at *D* is an interesting special case. For products like commercial timber with significant costs for the on-site act of harvesting, there are two general cases. In some cases there is an observable difference between the volumes and species of the remaining forest stocks standing on either side of the boundary at *D*. This difference is equal, in terms of the value of standing timber, to the on-site cost of harvesting the timber and preparing it for delivery to its market (including the costs of felling, bucking, skidding, and loading, but not including the cost of delivery itself). Of course, this on-site cost is large for timber. Therefore, the boundary of timber harvest activities may be easily recognized. One can stand near that point in the forest and observe the edge of the logging activity and the beginning of the undisturbed natural forest. This is a general case for temperate forests. These forests tend to be composed of a few species which tend to be clearcut or selectively harvested of large shares of their standing volume at one time.

---

9   Later, in chapter 7, we will discuss the rationale of those vertically integrated firms that leave a modicum of mature timber standing at their economic margins in the neighborhood of point *D*—even as they harvest other timber in this neighborhood. No one harvests beyond point *D* without either subsidies or other assistance borne by some different productive activity.

A second case is more common for the many tropical forests composed of mixed species and many age classes. In this case, logging at point D contains an element of hunting not unlike that for big game. Loggers search for the occasional groups of high-valued species and harvest only those. The rest of the natural forest around point D is composed of lower-valued species and it may be largely undisturbed, the decrease in standing volume surrounding the harvest site may be minimal, and the effect of logging on the local environment also may be minimal. This second case is also more representative for those other, non-timber, forest products that are simply gathered from the forest. These products have no substantial cutting and loading costs and the geographic limit of their extraction may not be so apparent to the eye.

In any event, the opportunity costs of the labor and capital used in extracting forest resources are important explanatory factors for the levels of deforestation and forest degradation, and both the extent of deforestation and the severity of forest degradation are less apparent in regions where the opportunity costs of extraction are greater. The previous paragraph explains this point for products with different harvest costs. The point also holds for the same product in two different regions. Consider Figure 2.4. Consider two regions that are similar in their agricultural and forest values, in their administration of property rights, and in the characteristics of their forests. They differ only in the opportunity costs of extraction from their forests. Loggers and harvesters of other forest resources extract from the forest out to the point where their compensation equals their opportunity cost. The lower their opportunity cost, the farther they are willing to go to extract products of the same value in the market. Therefore, the forest value gradient in the market with higher logging opportunity costs intersects the horizontal axis closer to the market at D_h, while the forest value gradient in the lower cost market intersects that axis farther to the right at D_l. Deforestation will be more extensive in the lower opportunity cost market, and we can conclude that the alternative opportunities for the employment of labor and capital, as they increase these opportunity costs, are critical determinants of the level of deforestation.

Forest degradation is also more severe in the lower cost market because smaller and lower quality products still offer sufficient reward to offset the lesser opportunities that forest users forego while extracting harvests from the open access lands in this market. Therefore, workers with lower wage opportunities extract more from any standing physical resource. The graphic below Figure 2.4 illustrates this point. The land between points B and D_l in the lower wage market is further degraded than the land between B and D_h in the higher wage market. Degradation may be so great that only scattered shrubs exist in the neighborhood of B in the lower wage market. Of course, the residual resource is larger at D_l than at B in both markets because the labor opportunities foregone are greater at D_l and the returns necessary to compensate that labor must also be greater.

Two examples, one from a developed country and one from a developing country, illustrate the generality of these characteristics of open access forests, deforestation, and forest degradation. In the rural areas of arid southeastern India,

24 The Pattern of Forestry in the Course of Economic Development

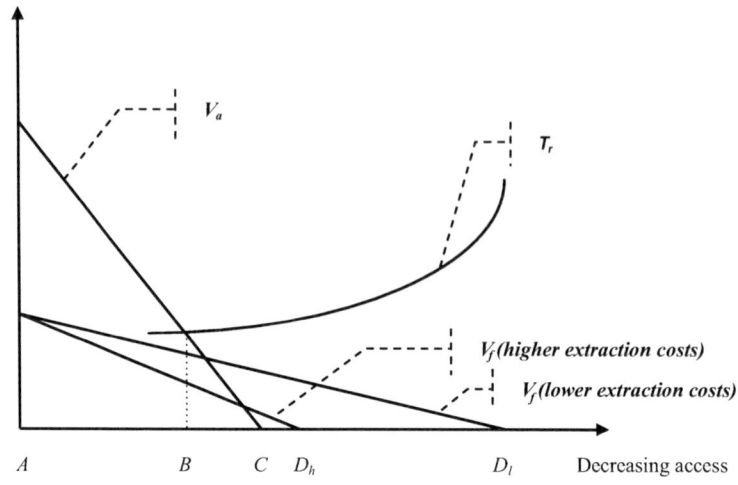

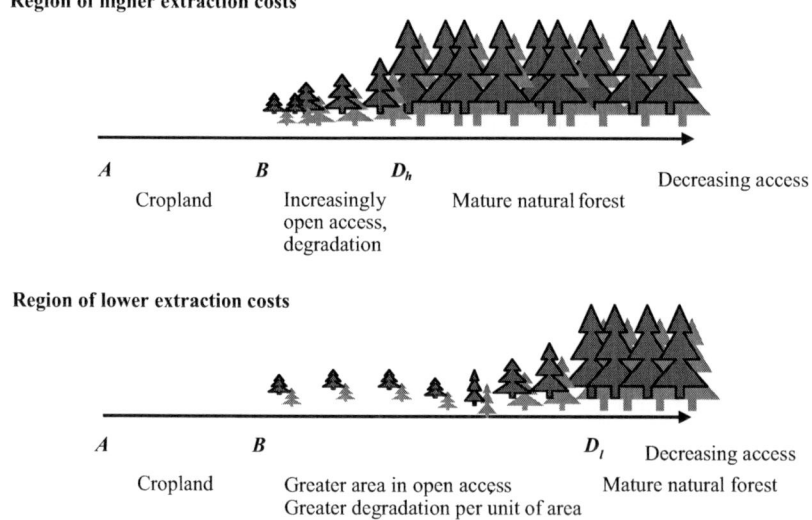

FIGURE 2.4 Labor and capital opportunity costs, deforestation and forest degradation

the population is very poor and its labor opportunities are most limited. Households collect fuelwood from woody bushes down to the size of an adult's index finger. The woody vegetation on the non-agricultural landscape is beyond degradation. It has disappeared almost entirely.

On the other hand, in rural southwestern Virginia in the United States, a few itinerant families collect holly and material from the growing tips of the evergreen forest for sale as Christmas greenery. These are poor families but their labor opportunities are nowhere near as scant as those in southeastern India. The "tipping" of the trees by these families degrades the standing forest in only a small degree and it does not measurably alter the growing stock. The degradation of the forest is not so easily noticeable. Nevertheless, it is an activity that often occurs without the permission of local landowners. This makes it a de facto open access activity. In the same region, others with more specialized skills collect ginseng, a prized forest herb. Virginia's ginseng crop of 6.5 tons was worth $1.8 million in 1991 (Hammett and Chamberlain 1998). Some ginseng collectors accept the risk associated with trespass and illegal collection from private forestlands while hoping to avoid detection. Their favorite harvest sites are closely held secrets, and the forest has been largely depleted of ginseng in many of its more productive and more accessible locations. In both southwestern Virginia examples the labor opportunities are greater than in southeastern India and the level of open access forest degradation is much less apparent. Nevertheless, at least a modicum of degradation does occur.[10]

Illegal logging, a second feature of open access: Fuelwood, Christmas greenery, and ginseng are non-timber forest products, but discussions of open access deforestation are most widely associated with timber as the primary product. Open access lands are a primary source of illegal logging and some amount of illegal logging occurs almost everywhere in the world. No formal ownership statement and no number of well-trained and well-motivated forest guards can fully prevent it. Pit sawyers in poor countries, for example, often pay little attention to the boundaries of forestlands that are the formal responsibility of the state forest agencies. They cut logs and saw lumber with a large degree of impunity because their two-person operations are so mobile and the forest area is so large that controlling their activity is nearly impossible even for the most motivated government foresters.

However, the general problem is often greater than a few pit sawyers. Growing global concern over illegal logging, associated with corruption in logging trade, has been a source of numerous national, regional, and international initiatives

---

10 In another example, McLean (2002) tells of the small town of Nakusp in the Kootenay region of British Columbia in Canada where the population grows 20-fold during prime mushroom season. Mushroom collection is a de facto open access activity conducted on the surrounding provincial forest lands. Individual collectors make as much as $40,000 annually from the activity and they guard the secret locations of their favorite productive sites as zealously as ginseng collectors in Virginia guard theirs.

Linde-Rahr (2005) provides an econometrically rigorous example for the extraction of non-market forest products in Vietnam. He observes that those households that are less well-endowed with other resources (i.e., poorer households) are more dependent on collection from the open access forest—and this forest remains an open access resource because its low value does not justify more favorable administration. In contrast, higher income households (whose labor opportunity costs, therefore, are greater) engage less often in the collection of non-market forest products and, when they do collect, their collection tends to be from more accessible forests with household specific user rights. Jodha (2000) draws similar observations from the state of Rajasthan in west central India.

in recent years. The control of this illegal activity has been on the agenda of the G7/G8 summits of the world's most developed nations since 1997, and G8 placed illegal logging on its list of priority actions in 2005.

Illegal logging may cost domestic national governments as much as US$15 billion annually in foregone taxes and timber royalties and it may be the source of one-quarter of all hardwood lumber traded globally (Sizer 2005, Seneca Creek 2004). For Brazil, Bolivia, Russia, Cambodia, Cameroon, Indonesia, Myanmar, and Papua New Guinea, illegal activity may account for 50 percent or more of all roundwood production (Contreras-Hermosilla 2001, Tacconi, Boscola, and Black 2003).[11]

These are developing countries or countries in economic transition—but illegal logging occurs in developed countries as well, although generally on a smaller scale. British Columbia alone suffers annual losses of US$200–300 million from timber theft and fraud (Smith 2002). In the United States, three percent of harvests from private land and 10 percent of harvests from the National Forests may be illegal. Landowners in the Appalachian region of the United States may lose $4 million each year to timber thieves and, in 2007, the non-profit Appalachian Roundtable was working with more than 50 families in that region who were victims to such losses (Timber 2007). Yet, in 2001, only three arrests were made for illegal logging anywhere in the United States (Mendoza 2003). A commercial logger who was also an elected member of the local government once showed me the site of his own illegal logging on public lands in northwestern Montana. As he explained, the key to his operation was the inability of the government foresters to monitor all the logging roads all the time. The U.S. government forestry agencies disapprove of these activities, but they cannot extend greater effort to prohibit them because their budgets are insufficient and the cost of enforcement would be much greater than the potential gain.

In general, illegal logging occurs where the log values are positive but where landowners perceive that the cost of enforcing their property rights exceeds the forest values at risk. In some developing countries, the risk to the logger may seem small when the return from a single illegal log can exceed several months' wages. The value of a single high-valued silk log in Sri Lanka may even exceed a year's wages for many rural farmers (Gunatilake 2007). Illegal logging is less frequent in developed countries only because the return to the illegal activity is less likely to be worth the risk: less risky and better paying alternatives to illegal activity exist for the skilled labor required to operate the heavy equipment that typifies logging operations in these countries.[12]

---

11 International market flows themselves provide evidence of illegal logging. Reported international shipments from many developing countries are less than the reported levels of log imports coming from those same countries, especially to the developed or rapidly developing countries of East Asia. The difference is unreported and generally illegal logging in the exporting country.

12 Our model implies a cost gradient associated with illegal activity. This idea is common in the economics of illegal activity and enforcement. Clarke et al. (1993) first explained its form for forestry.

Summary: To summarize the discussion to this point, the construct of Figure 2.1A conforms to the common description of any initial settlement. Trees impede agricultural development and the forest value gradient is very low. Net forest resource values are sufficiently low that point D, where the forest value gradient intersects the horizontal axis and forest values decline to zero, is to the left of point B where agricultural land use value declines to zero. This describes new settlements in the westward development of the Americas in the seventeenth to nineteenth centuries and the northward European settlement in southern Africa in the nineteenth century. It also describes, for example, continuing frontier settlement in many parts of South America, migrant settlement in Sumatra in Indonesia, new upland migration in the Philippines, and subsistence settlements in Zambia in the latter twentieth century.[13]

In other cases, described by stage II and Figure 2.3, agriculture no longer extends to the frontier of mature natural forests, point D is to the right of point C, and the land area between points B and D may be large and seriously degraded. This describes the poorest rural areas of many developing countries today, including for example regions as geographically disparate as the southern two-thirds of Malawi and portions of Tamil Nadu in India, of Nepal's hills, and of China's remote and arid Qinghai province. The positive net value of the original forest, together with its open access character, has assured removal of the best resources. Some degraded vegetation remains, and with time it may re-grow naturally. The poorest households will continue to exploit these degraded resources as the scattered vegetation grows to a minimum exploitable size or as its fruits begin to ripen.[14]

---

13 Numerous descriptions of this pattern exist. Richter (1966) provides a lucid description of this pattern of frontier settlement for the Ohio Valley of the United States in the early nineteenth century. He describes the daily life of families settling among the trees, building homesteads, girdling trees to create land for agricultural production, and sometimes abandoning the land for newer and more western frontiers. Chomitz and Gray (1996), Lopez (1998), Amacher et al. (1998), and Pfaff (1999) provide econometric measures of agricultural settlement and conversion of the modern forest frontier in Belize, Cote d'Ivoire, the Philippines, and Brazil, respectively. Heydir (1999) provides a description of historic forest use in Sumatra that extends into a description of settlement on the modern frontier. He observes that it is voluntary migrants, not government induced transmigrants, who are settling and converting Sumatra's forestland today. Alston, Libecap, and Mueller (2000) provide econometric evidence of the relationships between institutions, land use conflict and deforestation at Brazil's Amazon frontier.

14 Several recent empirical assessments establish this point about relative labor opportunities and forest use. Hyde and Seve (1993) develop an economic simulation for fuelwood collection beyond the frontier of agricultural land use that conforms with evidence for Malawi. Foster et al. (1998) and Amacher et al. (1999) provide econometric evidence of the effect of labor opportunities on forest use for India and Nepal, respectively. Foster et al. also develop the economic reasoning underlying the use of the open access region by lower wage households. Amacher et al. demonstrate the shift to increasing use of the open access resource as relative wages decline. Escobal and Aldana (2003) observe that only those forest users who find alternative employment unrelated to the forest manage to break the link between poverty and forest degradation in the Peruvian uplands. Young (2003) develops the more general relationship between property rights, labor opportunity, and deforestation, and supports it with econometric evidence from Brazil's Amazon. Tachibana, Nguyen, and Otsuka (2001) show that greater opportunity for agricultural labor deters deforestation in Vietnam.

28 The Pattern of Forestry in the Course of Economic Development

Stage II also describes timber harvest operations in regions like the Bolivian Amazon, Indonesia's East Kalimantan, and even the Appalachians and the southern Rocky Mountains of the United States, regions where logging operations are characterized by heavier equipment, and the skilled labor required to operate it is not so plentiful. These more capital-intensive logging operations may extend farther into the hinterlands than many small-scale loggers would go with their own private operations. The degradation in these regions is that which is conducted with heavy equipment on selected sites rather than the more complete removal of all vegetation across large expanses of land that we observe in examples like those where the labor opportunity costs are lower. For example, Long and Johnson (1981) estimate that fewer than 20 percent of all trees were removed during the early logging operations in Indonesia's East Kalimantan. Loggers harvested those high value logs that could be removed easily and floated to the ports for export. They left standing the lower-valued trees and those hardwoods that were too dense to float, and the untrained eye finds it difficult to distinguish the residual part of the harvested stand from the unharvested adjacent forest.

The use of the natural forest continues during this second stage of forest development, deforestation continues and the forest margin at point D slowly extends farther and farther from the market center. The delivered costs of forest products continue to rise. Nevertheless, the incentives of higher prices remain insufficient to induce even minimal tree planting or forest management and any attempt at forest management will be unsustainable. The only management on these lands is harvest scheduling to coincide with the use of the more expensive capital equipment necessary for some logging and wood processing operations.[15]

A Mature Forest Frontier, Stage III

Eventually, the frontier of economic activity at point D becomes sufficiently remote—and the delivered costs and local prices become great enough—to induce substitution. This occurs when the costs of removal and delivery to the market equal the backstop cost of the substitute. Substitution may take the form of new consumption alternatives to forest products—for example, brick, stone, and concrete block as substitutes for construction wood; or kerosene, LPG, and more efficient stoves as substitutes for fuelwood. Alternatively, the substitution may be production related—for example, permanent forest management on land with more secure rights and closer to the market as a substitute for the unmanaged products extracted from the open access natural forest. The gradual long upward trend in basic forest product prices comes to an end at this time.[16]

---

15 Both Berck (1979) and Johnson and Libecap (1980) show that the westward moving locus of timber harvests in nineteenth century United States followed this economic pattern. Hofstad (1997) relates a similar pattern of expanding forest extractions for charcoal in the vicinity of Dar es Salaam, Tanzania, and Chomitz and Griffiths (2001) describe it again for charcoal production to supply the population centers of Chad in central Africa.

16 von Amsberg (1996) also makes the conceptual case for prices having differential effects on natural and managed forests. The analytical evidence of rising prices inducing forest inves-

Of course, in conjunction with local economic development forest management itself displays increasing complexity at the margin of land use transition between the most intensive forestry and extensively managed agriculture. In modern forestry, we observe, first, site preparation and planting, and later, in some regions, pre-commercial thinning, fertilization, and still later, pruning. We also observe increasing distinction among the lands on which more intensive management activities occur (e.g., Yin and Sedjo 2001).

Very clearly, prices have not risen sufficiently to induce forest management in all of the markets of the modern world and many regions of the world have not yet attained the third stage of forest development. Nevertheless, many other regions have attained this stage—for industrial timber and also for other categories of forest or forest ecosystem services. That is, the evidence of sustained forest management is not ubiquitous—but it is not trivial either. Forest plantations cover more than 187 million hectares worldwide, and managed natural stands account for an unknown addition to that 187 million. Forest plantations account for more than four percent of the world's total forest area and more than 1.4 percent of the area in all land uses combined. The land area in plantations is growing by almost 4.5 million hectares annually, and plantations now supply approximately 22 percent of the world's industrial roundwood (FAO 2001).[17] For specifically tropical and sub-tropical plantations, Sedjo (1994) observes that the share of industrial wood production doubled from 1977 to 1992. Undoubtedly, that share has increased further since 1992 and, just as surely, it will continue to grow in the twenty-first century.

Consider two examples, China and Indonesia. China alone has 46 million hectares of forest plantations, including commercial timber plantations, plantations for other commercial uses of trees such as orchards, and plantations designed for non-commercial purposes such as erosion control. Its plantations account for 10

---

ment is limited—although the anecdotal evidence is extensive. Difficulties with data aggregation across various market conditions are part of the problem. In one exceptional assessment, Zhang, Uusivuori, and Kuuluvainen (2000) demonstrate that rising prices on the island province of Hainan in China induced both an increase in the land area in managed forests and a simultaneous decline in natural forest area—just as we have predicted. Several case studies generally designed for other purposes also illustrate the investment-inducing effect of rising prices. For example, Shively (1998) and also Garrity (1995) discuss the positive effect of rising prices (and increasing uncertainty) on smallholder tree planting in the Philippines, and Yin and Hyde (2000) show the positive effect of rising prices (and secure tenure) on planting trees for erosion control in China's North Central Plain. Amacher et al. (1993) show that Nepali households planted trees in a region of greater fuelwood prices while households in a neighboring region that experienced lower prices continued to collect from the natural forest. (These observations are consistent with Demsetz's [1967] fundamental argument that weak property rights (in forestry in our case) are associated with abundance and recognized rights to property improve with increasing scarcity.) In two examples of substitution on the consumption side, Chomitz and Griffiths (2001) and Hofstad (1997) observe that alternative energy sources and alternative cooking technologies constrained additional price increases and more distant extraction of both fuelwood in Chad and wood for charcoal production in Tanzania.

17 The data on forest plantations generally exclude agricultural tree crops such as orchards, palm plantations for oil and coconut, etc. China's data are the exception.

percent of China's standing forest volume. Commercial timber plantations account for slightly more than 50 percent of the 46 million hectares (PRC/SFA 2001).

Indonesia is often cited as an example of rapid deforestation and poor forest management—and this may be a fair characterization. Yet Indonesia has 2.6 million hectares of plantation forests plus another 14.5 million hectares in perennial tree crops growing products like palm oil, cloves, coconut, and rubber (known as "estate crops," GOI/MFEC 1997). Official industrial forest plantations comprise only 1.9 percent of Indonesia's forests. Yet trees in plantations of one sort or another account for over 17 million hectares, or 9.4 percent of Indonesia's total land area (GOI/MFEC 1997). Indonesia's forest plantations and its estate crops both produce managed substitutes for tree products that continue to grow in physically plentiful but economically less accessible natural forests.

In the general terms of Figures 2.1A and 2.3, these examples mean that the forest value gradient has continued to rise with the increase in harvest and delivery costs (from the dashed line to the new solid forest rent gradient in Figure 2.5) until, at some moment in time, it intersects the agricultural value gradient to the left of the intersection of agricultural value with the cost function for secure property rights.

We might call Figure 2.5 the illustration of a mature frontier. Localities and markets described by this figure are mature in the sense that sustainable forest activities (in region B'B") compete with some resource removal activities (at D). They are still frontiers because some unsustainable removal of natural stocks at the frontier remains competitive with the sustainable activities. In this case, the marginal cost of growing, harvesting, and delivering products from the managed forest equals the marginal cost of simply harvesting and delivering the same products from the more distant and less accessible mature natural forest. Harvests of comparable products from both forests obtain a comparable delivered price in the local market.[18]

The new managed forests may take the form of industrial timber plantations or of agroforestry or even of just a few trees growing around individual households, in home gardens, along fences and roadsides, and in village parks. The latter are generally not included in the common measures of the forest stock but their numbers and their economic importance can be large in almost any part of the world.

For example, the Southeast of England, including the city of London, is the most populated region of that country. It is also the most wooded and its wooded area is growing faster than that in any other region of the country. London

---

18 Obviously, marginal delivery costs are greater for the more distant natural forest. Marginal harvest costs may also be greater for the natural forest because the natural product is not as uniform in size or log form as the product of many managed forests. However, the natural forest incurs no growing costs. Therefore, the positive marginal cost of growing managed forests must offset its advantage in lower harvest and delivery costs.

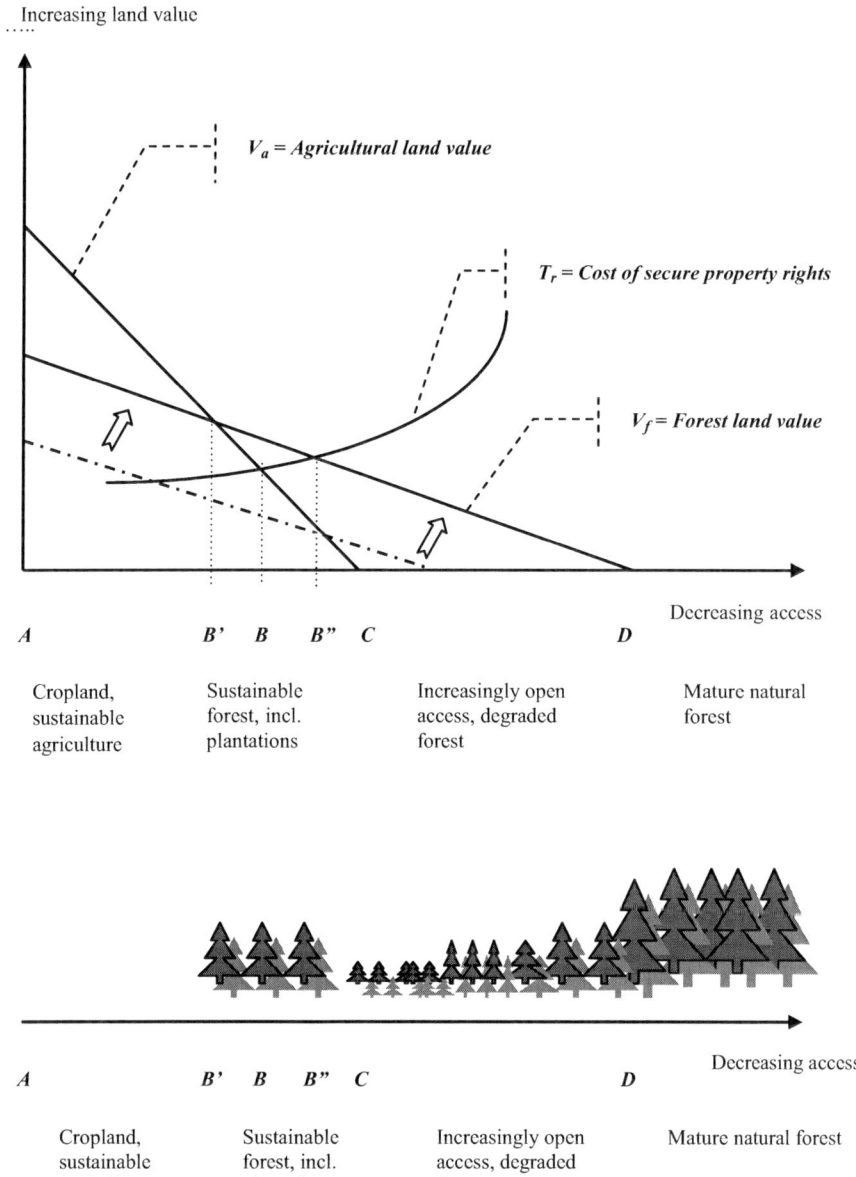

FIGURE 2.5 A mature forest frontier, stage III

alone has 65,000 stands of trees and woodlands covering almost 7,000 hectares. Almost 5,000 hectares are composed of woods larger than 9.5 ha. (UK/Forestry Commission 2001).

In the dry high plains of the Denver metropolitan area in the United States, for another example, trees comprise only seven percent of the land cover. They do not appear in the official U.S. forest inventory for the broader region, but they act as a $44 million regional storm-water management system, remove 2.2 million pounds of air pollutants annually, and save $50 per household in annual air conditioning costs (Stein 2001).

Throughout South and Southeast Asia, trees growing in plots that are too scattered and too sparse to meet the measurement standards for inclusion in official estimates of forest stock account for 65–85 percent of all market timber and fuelwood production (FAO RWEDP 2000). On the densely populated island of Java, they account for tree cover on as much as 47 percent of the total land area—well in excess of the 24 percent that comprises the government's official estimate of forestland.[19] (The additional 23 percent is not part of the official estimate because it is composed of scattered trees and stands rather than continuous forest and because its value is for shade, fuel, fruit, and estate crops rather than commercial timber.)

In El Salvador, following civil war in the 1980s, the natural forest may have covered only five percent of the landscape. With renewed economic and political stability, a new and somewhat managed forest has emerged. Shade coffee and orchards, hedgerow trees, urban tree cover, and forest regeneration on abandoned pastures contribute to a forest that covered 60 percent of the country by 2001. The new forests supply wood products, protect watersheds, and provide habitat for 520 bird species, 120 mammals, and 130 reptiles and amphibians (Hecht et al. 2006)

In Africa, trees growing on farm plots are major sources of fuelwood in Malawi (Hyde and Seve 1993) and timber production in Kenya (Scherr 1995). In fact, most farms contain some tree cover. Wood, Sebastian, and Scherr (2000) calculate that the majority of Latin American, African, and South and Southeast Asian farmland contains more than 10 percent tree cover and one-quarter of all agricultural land in these regions is under more than 30 percent tree cover.

Summary Observations and the Global Evidence

In all three stages of development, loggers concentrate their removals from the natural forest in the lands around a point like D in the figures. Those mature natural stocks growing closer to the market, or to the left of point D, were removed in earlier times. A mature natural forest of little current market value and no local subsistence value exists beyond D—farther into the hinterlands, into areas like swamps or mountains that are more difficult to penetrate. Sometimes these unmarketable forests are of negligible total area because the roads are good and the reach of neighboring markets with their own higher-valued agricultural, industrial, and commercial activity begins to overlap the market described by the figures. In this case the frontier may even disappear. Indeed, the land between two adjacent markets may not support any commercial forestry whatsoever if

---

19 The 47 percent is an estimate from personal communication with D. Garrity in May 2000. The official estimate of 24 percent is from GOI/MFEC (1998).

higher-valued land uses dominate. Regions that describe markets like these import their forest products.

Sometimes, however, the area of unmarketable natural forest beyond point D continues for great distances until it supports the largest share of reported physical stocks of forest. This is the case of the great non-commercial forests remaining in the world today—in Siberia, northern and interior Alaska, northern Canada, much of the Amazon, some of central Africa, most of Kalimantan in Indonesia, and large parts of Papua New Guinea, all regions where the area of mature natural forest dominates the area of actively managed forestry

Many countries contain multiple regions in different stages of forest development because most forest products are either bulky or perishable. They do not transport over land well. Therefore, their primary markets are geographically contained and the location of their next level of processing tends to be close to the source of primary production. Some regions contain no managed forests and their extractive forest activities occur only in the natural forest (stages I & II). In other regions of the same country, some firms and some households have planted trees and they manage some forests. The harvest operations in these regions occur in both managed stands and natural forests (stage III).

Consider timber production in the United States. We observe plantations and forest management in the coastal plain and the piedmont of the South and in coastal Pacific Northwest, but we also observe unmanaged natural forests farther inland and at higher elevations—and timber harvests occur simultaneously at both locations in these two regions. These regions are an example of stage III. We also observe timber harvesting in the Rocky Mountains and in drier eastern Oregon and Washington, but seldom from managed forests. There are no significant commercial timber plantations in these latter two regions. They are examples of stage II.

Java and the Kalimantan provinces of Indonesia provide a similar example for a developing country. The island of Java is an example of stage III. It contains both plantations and natural forests, and some level of harvest activity occurs in both. Indonesian Kalimantan, however, contains large expanses of natural forest and almost all timber harvested in the Kalimantan provinces originates from natural forests. The very few plantations that have begun to appear in Kalimantan in recent years suggest that this is a region in stage II, but one that is approaching maturity into stage III. Once more, uneconomic standing natural reserves remain in the less-roaded upland and inland areas of both Java and Kalimantan.

In sum, the three-stage characterization of forest development identifies three categories of forest: (1) managed forests including industrial forests and forest plantations, more scattered trees in residential areas, and well-managed agroforestry production all in the area identified by $B'B''$ in Figure 2.5; (2) degraded natural forests from point B in Figure 2.3 (or B'' in Figure 2.5) out to point D; and (3), a presently unmarketable mature natural forest beyond point D. For some purposes, it is convenient to separate the latter into two parts, including (4) the neighborhood of unsustainable current harvest activity from the mature natural forest centered around point *D*, as well as the presently unmarketable mature forest beyond it.[20]

Global evidence: The global forest inventory data organized by the Food and Agriculture Organization of the UN provide an impression of the magnitude of each of these four categories (FAO 2001). See Table 2.1. These are official data provided by the governments of the individual countries and they are physical estimates that were not designed for economic assessments such as ours. Nevertheless, some impressionistic judgments are possible.

FAO's estimate of the global forest is 3,869 million hectares. This estimate includes between 428 and 479 million hectares that are protected in parks and for wildlife preservation and other public values. The FAO estimate is a fair measure of total forest area but it is underrepresentative of the global impact of trees because the 3,869 million hectares includes very few of those trees that are outside continuous forests, in home gardens and backyards, along roads and city streets, in shelterbelts, and in agroforestry production systems.

About half of the total unprotected forest, or 1,686 million hectares, fall within officially recognized forest management plans. These 1,686 million hectares bear some comparability to our category of managed forests. Some additional forestland is managed, but not under an official government-recognized plan. On the other hand, many forest management plans recognized in the FAO estimate include a component of non-commercial forest. Others are simply an administrative response to a government requirement and are not reflective of any overt on-the-ground activity.[21] Therefore, all that is certain is that 187,000 hectares of forest plantations, approximately half of which are designated for timber

---

20 The three-stage characterization provides important insight to the ratio of harvests to growth. This ratio is often used without clear understanding of its implications. In fact, harvests always exceed growth during the first two stages of development because the harvest level is positive during these stages, yet growth is negligible in the remaining mature natural stands at the frontier and beyond. During the third stage of development, harvests approximate growth for the managed stands but the frontier remains a source of some harvests. Therefore, harvests from the combination of managed stands and the frontier must exceed growth (which very largely occurs in the managed forest) during the third stage as well.

Growth can exceed harvests only in regions where the measures of standing forest volume begin to include trees and forests grown for shade, parks, erosion control and other non-extractive purposes, or where abandoned agricultural land has begun to revert to forest. The volume of trees and forests growing under these latter conditions can be large. It has been sufficiently large that, in the United States for example, overall growth exceeded harvests for the last several decades of the twentieth century. Growth has begun to exceed harvests in China for the same reasons. Agricultural land use is declining, and the volume contained in areas under tree cover for non-commercial reasons is expanding rapidly.

21 For example, Stone (1998) tells of the eastern Amazon, where management plans required by Brazil's environmental agency (IBAMA) add logging operation costs of about US$1.30 per m^3 of harvested timber, and create employment for consulting foresters, but make little difference for the local forest environment. Bowles et al. (1998) conclude that, after all such adjustments are accounted for, less than 0.2 percent of tropical forests were managed in the late 1990s.

TABLE 2.1 The global forest

i) Managed forest	ii) Degraded natural forest	iii) Previously unmarketed mature natural forest	iv) Region of current extraction from the natural forest (a component of iii)	Total global forest area
1,686 million ha. under management plans, 187 million ha. of this in plantations.	? Total area unknown, but the degraded forest can be large. In India, for example, the official estimate for degraded forests is almost sixty percent of the total forest area.	775 million ha. (Bryant et al. 1997) Forty percent of the world's forest—1,548 million ha. (FAO 1998). [The FAO (2001) estimate is only 109 million ha. but this estimate seems to reflect land available for harvest rather than economic accessibility.]	155 million ha. currently threatened (Bryant et al. 1997). Current extraction occurs only on the share of this that is closest to the economic frontier.	3,869 million ha., 428–479 million ha. in protected park and wildlife areas

Note: The estimates in managed and in previously unmarketed mature natural forest sum to 2,461 million ha., or approximately 1,000 million ha. less than the FAO estimate of available global forest. That total is an underestimate by the amount of forest in unaccounted agroforestry, roadside trees, city parks, etc. Therefore, the only confident conclusion that can be taken from any of these numbers is that all four categories of forest are large on a global scale.

Source: FAO (2001), except where noted otherwise

production, fall within the economically managed category. The actual measure of economically managed forest is probably a great deal larger than 187,000 hectares but a good deal less than 1,686 million hectares.

One FAO (2001) document reports that only 109 million hectares, or three percent of the total that is not in protected forest, is an inaccessible residual comparable to our third forest category, the mature natural forest. However, FAO speculates that the proportion "available for wood supply" *increases* with distance from the transportation infrastructure! Surely this is incorrect. Surely the economic timber supply decreases as the forest becomes less accessible. Therefore, this FAO observation can only be meaningful if its real intention is to convey that those demands that compete with wood supply decline with increasing distance from roads and other means of easy transportation. We must search for a better estimate of the economically inaccessible forest.

Bryant, Nielsen, and Tangley (1997) estimate that one-fifth of the global forest, or approximately 775 million hectares, exists in large unbroken blocks of natural ecosystems. This characterization is more compatible with our third forest category. Yet even this 775 million may be an underestimate of the uneconomic residual mature natural forest. Even in the United States, for example, with its well-developed system of forest roads, between 20 and 30 percent of the total forest area is probably inaccessible for economic timber extraction.[22] An earlier FAO (1998) evaluation speculates that 40 percent of the world's forest (1,548 million ha.) is economically inaccessible. In sum, the economically inaccessible component of the global forest (our third category) is large—but probably smaller than the managed forest (our first category). Both may be larger than the FAO (2001) estimates in Table 2.1.

Bryant et al. estimate that 39 percent of the frontier forest, or 155 million hectares, is threatened. Logging is a threat to the natural forest but other, non–forest, land uses are also a threat. Therefore, the 155 million hectare estimate must be an upper bound for the fourth category, the area of current commercial harvest activity.

The degraded forest and the crucial roles of local institutions and rural labor opportunity: The land area in degraded and increasingly open access forest (our second category) is more difficult to estimate. It will be largest in those developing countries with both the smallest ratios of labor opportunity costs to forestland value and the least well-developed institutions for establishing and maintaining property rights. We previously discussed the effect of lower opportunity costs on forest extraction. With relatively lower labor opportunity costs, rural collectors of forest resources can justify greater personal expenditures of time in forest

---

22 Of 302 million hectares of forestland in the United States, 128 million are public lands (USDA Forest Service 2005). Only a fraction of these public lands, perhaps no more than one-half and very likely an even smaller fraction, could be accessible for purely commercial timber extraction without additional public financial assistance for road improvement or new road construction.

extraction and this allows them to conduct their extractive activities farther into the remote areas away from the market center. That is, the forest value function in the figures will be shallower and the boundary of mature natural forest (at D) will be farther from the local commercial market in these cases. The open access region will be heavily degraded precisely because the labor opportunities are low and the rewards from additional degradation exceed the alternative labor opportunities until the forest is heavily depleted. Furthermore, less developed regions and countries in general tend to have less completely developed institutions for the establishment and maintenance of property rights. Therefore, the transactions cost function associated with maintaining any particular level of forest protection will be higher in these cases. Taken together, smaller labor opportunity costs and greater transactions costs predict larger regions of open access and more completely degraded forests in these cases.

We can apply this reasoning to anticipate the countries with larger areas of open access degraded forest. General measures of wages or more general labor opportunity are not always available. However, lower wages undoubtedly occur where per capita incomes are also lower. Therefore, we can anticipate that countries with lower per capita incomes and lower per capita forest cover (which implies higher value for the remaining forest) are likely to possess larger areas of open access forest, some of it so severely degraded that it is no longer included in the official forest statistics. Countries described by the contrasting case, relatively higher per capita incomes and greater per capita forest cover, are likely to contain larger tracts of remaining natural forest. These latter are more likely to be in the first or second stage of forest development, their agricultural lands are more likely to abut the natural forest, and their regions of open access forest are likely to be less completely degraded. Of course, other factors are also at play in any relationship between income and forest cover. For example, Uruguay has a relatively high per capita income (US$9,420 in 2007) and a moderate-to-low per capita level of forest cover (0.4). Uruguay's proximity to large forests in neighboring Brazil and Argentina explain the difference. These two neighbor countries provide much of the forest resource that Uruguayans use for both commercial and non-commercial purposes.

Table 2.2 makes these comparisons for a number of South Asian, South American, and sub-Saharan African countries. In general:

- The first set of conditions, lower per capita incomes and lower per capita forest cover, describe the countries of South Asia. Open access is a critical issue for forest management in these countries and for India in particular. India's official statistics designate 75.5 million hectares, 23 percent of its total land base, as wasteland—which is a part of the open access area between agriculture and the natural forest. Degraded forest and grazing land and the area of shifting cultivation alone account for an additional 24 million hectares, most of which is also a part of the open access land in our categorization (NRSA 1995). India's official estimate of forestland, in contrast, is only 64.1 million hectares (FAO 2001).

TABLE 2.2 Predicting degraded open access forest

South Asia			South America			Sub-Saharan Africa		
Country	GDP/capita (US$-2004)[a]	Forest/capita (hectares)	Country	GDP/capita (US$-2004)[a]	Forest/capita (hectares)	Country	GDP/capita (US$-2004)[a]	Forest/capita (hectares)
Bangladesh	1,870	n.s.	Argentina	13,300	0.9	Botswana	9,950	7.8
Bhutan	290	1.5	Bolivia	2,270	6.5	Gabon	6,620	18.2
India	3,170	0.1	Brazil	8,200	3.2	Namibia	7,420	4.7
Nepal	1,490	0.2	Chile	10,870	1.0	Zimbabwe	2,070	12.0
Pakistan	2,230	n.s.	Colombia	7,260	1.2			
Sri Lanka	4,390	0.1	Ecuador	3,960	0.9	Angola	2,180	5.6
			Guyana	766	19.7	Cameroon	2,170	1.6
			Paraguay	4,810	4.4	Congo	710	7.7
			Peru	5,680	2.6	Cote d'Ivoire	1,550	0.5
			Suriname	4,100	34.0	Kenya	1,140	0.6
			Uruguay	9,420	0.4	Zambia	940	3.5
						Burundi	680	n.s.
						Ethiopia	760	0.1
						Nigeria	1,150	0.2
						Malawi	650	0.2
						Uganda	1,480	0.2

a = purchasing power parity
n.s. = not significant
Sources: The Economist (2006) for GDP/capita; FAO (2001) for forest area;

- The contrasting conditions of higher incomes and greater per capita forest cover describe the countries of South America—where the open access exploitation of the natural forest is an issue but, as anticipated, agricultural land often abuts the forest, wasteland is not a general problem of the same magnitude as in South Asia, and open access exploitation and forest degradation are less visibly recognizable than in South Asia.
- Some other regions of the world are more difficult to characterize. In sub-Saharan Africa, for example, countries such as Botswana, Gabon, and Namibia share labor and per capita forest cover conditions with the countries of South America. Their regions of open access forest are smaller. Ethiopia, Nigeria, and Malawi, in contrast, share labor and per capita forest cover conditions with the South Asian countries. Not surprisingly, their open access forest regions are larger and more seriously degraded.
- Areas of effective open access do exist in more developed countries but they are less easily recognized. Rural wages are higher and wages are a smaller share of the total costs of forest extraction. (Capital equipment is relatively more important.) Therefore, the forest value function is often steeper. Furthermore, the institutions establishing property rights tend to be more complete where local wealth and the per capita labor opportunities are greater. Therefore, the transactions cost function tends to be lower and its slope is shallower in these countries. The combined effect of the relatively steeper forest value function and the relatively lower transactions cost function allows for only a smaller region of open access. Thus, the level of forest degradation within developed countries is minimal simply because the rewards from further degradation are insufficient to justify the effort.

In sum, the global forest is composed of three large components: managed forests, degraded open access forests, and mature natural forests. The first and the third may be the larger two on a global scale but many countries contain some share of all three forests. While the degraded open access forest may be small in many developed countries, it can be very large in some poorer developing countries. Clearly, relative labor opportunities and relative transactions costs are determinants of the extent of the open access forest and of its degree of degradation. They are also important explanatory factors for the extent of extractive activities conducted at the boundary of the mature natural forest.

Preliminary policy insights: The crucial points are that:

- We can differentiate at least three categories of forest, and
- the appropriate measure of the forest depends on the policy objective.

The differentiation is crucial because the same market and policy factors can have opposite effects on the different categories of forest. Consider a few examples. Timber price incentives are an inducement to improve and expand the managed forests of regions in stage III, but they are also an incentive to harvest and extend the deforestation of natural forests. For regions in stages I and II, price incentives have only the latter, negative, impact on natural forests. There are no managed forests in these regions to take advantage of the price incentive.

Similarly, public cost-sharing and technical assistance programs for forest management are inducements to improve and expand the managed forests in regions in stage III, but they have no effect on regions in stage I or stage II, regions that contain natural forests but no managed forests. Agricultural policy incentives, another example, induce land conversion from the forest to agriculture for regions of new frontiers, stage I. They cause some agricultural conversion from previously degraded lands for regions in stage II—but whether these degraded lands are included in the official forest inventory depends on the local standards for forestry data. Agricultural incentives have little effect on the natural forests of mature regions in stage III, but they may cause agriculture to out-compete managed forests at their intensive margin (B' in the figures).

The importance of the distinctions among different measures of the forest stock can be illustrated in a comparison of policies intended either to improve carbon sequestration or to protect endangered habitat. All trees sequester carbon. Therefore, an accurate measure of the effect of a policy designed for this objective must include those managed trees contained in orchards, backyards, city parks, windbreaks, along roadsides, and in agroforestry cropping systems. These trees are not included in most official national forest statistics. Overlooking them grossly underestimates the total amount of carbon sequestered in trees. (An appendix to this chapter discusses the forest measurement problem further.)

On the other hand, most forest-based endangered habitat is contained in the remaining natural forest past point D in the figures. Measures of the forest that simply aggregate all three categories of forestland do not provide a good indication of the amount of remaining natural habitat, and policies designed to affect forests in general but which concentrate their effects on extending the area of managed forests or on improving the condition of the depleted open access area generally have little effect on endangered habitat.

In conclusion, the three-stage classification of forest development and the three distinctive categories of forest that emerge from it create a means for tracing both policies and spillovers from market activities onto the various components of the forest. This classification will become central to the assessment of direct forest policies, of the effects of market and policy spillovers from other sectors, and of modifications of the relevant institutions in chapters 4–6. It will also be instructive for the examination of forest properties and the behaviors of the various categories of forest landowners in chapters 7–10.

Modifications and Qualifications

The basic model of forest development is nearly complete. (Chapter 3 will complete it with a discussion of the effects of long-run changes in the forest and the economically adjacent agriculture and wood processing sectors.) Our examples should be convincing of its principles. Nevertheless, most of us recognize that transactions costs vary with differences in local institutions and, therefore, property rights are not as simple as our single transactions function T_r suggests. Furthermore, the environmental landscape around the market at point A is not homogeneous. This means that the three functions in the figures are not rigid and fixed, and that the land use margins we have described are not always sharp and clear.[23] We also know that forests contain multiple market values, as well as important non-market environmental services that the model seems to overlook altogether. We have alluded to these issues. This final section of the chapter extends the introduction to each of them.[24]

Multiple Market Values

Figures 2.1–2.5 are generalized representations. They may be understood most easily with reference to forests that produce a single product such as fuelwood or commercial timber of uniform species and age—but forests produce many products and even commercial timber comes in multiple species and grades, each with its own price. This means that the single forest value gradient in the figures is actually a series of gradients, one for each forest product and grade. Each gradient has its own characteristic slope and intercept.[25]

Consider three products from the Amazon frontier: the lower-valued timber species that are primary inputs for the local sawmill industry, more highly-valued mahogany (nara) that is also processed into lumber, and Brazil nuts, a highly-valued non-timber product of this forest.

The lower-valued species that comprise most of the logs processed locally might be represented by the value function V_{fl} in Figure 2.6. In most local markets of the Amazon these species are not sufficiently valuable to justify investment in their management. They have been removed to a point like D_l that identifies the general neighborhood of their current harvests.

Mahogany is higher valued. It is also rare, occurring in clumps known as "mancha" that occur in wide spatial variation but at a general rate of perhaps one mature

---

23 A series of papers by Albers (1996) and colleagues (Robinson, Williams, and Albers, 2002; Robinson, Albers, and Williams, 2008) demonstrate the conceptual importance of these differences.
24 The discussions of industrial and public ownership behavior in chapters 7 and 10 will introduce additional reasons why the actual land use margins are not as distinct as they appear in our figures. It will become clear that landowners and other users of the forest do not always remove resources all the way to the myopic open access economic margins of either the degraded forest or the natural forest frontier. The economic stock that remains standing in these cases becomes a crucial opportunity in times of greater uncertainty—as we will discuss in chapter 6. It will also become clear—in the appendix to chapter 3—that public landowners, following the rules of biological forest management, often subsidize harvests beyond the economic frontier.
25 Asner et al. (2005) describe the spatial distribution and selectivity of logging in eastern Brazil and Ruiz-Perez et al. (2001) describe the spatial distribution of non-timber forest products in the humid forest zone of Cameroon. Their observations of product diversity and specialization are consistent with the general pattern of value, accessibility, and secure property rights described by our model. Ruiz-Perez and Byron (1999) draw similar conclusions for nine specialized non-timber products in nine different developing countries.

42 The Pattern of Forestry in the Course of Economic Development

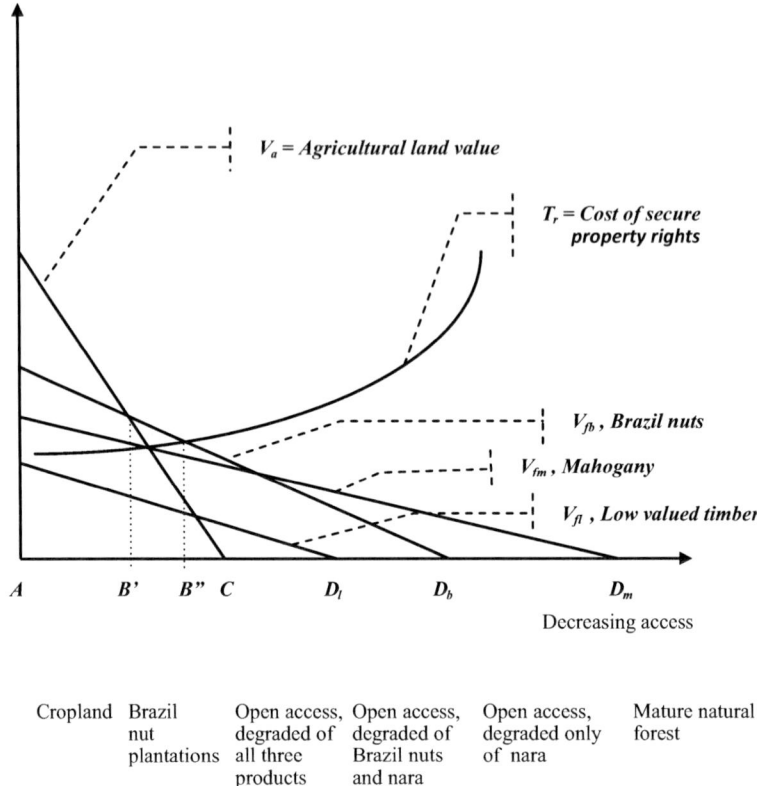

FIGURE 2.6 Multiple market values: Brazil nuts, mahogany, and lower-valued timber

tree per ten hectares. Sawmills employ scouts who spend months in the forest searching for these clumps of mahogany. (This scouting activity ranged far enough to make first contact with a previously uncontacted native group as recently as 1997.) Once the clumps are identified, sawmills build temporary road spikes into the mahogany growth sites just to harvest a few trees (Gullison et al. 1996). This implies a mahogany value function like V_{fm} in Figure 2.6. Its intercept with the vertical land value axis is above the intercept for lower-valued timber species but lower than the intercepts for agriculture or Brazil nuts, as mahogany plantations are a topic of only limited discussion. Mahogany prices are not yet sufficient to support them. However, mahogany harvests do extend to a point like D_m, well beyond the location of current logging for most lower-valued species at D_l.

Brazil nuts are interesting because they have been used as an example of a natural product that cannot be domesticated. Yet the market for Brazil nuts has risen sufficiently in recent years to induce minimal investment in a few successfully managed stands (Viana et al. 1996). This means that a function like V_{fb} represents

Brazil nuts. Managed stands of Brazil nuts occur in region $B'B''$, while harvests of the nuts from the natural forest still occur in the neighborhood of D_b. Since most of the forest has been harvested only to D_l, there is a standing forest beyond that point and that forest includes Brazil nut trees at D_b. Only occasional clumps of mahogany have been removed beyond this point.

While this discussion demonstrates that the concepts of Figures 2.1–2.5 hold in general, it also demonstrates the importance of clearly identifying the specific forest value gradient and the specific component of the full physical forest resource that is the appropriate target for any selective policy or management topic in question.

To make this point even clearer, consider the concepts of forest degradation and deforestation applied to this Amazon example. The land will have been totally deforested to point C, the fringe of agricultural use. Depending on one's measure of deforestation we might even say the land has been deforested all the way to D_l, the neighborhood of most current commercial timber removal—although a few stands closer to the market than D_l are managed for Brazil nuts ($B'B''$) and some non-commercial trees and woody shrubs will remain standing between points B_b'' and D_l. The land between D_l and D_m has been partially degraded of ripe Brazil nuts and fully depleted of mahogany—but certainly most of us would not consider the land between D_l and D_m deforested because the lower-valued species that make up the majority of the Amazon forest remain standing throughout this region.

The definitions of the forest value gradient and the terms "degradation" and "deforestation" become even more important when we address the modern policy issues of watershed protection, biodiversity and carbon sequestration in chapter 4. The measures of either degradation or deforestation will be very different for those discussions and still different yet for additional market-valued products of the forest.

Non-Market Values

Figures 2.1–2.5 tell the story of economic development with extractive forest resources. They appear to tell nothing about the important non-market values and non-consumptive ecosystem services of the forest. This is an important shortcoming that we will examine more thoroughly in discussions of public policy, public land, and the role of public regulation in chapters 4 and 10.

In fact, the formulation of Figures 2.1–2.5 can be very helpful when considering the allocation of forested sites between alternative land uses, including non-market land uses. It is helpful because the vertical distance between the forest value gradient and the transactions cost function identifies the contribution of market-valued products to the joint output of market and non-market valued production on the same site. This contribution is also identifies the market opportunity cost imposed by forest allocation solely to non-marketed goods and services.

For example, in Figure 2.7 on land identified by the hectare at j, $V_f(j)-T_p(j)$ is the market-valued opportunity cost of allocating this land to a non-market valued use. For efficient land allocation to incompatible non-market uses, the value of

44 The Pattern of Forestry in the Course of Economic Development

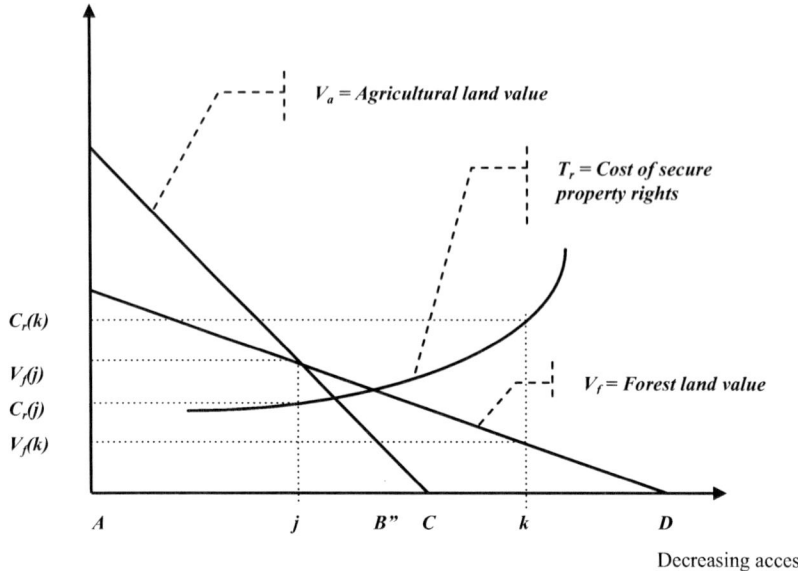

$V_f(j) - T_r(j)$ is the opportunity cost of converting land unit j to non-market valued use.

$V_f(k)$ is the contribution market-valued uses to joint market and non-market production on land unit k. Non-market valued uses must contribute a net amount of at least $T_r(k) - V_f(k)$ in order to justify joint production on the land.

FIGURE 2.7 Market-valued opportunity costs

any non-market land use must exceed this market value. That is, if the value of this hectare for producing timber is $V_f(j) - T_p(j)$ but the hectare's value for wilderness recreation or as habitat for endangered species, for example, is greater than $V_f(j) - T_p(j)$, then allocating the forest to the non-market valued use is efficient.

Alternatively, if the joint production of compatible market and non-market valued products like timber and important categories of wildlife is feasible on the same unit of land, then $V_f(j) - T_p(j)$ is the contribution of market-valued land uses to joint production. Beyond point B'', market-valued forest products are not economically viable unless the very same management activity also creates efficient joint products. This would be the case for the hectare at k if management for wildlife, for example, requires some level of timber extraction to improve the wildlife habitat and the resulting positive non-market wildlife value exceeds the net timber loss, $T_p(k) - V_f(k)$, incurred while protecting the land but harvesting that timber.

It would be easy to recommend public management of all land past point C in Figure 2.7 and to recommend management of most of this land for non-market environmental services. The problem is that the market values for extractive products are still positive on the lands between points C and D. Positive market values create an incentive for open access encroachment unless the responsible agency can absorb the costs of excluding unauthorized use of the forest—but we know that these costs exceed the revenues the agency might extract in contracts for market-valued uses. (The additional costs for the public land management agency would be at least $T_p(k)-V_f(k)$ for the k-th hectare.) Managers of forested national parks are acutely aware of the encroachment that occurs in such cases. Forested parks are safe from encroachment only for the region beyond point D.

Different Bundles of Property Rights

Some claims exist for virtually all land and all forest resources everywhere. The security of the claims varies, however, and, much as it is easier to conceptualize the forest value gradient with reference to a single product like commercial timber, it is also easier to conceptualize the transactions cost function with reference to a uniform bundle of rights that holds across an entire landscape. In fact, just as several products and values may originate from a hectare of forestland, land managers may choose from a wide array of configurations of property rights. Managers select the particular bundle of rights that fits the specific local situation. Complete and secure rights are not an efficient management choice for all lands, and the preferred bundle of rights depends on the local resources at risk, the institutional arrangements for their protection, and the choices available for their use.

For example, some residential housing compounds protect complete rights to all resources within them by erecting walls and positioning guards to effectively restrict unapproved entry. In other cases, such as agricultural cropland and forest plantations, the barrier may only be a simple wire fence. Trespass may occur at gaps in fences or anywhere the fence is weak or easy to cross. Monitoring the boundary of the property and repairing the fence may be a regular activity or only a task for seasons of slack labor—depending on the value of the crop or livestock within the fenced area and the risk of loss. In still other cases, only the rights to forage and fodder from the land may be protected, and they may be protected only by the periodic presence of those children who shepherd their families' grazing livestock. And even their presence while watching their grazing animals may not entirely prevent other activity on the same land, especially during times of day and seasons of the year when the shepherding children and their grazing livestock are somewhere else.

Obviously, the rights are less complete and secure in each of these three successive examples. Each property has its own bundle of rights and each bundle has its own cost function. The selected function is lower for each less complete bundle. As the appropriate cost function falls, as in Figure 2.8, only some more limited use of the land remains protected and the potential for encroachment increases.

The relative security of the property rights is crucial to the discussion of illegal logging in chapter 5. The completeness of the rights and the cost level associated with any set of rights are also central to modern discussions about decentralization and the transfer of rights from the central government's forestry agency to local governments and even to private land managers. In one form or another,

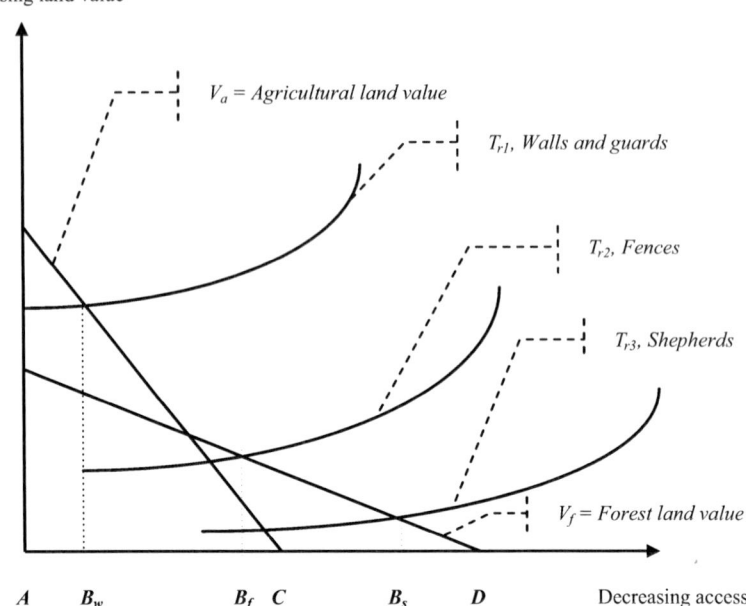

FIGURE 2.8 Property rights

this topic receives widespread attention—from "community involvement" in public forest management in the United States, "community management" in British Columbia, or "joint forest management" in India, to "community-based forest management" in preference to central government forest management in many developing countries, and to improved private incentives for land management in China and the former socialist states of Eastern Europe. Chapter 4 will reflect further on these policies and the potential advantages of transfers of property rights.

Variety in the Environmental Landscape

The final generalization of Figures 2.1–2.5 is a uniform environmental plain. Obviously real landscapes are not homogeneous—even for even-aged mature forests of single forest types. Basic roads and trails make access to some lands easier, better roads make access easier yet, and natural environmental barriers such as mountains, swamps, and canyons make access to other lands more difficult. Distinctions in soil quality also have a strong effect. The effects of these non-homogeneities might be illustrated by re-drawing Figure 2.5 from an aerial perspective. Economic activity remains centered on the market at a point such as A in Figure 2.9—with roads, mountains, and swamps now included in the landscape. We cannot display the transaction cost function from this perspective but the summary spatial distinctions from the discussion of Figures 2.1–2.5 remain intact.

Roads make access easier and, thereby, extend the boundaries of any land use class. Mountains and swamps make access more difficult and shrink the distance between the market and the boundary of each specific land use. Nevertheless, any transect emanating from the central market at point A shows the common sequence from higher-to-lower-valued land uses: in the case of this figure, first agriculture, then managed forest, then increasingly open access activities, then mature natural forest. The boundaries of these land uses in Figure 2.9 are comparable to points B', B'', C, and D, respectively, in Figure 2.5.

Variations in the physical quality of resources like land, water, and trees are a second qualification on homogeneity. Higher- and lower-valued resources do not always appear along uniform gradients emanating from the market center. They often occur in geographically discrete locations. Managers avoid and manage around the lower-valued resources and they invest in some greater level of protection for some higher-valued resources despite their geographic separation. Two good examples can be found in the history of claims for waterholes in the open range of nineteenth-century western America (Anderson and Hill 1975) and in customary claims of "tree tenure" for individual higher-valued fruit trees mixed in an open access forest around some traditional communities in some less developed parts of the world today (e.g., Peluso 1996; Fortmann and Bruce 1988). Figure 2.9 illustrates such non-homogeneities with higher-valued waterholes and fruit trees at points W_1 and W_2 and lower-valued swampland at S.

In sum, the environmental landscape is not homogeneous and land use boundaries shift with access and resource quality, but the characteristic pattern of land use remains unchanged from our fundamental description.

Conclusion

This chapter lays a foundation for the rest of the book. It describes three stages of economic development involving forestry. Natural forests interfere with new agricultural development during the first stage. During the second stage, natural forests no longer interfere with agricultural development, but they continue to be the sole source of forest products. Some level of unsustainable forest activity is inevitable in these two stages—so long as the maintenance of property rights imposes even a modest cost. The third stage occurs once removals from the natural forest create sufficient scarcity in the form of higher prices to justify establishing some form of property rights and at least a modest level of forest management.

The three stages of forest development imply at least three distinct categories of forest: managed trees and managed stands, increasingly open access and degraded forests, and a mature natural forest. For some purposes, a fourth category is also important. This fourth category is the neighborhood of current extractive activities at the frontier of the remaining mature natural forest of the third category. These three stages of development and four categories of forestland will provide focus and definition for the discussions of forest policy and landowner behavior in subsequent chapters.

48 The Pattern of Forestry in the Course of Economic Development

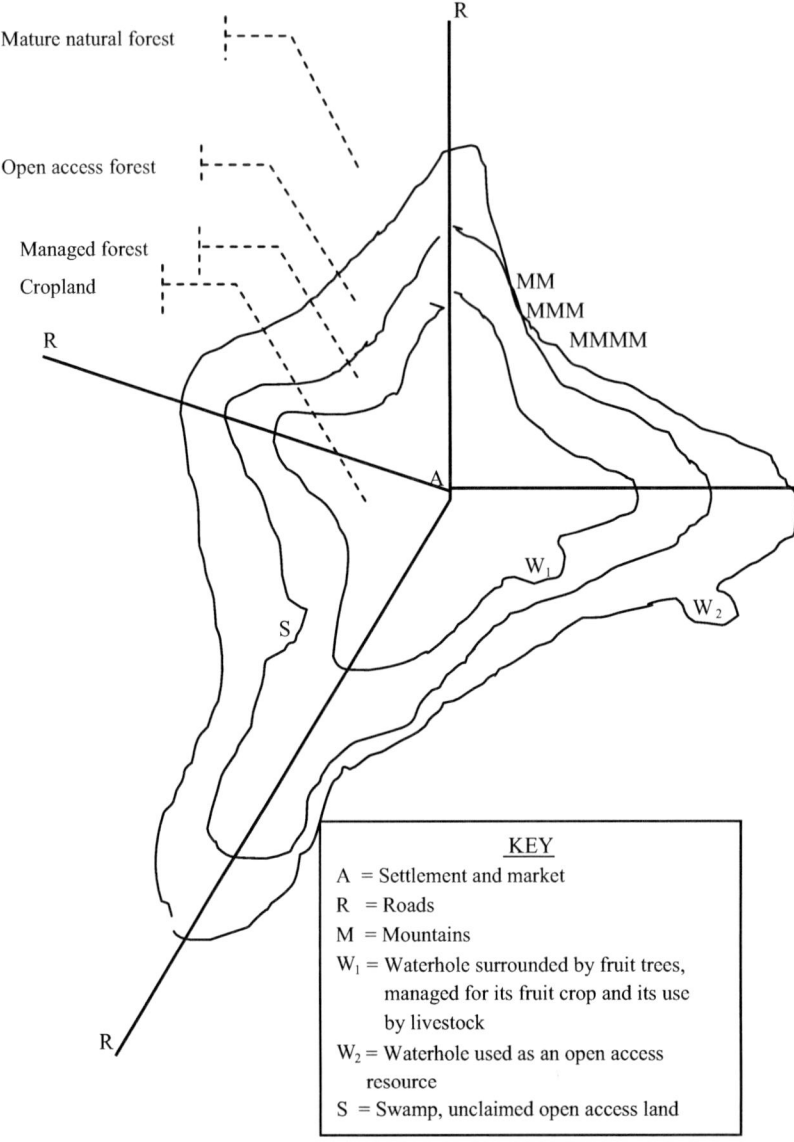

FIGURE 2.9 An aerial view, with roads and environmental barriers

The three stages of development are static snapshots in time. The next chapter completes this picture for a single regional market by introducing the longer-term effects of biological growth in the forest and technological improvements available to the regional economy.

We have also discussed the modifications to the model that are necessary to describe many specific local applications. The final result of both multiple products and grades and also natural physiographic differences is what Schneider et al. (2002) call an "unconsolidated" frontier. The three stages of forest development described in this chapter track the emergence of an unconsolidated frontier for which multiple forest values and multiple geographic features mean that the physical boundaries are not rigid and fixed. The boundaries for one economic activity, while following the three-stage model, are not the same for all other economic activities and some activities continue in deeper and more scattered locations into the natural forest far beyond the economic boundaries of others.[26]

Furthermore, these modifications—incorporating multiple forest products, non-market values, various configurations of property rights, and a heterogeneous landscape—alert us to differences in useful measures of forest data and different levels of forest degradation and deforestation. An appendix to this chapter examines the data issues more closely. Perhaps the crucial point to be taken from it is that the common national and global forestry data are not collected with reference to economic variables. They are physical data that fit neither our three-stage description of forest development nor most any other system of economic analysis. Therefore, the appendix makes an appeal for care in selecting appropriate data for any economic question in forestry and caution in accepting the economic, management, or policy conclusions that arise from many applications of national or global forestry data. Each particular analysis must consider its own objectives and the trees and forests that are relevant to it, and also the biases that are implicit when there are no alternatives to the commonly available physical data. Neither all forests nor all trees are relevant for all policy questions, and the official forestry data of most countries include some forests and trees that are irrelevant for many questions, but exclude others that are exceedingly relevant for different important questions.

Literature Cited

Albers, H. 1996. Modeling ecological constraints on tropical forest management: Special interdependence, uncertainty, and irreversibility. *Journal of Environmental Economics and Management.* 30: 73–94.

Alchian, A., and H. Demsetz. 1973. The property rights paradigm. *Journal of Economic History* 33: 17–27.

Alston, L., G. Libecap, and B. Mueller. 1999. *Titles, conflict, and land use: The Development of property rights and land reform on the Brazilian frontier.* Ann Arbor: University of Michigan Press.

Alston, L, G. Libecap, and B. Mueller. 2000. Land reform policies, the sources of violent conflict, and implications for deforestation in the Brazilian Amazon. *Journal of Environmental Economics and Management* 39(2): 162–188.

Amacher, G., W. Hyde, W. Cruz, and D. Grebner. 1998. Environmental motivations for migration: Population pressure, poverty and deforestation in the Philippines. *Land Economics* 74(1): 92–101.

Amacher, G., W. Hyde, and B. Joshee. 1993. Joint production and consumption in traditional households: Fuelwood and agricultural residues in two districts of Nepal. *Journal of Development Studies* 30(1): 206–225.

---

26 See Bowles et al. (1998) and Deininger and Minten (2002) for additional discussion of this point.

Amacher, G., W. Hyde, and K. Kanel. 1999. Nepali fuelwood consumption and production: Regional and household distinctions, substitution, and successful intervention. *Journal of Development Studies* 35(4): 138–163.

Anderson, T., and P. Hill. 1975. The evolution of property rights: A study of the American West. *Journal of Law and Economics* 18(2): 163–179.

Angelsen, A. 2007. *Forest cover change in space and time: Combining von Thunen and forest transition theories*. Washington, DC: World Bank.

Asner, G., D. Knapp, G. Broadbent, P. Oliviera, M. Keller, and J. Silva. 2005. Selective logging in the Brazilian Amazon. *Science* 310: 480–482.

Barnett, H. J., and C. Morse 1963. *Scarcity and growth: The economics of Natural resource availability*. Baltimore, MD: The Johns Hopkins University Press.

Berck, P. 1979. The economics of timber: A renewable resource in the long run. *Bell Journal of Economics* 10(1): 447–462.

Boeke, J. 1948. *The interests of the voiceless Far East*. Leiden, The Netherlands: Universitare Pers.

Boeke, J. 1953. *Economics and economic policy of dual societies*. New York: Institute of Pacific Relations.

Bowles, I., R. Rice, R. Mittermeier, and G. da Fonseca. 1998. Logging and tropical forest conservation. *Science* 280: 1899–1900.

Bryant, D., D. Nielsen, and L. Tangley. 1997. *The last frontier forests: Ecosystems and economies on the edge*. Washington, DC: World Resources Institute.

Chomitz, K., and C. Griffiths. 2001. An economic analysis and simulation of woodfuel management in the Sahel. *Environment and Resource Economics* 19: 285–304.

Clarke, H. R., W. J. Reed, and R. M. Shrestha. 1993. Optimal enforcement of property rights on developing country forests subject to illegal logging. *Resource and Energy Economics* 15: 271–293.

Chomitz, K., and D. Gray. 1996. Roads, land use, and deforestation: A spatial model applied to Belize. *World Bank Economic Review* 10(3): 487–512.

Coase, R. 1937. The nature of the firm. *Economica* (New Series) 16: 386–405.

Contreras-Hermosilla, A. 2001. *Forest law compliance: An overview*. Rome: Food and Agricultural Organization of the UN.

Demsetz, H. 1967. Toward a theory of property rights. *American Economic Review* 57(2): 347–359.

Deininger, K., and R. Minten. 2002. Determinants of deforestation and the economics of protection: An application to Mexico. *American Journal of Agricultural Economics* 84(3): 943–960.

Escobal, J., and U. Aldana. 2003. Are nontimber forest products the antidotes to rainforest degradation in Madre de Dios, Peru. *World Development* 31(11): 1873–1877.

Finish Forestry Research Institute. 2000. *Finnish statistical yearbook of forestry*. Helsinki: Finish Forestry Research Institute.

Food and Agriculture Organization of the United Nations (FAO). 1995. *Forest resource assessment 1990: Global synthesis*. Rome: Food and Agriculture Organization of the United Nations.

Food and Agricultural Organization of the United Nations. 1998. Global fibre supply model. http://www.fao.org/docrep/006/x0105e/x0105e00.htm (accessed April 7, 2004).

Food and Agriculture Organization of the United Nations. 2001a. *Global forest resources assessment 2000*. Forestry paper 140.Rome: Food and Agriculture Organization of the United Nations.

Food and Agriculture Organization of the United Nations. 2001b. *Comparison of forest area and forest area change estimates derived from FRA 1990 and FRA 2000*. Forest Resources Assessment Programme working paper 59. Rome: Food and Agriculture Organization of the United Nations.

Fortmann, L., and J. Bruce (Eds.). 1988. *Whose trees? Proprietary dimensions of forestry*. Boulder, CO: Westview Press.

Foster, A., M. Rosenzweig, and J. Behrman. 1997. *Population and deforestation: management of village common land in India*. Unpublished manuscript, Department of Economics, University of Pennsylvania, Philadelphia.

Furnival, J. 1939. *Netherlands India: A study of rural economy*. Cambridge, UK: Cambridge University Press.

Furnival, J. 1948. *Colonial policy and practice: A comparative study of Burma and Netherlands India*. Cambridge, UK: Cambridge University Press.

Garrity, D. 1995. Agroforestation: Getting smallholders involved in reforestation—market driven smallholder timber production on the frontier. In A. Tampubolon, A. Otsamo, J. Kuusipalo, and H. Jaskari (eds.), *From grassland to forest: Profitable and sustainable reforestation of Alang-Alang Grasslands in Indonesia.* Jakarta, Indonesia: Enso Forest Development Oy Ltd., pp. 109–112.

Ginsberg, N. 1973. From colonialism to national development: Geographical perspectives on patterns and policies. *Annals of the Association of Geographers* 63(1): 1–21.

Government of Indonesia, Ministry of Forestry and Estate Crops. 1997. *Statistical estate crops of Indonesia, 1996–98.* Jakarta, Indonesia: DG of Estate Crops.

Gullison, R, S. Panfil, J. Strouse, and S. Hubbell. 1996. Ecology and management of mahogany (Swietenia macrophylla King) in the Chimanes Forest, Beni, Bolivia. *Botanical Journal of the Linnean Society* 122(1): 9–34.

Gunatilake, H. 2007. *Efficient technology and the conservation of natural forests: Evidence from Sri Lanka.* ERD working paper 105, Asian Development Bank.

Hammett, A., and J. Chamberlain. 1998. Sustainable use of non-traditional forest products: Alternative forest-based income opportunities. *Proceedings, Natural Resource Income Opportunities on Private Lands.* pp. 141–147.

Hecht, S., S. Kandel, I. Gomes, N. Cuellar, and H. Rosa. 2006. Globalization, forest recovery, and environmental politics in El Salvador. *World Development* 34(2): 308–323.

Heydir, L. 1999. Population-environment dynamics in Lahat: Deforestation in a regency of South Sumatra province, Indonesia. In B. Baudot and W. Moomaw, eds., *People and their planet.* New York: St. Martin's Press, pp. 91–107.

Hofstad, O. 1997. Deforestation by charcoal supply to Dar es Salaam. *Journal of Environmental Economics and Management* 33(1): 17–32.

Hyde, W., and J. Seve. 1993. The economic role of wood products in tropical deforestation: The severe experience of Malawi. *Forest Ecology and Management* 57(2): 283–300.

Jodha, N. 2000. Common property resources and the dynamics of rural poverty: Field evidence from the dry regions of India. In W. Hyde, G. Amacher, and colleagues, *Economics of forestry and rural development: An empirical introduction from Asia.* Ann Arbor: University of Michigan Press, pp. 181–202.

Johnson, R., and G. Libecap 1980. Efficient markets and Great lakes timber: A conservation issue reexamined. *Explorations of Economic History* 17: 372–385.

Lanly, J-P. 2003. *Deforestation and forest degradation factors.* Paper presented at XII World Forestry Congress. Quebec City, Canada. September 23.

Linde-Rahr, M. 2005. Extractive non-timber forestry and agriculture in rural Vietnam. *Environment and Development Economics* 10: 363–379.

Long, A., and N. Johnson. 1981. Forest plantations in Kalimantan, Indonesia. F. Mergen, ed., *International Symposium on Tropical Forests.* New Haven, CT: Yale School of Forestry and Environmental Studies.

Lopez, R. 1998. The tragedy of the commons in Cote d'Ivoire agriculture: Empirical evidence and implications for evaluating trade policies. *World Bank Economic Review* 12(1): 105–132.

McLean, S. 2002. *Welcome home: Travels in Smalltown Canada.* New York: Penguin.

Mendoza, M. 2003. Timber thieves in the U.S. saw forests for the trees. *Denver Post* (May 18) p. 9A.

Muller, D., and M. Zeller. 2002. Land use dynamics in the Central Highlands of Vietnam: A spatial model combining village survey data with satellite imagery interpretation. *Agricultural Economics* 27(3): 333–354.

Nair, P. K. R. 1993. *An Introduction to Agroforestry.* Dordrecht, The Netherlands: Kluwer Academic.

National Remote Sensing Agency (NRSA). 1995. *Report on area statistics of land use/land cover generated by using remote sensing techniques.* Hyderabad, India: Department of Space, National Remote Sensing Agency.

North, D. 1990. *Institutions, institutional change and economic performance.* Cambridge, UK: Cambridge University Press.

Peluso, N. 1996. Fruit trees and family trees in an anthropogenic forest: Ethics, access, property zones, and environmental change in Indonesia. *Comparative Studies of Society and History* 38: 510–548.

People's Republic of China, State Forest Administration. 2001. *China's Forest resource statistics.* Beijing: China's Forestry Press.

Pfaff, A. 1999. What drives deforestation in the Brazilian Amazon? Evidence from satellite and socioeconomic data. *Journal of Environmental Economics and Management* 37: 23–43.

Richter, C. 1966. *The trees*. Athens: Ohio University Press.

Robinson, E. 2008. India's disappearing common lands: Fuzzy boundaries, encroachment, and evolving property rights. *Land Economics* 84: 409–422.

Robinson, E., H. Albers, and J. Williams. 2008. Spacial and temporal modeling of community non-timber forest extraction. *Journal of Environmental Economics and Management* 56: 234–245.

Robinson, E., J. Williams, and H. Albers. 2002. The influence of markets and policy on special patterns of non-timber forest product extraction. *Land Economics* 78: 260–271.

Ruiz-Perez, M. O., Ndoye, A. Eyebe, and A. Puntodewo. 2001. *Spatial characterization of non-timber forest products markets in the humid forest zone of Cameroon*. Unpublished manuscript, Center for International Forestry Research, Bogor, Indonesia.

Ruiz-Perez, M., and N. Byron. 1999. A methodology to analyze divergent case studies of non-timber forest products and their development potential. *Forest Science* 45(1): 1–14.

Samuelson, P. 1983. Thunen at two hundred. *Journal of Economic Literature* 21(4): 1468–1488.

Scherr, S. 1995. Economic factors in farmer adoption of agroforestry: Patterns observed in western Kenya. *World Development* 23(5): 787–804.

Schneider, R., E. Arima, A. Verissimo, C. Souza, and P. Barreto. 2002. *Sustainable Amazon*. World Bank technical paper no. 515. Washington, DC: World Bank.

Sedjo, R. 1994. *The potential of high-yield plantation forestry for meeting timber needs: Recent performance and future potentials*. Resources for the Future discussion paper 95-08. Washington, DC: Resources for the Future.

Seneca Creek Associates, LLC and Wood Resources International. 2004. *"Illegal" logging and global wood markets: The competitive impacts on the U.S. wood products industry*. A report prepared for the American Forest and Paper Association. Poolesville, MD: Seneca Creek Associates, LLC and Wood Resources International.

Shively, G. 1998. Economic policies and the environment: Tree planting on low income farms in the Philippines. *Environment and Development Economics* 3(1): 83–104.

Sizer, N. 2005. Halting the theft of Asia's forests. *Far Eastern Economic Review* 168(5): 50–53.

Smith, W. 2002. The global problem of illegal logging. *Tropical Forest Update*. 12(1): 3–5.

Stevens, S. 1988. Sacred and profaned Himalayas. *Natural History*. 97(1): 26–35.

Stein, T. 2001. Savings grow on trees along Front Range. *Denver Post* (April 18) 1A, 9A.

Stone, S. 1997. Economic trends in the timber industry of Amazonia: Survey results from Para state, 1990–95. *Journal of Developing Areas* 32: 97–122.

Stone, S. 1998. The timber industry along an aging frontier: The case of Paragominas (1990–1995). *World Development* 26(3): 433–448.

Tacconi, L., M. Boscola, and D. Black. 2003. *National and international policies to control illegal forest activities: A report to the Ministry of Foreign Affairs, Government of Japan*. Bogor, Indonesia: Center for International Forestry Research.

Tachibana, T., T. Nguyen, and K. Otsuka. 2001. Agricultural intensification versus extensification: A case study of deforestation in the northern-hill region of Vietnam. *Journal of Environmental Economics and Management* 41(1): 44–69.

Templeton, S., and S. Scherr. 1999. Effects of demographic and related microeconomic change on land quality in hills and mountains of developing countries. *World Development* 27(6): 903–918.

Timber thieves make away with leafy bounty. 2007. (December 30, 2007). http://www.cnn.com/2007/US/12/30/stealing.trees.ap/index.html?eref=rss_topstories (accessed January 15, 2008).

Tomich, T., M. van Noordwijk, S. Budidarsono, A. Gillison, T. Kusumanto, D. Murdiyarso, F. Stolle, and A. Fagi,(eds.). 1998. *Alternatives to slash-and-burn in Indonesia: Summary report and synthesis of phase II. ASB-Indonesia Report No. 8*. Bogor, Indonesia: ASB-Indonesia and International Center for Research in Agroforestry.

Tomich, T., M van Noordwijk, D. Murdiyarso, F. Stolle, and A. Fagi. 2001. Agricultural intensification, deforestation, and the environment. In D. Lee and C. Barrett (eds.) *Tradeoffs or synergies? Agricultural intensification, economic development, and the environment*. Wallingford, UK: CAP International. 221–244.

UK/Forestry Commission. 2001. Forestry Statistics 2001. Edinburgh. Also http://www.forestry.gov.uk/pdf/forestrystatistics2001.pdf/$FILE/forestrystatistics2001.pdf
USDA Forest Service. 2005. USDA Forest Service Forest Inventory and Analysis. http://apps.fs.fed.us/fiadb-downloads/datamart.html accessed March 5, 2008).
Viana, V. M., R. Mello, L. deMorais, and N. Mendes. 1996. *Ecology and management of Brazil nut populations in extractive reserves in Xapuri, Acre.* Unpublished research paper. Department of Ecology, University of Sao Paolo.
von Amsberg, J. 1996. *Economic parameters of deforestation.* World Bank Policy Research Working Paper no. 1350. Washington, DC: World Bank.
Wood, S., K. Sebastian, and S. Scherr. 2000. *Pilot assessment of global ecosystems: Agroecosystems.* Washington, DC: World Resources Institute and International Food Policy Research Institute.
World Bank. 2003. *Sustaining forests: A World Bank strategy.* Washington, DC: The World Bank.
Yin, R., and W. Hyde. 2000. The impact of agroforestry on agricultural productivity: The case of northern China. *Agroforestry Systems* 50: 179–194.
Yin, R., and R. Sedjo. 2001. Is this the age of intensive management? A study of Loblolly pine on Georgia's Piedmont. *Journal of Forestry* 99(12): 10–17.
Young, C. 2003. Land tenure, poverty and deforestation in the Brazilian Amazon. Unpublished research paper available from the author at young@ie.ufrj.br.
Zhang, Y., J. Uusivuori, and Y. Kuuluvainen. 2000. Econometric analysis of the causes of forest land use changes in Hainan, China. *Canadian Journal of Forest Research* 30(1): 1–9.

Appendix 2A:
Forestry Data

The fundamental data for economic analysis refer to prices and quantities. In addition, economic analyses of production in natural resource systems such as forests make extensive use of measures of the resource stock because today's stock can be considered an inventory stored in a natural warehouse where it is available for market exchange at some future date. This appendix reviews each of these: price, quantity, and stock data.

Price and Quantity Data

The ideal price and quantity data would be based on complete markets and full reporting of market transactions for all forest-based goods and services for which we might have analytical interest. Obviously, this ideal is a long way from reality. In fact, forests produce many non-market outputs for which there is very little transactions evidence. Widespread data exist for only one of the many commercial forest products—timber—and even timber transactions data have their weaknesses.

The most common transactions evidence for forestry refers to commercial wood at one of three points of exchange.

- "Stumpage" refers to standing trees ready for harvest, and also to their price.
- "Logs" and "pulpwood" refer to stumpage that has been felled, bucked, skidded to the roadside, and stacked ready for loading and delivery to a sawmill, veneer or plywood mill, or pulpmill. The term "logs" can also refer to delivered logs. The distinction between logs stacked at the roadside and logs delivered to the mill is clear to the logger and to the mill manager but it is not

always clear in the analytical literature. (Another term, "roundwood" is the volumetric combination of logs and pulpwood.)
- "Sawnwood" and "pulp" refer to the first processed intermediate outputs in the production of either lumber or paper. "Market pulp" refers to that which is sold to other manufacturers of paper, as opposed to that which remains in the production process of the originating vertically integrated pulp and paper manufacturer.

Therefore, the terms stumpage, logs or pulpwood, and sawnwood or market pulp each refer to an intermediate product in the manufacture of either lumber and wood products or paper and paperboard and each implicitly refers to the location of its own market exchange.

Problem

Uniform points of market exchange are important for the comparison of any two similar products across markets or over time. The location of market exchange is especially important for stumpage and logs or pulpwood because transportation costs comprise the difference in any market comparison of these products and transportation costs are a large share of their delivered price, an unusually large share when compared with most other goods and services. Consider, for example, log prices in three different markets described by the three stages of forest development in Figures 2.1–2.5. Logs from the natural forest sold at the roadside near points like D in all three figures attract approximately the same low price. However, these logs would sell at very different delivered values at the mill because (other things being equal) the distance to the mill and, therefore, the transportation cost, would be greater in the developing frontier market of Figure 2.3 than in the new frontier market of Figure 2.1A, and greater still in the marketplace of the mature economy described by Figure 2.5.

Market comparisons in mature forest economies are additionally complex because stumpage or logs at the roadside in managed forests in the vicinity of $B'B''$ are worth more to the mill than stumpage or logs at the natural forest frontier at D. Yet there may be no physical difference between logs or stumpage from the two points. The difference in log price at the two locations is due to the increased delivery cost for logs or stumpage from the natural forest, and any market comparison that ignores this difference in delivery cost is useless to both landowners and mill managers. Furthermore, any average of the prices from the two locations is deceptive because it represents the products of neither of the two forests.

Therefore, simple comparisons of either log or stumpage prices in different markets at any moment in time, or within any particular market over time, are not very meaningful as indicators of the scarcity value of wood. Meaningful comparisons require an adjustment for location (delivery costs). Comparisons of pulpwood prices suffer the same problem and require a similar adjustment.

First Solution

The best solution to this problem would be to use delivered prices for logs or pulpwood at the mill, the prices at point A in Figures 2.1–2.5. Delivered prices represent comparable products at comparable locations across markets and over time. Because they include transportation costs, they are true measures of the increasing physical and economic scarcity implied by more distant locations and more difficult harvest conditions. However, in most countries, market prices are quoted either for standing timber on the stump or for logs prepared for loading at the roadside. Market reports of delivered prices are much less common.

Just how important is this problem? We can examine its empirical relevance by finding those locations where price data exist for both stumpage or roadside and delivered logs and observing the comparative movement of the two price series over time. After correcting for time lags between stumpage or roadside offerings and mill delivery, locational distinctions should account for the entire remaining difference. Since the roadside and stumpage locations vary with time as harvest locations continually shift, the stumpage or roadside price series and the delivered price series should not move together over time.

However, if the two price series unexpectedly do move together, then changing differences in location and delivery costs are not critical indicators of the changing relative scarcity of the basic wood resource, and other unidentified costs must be the dominant determinants of the prices of both delivered logs and stumpage. In this case, we could use either roadside log or stumpage prices as proxies for delivered log prices in analyses of relative scarcity.

Because series of delivered prices were unavailable, Toppinen (1997), Jung, Krutilla, and Boyd (1997), and Sun and Zhang (2006) tested the hypothesis that these prices do move together by comparing intertemporal movements of stumpage and lumber prices in Finland, the U.S. Pacific Northwest, and the U.S. South, respectively. (Recall that lumber prices are one production stage beyond delivered log prices.) None of these three regional assessments can conclude that stumpage prices are significant predictors of lumber prices.[27] Therefore, since sawmill costs and lumber prices must follow similar patterns over the shorter periods of time in these assessments, then variation in location or variation in harvest and delivery costs (which are the residual after sawmill costs are deducted) must be an important source of the difference, and the problem remains as we have stated it.

---

27 More precisely, Toppinen observes that neither export lumber prices nor domestic sawlog prices are statistically significant predictors of the other in Finland—regardless of adjustments for various time lags in the independent variable. (Exports consume approximately 70 percent of Finland's lumber production ([FRI 2000]).) Jung et al. observe that neither stumpage nor log or lumber prices move together over time in the U.S. Pacific Northwest. Sun and Zhang observe that neither stumpage nor delivered sawlog prices move together in the U.S. South. (In econometric terms, no two of these prices are cointegrated.) Buongiorno (personal communication, March 14, 2000) similarly observes that stumpage prices are not significant predictors of either current or future lumber prices in the U.S. Pacific Northwest.

Barnett and Morse (1963) also identify the location and delivery cost problem in their classic study of natural resource scarcity in the United States. They obtained price series for other primary resources but, for forestry, they were forced to rely on observations of lumber and pulpwood prices. They observe increasing relative lumber prices over a period of more than a century but they hesitate to conclude that this implies increasing scarcity for the basic wood resource because (1) they also observe *decreasing* relative pulpwood prices, yet pulpwood prices are also affected by the delivery cost problem and because (2) lumber is a secondary product that incorporates other wood processing costs that are not associated with the primary product. Therefore, Barnett and Morse were unable to collect prices for wood as a raw material that were comparable over time or across space and, as a result, they were unable to solve the problem.[28]

Second Solution

How can this problem be solved in the absence of either delivered prices or reasonable proxies for delivered prices? An obvious second solution would be to add estimated harvest and delivery costs to reported stumpage prices to obtain estimates of log or pulpwood prices at the mill. This solution requires many independent calculations, one for each stumpage sale location in each market in each period in time. It may be an acceptable solution for local case studies and it is a common calculation for consulting foresters, mill managers, and loggers as they sort out conditions in their own local markets. However, the large number of independent calculations makes this approach unacceptable for most broad scale policy analyses.

In fact, the search for a fully satisfactory means of making broad scale intertemporal or cross-sectional market comparisons of either market prices or improved measures of economic scarcity remains an elusive quest for forest policy analysis.

Measures of Physical Stocks

Individual landowners conduct inventories of their own forest stocks, but their information is proprietary and generally unavailable for the planning activities of other landowners or mill managers or for policy analysis. Most governments conduct broader regional or national estimates of forest stocks and their information is available for public use. These government survey efforts are called forest resource assessments (FRAs). Like the market price and quantity data, FRAs have a strong commercial timber orientation.[29]

FRAs are generally collected with reference to land use and one or more of

---

28 Lyon (1981) also examined the increasing price of lumber in the United States and he relates it, as we do in this book, to mining the natural forest over time.
29 Janz and Persson (2002) review the problems of forest resource assessment from the perspective of those who specialize in conducting these assessments. Their observations are largely consistent with those of this appendix.

TABLE 2A.1 Characteristic thresholds for national definitions of forest stock

Forest attribute	Number of national definitions reliant on this attribute	Least min. value of attribute		Greatest min. value of attribute		Mode	Mean	FAO's FRA 2000
		Value	Country	Value	Country			
Area (ha.)	40	0.01	Czech Rep.	100	PNG	0.1 or 0.5	3.4	0.5
Strip width (m.)	20	9	Belgium	50	Taiwan	20	27.6	20
Tree height (m.)	33	1.3	Estonia	15	Zimbabwe	5	5.7	5
Crown cover (%)	54	1	Iran	80	Zimbabwe	10	27.6	10

Source: Compiled from Lund (2000)

four attributes of forest cover: area, strip width, tree height, and crown cover.[30] Table 2A.1 shows the number of countries in a 128-country sample that rely on a threshold measure for one or more of these attributes to guide the country's own official definition of "forest".[31] The table also shows the range of threshold values for each attribute. The minimum area for forest classification in Papua New Guinea is 10,000 times the minimum area for the Czech Republic. The minimum percentage crown cover for forest classification in Zimbabwe is eighty times the minimum crown cover in Iran.[32] Clearly, national guidelines vary widely and meaningful international comparisons must remain uncertain at best. The Food and Agriculture Organization of the United Nations is the international repository for data from the national FRAs. FAO reports the national data as it receives them (FAO 2001). Therefore, meaningful international comparisons based on these summary data must also remain uncertain.

Furthermore, attempts to standardize the various international data according to some recommended threshold values—such as those recommended by FAO in the final column of the table—are exceedingly difficult. Sixty-five of the 128

---

30 This paragraph and the next two draw on personal discussion with G. Lund and his survey paper (Lund 2000).
31 Some countries use altogether different guidelines. For example, the official Philippine forest statistics include any land of greater than eighteen percent slope. Some Nordic countries consider forests to be any land capable of producing one cubic meter of wood per annum. Moreover, most countries exclude from their forest resource assessments (FRAs) any agricultural land with tree cover—even if it satisfies the official threshold definition for forest. This means that horticultural crops, palm crops, some small woodlots, and trees incorporated in most agroforestry systems are not included in the FRAs of most countries.
32 Different government agencies within a country can add to the confusion. For example, as many as ten U.S. federal government agencies each use different attributes and different minimum standards in their own definitions of the "forest".

national FRAs in FAO's sample did not satisfy even one of FAO's threshold values in 2002. For this reason FAO is in the process of incorporating satellite imagery, expert opinion, and other published scientific analyses, along with the national FRA data, to create a standardized set of national forest accounts. Its progress has been used in FAO's own international comparison of forest area between 1990 and 2000 (FAO 2001 ch. 46, 2001a). However, the general effort to create comparable international accounts remains a work in progress.

Within each country, forest inventory specialists use the national threshold values, together with techniques ranging from on-the-ground sampling to satellite imagery, as the bases for estimates of standing forest area or volume at a moment in time. They generally prefer to re-apply the same standards (preferably on the same sample plots) again when they periodically renew their estimates at later dates. In this way they create continuous forest inventories (CFIs) of comparable physical data. FRAs are expensive. Therefore, they are conducted infrequently and no country's CFI consists of more than a few periodic observations. The Nordic countries began conducting official inventories in the 1920s, and they had completed seven or eight FRAs by the year 2004. The United States began conducting official forest inventories in the 1930s and it had completed five FRAs by year 2000. (The U.S. Forest Service now compiles annual updates.) Italy has completed only one, in the 1980s. China has conducted six, but it changed the standard for forest cover in the fifth. Therefore, China's fifth and sixth FRAs are not strictly comparable with the first four. Global concerns with deforestation tend to focus on the tropics. Yet only 21 of ninety tropical countries had conducted more than one FRA by the year 2000.[33]

Problem

Clearly, comparative rates of change in deforestation, or growing stock, or anything else, must be accepted only with a fair dose of thoughtfulness and caution when (1) the bases of the national data used in the comparisons are so widely divergent and (2) the number of intertemporal observations is so sparse. Meaningful estimates are all the more difficult for the many countries that can supply FAO with only one or a very few temporal estimates of their forest inventories.

Unfortunately for economic analysis, this is not the end of the problems with forest inventory data. There is a third problem: the physical data in these FRAs are not intended to represent economic stocks. In fact, they are poor measures of a country's economic forest stock even at the date of any particular FRA. Moreover, the unchanging minimum thresholds for estimating physical stocks means that the CFIs fail to incorporate the long-term adaptive technological responses of dynamic economic systems. Therefore, the difference between physical and economic measures of forest stock changes with each new inventory. Consider each of these points in turn.

---

33 Morris, Mulcauley, and Sedjo (2010) make many of these same points.

The errors of using physical inventory data to represent static economic stocks can be visualized with reference to the mature forest economy of Figure 2.5. First, the physical inventory includes all of the mature natural forest beyond the frontier at point D that satisfies the national measurement threshold. Second, the physical inventory also includes any degraded forest on the land between B'' and D that satisfies the country's threshold standard for its "forest". The largest shares of stocks in both of these categories of forest are uneconomic. Third, physical inventories do include plantations and managed stands in the region $B'B''$—and these are economic, of course. However, farm lands with small forest plots, shelterbelts, and agroforestry trees are generally excluded, as are trees in urban parks and residential areas—but these too are economic producers of some forest products and ecosystem services.

When all of these reasons are taken together, both the sign and the magnitude of the implicit error when the official national forest inventories are used as economic measures must vary from market to market and country to country, and also from time to time in any single market or country. The variation in the magnitude of the error can be critical in some cases. For example, farm and household grown trees that do not appear in the official inventories account for 65–85 percent of all marketed fuelwood in South and Southeast Asia (FAO RWEDP 2000). This means that the official measures of forest inventory substantially *under*estimate economic supply in these countries.[34] In contrast, for countries like the United States, Canada, Cameroon, Brazil, and Russia with large hinterlands of uneconomic physical stock, the official measures of physical stock seriously *over*estimate economic supply.

The generally unchanging nature of the threshold standards for CFIs compounds the uncertain difference between physical and economic stocks. Unchanging physical standards fail any test of economic usefulness because economic inventories expand and contract as prices rise and fall. For example, stumpage prices more than doubled in the Douglas fir region of the U.S. Pacific Northwest in the early 1990s. Surely loggers were willing to go farther into previously inaccessible areas for timber that was now much more valuable than it had been. In British Columbia, the very short-term price elasticity of supply may be as great as 7–8, which means that loggers do rapidly extend their harvest operations to many previously uneconomic stocks in response even to small increases in price.[35]

Economic forest inventories also expand as logging and processing technologies improve and as roads extend farther into the frontier. To grasp the importance of this point, consider the example of technological change in logging in the United States. When the United States conducted its first official inventory in the 1930s, the minimum-sized log acceptable at mills in some parts of the American West was 32 feet (9.75 meters) in length. By the 1980s, eight-foot (2.4 meter)

---

34 The means for gathering data about scattered trees outside of forests are known and a few local forest inventories do include this information (Singh 2001, Pandey 2000).
35 Personal communication, D. Haley, University of British. Columbia, July 15, 2001.

60 The Pattern of Forestry in the Course of Economic Development

logs were processed in all parts of the country. Yet the official U.S. standard for the "commercial" forest inventory has not changed. In the shorter period of time from 1970 to the present, the economic status of at least three species—aspen, alder, and western hemlock—has changed from unproductive weed species to valuable timber species. Yet the standards for incorporating measures of their volumes and areas are unchanged within the official U.S. forest inventories over this 40-year period.

The final problem is that these physical estimates of forest stock were not designed to indicate anything about other, non-timber, values of the forest.[36] The point should be clear for watershed values, forest recreation, wildlife and fish, but the point is also valid for values like fuelwood, the protection of biodiversity, and carbon sequestration that depend on the trees themselves. Much of what becomes fuelwood comes from trees and woody plants that are too small to be included in estimates of forest volume. Critical biodiversity is most closely tied to those selective areas within the official measures of the forest that provide the habitat for unique species. Many of these are in undisturbed areas of the natural forest beyond the frontier and to the right of point D in the figures. The official measures of the forest are poor proxies for biodiverse habitat to the extent that these measures include large shares of managed and degraded forest, and natural forest too, that are less supportive of endangered biodiversity.

On the other hand, all trees sequester carbon, including fruit trees and shade trees in residential areas and trees along the edges of fields and roadsides in more rural areas, as well as those trees in forested regions that have not attained the minimum threshold physical measures necessary for inclusion in the national FRAs. The state of New Jersey in the United States and the island of Java in Indonesia illustrate this point. New Jersey and Java are heavily populated and neither is generally considered to be heavily forested. Yet their lands support approximately 70 percent tree cover although only 38 and 24 percent, respectively, of the land in these two cases satisfies the minimum standards for inclusion in official forest inventories (USDA Forest Service 2005; GOI/MFEC 1998). In another example, blocks of forest greater than one hectare in area, yet outside the official recorded forest inventory account for thirteen percent of India's forest. Forested areas of less than one hectare create another increment. In some of India's states the accumulation of these small blocks is greater than the official recorded forest inventory.[37] The discrepancy between the actual area under tree cover and the official forest stock varies from New Jersey's, Java's, and India's experiences for other parts of these same three countries as well as for other countries of the world. It should be clear that analyses of carbon sequestration that overlook the tree cover that is

---

36 Forest inventory specialists began discussing multiple resource assessments (MRAs) in the 1980s. MRAs incorporate evidence on non-timber as well as timber resources. By 2003, only six countries had conducted even one MRA and timber remains the overwhelming focus even in these six cases.

37 Personal communication, J. K. Rawat, Director, Forest Survey of India, September 21, 2003.

left unaccounted in official measures of forest inventories must be underestimates, perhaps gross underestimates. Estimates of the change between any two periods may be even less accurate because the largest growth in tree volume may occur in those trees that fail to appear in the official inventories.

Thus, official measures of standing forest stock are poor indicators of trends in fuelwood, biodiversity, and carbon sequestration. Even the direction of bias in their indications of these trends must be in doubt. In sum and in general, official measures of the standing forest stock are only indicative as physical measures of industrial-sized timber. They are not reliable measures of economic value.

Solution

The preferred solution for any economic analysis would be to revise the estimates of physical stock to incorporate economic as well as physical characteristics of supply and to measure supply with a specific social value in mind.

Krutilla, Xu, Barnes, and Hyde (2000) provide an example—and an indication of the error due to using unrevised measures of the physical stock. Krutilla et al. were interested in the economic supply of raw material for all consumable forest products. They began with satellite imagery as a basis for calculating total forest biomass and corrected for the difference between physical stock and economic supply by adjusting the measure of biomass for road density, terrain, and moisture. (Forests in steeper and wetter areas are generally less accessible to human consumption.) Their corrections explain 40 percent of the variation in stocks for 300-mile (480 kilometer) radii around 31 cities in the eight tropical countries of their assessment. General policy and institutional differences between the eight countries explain no more than 20 percent of the variation. The obvious point is that measurement, infrastructure, and terrain, are critical. The effect of those variables dominates the effect of all elements of policy and all institutional differences between the countries in the sample.

This first solution is more appropriate for addressing questions of forest management or for analyses of policy within a single market or small region. In fact, consulting foresters in the U.S. South often maintain data on timber sales distinguished by local features like terrain and road access. For more geographically aggregate analyses, such as the Krutilla et al. assessment of 31 regional markets in multiple countries, these modifications would be expensive and time consuming.

A second alternative that might be more suitable for broader policy analyses would be to separate the physical forestry data by country groups that reflect economically comparable inventory standards, and then to assess each group independently. One group might contain all regions or countries for which the entire physical inventory is economic for the product in question, while additional groups might reflect decreasing ratios of economic to physical inventories.

This second alternative also requires an underlying assessment of the geographic and physical sensitivity of each inventory to the critical economic variables in order to make the country assignments to each group. While this approach might not be as expensive and time consuming as the first alternative, it would still consume significant analytical resources. Furthermore, the underlying assessment would have to be repeated for each new periodic inventory even within one region or country because the local economic conditions and the local infrastructure change within the six- to ten-year period between most national forest inventories—even while it is the nature of CFIs to report forest stock subject to unchanging physical standards.

A third approach for broader policy analyses would be to argue from rigorous theory and good intuition supported with empirical evidence from appropriate regional examples. Fortunately, numerous regional case studies do exist. Many are cited in this book. The selection of the appropriate cases for any particular question depends on the comparability of general human demands on the forest as well as the economic comparability of the minimum standards for physical forest inventories. Finally, this approach also requires recognition that the economic standards of forest volume and forest area vary, even within a country, according to the policy or management issue in question. That is, as we have seen, even within any one country the economic forest inventory for assessments of timber are different from those for assessments of biodiversity and different still for assessments of carbon sequestration, etc.

A final approach, useful for some international comparisons, would be to rely on rates of change in national measures of forest stocks. FAO (2001a) and Bhattarai and Hammig (2001) use this approach in their comparisons of rates of deforestation. Rates of change are not labeled in physical units. Therefore, this approach only requires comparable measures for two moments in time for each country in the sample. It does not require comparable physical standards across countries. It will be successful to the extent that the international sample includes a broad sample of countries that have important forests and that each of these countries have completed two official FRAs. It will still be biased to the extent that the rate of change in the unmeasured inventory of trees in homesites and parks, along fencerows and roadsides, and in agroforestry systems is different for different countries in the sample.

In sum, FAO and the forestry agencies of several countries work continuously to upgrade national and global forest inventories. However, given the changing nature of economic value in the forest, a satisfying economic inventory is probably an unrealistic goal. Regular improvement in the physical data is a more realistic objective for these agencies. Indeed, that is their objective. This means, however, that caution must continue to be the watchword when using forest inventory data in economic analyses. When using these data, analysts have an obligation to understand the data and its strengths and weaknesses for their own purposes. They also have an obligation to inform their audiences of the expected directions of bias in their applications of these data and the increased uncertainty in their results caused by data that were not designed to measure the questions addressed by their analyses.

Literature Cited

Barnett, H. J., and C. Morse. 1963. Scarcity and growth: The economics of natural resource availability. Baltimore, MD: The Johns Hopkins University Press.

Bhattarai, M., and M. Hammig. 2001. Institutions and the environmental Kuznets curve for deforestation: A crosscountry analysis for Latin America, Africa and Asia. *World Development* 29(6): 995–1010.

Food and Agriculture Organization of the United Nations. 2001a. *Global forest resources assessment 2000.* Forestry paper 140. Rome: Food and Agriculture Organization of the United Nations.

Food and Agriculture Organization of the United Nations. 2001b. *Comparison of forest area and forest area change estimates derived from FRA 1990 and FRA 2000.* Forest Resources Assessment Programme working paper 59. Rome: Food and Agriculture Organization of the United Nations.

Government of Indonesia, Ministry of Forestry and Estate Crops (GOI/MFEC). 1998. *1997–1998 Forest utilization statistical yearbook.* Jakarta: DG of Forest Utilization.

Krutilla, K., J. Xu, D. Barnes, and W. Hyde. 2000. Estimates of economic supply from physical measures of the forest stock: An example from eight developing countries. In W. Hyde, G. Amacher, and colleagues, *Economics of forestry and rural development: An empirical introduction from Asia.* Ann Arbor: University of Michigan Press. pp. 103–120.

Lyon, K. 1981. Mining of the forest and the time path of the price of timber. *Journal of Environmental Economics and Management* 8(4): 330–334.

Morris, D., M. Mulcauley, and R. Sedjo. 2010. Why we need accurate maps of the world's forests. *Resources* 174: 25–28.

Janz, K., and R. Persson. 2002. *How to know more about forests? Supply and use of information for forest policy.* CIFOR Occasional paper no. 36. Bogor, Indonesia: Center for Forestry Research.

Jung, C., K. Krutilla, and R. Boyd. 1997. Aggregation bias in natural resource price composites: the forestry case. *Resource and Energy Economics* 20(1): 65–73.

Lund, H. G. (coord.) 2000. Definitions of forest, deforestation, afforestation, and reforestation. Manassas, VA: Forest Information Services. http://home.comcast.net/~gyde/DEFpaper.htm (accessed October 9, 2008).

Pandey, D. 2000. Methodologies for estimating wood resources in South Asia. *Wood Energy News* 15(1). FAO/RWEDP.

Singh, K. 2001. *Guidelines on national inventory of village forests.* Bogor, Indonesia: Center for International Forestry Research.

Sun, C., and D. Zhang. 2006. Timber harvesting margins in the southern United States: temporal and spacial analysis. *Forest Science* 52(3): 273–280.

Toppinnen, A. 1997. Testing for Granger causality in the Finnish roundwood market. *Silva Fennica* 31(2): 235–232.

USDA Forest Service. 2005. Forest Inventory and Analysis webpage: http://fia.fs.fed.us.(accessed March 5, 2008).

3
FOREST DEVELOPMENT IN THE LONG RUN

When foresters discuss the effect of long time periods on the resource, they focus on biological growth. They think about harvesting and then regeneration and management over the next full timber rotation, the period over which all elements of the forest are renewable. This coincides with the economist's view of the long run as the period when all factors of production are variable—and the tree itself can be replaced. Both perspectives mean that the long run is a period associated with the age of a mature tree or forest stand, a tree or stand that is ready for harvest and replacement in the ground with new seedlings. This period can be as short as 3–5 years for a few fuelwood and agroforestry species although it is generally much longer, ranging from twenty to over 75 years for final harvests of many commercial timber species, and some forests require well over 100 years to attain the maturity required for certain unique varieties of biodiversity.

The concept of sustainability arises whenever we think about such long periods. As it does, foresters quickly remind us of the difference between "first growth" and "second growth" forests and they note the impact of this difference on sustainable harvest flows. That is, initial harvests are taken from "old growth" forests on the frontier. Nature determined the state of these forests, while prices and the economic costs of access and extraction determine when mankind chooses to harvest them. The trees in these first forests tend to be old and the volumes per tree and per timberstand tend to be large.

Second forests develop from the new growth that follows these initial harvests. They occur in locations that are already accessible and some of these second forests are managed. Others are the product of natural regeneration in regions of limited management or even open access. Prices, management costs (for the managed stands), and rates of biological growth, as well as the costs of access and removal, determine the younger harvest ages and smaller harvest volumes from these second forests. The potential difference in harvest volumes between the first and second forests causes foresters to reflect on the growth rate of the second forest and the potentially unsustainable nature of the larger harvest flows generated from the first forest, on adjustment periods for conversion to the smaller harvest flows from the second forest, and on regulatory controls or "allowable" harvest levels that would permit a managed period of adjustment to these smaller flows.

As economists reflect on longer periods of productive activity, they anticipate the important effect of technological change. The impact of technological change can be immense over periods as long as a commercial timber rotation. For example, for many decades in the twentieth century the United States experienced an economy-wide average annual rate of technical change in the neighborhood of 2.5 percent.[1] At this rate, productivity doubles approximately every 29 years. Applying this rate to commercial timber would mean that the consumer demands of one year could be satisfied with half the physical harvest level or with harvests from approximately half as much forestland in the twenty-ninth year of a timber rotation. Therefore, technological change can have a tremendous favorable effect on resource scarcity and forest sustainability.

Indeed, technological change has had a remarkable effect on global wood consumption in recent years. Paper production serves as an example. Global consumption of industrial roundwood by the pulp and paper industry grew only ten percent between 1980 and 2000 while the global production of paper products grew much more rapidly. The production of wood pulp increased by more than 45 percent and the global production of paper and paperboard increased by 90 percent over the same period (FAO/UN 2005). That is, the industry's output of paper and paperboard increased 4.5 and 9 times, respectively, as rapidly as its increase in demand for the wood input.

When we think of technological change in forestry we often reflect on the Green Revolution in agriculture. We know how the Green Revolution increased agricultural yields and we are aware of the potential for various seed selection, tree breeding, and even bioengineering programs to create very large improvements in forest growth and yield. Technological change can also affect forest

---

[1] Solow (1957) and Denison (1962, 1967) are the classic studies for technological change and economic growth in the United States. Solow observes that 87.5 percent of per capita income growth before 1955 was due to technical change, and only 12.5 percent was due to capital deepening. Solow's estimate and the economy-wide estimate of 2.5 percent are for process-oriented technological change. Process-oriented technological change has to do with improvements in existing products. A second variety of technological change, new product technological change, contributes an ever larger share to economic growth over time. However, no good method has been developed for estimating product-oriented technological change. Therefore, the 2.5 percent measure must be an underestimate of the full contribution of all technological change to U.S. economic growth.

Scherer (1984) provides a complete survey of the topic through 1980. Parry (1999) provides a more recent assessment of technological change in natural resource industries (although he confuses the logging and the sawmill industries in his discussion of forest products).

management by improving the silvicultural activities associated with regeneration, fertilization, weeding, thinning, etc. Improvements in the tree and the forest stand are one potential affect of technological change in forestry. Technological change also affects the processing industry's preferences among tree species and the volume of recoverable product per harvested tree. Technological change can even affect the selection of lands allocated to forest production.

The first section of this chapter introduces the topic of technological change from the illustrative experience of the U.S. South. Forestry in the South began as a relatively uncomplicated extractive industry but forestry in the modern South is perhaps more technologically advanced than in any other region in the world. The South's history over the past century illustrates the three-stage pattern of forest development discussed in the previous chapter. It also illustrates the crucial impact of technological change and, among the varieties of technological change, a slower pace of technological improvements in forest management and timber growth and yield compared with a more rapid pace of improvements in logging and wood processing.

The chapter continues with a discussion of the specific varieties of technological change that affect forests and forestry and with an examination of the effects of each within the three stages of forest development. Finally, the chapter closes with a review of the historical rates of technological change in U.S. forestry and also a comparison of the rates for forestry with those for other sectors of the U.S. economy. Clearly, technological change can improve forest management and timber growth, but it is an unusual characteristic of forestry that technical change typically has had a larger effect on the location of timber harvests and on recovered harvest volumes than on the physical growth and yield of the forest itself. We will show how the three-stage pattern of forest development anticipates this result and allows us to understand the circumstances under which improvements in forest management itself will eventually become more important.

Illustrations from the United States, and from the U.S. South in particular, are convenient for this discussion because the history of U.S. forestry is well known, its statistical record is detailed, and the supporting analytical literature on technological change in the United States is also well developed. Nevertheless, these illustrations are general—and relevant for forestry in other regions of the world as well.

The characteristic effects of technological change that are apparent for commercial forestry are also applicable to other market-valued forest products—just as the lessons of chapter 2 were applicable to fuelwood and charcoal, latex, mushrooms, fruits and nuts, etc., as well as to industrial wood. Timber simply provides a convenient single product focus for the discussion. Its data are more complete than the data for other forest products. Furthermore, timber is a product that often grows in managed stands. Therefore, its representation covers all three stages of forest development.

The discussion of technological change and the economic long run in this chapter completes the basic forestry model. Each more specialized discussion in the remainder of the book will rely on this model for a part of its formulation. This chapter closes with a summary of the full model's implications for the economics of forestry. Just what is unique about this subject matter and what do its special characteristics suggest for management decisions and management behavior?

Two appendices follow the chapter. They review the traditional economic and biological models of forest management and describe their context within the three-stage spatial model. The traditional economic, or Faustmann, model applies to the important but specialized case of uniform plantation management. The fundamental biological model, also known for the "allowable cut" calculation and sometimes identified as the Hanzlik model, is seldom explained in textbooks on forest economics. Its absence means that economists seldom gain a fair understanding of the management perspectives of professional foresters and government forestry agencies. As a result, we often fail to grasp the different meanings that foresters and policymakers apply to terms like "allowable cut" and "rate of return," and we fail to comprehend the implications of these differences for timber supply. The explanation in the second appendix is an attempt to improve communication on these crucial forestry topics.[2]

Technological Change: The Experience of the U.S. South[3]

We can begin by identifying some common distinctions within the forestry sector of any economy. The owners of forestland are generally categorized as industrial (or owners of integrated forest and wood processing operations), non-industrial private including farmers and other generally smaller landowners, and the public.[4] The forest products industry is further divided into logging, lumber and wood products (including sawmills and producers of plywood and other wood-based panel products), furniture, and paper and allied products. Of course, the logging industry is a feeder industry for the other three industry categories. The furniture category includes metal furniture, metal furniture components, mattresses, and upholstery as well as wooden furniture, and only the latter is truly a forest product. These distinctions are common to most international statistical reporting for forestland area, standing forest volume and harvest levels, and also for the reporting of both industrial materials and products.[5] The discussion of the rest of this chapter and the rest of the book remains consistent with these distinctions.

---

2 It might be useful to convert our graphic description of the 3-stage model into a formal (optimal control) mathematical model as well. One merit of this exercise would be the ability to show the clear mathematical distinctions between this model and the Faustmann and allowable cut models. This challenging exercise is left as an exercise for a better mathematician!

3 This section relies on Hyde and Stuart (1999), revised with more recent data from the USDA Forest Service, U.S. Census Bureau, and the Food and Agriculture Organization of the UN.

4 Later, chapter 8 will introduce a fourth class of landowners, large institutional landowners who, like non-industrial landowners, own forestland but, unlike industrial landowners, do not their own processing facilities.

5 This statistical reporting system is known as the International Standard Industrial Classification or ISIC. In the United States it is simply SIC. Canada, Mexico, and the United States converted to a related system known as the North American Industrial Classification System (NAICS) in 1997, and revised this system again in 2002.

The U.S. South is the home of the most technologically advanced forestry sector in the world. Forests cover 86 million hectares or 55 percent of the southern landscape. They account for approximately two percent of the world's total forest but southern forests include 15 million hectares of plantations, or eight percent of the global total (USDA Forest Service 2005; FAO/UN 2001). The South's shares of the world's commercial forests and of those plantations strictly devoted to commercial timber production are larger yet. Southern forests produced 15 percent of the world's harvest of industrial wood in the early years of the twenty-first century. Most of their production supplies the South's fourteen thousand mills and an untold number of furniture manufacturers, including some of the largest and most technologically advanced sawmill, plywood, and pulp and paper facilities in the world. At least 40 southern pulp and paper facilities, for example, exceed 500,000 (short) tons in annual capacity and two exceed one million tons (CPBIS 2005). The two larger components of the forest products industry, lumber and wood products and paper and allied products, combine to form the largest single industry in the South today. A third component, furniture, is larger in the South than in any other region of the United States. The lumber and wood products and paper and allied products industries employed almost 400,000 workers and shipped products with an annual value of almost one hundred billion dollars in 2002 (U.S. Census Bureau 2006). The industry exports most of its products to markets in other parts of the United States and the world.

The South's Four Forests

The forestry sector was not always so important to the South's economy, and southern forestry was not always at the forefront of global forest technology. Southern forestry evolved from an extractive activity to include a timber management component over the course of the twentieth century.

Foresters speak of the South's four forests. The First Forest was the mature natural forest famous for its tall stands of clear longleaf pine. The initial harvests from this forest followed the pattern of agricultural development—expanding inland from the Atlantic and Gulf coasts with the conversion of forestland for the agricultural production of cotton, tobacco, and rice. The labor supply for the early forest industry was also associated with agriculture as most loggers were farm workers in search of supplementary income.

The early southern industry relied on the region's rivers to transport logs to sawmills along the coast. These coastal mills were oriented to export markets. The development of railroad logging changed this orientation as the railroads provided overland access to the markets of the U.S. Northeast and Midwest, as well as penetration into previously inaccessible regions of the South that subsequently became inland logging and sawmill centers.

Rail construction itself was a large share of the total cost of the logging operation and, once the rails were in place, rerouting them was expensive. The fixed costs of rail construction created a strong incentive to recover as much value per rail mile as possible. Concentrated and integrated logging and sawmill operations were the result. These operations developed their own demands for a reliable and permanent labor force, and a pattern of company towns emerged. Meanwhile, land ownership also became more concentrated—with 925 landowners accounting for more than one-half of the region's standing timber in 1919. Just 67 landowners accounted for half of that half.

By 1919, the cumulative effect of prior timber harvests accounted for more than forty million hectares of cutover forestland across the South. The land area in agriculture use peaked about this time and the demand to convert more of the original natural forest to agriculture declined. Southern lumber production declined as a result. In the 1920s, after approximately two decades of operation, many larger sawmill operators "cut and ran." They liquidated holdings, closed their mills, and moved to the U.S. West. *Ad valorem* taxes on the standing forest also encouraged rapid harvesting and discouraged reforestation, thereby reinforcing the cut and run pattern of early southern forestry.

The cutover lands did not remain in that condition for long. Southern pines are aggressive pioneering species and the southern pine forest regenerated naturally on abandoned land. A resulting natural crop of "old field pine" became the South's Second Forest. It became an important source of raw material for the sawmills of the 1930s and early 1940s and also for the emerging pulp and paper industry.

The gasoline engine and the electric motor changed the lumber industry again in the 1930s and 1940s. The region's sawmills took advantage of both technologies, of the labor force that was left behind when the large concentrated operations departed the region, and of the more dispersed forest that remained after railroad logging had run its course. Trucks replaced railroads for hauling logs and they were able to open many smaller pockets of timber that had been inaccessible and unprofitable for railroad logging. Logging operations themselves became smaller and more versatile. With the complementary development of more mobile electric sawmills, the South continued to supply approximately 35 percent of total U.S. lumber production even through the Great Depression of the 1930s.

Meanwhile, an emerging pulp and paper industry also drew on the South's Second Forest. The new sulfate pulping technology made it possible to produce pulp from resinous southern pines. The Roanoke Rapids mill opened in 1909 in Virginia's coastal plain. Fifteen more mills followed through the 1930s, all in the coastal plain and all manufactured kraft paper. (The South's first newsprint mill did not begin operation until 1940—after the industry learned that young southern pines are free of resin.) Many of these kraft mills relied on sawmill residue for a significant share of their fiber supply but their source for this input diminished as the lumber industry declined. Even these early paper mills were large capital-intensive operations and, unlike sawmills, they could not be transported to regions of greater supply once their local sources of fiber ran out. A sustained local supply was vital to their continued operation.

Indeed, resource supply was becoming an issue for the forest industry in general. Supply uncertainty had been a force behind the creation of the U.S. Forest Service and the National Forest System of public lands in the 1890s. With the successful protection of public forests, professional foresters in the 1930s turned their attention to the private lands. They were concerned that private forests were poorly managed and that poor management now would be a prelude to future timber supply shortages. Meanwhile, a few larger industrial landowners began reforesting their own lands (Urania Lumber Co. in Louisiana, Chesapeake Co. in Virginia, Crossett in Arkansas, Great Southern-Gaylord-Crown first in Louisiana then extending to Mississippi). Their experience undoubtedly helped divert public scrutiny away from the industrial forestlands, allowing it to shift toward small private landowners. Government regulation received some consideration as a means to insure reforestation and, therefore, a sustained timber supply, but regulation of many thousands of scattered independent private landowners would have been difficult.

Federal financial and technical assistance to landowners became the preferred alternative. Fire control was a prominent component of the early assistance programs. Once fire was controlled, large areas of abandoned agricultural land and cutover timberland regenerated naturally to pine and mixed pine-hardwood forests. Professional foresters in the United States still consider fire control in the South during this period as one of their great successes.

The Third and Fourth Forests

The South's Third and Fourth Forests are known for their plantations. The plantations of the Third Forest were established with wild seed stock in the years following World War II. Those in the Fourth Forest, beginning in the mid-1960s, were planted with selectively improved seedlings. Of course, plantations have never accounted for the majority of the southern forest. Plantations were unnecessary while there was a plentiful supply of natural timber and, even today, plantations only account for 17 percent of all southern forestland. The inventory of naturally growing timber remains substantial—although it is not always composed of preferred species in advantageous locations.

Following World War II, the larger sawmills continued the exodus from the South that began in the 1920s and the southern pulp and papermills continued their search for fiber. Smaller sawmills continued to replace their larger predecessors. These smaller mills were mobile operations, often composed of little more than a circular headsaw powered by a farm tractor. They took advantage of remaining pockets of timber wherever they could be found, and of off-season farm labor.

Meanwhile, pulp and paper gradually replaced lumber as the dominant component of the southern wood products industry. The value-added in pulp and paper exceeded value-added in sawmills and planing mills by the early 1950s, and the southern pulp and paper industry has continued to expand to this day.

Labor scarcity drove the subsequent technological adjustment. Agriculture mechanized, the aggregate regional demand for labor declined, and much of the rural population emigrated. The forest industry increased wages in order to maintain its work force and higher wages became an incentive to mechanize. Mechanized logging became more competitive in the 1950s and 1960s.

It helped that pulpwood demand was outpacing sawtimber demand. Pulpwood is smaller and lighter than sawtimber, and lighter and more uniform products are easier to mechanize. Even the new studmills and plywood mills that began to appear in the 1960s relied on smaller raw material. New capital investment in logging increased fourfold in real terms between 1954 and 1972. It doubled in sawmills and planning mills, increased twelve-fold in plywood and structural members, and doubled in paper and allied products. Labor productivity, a measure of increasing mechanization, more than doubled in sawmills and planning mills and more than tripled in logging, in plywood, and in pulp and paper between 1954 and 1972.

Mechanization continued at a rapid pace over the last quarter of the twentieth century. The level of new capital investment in lumber and wood products first increased, and then tapered off at a level that is still 50 percent greater in real terms than it was in the 1960s. New investment continued at a rapid rate, first in plywood, and then in reconstituted wood products as the construction industry began shifting away from plywood as a structural material, and as new substitutes like oriented strand board (OSB) became the means to utilize the plentiful upland hardwood resource. The seven large new OSB mills that began operation in western Virginia and West Virginia in the late 1990s and first years of the twenty-first century illustrate this trend.

The pace of mechanization in the paper industry has been even more rapid. Annual investment in new capital more than tripled in real terms between 1972 and 2002 and the value of shipments almost tripled—while employment decreased to 60 percent of its 1972 level. Labor productivity increased by more than 70 percent in real value per unit of cost. Meanwhile, this industry, like the plywood industry, continues to make increasing use of hardwoods and the previously less utilized upland forest resource.

This long period of increasing capital investment and mechanization means that much of the forest products sector's investment is now concentrated in its plant and equipment. The return on this capital has become a crucial indicator of industry success. The high fixed costs and low variable costs associated with operating this capital, particularly for pulp and paper operations, place a premium on uninterrupted operation. Therefore, we expect the forest products industry in general, and pulp and paper operations in particular, to have an incentive to maintain limited forest resources under their own control as guarantors of uninterrupted production for their mills.

Indeed, the industry does own forestland and the industry has been at the forefront in introducing improvements in forest growth and yield on its lands. It introduced the first nurseries in the 1960s to grow the improved seedlings identified with the Fourth Forest. Some firms began participating in university research cooperatives at about the same time. Certain of these firms, and others as well, began their own independent forest management research programs. More recently, a few industry-owned fiber farms began to appear in the 1970s. These incorporated fertilization and irrigation into their management regimes in order to improve further on fast growth for high yields of uniform fiber.

Nevertheless, forest research expenditures and forest plantations remain a very small part of the industry's total financial investment and harvests from the industry's plantations are only sufficient to insure an uninterrupted flow of fiber in the most competitive times.[6] Only 45 percent of the industry's lands were in forest plantations, and the total of all industry lands was less than 20 percent of the total southern forest by late twentieth century.[7] In the early twenty-first century, two-thirds of all timber harvested in the South still originates from non-industrial private forestlands, approximately 90 percent of which are still managed extensively if at all.[8] These non-industrial private forests may not be at a "frontier" in the colonial development sense of that word. They are, however, at an economic frontier in the sense that the financial return on many of these lands is close to zero and they seldom have higher-valued competitive uses.

Clearly, the economic incentives for intensive forest management, while expanding, are not overwhelming—even in this very productive and technologically advanced region. Forestry has prospered in the South, even as the geographic locus of timber harvests has shifted inland and upland with time. Two competing sources of fiber, debris from logging and processing and 50 million hectares of hardwood and mixed pine-hardwood forest, including nearly 30 million hectares of unmanaged and lower-valued upland hardwoods, continue to offer inexpensive substitutes for the higher cost fiber from more intensively managed pine plantations. The pulpmills have additional sources of fiber in the form of sawmill waste, recycled paper, and old corrugated paperboard. The South's largely unmanaged non-industrial forests and these substitute sources of fiber satisfy more than 60 percent of the southern industry's demand for raw material

---

6 A National Science Foundation (1981) survey finds that the paper and allied products and wood and wood products industries were eighth and tenth among fourteen two-digit SIC industries in research expenditures as a share of total revenues. The lumber and wood products industry spent less than 0.8 percent of its revenues on research and development over the twenty-year period from 1960 to 1980. The paper and allied products industry spent less than 1.1 percent of its revenues on R&D over the same period. These are old data, but the research shares and the rankings among SIC two-digit industries were steady over the years before 1980 and they probably have not changed much since then.

7 By the earliest years of the twenty-first century the southern industry had divested itself of much of its forest land. Institutional landowners now own or manage most of these lands. Chapter 8 discusses this most recent trend in forest landownership.

8 Alig et al. (2001) project southern forest land use through the year 2020. They foresee no change in these conditions over this period. The U.S. South is not unique in this respect. Hoen, Eid, and Ok (2001) examine production possibilities and yields for Norway, another technologically advanced forest economy. They observe that natural regeneration accounts for the majority of Norway's stocking today and they anticipate that, even with a rising rate of return to forest production, the area under natural regeneration will actually increase in the next decade.

even today, and they are the reason that forest management itself has not become a more widespread and more intensive activity.

More General Observations

The forest industry in other regions of the world has followed a similar pattern of development and a similar pattern in its adoption of new technologies.[9] Some parts of the world, such as the southern regions of the Nordic countries, are at least as technologically advanced as the U.S. South in their plantation management, their logging operations, and their processing facilities. Other parts of the world fall within earlier stages in the patterns of forest development and the adoption of new technologies. Those who live in regions within the earlier stages of development may or may not be aware of the latest technologies adopted in regions like the U.S. South and parts of the Nordic countries. The different mix of economic incentives prevailing in these other regions explains local reluctance to adopt some technologies that are known but are not yet appropriate for regions in the first two stages of forest development.

In the U.S. South, the Nordic countries, and in all other regions of the world, the local adoption of improvements in harvesting and wood processing technologies preceded the adoption of most improvements in forest management itself. This adoption sequence occurred because, even today, the natural forest alternative to fiber from managed forests still provides most of the raw material for the wood processing industries in virtually all regions of the world. Improvements in forest management tend to be adopted only after lower cost sources of fiber have been depleted. Therefore, we can anticipate that the rates of advance in harvest and wood processing technologies will continue to outpace the rate of change in forest management itself in regions where an accessible volume of natural forest remains. However, we can also anticipate that the relatively low rate of technological advance in forest management will increase as more of the world moves into the third stage of forest development and managed forests become more competitive with commercial production from natural forests.

Technological Change and Biological Growth: Long-term Effects on Forestry

We can generalize the experience of the U.S. South with reference to the three-stage pattern of forest development and figures like those in chapter 2. As we do, it will be useful to distinguish the effects on the forest from technological change in the adjacent agriculture sector of the economy as well as the effects of technological change in the forestry sector itself. Technological change in

---

9 Stone (1997, 1998) reflects on a similar pattern of development and technological change, although for a shorter period, for the forest industry of the eastern Amazon. Chapter 7 will review the detail of Stone's observations with special attention on the wood processing industry.

agriculture affects the demand for forest land. Three categories of technological change in forestry; improvements in wood utilization at the mill, in logging, and in forest management itself; have their own effects on the forest. In addition, improvements in the rural infrastructure also have an effect. These different categories of technological change can have different, and sometimes opposing, effects on the forest—even within the same region and during the same stage of forest development.

Agricultural Technologies

Agricultural technologies affect forestry by altering agriculture's competitive advantage for marginal lands. The effects of agricultural technologies on forestry differ, however, with the different stages of forest development.

Improved agricultural technologies increase agricultural productivity. This raises the net agricultural value function, as from V_a to $V_{a'}$ in Figure 3.1A. For the first stage of forest development, while the forest still interferes with agricultural activity, the margin of productive agriculture at point B is beyond the margin of extractive forest activity at D. Land is a plentiful resource, relatively cheaper than agricultural labor or capital which has to be transported to the less accessible productive site. Therefore, in this first stage, most successful new agricultural technologies must be relatively land-using and labor or capital-saving. They shift the intersection of the net agricultural value function with the transactions cost

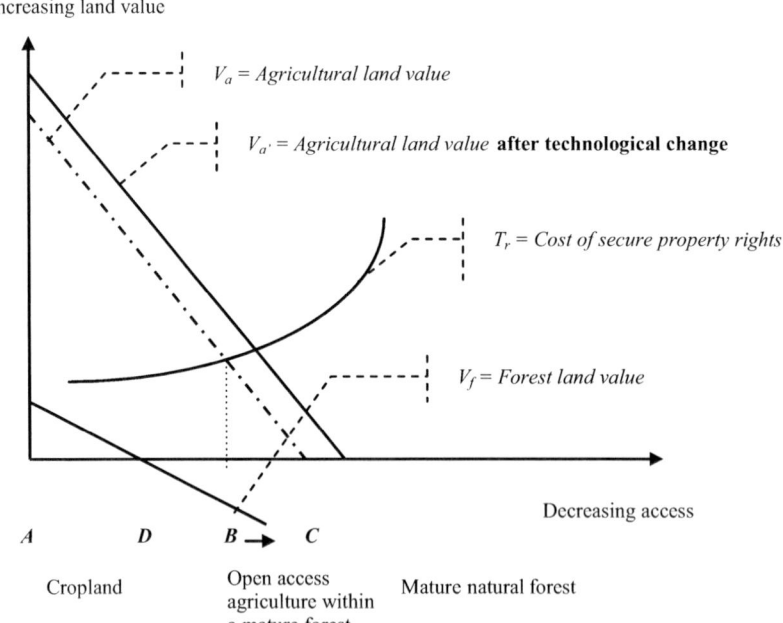

FIGURE 3.1A Agricultural technologies on new frontiers

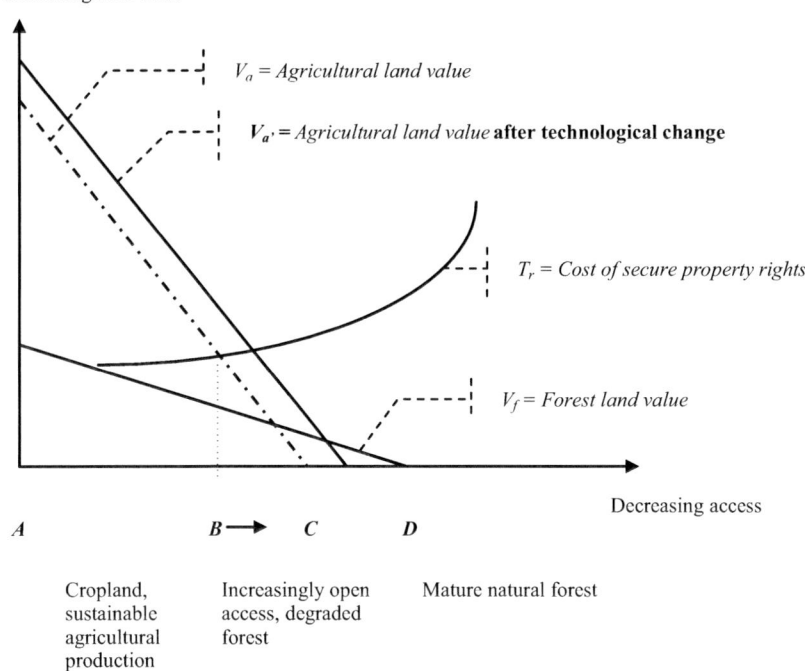

FIGURE 3.1B Agricultural technologies on developing frontiers

function to the right of their previous intersection at B and expand agricultural use farther into land that is still under natural tree cover. The bold arrow below the figure illustrates this effect.[10]

Many new agricultural technologies also encourage agricultural expansion during the second stage of forest development. Land remains a plentiful resource for agriculture, but extractive forest activity now extends past the limit of productive agricultural activity. New agricultural technologies cause the intersection of the agricultural value function with the transactions cost function to shift to the right once more. However, this time agricultural land use expands into the area of previously degraded forestland—between points B and C in Figure 3.1B.

---

10 Cattaneo (2005) provides an empirical example from Brazil's Amazon. Innovation in livestock practices within the Amazon, a region in the first stage of forest development, induced land conversion from forest to agriculture—just as our analysis anticipates. However, Cattaneo also demonstrates the greater complexity that arises when the analysis moves beyond one region to include trade between regions. He observes that agricultural innovation in other regions increased productivity in those regions and drew labor away from the Amazon, thereby decreasing deforestation. In sum, agricultural innovation in the Amazon and agricultural innovation away from the Amazon had opposite effects on deforestation in the Amazon. Chapter 6 will extend the regional analysis to include multiple regions and the effects of trade between them. The discussion of another example from Brazil's Amazon (Young 1996) will further illustrate the complexities that may arise.

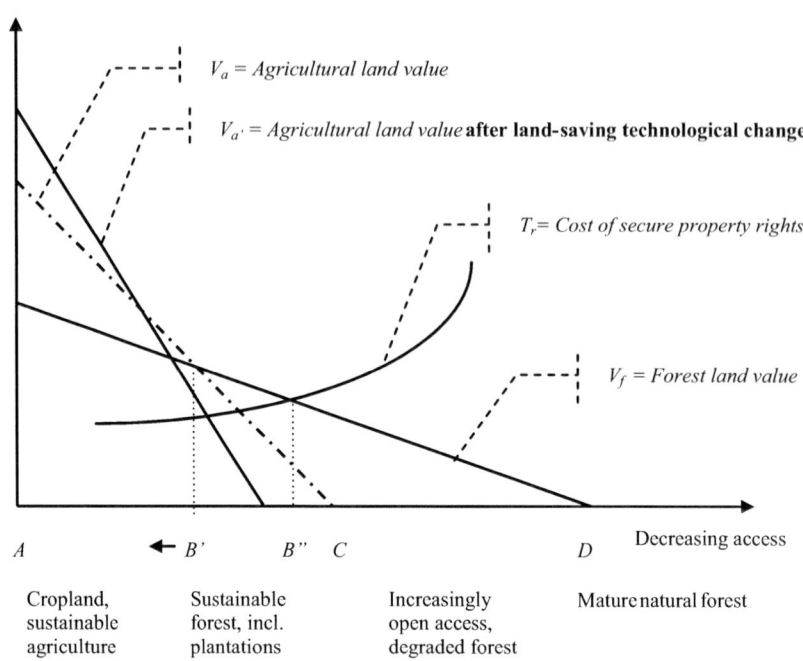

FIGURE 3.1C-I Land-saving agricultural technologies on mature frontiers

Once more, the bold arrow below the figure illustrates the land use shift. Agricultural technologies no longer have an effect on the mature natural forest in regions characterized by this stage of development.

In the third stage of forest development, forest management and agriculture are competitive at what is the extensive margin for agriculture but the intensive margin for managed forestry. Land is no longer a plentiful resource for either agriculture or forestry and agriculture's higher relative value in this stage provides an incentive for labor- and capital-using and land-saving agricultural technologies like mechanization and high yield crop varieties. The incentive for agricultural technologies in general has been present all along but the new agricultural competition with managed forests introduces the incentive for agricultural technologies that have a land-saving effect. See Figure 3.1C-I. As land-saving technologies shrink the base of agricultural land use, managed forestry wins the decreasing competition with agriculture for some lands at the shared land use margin at B'. This margin shifts to the left as managed forests expand into lands previously devoted to agriculture.

In fact, land-using agricultural technologies did dominate the North American experience during the pioneer years of western expansion.[11] Subsequently,

---

11 See North (1966) for the classic description of early U.S. economic growth.

land-saving technologies dominated during most of the twentieth century, and American agricultural land use has remained relatively constant since the 1930s as a result—even as aggregate agricultural production has increased. For the U.S. South we noted that agricultural mechanization in the 1950s caused a concentration in agricultural land use and a decrease in the demand for farm labor. Emigration from the region in the 1960s was one result. The return of some untended old fields to forest cover was another. The land area in farm forestry also expanded—by more than 50 percent in the 1960s and 1970s (USDA Forest Service 2005).

Nevertheless, not all new agricultural technologies arise from the competition between agriculture and forestry. By the 1970s, the South was in the third stage of forest development, and some technologies, such as wetland drainage, that began to arise in response to other incentives were not land-saving. Technologies that are not land-saving have a contrasting effect, improving the competitive position of agriculture relative to forest management and inducing the conversion of some forestland to agriculture. These technologies cause forestry's competitive margin with agriculture at B' to shift to the right—as the bold arrow below Figure 3.1C-II indicates.

Thus, for regions in the third stage of mature forest development, the net effect of new agricultural technologies on the shared margin of agriculture and

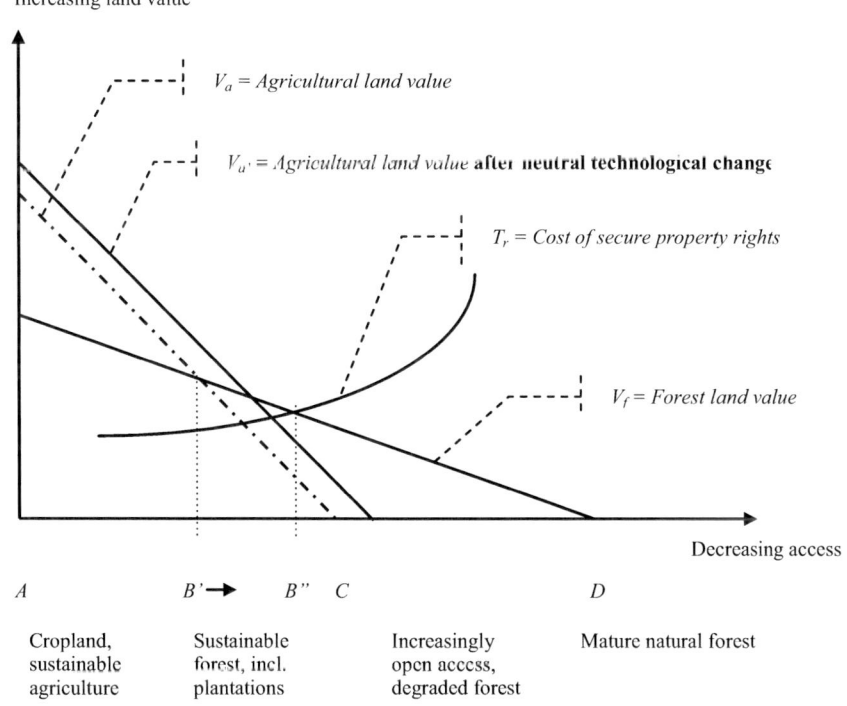

FIGURE 3.1C-II Neutral and land-using agricultural technologies on mature frontiers

managed forests is indeterminate, and it may change from time to time depending on the relative prices of land, labor, and capital, and the relative land-saving or land-using character of the latest successful new technology. New agricultural technologies during this stage have no effect, however, on either the degraded open access forest between B'' and D or the remaining mature natural forest beyond point D.

In sum, the effect of new agricultural technologies is complex. Agricultural technologies affect the natural forest in some cases, managed forests in others, and have little effect on either in still others. The effect in each case depends on the stage of regional forest development and on the nature of the agricultural technology itself. In the broader scheme of things, when trade between regions also enters the problem, we will find (in chapter 6) that local comparative advantage and the elasticity of demand for additional agricultural production are also important determinants of the final effects on the forest.[12]

A Developing Rural Infrastructure, Logging and Wood Utilization Technologies and, Ultimately, New Forest Management Technologies

Standing forests are a deterrent to many productive activities at new frontiers. Therefore, technological improvements that reduce their deterrent effect attract new settlers and increased extractive activity. Two categories of improvements come to mind, improvements in the regional infrastructure and improvements in logging technologies. Improvements in the regional infrastructure, especially new or better roads and more extensive road networks, permit greater access for agricultural development as well as for the various activities that exploit the natural forest. They reduce transportation costs and, thereby, shift the net value functions V_a and V_f in Figure 3.2A. New or improved road construction may extend to previously less-roaded areas and these tend to be farther from the more concentrated development that exists in towns and market centers. Therefore, the transportation cost savings from new roads may be greater at greater distance from the market. This means that the shifts in V_a and V_f may be larger for less accessible lands and the margins of both agricultural production at B and C and forest development at D will extend to the right in our figures. In this event, improvements in the regional transportation network allow agriculture to expand farther into the open access area of degraded forest and they allow extractive forest activities to extend farther into the mature natural forest.

Other improvements in the regional infrastructure, such as improvements in communication networks and improvements in the institutions of governance can also be explained in terms of better information for land management and

---

12 Tachibana, Nguyen, and Otsuga (2001) and Cattaneo (2005) demonstrate the range of possibilities in their examinations of alternative sources and effects of agricultural intensification in Vietnam and Brazil, respectively. Kaimowitz and Angelsen (1998) also introduce a range of examples.

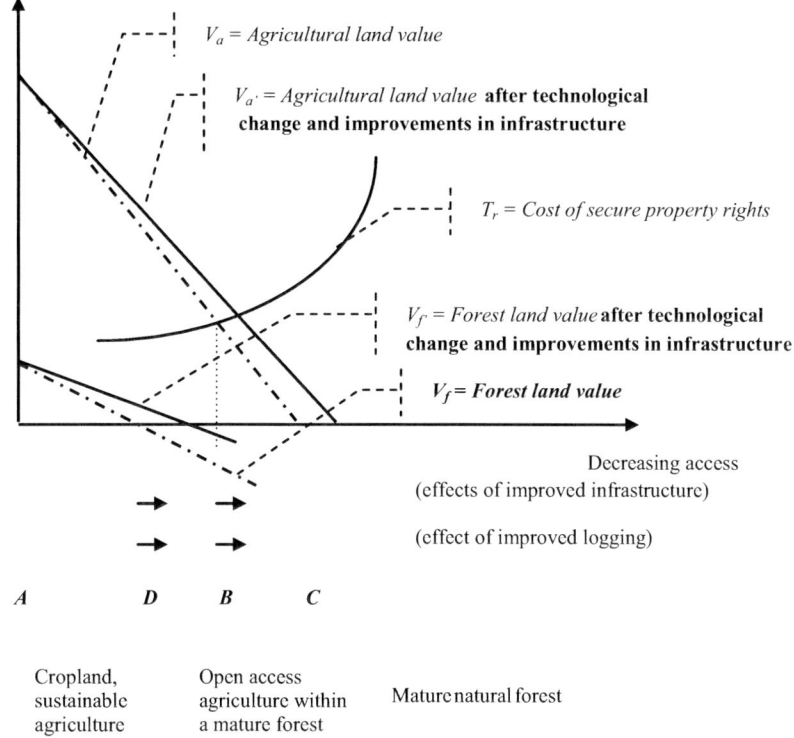

FIGURE 3.2A Improved rural infrastructure and logging technologies on new frontiers

similar shifts in V_a and V_f or, alternatively, in terms of decreases in transactions costs T_r.[13] Both explanations have the same marginal effect on new frontiers, shifting the intersection at B to the right and adding to the movement of agricultural activity into the area of the open access natural forest. (We will return to improvements in roads and governance again in the discussion of forest policy in chapter 4.)

---

13 Global Positioning Systems (GPS) and cell phones are two examples. GPS provides the means to quickly and inexpensively verify environmental performance that was previously confirmed only with extensive field sampling. These two technologies also provide the means to discover and track trespass in the forest. For example, park managers in some African countries use both technologies to deter the poaching of wildlife, and villagers in Cameroon use their mobile phones to report violations of rural property rights. As the use of cell phones has increased, the response time required for identifying violations of rural property rights has declined and violations themselves have declined (J. Xu, personal communication, July 1, 2004; *Economist* 2004; J. Vincent, personal communication, May 20, 2004; Global Forest Watch 2005). In terms of our figures, technological improvements in rural communication decrease the cost of monitoring and enforcing property rights and, therefore, shift the transactions cost function T_r downward and decrease the area of open access activity.

Improvements in logging technologies, ranging from better chainsaws to improved heavy equipment like skidders, reduce the cost of removing the natural forest cover. They shift the net value function for timber outward along its horizontal axis—similar to the effect of improved roads. This shifts the margin of forest extraction at D to the right. For regions at the very outset of human development (point D is to the left of point B), new logging technologies help remove the forest from its interference with agriculture. In this case, the new technologies assist agricultural conversion of the forest and also help expand the range of extraction from the natural forest. For regions that are only a little more developed (point D is between points B and C), the forest no longer interferes with agricultural production and the only effect of new logging technologies is to extend extractive activities farther into the natural forest, thereby increasing the area of open access.

Developing forest frontiers: Improvements in the regional infrastructure and in logging technologies have comparable effects on regions in the second stage of forest development—as Figure 3.2B illustrates. Improvements such as roads increase the net values of both agricultural and forest land use, extending agricultural activity into the degraded open access territory to the right of point B and extending extractive forestry activities farther into the mature natural forest to the right of D. Other improvements in the infrastructure, such as improvements in communication and in governance, also enable agriculture's extension into the degraded forest to the right of B.

These technologies also have a second effect, increasing the productive recovery of material extracted from the forest and making it profitable to return and remove additional woody material from the previously logged areas of the depleted forest between points B and D. We observed this effect in the U.S. South in the 1930s when trucks replaced railroads and loggers began entering the smaller localized forested coves that had been inaccessible previously. The bold arrows in the graphic below Figure 3.2B identify this effect.

At some point, the wood processing sector recognizes the cost advantage originating from its own improved utilization of delivered wood, as well as from the increased utilization of lower quality material from the forest. Typically, these advantages occur as the traditional resource supply begins to decline late in the second stage of forest development. Pulpmills in the U.S. South, for example, recognized potential shortages of fiber supply in the 1930s and 1940s and began recycling residue fiber from sawmills. Later they learned to use less resinous young pines in the manufacture of newsprint, and still later they developed the technology for producing pulp from hardwood species. Later yet, the new OSB technology for wood-based panels began to utilize upland hardwoods, often from previously logged forests.

As the recovery of mill waste and material from the degraded forest increases, the new value of these previously uneconomic resources adds another increment to the value of the lands on which they grow. In terms of Figure 3.2B, the net

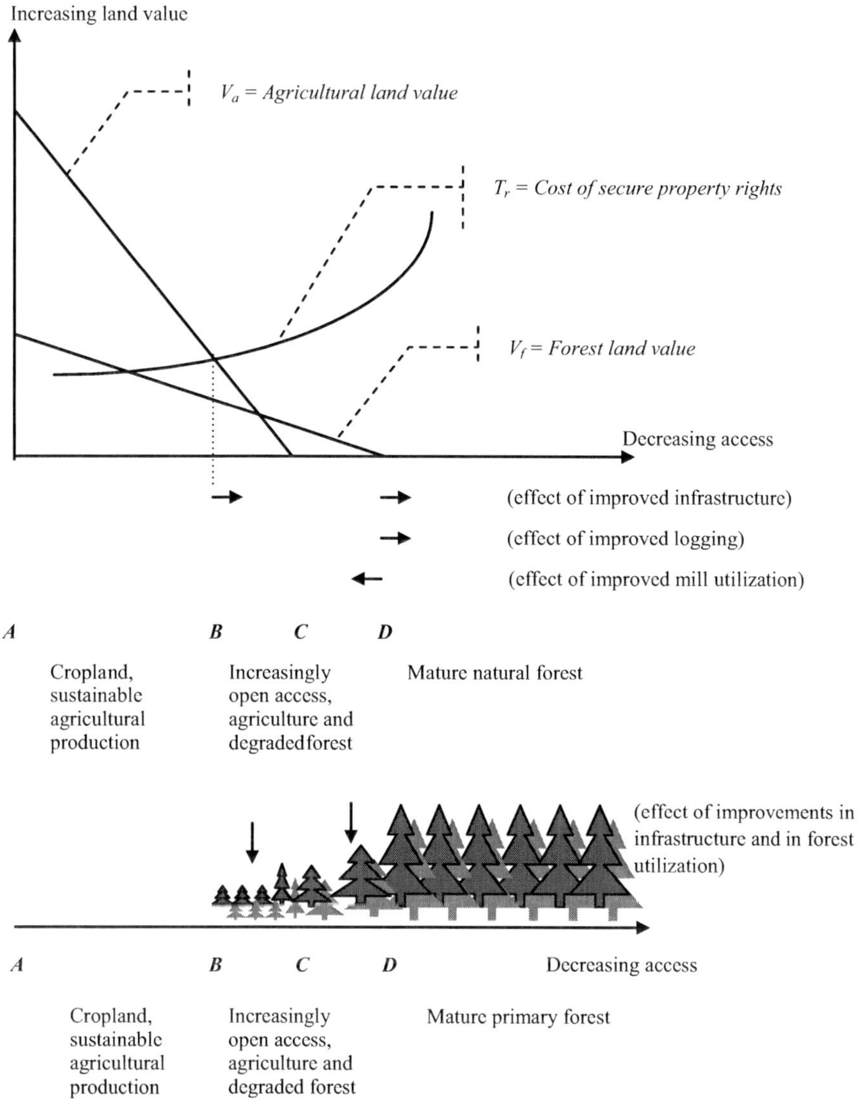

FIGURE 3.2B Improved infrastructure, logging, and forest utilization on developing frontiers

forest value function V_f shifts upward. (In order to reduce clutter, Figure 3.2B no longer displays the shifts in V_a and V_f.)

Does improved forest utilization shift the forest value function outward too? Probably. But improved utilization also has a forest conserving effect, substituting for some amount of wood that would have been delivered to the mill in the absence of the new technology and decreasing the overall rate of harvest at the natural forest frontier. The combined result from these two opposing effects depends on the mill demand and resource supply elasticities. The latter, forest conserving, effect can be large and surely it is the dominant effect in many cases. In one impressive example, the introduction of truss frame housing construction in the United States saved an amount of wood equivalent to the potential annual harvest from all de facto Wilderness areas.[14] In a developing country example, Gunatilake and Gunaratne (2010) calculate that by retiring inefficient sawmills and shifting Sri Lankan production in the direction of that country's more efficient mills would reduce sawmill demand and replace or delay annual harvests on more than 3.6 million hectares of that country's remaining natural forest.[15]

In sum, during the second stage of forest development the combined net effects of improved infrastructure, improved logging, and improved forest utilization extend the margin of managed agricultural land use and decrease the remaining depleted forest stock on the open access lands. In terms of our figures, they extend point B to the right and decrease the forest stock between points B and D. Their net effect on the natural forest frontier at point D is uncertain.

Mature forest frontiers: Similar effects of improved infrastructure and improved logging and forest utilization continue into the third stage of forest development.

In addition, both improved infrastructure and improved logging and forest utilization affect the managed forests that begin appearing in this third stage of development. Forest land has become a relatively scarce resource and, for the first time, improvements in forest management like thinning, weeding and fertilization, and biological technologies like improved seed stock can have an effect. They too register their effect on the managed forests between B' and B'' in Figure 3.2C.

All three improvements (infrastructure, logging and utilization, and forest management) shift the net forest value function V_f upward. The increase in managed forest value enhances the ability of intensive forest management to compete with extensive agriculture—and this extends forestry's intensive land use margin at B' to the left as forestry competes favorably with some agricultural activities. Improvements in forest biology and management may also extend the margin of extensive management at B'', as when these new technologies become so profitable that they induce the removal of some natural forest cover simply to provide space for newly profitable intensive management. The conversion of some bottomlands in the U.S. South from their original hardwood forests into pine production and the conversion of some natural forestland in Indonesia to palm plantations for oil production are examples.

---

14 Specifically, the 28.6 million acres proposed for wilderness withdrawal in the policy proposition known as RARE II. (U.S. Forest Service calculation provided to R. Buckman by H. G. Wahlgren, February 15, 1989.)
15 Kneese and Schultz (1975) first identified the more general environment-saving characteristic of technological change. Smith (2008) provides a more recent reflection on the general topic.

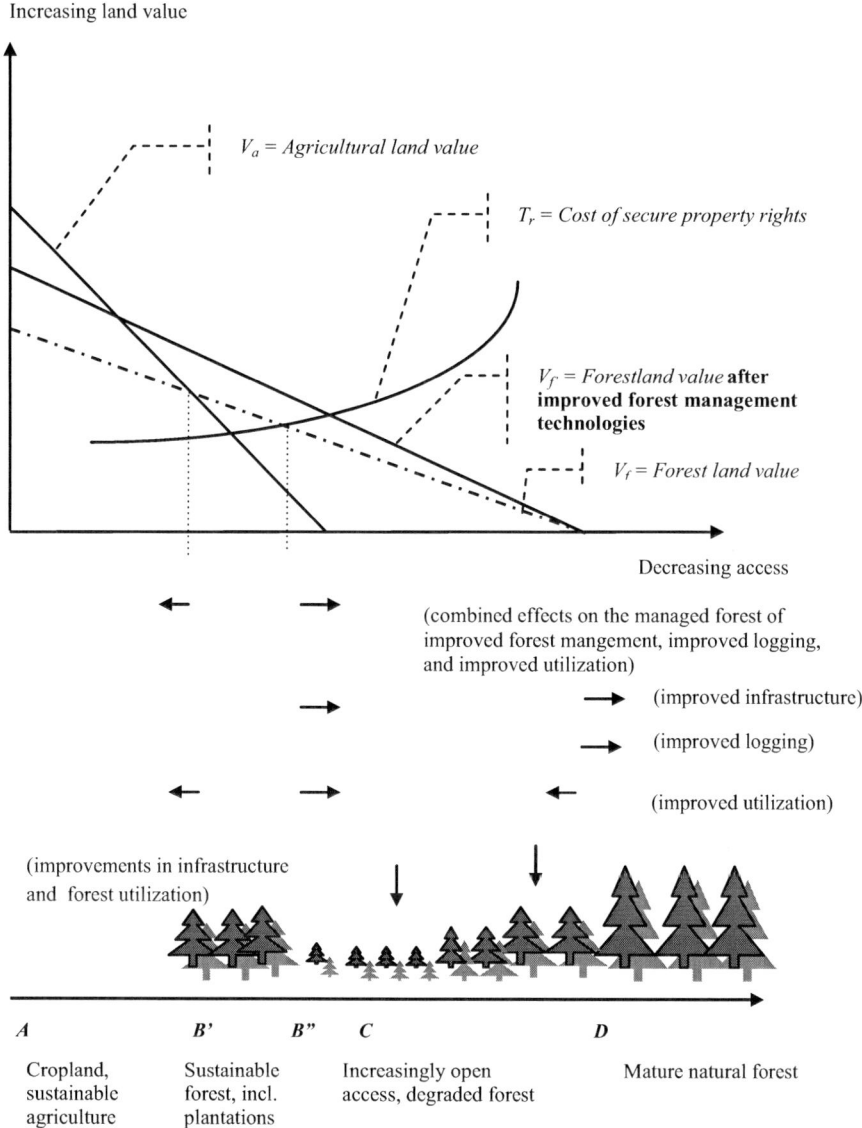

FIGURE 3.2C Improved infrastructure, logging, and forest utilization on mature frontiers

Once more, in stage III as in the stage II, new technologies that increase the utilization of raw material from the managed forests and increase the recovery of raw material from the natural forest tend to substitute for some amount of more remote extraction from the natural forest. They increase the share of production from managed forests and decrease the rate of extraction from the natural forest, but they will not replace natural forest extraction altogether.

This conceptual organization and the delayed implementation of new biological and forest management technologies are entirely consistent with our observations from the U.S. South. Southern forestry incorporated many improvements in infrastructure, logging technology, and mill technology before World War II, but it did not begin introducing improvements in forest management until the Third Forest. Only then did forest plantations and forest management become competitive with extraction from natural forests. Furthermore, even for the South's Fourth Forest and even today, the most intensive management practices occur only on a few hectares at forestry's intensive margin. Most of these lands occur in the southern coastal plain where the land tends to be more productive and the access to the region's capital-intensive pulp and papermills is better. Numerous additional lands are planted and then left to grow in a minimal regime of extensive management. Still more regenerate naturally and remain as essentially unmanaged private forests until they are ready to harvest. These observations are consistent with a net forest value function that shifts upward at the intensive margin at B' in the figures, but which is relatively unchanged at the extensive margin of managed forestland at B''.

Finally, even today we observe that southern pulp and papermills obtain more than 60 percent of their fiber input from recycled material, including the fibrous debris of other wood processing activities, and we also observe almost thirty million hectares of upland hardwoods remain undeveloped, plus almost as much unmanaged bottomland hardwood and mixed pine-hardwood forest—well over half of the South's total forest area (USDA Forest Service 2005). These natural forests still support a limited amount of harvesting. The natural forest frontier remains an inexpensive alternative to some more expensive managed forest technologies and that frontier is still an important source of wood fiber even for today's mature and technologically advanced southern forest industry.

The availability of substitute material explains why the impact of biological improvement in forestry has not been as remarkable as the Green Revolution has been for agriculture. The wild crop component of all agricultural production is very small. Agriculture has to plant, manage, and grow the vast majority of its product. Forestry does not. The *physical potential* of biological gains may be as promising for forestry as for agriculture. The *economic* returns to biological improvements for forestry will be less impressive and their adoption will be less widespread as long as the South's, and the world's, wood processing industry continues to have access to substitute sources of naturally growing fiber. This means that the potential for the adoption of further improvements in logging and mill utilization technologies will generally continue to be more promising than the potential of further improvements in forest biology and management—although the latter are growing in importance in those parts of the world that are in the third stage of forest development.

Summary

Tables 3.1A and 3.1B summarize the impacts of technological change in two different formats. Table 3.1A traces the effects from each category of technological change through each stage of development and on to its marginal impact on land

TABLE 3.1A Categories of technological and institutional change and impacts on land use

Category of change	Stage of forest development and resulting shift in marginal land use
1. Agriculture	I: Extend cropland into the open access region (B shifts right) and extend livestock grazing into the natural forest (D shifts right)
	II: Extend agricultural activities into the depleted open access forest (B shifts right)
	III: Uncertain agricultural effect on the intensive margin of forest management (B')
2. Infrastructure	I & II: Expand agricultural activities and forest extraction (B and D shift right)
	II & III: Increase recovery from the open access forest (the regions of BD or $B''D$)
	III: Expand the area of forest management activities and extend all activities in the natural forest (B'' and D shift right)
3. Logging	I: Decrease forest interference with agriculture, thereby increasing agricultural land conversion (B and D shift right)
	I, II & III: Increase recovery from the open access forest (the regions of BD or $B''D$)
	II & III: Net expansion into the mature natural forest (D shifts right)
	III: Extend both the intensive and extensive margins of forest management (B' shifts left and B'' shifts right)
4. Forest utilization	I: Decrease forest interference with agriculture, thereby increasing agricultural land conversion (B shifts the right)
	I, II & III: Increase recovery from the open access forest (the regions of BD or $B''D$)
	I, II & III: Conserving effect on the mature natural forest (D shifts left)
	III: Extend both the intensive and extensive margins of forest management (B' shifts left and B'' shifts right)
5. Forest management	I & II: No effect
	III: Extend both the intensive and extensive margins of forest management (B' shifts left and B'' shifts right)

use. Table 3.1B traces effects in the other direction, from the marginal land use impacts in each stage of forest development to the technological sources of these effects.

Technological change plays a crucial role in forest development, especially since forest production periods tend to be long enough that even small rates of annual technological change accumulate into both large impacts on productivity and important impacts on land use at the margin. However, we have seen that the role of technological change is complex, and it is complex, at least partly, because of forestry's unusual three or four margins of productive activity. Its

TABLE 3.1B Stages of forest development and causes of marginal land use change

Stage of forest development	Impact	Marginal shift	Sources or categories of change
I. New frontier	1. Extend cropland conversion	D shifts right	Agriculture, infrastructure, utilization
	2. Expand other agricultural activities into the natural forest	B shifts right	Agriculture, infrastructure, logging
II. Developing frontier	1. Extend agricultural land conversion into the depleted open access forest	B shifts right	Agriculture, infrastructure
	2. Increase recovery from the depleted open access forest	Forest stock decreases in BD	Infrastructure, logging, forest utilization
	3. Uncertain harvest expansion into the mature natural forest	D uncertain	Infrastructure, logging, forest utilization
III. Mature frontier	1. Increase recovery from the depleted open access forest	Forest stock decreases in B"D	Infrastructure, forest utilization
	2. Extend harvests into the mature natural forest	D uncertain	Infrastructure, logging, forest utilization
	3a. Extend extensive margin of managed forest	B" shifts right	Infrastructure, forest utilization, forest management
	3b. Uncertain effect on intensive margin of managed forest	B' uncertain	Agriculture uncertain, others extend B' to the left

effect depends on the particular technology and the local relative scarcities. Technological change in different components of the forest industry has different effects on the forest itself and at different stages of the industry's development and one category of technological change, improvement in forest management, does not even begin to have an economic impact until the third stage of forest development.

The later adoption of improvements in forest management highlights two additional points of interest. First, we often think of biological growth when we think about "renewable" resources like forests. However, the other categories of technological change will be more important than biological growth and improvements in forest management—at least for the first two stages of forest development. And second, during the third stage of forest development, rising prices and all categories of technological change induce increases in the share of production originating from managed forests. As this share increases, the share of production from the natural forest must decline. Therefore, in the third stage of forest development, technological change can help to control the rate of deforestation and improve the protection of the natural forest. Since managed forests provide an increasing share of global production today, modern technological change must be increasingly responsible for containing, if not overcoming, the rate of deforestation.

In sum, the role of technological change is important to any clear understanding of forest production and forest land use over time. However, we can be certain that the net effect of all technological change on any single forested location is an empirical question that depends on characteristics of the specific technological improvement itself as well as on the demand for the local forest product and the conditions of the local forest. We can expect to find a diversity of examples as we observe different cases around the world. In fact, the analytical evidence on technological change in forestry is sparse and empirical assessments of the effects of technological change on the margins of land use are non-existent. The next section of the chapter reviews the implications of that U.S. evidence that is available.

Empirical Evidence on Rates of Technological Change

Observed rates of technological change provide evidence that these arguments are not just theoretical and speculative. The long-run economy-wide rate of technological change for the United States has been in the neighborhood of 2–3 percent per annum over the course of the last 70 years. Technological change in any one sector like forestry can exceed the economy-wide rate only occasionally and for a few years, during short periods of widespread adoption of particularly rewarding innovations. Technological change decreases the costs of production and the adoption of a new technology reduces the opportunity for further cost savings and technological change. This means that, as change occurs relatively more rapidly in one sector, that sector's relative costs decline, and the incentives for adopting further improvements in that sector decline relative to the incentives for change in competing sectors. As the incentives decline, the first sector's rate of change must eventually decline to a rate that is comparable to the economy-wide rate. On the other hand, technological change in one sector can occur at a long-run rate that is less than the economy-wide rate if that sector is declining in relative importance in the aggregate economy.

This reasoning suggests a preliminary inquiry of forest products' share of longer-term economic activity. In fact, the wood processing industry's share of the U.S. economy declined from 1.75 percent in 1977 to 1.06 percent in 2002 and it has probably continued to decline as the U.S. housing market has stagnated since then.[16] Therefore, it is not surprising that technological change in the lumber and paper industries was in the range of 1.5 to 1.9 percent for the same period,

---

16 Calculated as the sum of value-added by the wood products, furniture, and paper industries divided by the gross national product (http://www.bea.gov/industry/gdpbyind_data.htm, accessed July 21, 2007).

a range that is on the low side of the economy-wide rate of technical change (Stier and Bengston 1992).[17]

The share of forest management in the full economy is difficult to calculate because the statistical records of timber revenues and of expenditures on forest management are either incomplete or not readily available for many years. However, we do know that the annual rate of technological change in southern forest management from 1950 to 1990 was less than 0.6 percent, a rate that is much smaller than either the economy-wide rate or the rate experienced by the wood processing industries (Newman 1991).[18] Moreover, technological change in forest management has probably been even slower for those other regions of the United States where the experience with forest plantations is more recent and more limited than in the South.

A slower rate of technological change in forest management means that cost reductions in primary forest production have not kept pace with cost reductions in other economic activities, and that wood fiber became *relatively* more expensive over the course of the twentieth century. In fact, there is some indication of long-term increases in relative prices in delivered logs and lumber (Barnett and Morse 1963; Ruttan and Callahan 1962; Phelps 1975; Olson 1971). Increasingly expensive wood could suggest a developing problem with resource supply at the mills and this should mean increasing incentives for wood-saving technological change at the mills. Indeed, within the last 40 years, technologies like computer programs to determine the configuration of saw slices increased physical yield in sawmills by 10 percent, powered back-up rollers increased plywood production by 17 percent, and various environmental improvements, while not increasing production, have enabled pulpmills to limit deleterious chemical discharges (Hyde, Newman, and Seldon 1992; Horvath 1980). Some of these, for example, improved watering and pressing technologies, have increased paper production as well as improved environmental quality (Damani 2004).

However, the question is not whether wood processing technologies have improved. They have. The question is whether improvements in the use of wood occurred more or less rapidly than either improvements in the use of other (labor and manufactured capital) inputs to wood processing or improvements in other sectors of the economy. The most direct comparison would be between different inputs in the same industry. That is, if wood inputs were increasingly scarce relative to labor and capital inputs in the wood processing industries, then we would expect a greater rate of technological change in wood-saving technologies than in labor- or capital-saving technologies.

---

17 Various assessments of the Canadian lumber industry report even smaller rates of technological change, ranging from negative to + 0.61 (summarized and re-appraised in Nagubadi and Zhang (2006) and Zhang and Nagubadi [2006]).
18 Even at this 0.6 percent rate, production doubles every 120 years—or for each passing 120 years only half as much land is required to produce the original volume of wood.

In fact, Steir and Bengston's (1992) survey of 24 empirical assessments from various U.S. and Canadian wood products industries concludes that technological change was predominantly labor-saving and capital-using.[19] This observation is consistent with the overall U.S. experience since World War II. Stier and Bengston find little evidence of wood-saving technical change and many of the assessments they review observed a wood-using bias. Thus, we can conclude that, while new wood-saving technologies certainly have been introduced throughout the wood product industries, the rate of their introduction has been relatively slower than for that of new labor-and capital-saving technologies.

There is only one possible explanation. As the relative cost of wood rose and the supply of wood fiber became a more important issue for the forest products industry in the United States, this increasing cost, together with a concern for future raw material supply, encouraged some investment in forest plantations and a certain level of innovation in forest management itself. Nevertheless, labor costs in particular remain *relatively* greater burdens for the wood processing industries. While the relative costs of woody raw material may have risen, the availability of substitute products like brick and concrete for construction and substitute inputs like woody debris, other fibrous waste, and alternative sources of primary natural fiber for the wood processing industries have kept the relative cost of raw material from rising more rapidly and, thereby, have constrained the incentive for more rapid adoption of new biological and forest management technologies.

Conclusions: The Unique Characteristics of Forestry

This completes the description of the patterns of forest development. Chapter 2 traced three stages of forest development. Chapter 3 has introduced the effects of technological change on each of those stages. Technological change is crucial because even slower rates of technological improvement have immense impacts on the demands for and sources of supply of primary forest products over periods as long as the growing period for many forests products.

Five categories of technological change are important for forestry: (1) agricultural technologies which can affect the boundaries of agricultural and forest land use and, therefore, the land used for forest production and the volume of production itself in some stages of forest development; (2) the improvements in the regional infrastructure that enhance access to the forest resource, and technological change in (3) logging and (4) mill utilization of the forest product and in (5) forest management itself. The latter four affect productivity on existing forestland and also at the margins of forestland use. Forest managers and forest policy analysts often focus on the fifth category, forest management. Yet, the first four typically have had greater effects on the margins of forest activity. They are responsible for entrance to previously inaccessible natural forests, for

---

19 Bengston and Gregerson (1992) and Smith and Munn (1998) draw similar conclusions.

the use of lower quality species and smaller size classes, for increased utilization of any given species and size class, and for the increased substitution of fibrous debris from other wood processing activities for uncut "virgin" logs in products like wood-based panels and paper. These have been the greater sources of new fiber and of increases in the production of wood products over the history of the forestry in most countries.

In fact, we have seen that improvements in forest management can have no effect whatsoever until the third stage of forest development when managed forests and forest plantations begin to appear. Less expensive substitute fibers have always been available from the natural forest. Even in this final stage of development, the products of improvements in forest management must compete with the products arising from those other varieties of technological change affecting forestry and from the availability of substitute raw material from those other sources. These alternatives constrain the rising price of woody material from the forest and the incentive to adopt innovations in forest management.

What general conclusions can we draw from the complete three-stage pattern of forest development—aside from the importance of technological change and alternative sources of woody material? The conclusion to chapter 2 summarized the most basic lessons about three distinct categories of managed, increasingly open access and degraded land and forest, and mature natural forests; and the important roles of property rights and relative wages in determining the geographic extent of each of these categories.

An alternative way to summarize these same conclusions would be to identify the characteristics that distinguish forestry from other categories of economic inquiry. Forestry's primary distinctions are its three margins of land use and its additional margin of extraction from the degraded natural forest. Those are

1. the intensive and
2. extensive margins of managed forests,
3. the accessible frontier of mature natural forest, and
4. the margin of lower quality material available from the previously logged area of degraded open access forests.

These four margins compare with the two margins common to most economic activity and they add considerably to the complexity of economic analysis in forestry. One appendix to this chapter describes the classic economic model for forestry. That model is based on a fixed area of land within the first two margins. It disregards both adjustments in the intensive and extensive margins of managed land as well as the mature natural forest and the area of open access. A second appendix describes the standard biological model for forests. Planning based on this model fails as well, although because of its different focus. The biological model overlooks the first and fourth margins and misunderstands the limits of the third.

The third and fourth margins are the only sources of forest products during the first and second stages of forest development and they remain important sources even when managed forests begin appearing in the mature forest economies of the final stage of development. These two margins make forestry comparable to hardrock minerals as an economic resource. However, hardrock minerals tend to be concentrated in location and greater in value per unit of area, while primary forest products are comparatively lower-valued and dispersed across the landscape. These differences mean that the economic justifications for property rights are easier to establish for hardrock minerals than for forestry and that open access depletion is a more common consequence in forestry than in hardrock minerals.

It is true that forests are biologically renewable resources in the sense that they grow and then, after harvesting, they recover and grow again. To the extent that forests recover rapidly and grow voluntarily, the economic activity of forestry is like ocean fisheries. That is, forestry and fisheries share the characteristic (fourth) margin of depleted but naturally regrowing stocks.

We must add a caution, however, against making too much of forestry's biological renewability. Forests can be managed, and *some* forests are managed as economically renewable resources during the third stage of forest development. In this case and for the managed share of production, forestry as an economic activity is comparable to livestock. That is, the same fundamental resource is both a managed and growing input in one period (a juvenile tree or a calf) and a harvestable product in a future period (a mature tree or a steer ready for slaughter). Until this final stage of forest development, however, biological growth has only a limited role in explaining forest productivity.

Foresters often continue the argument about renewability with stress on the long-term nature of the activity. It is true that managed forests may require 20 or 50 or sometimes even 100 years of economic oversight. Nevertheless, there are two reasons for caution with this argument as well. First, this long production period is decreasing with improvements in forest management. In fact, the time period for fiber plantations can be less than 20 years, and for some agroforestry species and products the growth period can be as short as three or four years. Second, managed forestry is not unusual in its longer term of productive activity. Many manufacturing activities and even retail trade and agriculture make use of long-lived fixed capital inputs.

What is unusual for forest management is the large share of total investment that is tied up in fixed capital—in the tree itself. This means that uncertainty about markets in the distant future and the flexibility of the forest industry to adjust to that uncertainty are crucial variables when assessing the demand for managed forest products. It also means that the flexibility of that capital investment in the tree to satisfy a diversity of future demands is a key element in long-term forest management decisions. Technological change expands the flexibility of the mill and it may also expand the diversity of uses for the tree itself. It is crucial for a good understanding of forestry in the economic long run.

Literature Cited

Alig, R., D. Adams, J. Mills, R. Haynes, P. Ince, and R. Moulton. 2001. Alternative projections of the impacts of private investment on southern forests: a comparison of two large-scale models of the United States. *Silva Fennica* 35(3): 265–276.

Barnett, H., and C. Morse. 1963. *Scarcity and growth*. Baltimore, MD: Johns Hopkins University Press for Resources for the Future.

Bengston, D., and H. Gregersen. 1992. Technical change in the forest-based sector. In P. Nemetz, ed., *Emerging issues in forest policy*. Vancouver, Canada: University of British Columbia Press, 187–211.

Cattaneo, A. 2005. Inter-regional innovation in Brazilian agriculture and deforestation in the Amazon: Income and environment in the balance. *Environment and Development Economics* 10: 485–511.

CPBIS (Center for Paper Business and Industry Studies). 2005. http://www.cpbis.gatech.edu/research/projects/gasification/webtool/Main.php (accessed June 17, 2008).

Damani, P. 2004. *Vertical integration in the American pulp and paper industry*. Unpublished M.S. thesis. Center for Paper Business and Industry Studies, Georgia Institute of Technology, Atlanta.

Denison, E. 1962. *The sources of economic growth in the United States and the alternatives before us*. New York: Committee for Economic Development

Denison, E. 1967. *Why growth rates differ*. Washington, DC: Brookings Institution.

Economist. 2004, June 12. Hunter-programmers 371(8379): 79.

FAO/UN (Food and Agriculture Organization of the United Nations). 2001. Global forest resources assessment 2000. FAO Forestry paper 140. Rome: Food and Agriculture Organization of the United Nations

FAO/UN (Food and Agriculture Organization of the United Nations). 2005. Yearbook of forest products, FAOSTAT statistics database. http://apps.fao.org/ (accessed June 17, 2008)

Global Forest Watch. 2005. *Interactive forestry atlas of Cameroon* (version 1). Washington, DC: WRI-MINEF.

Gunatilake, H., and C. Gunaratne. 2010. Technical efficiency of sawmilling and the conservation of natural forests: evidence from Sri Lanka. *Journal of Natural Resources Policy Research* 2(2): 149-169.

Hoen, H., T. Eid, and P. Okseter. 2001. Timber production possibilities and capital yields from the Norwegian forest area. *Silva Fennica* 35(3): 249–264.

Horvath, G. 1980. Lumber, pulp, and paper. Ion J. Ullman, ed., *The improvement of productivity: Myths and realities*. New York: Praeger, pp. 158–174.

Hyde, W. D., Newman, and B. Seldon. 1992. *The economic benefits of forestry research*. Ames: Iowa State University Press.

Hyde, W., and W. Stuart. 1999. The US South. In B. Wilson, G. van Kooten, I. Vertinsky, and L. Arthur, eds., *Forest policy: International case studies*. New York: CABI International, pp. 23-46.

Kaimowitz. D., and A. Angelsen. 1998. *Economic models of tropical deforestation—A review*. Bogor, Indonesia: Center for International Forestry Research.

Kneese, A., and C. Schultz. 1975. *Pollution, prices, and public policy*. Washington, DC: The Brookings Institution.

National Science Foundation. 1981. *Research and development in industry*. NSF 82–317. Washington, DC: National Science Foundation.

Nagubadi, R., and D. Zhang. 2006. Production structure and input substitution in Canadian sawmill and wood preservative industry. *Canadian Journal of Forest Research* 36: 3007–3014.

Newman, D. 1991. Changes in southern softwood productivity: A modified production function analysis. *Canadian Journal of Forest Research* 21(8): 1278–1287.

North, D. 1966. *The economic growth of the United States*. New York: W.W. Norton.

Olson, S. 1971. *The depletion myth: A history of railroad use of timber*. Cambridge, MA: Harvard University Press.

Parry, I. 1999. Productivity trends in the natural resource industries: A cross cutting analysis. In Simpson, D. (ed.), Productivity in natural resource Industries. Washington, DC: Resources for the Future, pp. 175–204.

Phelps, R. 1975. *The demand and price situation for forest products, 1974–75.* USDA Misc. Publ. 1315. Washington, DC: USDA.
Ruttan, V., and J. Callahan. 1962. Resource inputs and output growth. *Forest Science* 8(1): 68–82.
Scherer, F. 1984. *Innovation and growth.* Cambridge, MA: MIT.
Smith, V. 2008. Reflections on the literature. *Review of Environmental Economics and Policy* 2(1): 130–145.
Smith, P., and I. Munn. 1998. Regional logging function analysis of the logging industry in the Pacific Northwest and Southeast. *Forest Science* 44(4): 517–525.
Solow, R. 1957. Technical change and the aggregate production function. *Review of Economics and Statistics* 39(3): 312–320.
Steir, J., and D. Bengston. 1992. Technical change in the North American forestry sector: a review. *Forest Science* 38(1): 134–159.
Stone, S. 1997. Evolution of the timber industry along an aging frontier: the case of the Paragominas (1990–1995). *World Development* 26(3): 433–448.
Stone, S. 1998. Economic trends in the timber industry of Amazonia: Survey results from Para State, 1990–1995. *Journal of Developing Areas* 32: 97–122.
Tachibana, T., T. Nguyen, and K. Otsuga. 2001. Agricultural intensification versus extensification: A case study of deforestation in the northern-hill region of Vietnam. *Journal of Environmental Economics and Management* 41: 44–69.
U.S. Census Bureau. 2006. 2002 Economic census, Industry series reports. http://www.census.gov/econ/census02/guide/INDRPT31.HTM (accessed June 17, 2008)
USDA Forest Service. 2005. Forest inventory and analysis webpage: http://fia.fs.fed.us (accessed March 5, 2008).
Young, C. 1996. Economic adjustment policies and the environment: A case study of Brazil. Unpublished PhD thesis. Department of Economics, University of London.
Zhang, D., and R. Nagubadi. 2006. Total factor productivity growth in the sawmill and wood preservation industry in the United States and Canada: a comparative study. *Forest Science* 52(5): 511–521.

Appendix 3A: The Faustmann Model

This appendix describes the standard economic (theory of the firm) model for a managed stand of timber, also known as the Faustmann model for Martin Faustmann, who first described it (Faustmann 1849).

The model features constant returns to scale on a fixed area of timberland on which trees grow from the time they emerge as seedlings to the time they attain an economically mature harvest age. The economic literature discusses this model as if its application is general. In fact, it is an appropriate model, without restriction, only for an important subset of commercial forest plantations, which are themselves a subset of all forests and all sources of timber production. We will see that two straightforward, but uncommon, modifications are necessary before the Faustmann model becomes applicable for the remaining categories of forest plantations and for managed natural timberstands. The Faustmann model provides no insight for marginal land use shifts as it holds constant the land area under management, and it is entirely inapplicable for the bulk of all economic activity that occurs in the natural forest. It is limited further by its implicit assumption of no relevant technological change—or no decrease in any of the costs of plantation management over the period of inquiry.

The Basic Model[20]

The model begins with a fixed base of undeveloped bare land—which means that it implicitly assumes the absence of scale economies in land use. (This assumption is not universally valid, but it is valid for a wide range of cases. Chapter 9 will consider the exception within the discussion of non-industrial private land managers.) The objective is to obtain the maximum financial return from growing a crop or successive crops of timber on this land base.

Production Q is a function of time T and silvicultural effort E. Silvicultural effort includes both the labor and capital components of timber management activities like site preparation, planting, thinning, and forest protection. Many versions of the Faustmann model restrict these silvicultural applications to a single initial event. This restriction allows the model to capture the most important costs (site preparation and planting) for many commercial forest plantations, and it simplifies the derivation of the optimality conditions.[21]

Production follows the logistic or sigmoid path over time common to biological growth. It is a concave function of silvicultural effort. Therefore,

$$Q = Q(T, E) \quad (3a.1)$$
$$Q_T > 0 \qquad \text{for } 0 < Q < T_x$$
$$Q_{TT} > 0 \qquad \text{for } T \le T_i$$
$$Q_{TT} \le 0 \qquad \text{for } T > T_i$$
$$Q_E > 0, \; Q_{EE} < 0$$
$$Q_{TE} = Q_{ET} > 0$$

where the subscripts denote first and second derivatives.

The eventual product sells as standing timber, or stumpage, which obtains an expected price p at some time after the timberstand, attains marketable size.[22] The production costs are simply the cost per unit of silvicultural effort w plus the periodic rental value for land use R. Finally, the land manager's opportunity cost of capital r is introduced in order to convert all expenditures and expected revenues to a common period of reference.

The Complete Economic Model

The full economic model is captured in the maximization of the sum of discounted expected revenues, net of both silvicultural costs and the periodic land rent R.

---

20 See Gaffney (1957), Bentley and Teeguarden (1965), Samuelson (1976), and Chang (1983, 1998) for the classic discussions of the Faustmann model and the various related formulations of the economic maturity problem for forestry.

21 Multiple discrete silvicultural inputs create multiple discrete shifts in the production function and multiple solutions to the optimality problem. The problem is not conducive to easy algebraic manipulation, but large computer simulations handle it with ease.

22 In some cases, the timber is sold as logs cut and delivered to a location agreed upon in the timber sale contract. In this event, the expected price is the delivered log price less the costs of felling, bucking, skidding, loading, and hauling to the agreed location.

$$V = \max_{T,E}[pQ(T,E)e^{-rt} - wE - R\int_0^T e^{-rt}dt] \quad (3a.2)$$

If both the product and factor markets are perfectly competitive, and silvicultural costs include returns to entrepreneurial skills as well as to the labor and capital components of timber management, then rent accounts for the entire residual between optimally determined costs and expected revenues. The only rents are locational, and rent is zero at the extensive margin of economically productive timberland.

Since the land rent is generally unknown, an alternative form of the model is more common. This alternative anticipates that timber production is the highest and best use of the land, both now and in the future. In this case, the maximum return is that obtained from a perpetual series of timber crops ("rotations" in forestry terminology).

$$V = \max_{T,E}[pQ(T,E)e^{-rT} - wE](1 + e^{-rT} + ... + e^{-nrT} + ...)$$

$$= \max_{T,E}[pQ(T,E)e^{-rT} - wE](1 - e^{-rT})^{-1} \quad (3a.3)$$

The first and all subsequent rotations and their silvicultural applications are identical because all terms except T and E are unchanging and the periodic effects of T and E on Q do not change from one rotation to the next. Eq. (3a.3) is known as the Faustmann equation. rV, the periodic rental return calculated from this form, is known as the "soil expectation" or "land expectation" value, or just SEV or LEV.

Samuelson (1976) proved the identity of eq. (3a.3) with eq. (3a.2) for both a single optimally selected rotation and for an infinite series of rotations. His proof follows from the argument that if rent accounts for the entire residual between optimally determined revenues and costs in eq. (3a.2), then $V(1-e^{-rT}) = R\int_0^T e^{-rT}dt$.

Optimality Conditions

The necessary conditions for a maximum are:

$$V_T = \frac{[pQ_T(1-e^{-rT}) - r(pQ - wE)]e^{-rT}}{(1-e^{-rT})^2} = 0 \text{ and} \quad (3a.4)$$

$$V_E = \frac{pQ_E e^{-rT} - w}{1-e^{-rT}} = 0 \quad (3a.5)$$

where the subscripts continue to denote partial derivatives. The sufficient conditions are that

$$V_{TT} = \frac{[pQ_{TT}(1-e^{-rT})^2 - 2rpQ_T(1-e^{-rT}) + r^2(pQ - wE)]e^{-rT}}{(1-e^{-rT})^2} \leq 0, \quad (3a.6)$$

$$V_{EE} = \frac{pQ_{EE}e^{rT}}{1-e^{-rT}} \leq 0, \quad (3b.7)$$

and $V_{TT}V_{EE} > (V_{ET})^2$, where $V_{ET} = V_{TE} = \dfrac{[pQ_{TE}(1-e^{-rT}) - r(pQ_E - w)]e^{-rT}}{(1-e^{-rT})^2}$.

Conditions (3a.6) and (3a.7) are non-positive because the biological growth function Q is concave with respect to E and T over its relevant range.

We can use Cramer's Rule to determine the optimal input levels, T^* and E^*.

$$T^* : Q_T = \dfrac{r(Q - \dfrac{w}{p}E)}{1-e^{-rT}} \quad \text{and} \quad (3a.8)$$

$$E^* : Q_E = \dfrac{w}{p}e^{rT} \quad (3a.9)$$

In words, the rotation age is optimal when the value of additional growth is just equal to the revenues foregone by delaying the harvest minus the gain from delaying the initial regeneration costs for the next rotation. The level of silvicultural effort is optimal when the value of the incremental timber yield at harvest time due to an additional unit of silvicultural input is just equal to the cost of that unit of input compounded to the time of harvest.

Variations

At the outset of this appendix we noted that the Faustmann model is appropriate for a select set of managed timber plantations and that two uncommon but straightforward modifications can improve its relevance for industrial timber plantations and for managed natural stands.

Industrial timber plantations are planted (rather than naturally regenerated) timberstands that are owned by vertically integrated wood product enterprises. The primary objective of these plantations is to provide a secure source of raw material for the industrial enterprise's wood processing facilities—within the context of overall enterprise profitability. Reliable flows of woody raw material are crucial for those less mobile and more-capital intensive enterprises that operate in regions of competitive wood markets. Pulpmills in the U.S. Southeast, for example, often own or control the management of a limited area in forest plantations and they often manage their plantations more intensively than neighboring non-integrated private forest landowners. Their behavior can be explained within the Faustmann formulation by simply adding a premium to the expected price term p in eq. (3a.3). The premium is the amount the industrial enterprise is willing to pay to insure the flow of wood that enables its capital-intensive processing facilities to operate without interruption.

Over 90 percent of the world's managed forests are natural stands—not plantations (Table 2.1.) For some of these, natural regeneration rapidly follows the harvest operation—either because the local natural conditions are conducive to rapid regeneration or because the selective harvest activities conducted on the previous mature timberstand are designed to encourage rapid regeneration. In others, the recently harvested land is almost immediately replanted with seedlings that may already be as much as three years old. Very many other cases, however, rely on natural regeneration, which can be much slower. In some of the better public timberlands in the highly productive Douglas fir region of the U.S. Pacific Northwest, for example, natural regeneration has taken as much as 25 years, a period longer than one-third of the optimal economic timber rotation (Hyde 1980). Cases like these, where the regeneration period is significant, can be accommodated within the Faustmann formulation by simply adding an expected regeneration lag (either negative as for three-year-old seedlings or positive and perhaps even very large as on some public lands in the Douglas fir region) to the time input in eq. (3a.3).

These differences in forest regeneration are, in other terms, differences in the applied technology. Applications of the Faustmann model can adjust for other technological differences as it adjusts for these differences in regeneration, by changing the underlying production term, $Q(t)$ and the associated stream of silvicultural inputs wE.

Neither the premium for a reliable timber supply nor the regeneration lag or other technological adjustments is widely discussed in the economic literature of forestry. Both can lead to significant adjustments in optimally determined rotation ages and levels of silvicultural effort, as well as in optimal harvest flows from the managed forestlands.[23] Their inclusion would improve the predictive ability of many economic models.

Special Cases

The forest economics literature considers numerous special cases: prices rising over time and the related idea of wood quality increasing with age, improved silvicultural treatments, uncertainty in prices and yields and the related impact of natural hazards on the forest crop, etc.[24]

Three special cases receive widespread attention. The first is a novelty, but its recognition is widespread. When regeneration is natural, then initial silvicultural costs are zero and a simple manipulation of eq. (3a.8) shows that the cost of capital is the only determinant of the optimal timber rotation. Price has no role!

$$\frac{Q_T}{Q} = \frac{r}{1-e^{-rt}} \qquad (3a.10)$$

This is a common formulation in the forestry literature but its practical application is limited because managed stands almost always require some minimal level of periodic inputs.

---

23 As the forest industry generally limits harvests from its own plantations until times of tight markets, the age and volume of harvested timberstands are sometimes much greater than the optimal age in the simplest Faustmann models. A senior industrial manager pointed out that this also means that individual trees are larger in diameter and their larger diameter makes them useful in a greater range of production processes. Since he could not predict the future market for different sizes of sawlogs, he continued by explaining that having a few stands of larger trees available assures his company's ability to compete in a wider variety of market situations.
24 Newman (1988) reviews this literature.

The second special case is simply an acknowledgement that the standard Faustmann formulation refers to "even-aged" stands. Even-aged stands are composed of trees of one or only a small number of well-spaced age classes. They are characteristic of the many conifers and other pioneer species that comprise most of the commercial timber in the temperate world. Uneven-aged stands are composed of trees of several ages according to the more random periodicity of their establishment within the stand. They are characteristic of many temperate hardwood forests and most tropical forests. Even-aged stands are often clearcut when the entire stand achieves maturity, while uneven-aged stands are generally harvested selectively for specific mature trees and commercial species within the stand—although we observe variations in these harvest practices for both even- and uneven-aged stands.[25]

The third special case is an attempt to adjust the Faustmann formulation to incorporate non-timber values. The earliest attempts associated increasing non-timber values with increasing age of the timber stand. Therefore, they argue for longer optimal timber rotations (Hartman 1975; Calish, Teeguarden, and Fight 1978). Subsequent formulations recognize that non-timber values tend to be greater where a diversity of forest conditions prevail; including mature old stands but also including open meadows, scenic vistas, and an abundance of the forest edge that provides habitat for so many wildlife species. Bowes (1983), Bowes and Krutilla (1989) and Swallow, Parks, and Wear (1990) show that optimizing under these conditions requires a multiple stand model (rather than the single stand in the classic Faustmann formulation and Hartman's modification of it). These conditions also require careful specification of the important local values and their relationships to forest management. In this case, non-timber values do not necessarily increase with stand age, and maximizing across all locally relevant timber and non-timber values is likely to decrease the optimal age for some timber stands while increasing the optimal age for others.

Comments on Appropriate Context

After careful modifications to match the objectives of each class of landowners and each land productivity class, the Faustmann model is generally descriptive of many management decisions within a fixed land base of forest plantations or managed natural stands; that is; an unchanging piece of land within the territory between B' and B'' in the third stage of forest development.[26]

---

25 Chang (1998) discusses a generalized uneven-aged Faustmann model.
26 The restriction to a fixed land base is critical—and also unreasonable for wide variations in timber price. It can be shown, by taking the derivative of volume with respect to price from eq. (3b.3), that the Faustmann equation yields a backward bending timber supply curve (Clark 1976; Hyde 1980). In fact, this will not occur, because large increases in timber price would cause forest managers to increase silvicultural inputs on existing forestland and also add land to the area under management. Significant decreases in price would cause the opposite.

However, plantations only account for 187 million hectares or five percent of the world's forests. Even in the United States, they only account for 16 million hectares or 1.8 percent of all forests (FAO 2001). Furthermore, even with careful modification, each landowner class has additional motivations and opportunities that are not fully captured in the long-run profitability objective of the Faustmann model. Small private farm landowners are especially responsive to short-term price fluctuations and also to more immediate personal financial needs. Other small private landowners may be more responsive to a range of non-market values. Industrial landowners may be more responsive to internal demands for raw material or to short-term corporate cash flows. That is, early timber harvests may insure continuous operation of a firm's mills and they can offer the firm a means to increase its cash flow when the firm needs to increase funds for the payment of dividends to its stockholders or during those times when its integrated wood processing operations require additional financing. Institutional forest landowners, those large landowners without integrated wood processing facilities, may have the least constrained profit motives for their forestlands, but even they are concerned with the flow of earnings from their forest properties within the broader context of diversified financial portfolios. Chapters 7–9 will examine each of these landowner classes and their management objectives in more detail.

Literature Cited

Bentley, W., and D. Teeguarden. 1965. Financial maturity: A theoretical review. *Forest Science* 11(1): 76–87.

Bowes, M. 1983. *Economic foundations of public forestland management.* Quality of the Environment discussion paper D-104. Washington, DC: Resources for the Future.

Bowes, M., and J. Krutilla. 1989. *Multiple-use management: The economics of public forestlands.* Washington, DC: Resources for the Future.

Calish, S., D. Teeguarden, and R. Fight. 1978. How do nontimber values affect Douglas-fir rotations? *Journal of Forestry* 76: 217–221.

Chang, S. J. 1983. Rotation age, management intensity, and the economic factors of timber production. *Forest Science* 29(2): 267–278.

Chang, S. J. 1998. A generalized Faustmann model for the determination of optimal harvest age. *Canadian Journal for Forest Research* 28: 652–659.

Clark, C. 1976. *Mathematical bioeconomics: The optimal management of renewable resources.* New York: John Wiley and Sons.

Faustmann, M. 1849. On the determination of the value which forest land an immature stands possess for forestry. In M. Gane, ed., *Institute Paper 42* (1968). Commonwealth Forestry Institute, Oxford University.

Food and Agriculture Organization of the United Nations (FAO). 2001. *Global forest resources assessment 2000.* Forestry paper 140. Rome: FAO

Gaffney, M. 1957. *Concepts of financial maturity of timber and other assets.* Economics Information Series no. 62. Unpublished manuscript, North Carolina State College, Raleigh.

Hartman, R. 1975. The harvesting decision when the standing forest has value. *Economic Inquiry* 14(1): 52–58.

Hyde, W. 1980. *Timber supply, land allocation, and economic efficiency.* Baltimore, MD: Johns Hopkins University for Resources for the Future

Newman, D. 1988. The optimal forest rotation: a discussion and annotated bibliography. General Technical Bulletin SE-48. Asheville, NC: USDA Forest Service Southeastern Forest Experiment Station.

Samuelson, P. 1976. Economics of forestry in an evolving society. *Economic Inquiry* 14(4): 476–492.
Swallow, S., P. Parks, and D. Wear. 1990. Policy relevant nonconvexities in the production of multiple forest benefits. *Journal of Environmental Economics and Management* 19: 264–280.

Appendix 3B: Allowable Cut

This appendix describes the standard biological model for determining timber harvest levels, identifies its extension to timber management, and discusses its departures from economic efficiency. The model is known variously by the terms "allowable cut" or "allowable annual harvest", and for its stated objective to maximize a version of sustainable physical yield. It is also known as the Hanzlik model—for the author of its original mathematical form (Hanzlik 1922).[27] Some form of this model is common to the planning activities of most public forestry agencies around the world.[28] Some industrial forestry operations follow variations of it as well.

The model's departures from economic efficiency are not trivial. Clawson (1976) estimates, for the 1970s, that departures from efficient harvest levels for the U.S. National Forest System cost approximately US$600 million in foregone annual timber revenues. At about the same time and also for the U.S. National Forest System, Hyde (1980) estimates that departures from economic land use in the Douglas fir region of the U.S. Pacific Northwest extended the timber harvest area by one-fourth—reducing the de facto forested roadless area to less than one-third of its efficient level—while decreasing the economic harvest level by as much as 70 percent.

The Basic Model

The model begins with a fixed land base divided into two components, one actively managed or "regulated," and the other containing a standing unmanaged natural forest of mature timber or "old growth." The allowable cut objective is to obtain an "even flow" of annual physical harvests from this land base, while incrementally converting the unregulated component into the expanding regulated forest. Fundamentally, the allowable cut model is a biological maximization model with an additional term for old growth "conversion". It has no explicit economic rationale or content, although foresters derive what they call a rate of return from its use.

As for the Faustmann model, the allowable cut model can be described for either "even-aged" or "uneven-aged" stands. This appendix describes the better-

---

27 The term "allowable cut" has at least three different meanings—referring to the optimization model itself, the long-term planned harvest levels that arise from its applications, and the formally approved planned annual harvest level for a particular year. The third may or may not be based on an allowable cut optimization model. Clearly, the three can be very different and the potential for miscommunication is great unless the specific use of the term is clear.
28 Indeed, I have not discovered an exception.

known even-aged model and leaves it for interested readers to examine the literature for uneven-aged models (e.g., Davis and Johnson 1987).

The allowable harvest volume AC for any year i is obtained by (1) maximizing annual harvests from A_m initially managed, or regulated, hectares where cumulative stand volume Q is a logistic function of time T, while (2) harvesting and regenerating the remaining A_g hectares of unregulated and mature forest in equal annual segments throughout a period of years T_c. The entire forest ($A_m + A_g$) becomes one single fully regulated unit at the conclusion of the conversion period. The model assumes that growth on the unregulated component is stagnant at a volume of G per hectare and that successful reforestation follows immediately after harvests from both the regulated component and the newly harvested, previously unregulated, components.

$$AC_i^1 = \max_T A_m[Q(T)/T] + A_g(G/T_c) \tag{3b.1}$$

for $i = 1, 2, \ldots, T_c$ where $T_c \leq T^\star$.

The only decision variable is the length of the conversion period T_c for the unregulated component. In many applications the conversion period is set equal to the optimal rotation age and determined simultaneously with it. In other applications it is exogenous. Foresters know the optimal harvest age on the regulated component as the age when growth achieves its maximum average annual physical rate or the age of "culmination of mean annual increment" (CMAI). CMAI and the Faustmann-determined economic harvest age are identical when regeneration is immediate and all costs and the interest rate are zero. (This widely recognized case may be mathematically interesting but it is an entirely unlikely real world event.)

Once the optimal harvest age $T^\star$ is determined, the regulated component is divided into $T^\star$ equal segments, one in each age class between zero and that optimal age. The unregulated component is divided into one segment for each year in a conversion period. If the conversion period $T_c = nT^\star$ where n is a positive integer, then the second term on the right-hand-side (3b.1) disappears at the end of the conversion period, the entire forest becomes regulated, and a steady state develops in which annual harvests equal annual growth at its maximum average level.

$$AC_i^2 = \max_T (A_m + A_g)[Q(T)/T] \tag{3b.2}$$

Most applications of the allowable cut model focus on eq. (3b.1) for the practical reason that timber rotations and forest conversion periods are so very long that a fully regulated forest and the application of eq. (3b.2) only occurs well into the future. Furthermore, most foresters recognize that conditions change over such long periods, and periodic changes that affect any of the model's parameters require revisions in the allowable cut calculation.

Qualifications and Variations

The first variation in eq. (3b.1) refers to the distinction between volume control and area control. Volume control features equal annual harvest levels regardless of their areas of origin. Area control features equal annual harvest areas regardless of the volumes extracted. Volume and area control yield identical annual harvest volumes when all lands are of identical physical productivity. The distinction becomes important when the regulated component includes lands of varying productive potentials and the unregulated component includes old growth stands of various densities and, therefore, various volumes per hectare. Practical applications of eq. (3b.1) can be specified for either area or volume control.

A second variation occurs because some managers have a preference for a specific product like sawtimber, or even sawtimber of a particular size or quality. One representation of this preference can be described mathematically with a constraint on the basic equation

$$AC_i^3 = \max_T \{A_m[Q(T)/T] + A_g(G/T_c) + \lambda(T - T_k)\} \quad (3b.3)$$

where λ is a Lagrangean multiplier and T_k is the age at which trees in the regulated component attain the minimum acceptable size.

More generally, however, forest managers account for minimum size objectives by defining Q such that it only measures volume greater than the minimum standard. In this case $Q(T)=0$ for all $T<T_k$. Subsequent to T_k, the specification of Q takes on positive value and it grows from this base volume of, say, sawtimber size $Q(T_k)$, according to the standard logistic form for biological growth.

Of course, this objective and its allowable cut formulation disregard market evidence that the same forest may produce greater volume or value for some different product. It also disregards evidence that minimum size standards, even for the preferred product, have adjusted with time and technical change. For example, public forest managers in the United States persisted for many years in measuring sawtimber as a product from trees of no less than 11½ inches in diameter despite evidence that many sawmills cut lumber from trees as small as four inches in diameter. Sullivan, Bell, and Usher (1975) show that a simple measurement change from the 11½ inch standard to a five inch standard would have altered the annual allowable cut of the U.S. National Forest System in the mid-1970s by as much as 30 percent.

Finally, some forest managers introduce a third variation in the allowable cut model in order to address the decline in harvests that occurs at the end of the conversion period. This decline is known as "falldown". It occurs because the last hectare of unregulated old growth contains older trees and more volume (G) than the mature volume on a hectare of fully regulated forest [$Q(T^*)$] harvested at CMAI. Managers concerned with this problem respond by extending the conversion period, thereby decreasing the level of current annual harvests from the unregulated mature forest. This reduces the eventual falldown and creates a more "even flow" of annual timber harvests over the long run.

Of course, this extension of the conversion period imposes large losses in terms of foregone current harvest volume and value. Hyde (1980) calculates that the extension of the conversion period dissipates the entire timber value in at least one high profile example in the United States. The concern with falldown also disregards all reasonable expectations that changing market conditions will induce various long-run adjustments in economic measures of the forest inventory as well as in the economic land base for timber production.

Forest Management and Financial Returns

The allowable cut model provides no basis for judging forest investments. Nevertheless, such judgments are made—by creating a relationship between the implicit cash flows, regardless of their source, and forest investments, regardless of their allocation.

For example, consider an investment in reforestation on a unit of previously fallow land. Reforestation adds A_f hectares to the regulated component in eq. (3b.1) such that

$$AC_i^4 = \max_T \{(A_m + A_f)[Q(T)/T] + [A_g - A_f Q(T)/G](G/T_c)\} \qquad (3b.4)$$

Annual growth on the newly reforested lands, the first term on the right-hand-side (RH), justifies an increase in the allowable harvest level—but the new seedlings on these lands will have no marketable volume for many years. Therefore, the actual increase in harvests must be drawn from the standing inventory of old growth, the second RH term. Annual harvests from the old growth component must continue at the greater level for each year until year $T^\star$ when trees on the newly reforested land attain harvestable size. The harvest increase due to this decision to tie two unrelated activities, reforesting some lands and harvesting others, is known as the "allowable cut effect" or "ACE".

Forest planners calculate the return r on this investment in forest management as the return that equates the value of the discounted sum of increments in annual harvests from eq. (3b.4) with the initial reforestation costs C.

That is,

$$\sum_{i=1}^{T_c} p \frac{AC_i^4 - AC_i^1}{(1+r)^i} = C \qquad (3b.5)$$

where p is the stumpage price.

Calculations similar to eqs. (3b.4) and (3b.5) can be made for silvicultural investments such as thinning or fertilizing that are intermediate to any harvest activity. In this case, the term for reforested lands in eq. (3b.4) drops out and production Q in the first RH term becomes a function of activities like thinning and fertilizing as well as time. The increase in annual harvests is justified by the additional growth on the currently managed lands (A_m)—instead of reforestation and growth on new lands (A_f) as in the example of eq. (3b.5).

Annual returns as great as 390 percent have been calculated in this manner (USDA FS 1969). Obviously, they are an artifact of combining two independent decisions, financially profitable old growth harvests on some lands and new investments of undetermined profitability on other lands. It should be just as obvious that rates of return from such calculations are entirely misleading when used to compare the viability of alternative forest investments or the viability of forest investment with the activities of other, non-forest, enterprises that rely on standard accounting practices.[29]

The Economic Context

The simple biological maximization model, the first RH term in eq. (3b.1), has its roots in the old German concept of a continuous forest, or *dauerwald*. This model may have been appropriate for the owners of landed estates in northern Germany in the early eighteenth century.[30] The local economy was stable and largely closed to external influence. Its technology rewarded craftsmanship more than innovation and the undeveloped condition of the regional transportation and communication networks, reinforced by high customs barriers, acted to restrict external market exchange. Small and self-sufficient political and economic units were the rule.

The relative value of wood was high. Accordingly, productive timberland had a high value and it was located closer to the center of the economic unit than it is today. Moreover, sound ecological justification may have accompanied the simple biological model as rapid natural regeneration was the norm with the local species and silvicultural investments and other overt timber management may have been unnecessary because a vigorous understory was always ready to replace mature trees after the latter were harvested. A steady state forest may have been a reasonable characterization.[31]

The first professional foresters served the owners of the large German estates of the eighteenth and nineteenth centuries. These estate owners used their lands as biological preserves, and the great stags they protected, and hunted, are legendary. Timber production was not an important objective for these estate owners and they excluded the public from most uses of the forest. For them and their estates, biological maximization of the forest may have been reasonable.

However, if *dauerwald* and the biological maximization model were justifiable from these perspectives, their justification clearly failed by the middle of the nineteenth century when public objections to the laws and behavior related

---

29 Hyde (1980) provides a numerical example that shows how a two percent financial return on sixty-year rotations of Douglas fir equates with a ninety percent allowable cut rate of return.
30 The first attempts to regulate timber harvests in Germany may have occurred as early as the thirteenth century.
31 See Behan (1975), Gould (1962), and Raup (1964). Von Thunen (1826) himself identified forestry as an inner ring in the geography of concentric economic rings surrounding a local market. In terms of our model in chapters 2 and 3, forest production occurred closer to the market and other lower-valued economic activities occurred on lands with established property rights farther from the productive forest. Open access lands and the natural forest occurred farther yet from the center. Therefore, the open access lands were unimportant to the early applications of *dauerwald*. (In terms of our model from chapter 2, forest development was in its mature third stage.) Spurr (1964) describes the ecological conditions that justified *dauerwald*.

to forest estates were a primary cause of the Revolution of 1848.[32] In any event, *dauerwald* is not a relevant condition today. Local economies are no longer closed and stable, and modern forest production originates either from mature trees on natural forest frontiers or from intentional investments in managed forests—neither of which the basic allowable cut model describes adequately.

Nevertheless, it is true that even today some large profit-oriented industrial enterprises rely on their own variations of the allowable cut model. Their requirements for an annual source of funds to run their operations and also for stockholder dividends make the cash flow feature of this model attractive. Their concern for a secure source of raw material makes a tie between harvests and investments attractive. Furthermore, the generally higher biological productivity of industrial plantation lands saves these commercial enterprises from falling into the worst trap of the allowable cut model—using revenues from productive lands to support reforestation and management on lands of such poor quality that they generate virtually no independent returns.

However, it is also true that even these industrial enterprises make exceptions. A quick examination of the annual financial reports of any firm that purports to follow the principles of the allowable cut model will show that it often sells additional timber (and some firms sell forestland as well) whenever the market for its final product suffers a downturn or whenever it needs financial resources for other investments. This means that the firm treats its forest stock as a store of wealth and that economic behavior rather than biological maximization is its primary operational rule—regardless of the contrary evidence suggested by its references to the allowable cut model.

Of course, many other firms, and especially those that are less privileged with holdings of high quality forestland, operate closer to economic principles and make less pretense of following corporate rules for biological maximization on their forests. They may be concerned with insuring the flow of raw material to their mills, but the accounting principles for their timberlands divisions are more likely to tie forest investments directly to the expected returns on each land unit. (Chapter 7 will examine the behavior of industrial landowners in more detail.)

If variations in the allowable cut model are not reliable predictors of management behavior on industrial plantations, then what can we say about its application in other situations? We can hardly expect farm foresters and other owners of small woodlands to operate on a rigid sustainable biological basis. Their individual land holdings are generally too small to provide a regular harvest flow, and their financial needs also tend to be irregular. (Chapter 9 will examine the behavior of these landowners in more detail.)

---

32 Visit the museum at Hunting Castle in Kranichstein, near Darmstadt, Germany, the hometown of Martin Faustmann, for evidence of the relationship between the management of forest estates and the Revolution of 1848.

The remaining large category of land is the frontier forest generally managed or at least administered by national governments. The forest plans of national governments around the world—or their forestry agencies—generally do follow the principles of allowable cut. The argument given for following these principles is that they insure larger sustainable harvest flows than would be obtained under any other principle, including a principle of economic operation.

Consider this argument carefully. It is true that biological yields are greater than economic yields under select conditions. That is, physical yields derived from biological maximization are greater than physical yields derived from financial maximization *for a stand of naturally regenerated and naturally growing timber on a fixed and well-defined unit of land*. However, this condition excludes two important production possibilities: (1) financially justifiable more intensive management and its greater yields on some lands, and (2) other lands which the practice of biological maximization forces into delayed harvests and, therefore, into an overall uneconomic class, lands that would have been economic to manage and harvest under criteria that are different from the allowable cut criteria but which, in independent calculation, report net financial losses to maximization under biological rules.[33]

This restrictive set of conditions is implicit in the argument, but it is seldom recognized. The evidence that the allowable cut model allows larger sustainable harvest flows is inconclusive once these conditions are removed: biological yields could be either greater or smaller than economic yields for any given land unit. An accurate assessment in the absence of the allowable cut restrictions would depend on local market conditions and local land productivity. In any event, for public management which is generally at the geographic frontier of economic activity, biological maximization forces some lands out of the productive class into a class that is uneconomic to harvest. Even if these lands remain in the allowable cut base and even if the government agency can enforce the property rights, no logger would harvest these lands without external financial assistance. Expenditure for this purpose is one reason that some calculate that the U.S. National Forest System as a whole has failed to obtain a positive financial flow from its timber management for any year of its history (Barlow and Helfand 1980; Barlow et al. 1980; Forest Sense 2001; Zimmermann and Collier 2004). In the absence of external financial assistance from the U.S. Treasury, some of the allowable cut land base under public management would be removed from production and aggregate physical yields would decline, contrary to the argument that biological maximization insures the larger yields.

---

33 Hyde (1980, ch. 2) provides a diagram and more complete discussion of the reasons that biological yields or yields determined by the allowable cut model are not necessarily larger than economic yields.

Literature Cited

Barlow, T., and G. Helfand. 1980. Timber giveaway—a dialogue. *The Living Wilderness* 44: 38–39.

Barlow, T, G. Helfand, T. Orr, and T. Stoel. 1980. *Giving away the national forests: An analysis of U.S. Forest Service timber sales below cost*. Washington, DC. Natural Resources Defense Council.

Behan, R. 1975. Forestry and the end of innocence. *American Forests* 81(5):16–19.

Clawson, M. 1976. The national forests. *Science* 191(4227): 762–767.

Forest Sense. 2001. Forest service lost $407 million selling trees in 1998. *Forest Sense* 3(2):1

Gould, E. 1962. *Forestry and recreation*. Harvard Forest Papers No. 6. Petersham, MA: Harvard Forest

Davis, L., and K. N. Johnson. 1987. *Forest management*. New York: McGraw-Hill.

Hanzlik, E. 1922. Determination of the annual cut on a sustained basis for virgin American forests. *Journal of Forestry* 20(5): 611–625.

Hyde, W. 1980. *Timber supply, land allocation, and economic efficiency*. Baltimore, MD: Johns Hopkins University Press for Resources for the Future.

Raup, H. 1964. Some problems in ecological theory and their relation to conservation. *Journal of Ecology* 52 (Supplement).

Spurr, S. 1964. *Forest ecology*. New York: Ronald Press.

Sullivan, R, E. Bell, and J. Usher. 1975. *Information relating to RPAT timber policy issue #1*. Washington, DC: USDA Forest Service Timber Harvest Issues Study Team.

USDA Forest Service. 1969. *Douglas fir supply study*. Portland, OR: Pacific Northwest Forest and Range Experiment Station.

Von Thunen, J. 1826. Der Isolierte Staat in Bezeihung auf Landwirtschaft und Nationalokonomie [The Isolated State]. Berlin: Schmaucher Zarchlin. Translated by Carla M. Wartenberg. Edited with an introduction by Peter Hall. New York: Pergamon Press, 1966.

Zimmermann, E., and S. Collier. 2004. *Road wrecked: Why the $10 billion Forest Service road maintenance backlog is bad for taxpayers*. Washington, DC: Taxpayers for Common Sense.

4
FOREST POLICY

The last two chapters examined the principles of market operation in forestry. Many forest activities, however, do not occur in entirely free, complete, and competitive markets. Some occur entirely outside the market—as some forms of forest recreation are neither bought nor sold. Some marketed activities in the forest have unmarketed effects outside the forest—as timber harvests can cause erosion and sedimentation in downstream watercourses and riparian lands. Furthermore, some marketed activities external to the forest can have effects on the forest—as agricultural development can be a source of deforestation and the loss of non-market ecosystem services such as biodiversity and carbon sequestration that originate in the forest. Finally, market-based activities in the forest (and elsewhere) must carry a burden in addition to the price collected by their sellers in order to provide their share of the financial support for general public services like roads and schools and the police.

Governments intervene in cases like these where markets alone fail to create socially desired levels of production and allocation. They also intervene to promote employment and regional development. In forestry, governments have often intervened to insure a long-run timber supply. The interventions take the form of various taxes, financial incentives, and regulations, as well as public ownership and management of the resource. For all of these reasons an understanding of competitive markets alone does not provide a sufficient description of human activity in the forest.

This chapter reflects on the various categories of public market interventions. As it demonstrates the modifying effects that government policies have on the forest in each stage of development, it can help us understand the relative successes that some policies have in accomplishing their broader social objectives. It can also help anticipate the merits of alternative public policies that may be proposed in the future. Finally, and not least importantly, it can help us understand the failures of still other policies.

The two fundamental approaches for examinations of policy begin either with the categories of policy instruments or with the specialized policy objectives themselves. This chapter builds from a discussion of the instruments that have been most common in the last century of forest history. As the chapter progresses, it identifies the policy objectives often associated with each instrument and illustrates the discussion with examples. An appendix shifts the focus from the instruments of policy to the objectives and examines five widely discussed policy objectives of modern forestry and speculates on the best means for accomplishing each.

The instruments of direct forest policy are the taxes or subsidies and the regulations or physical standards imposed on activities that occur on private forestlands. We examine all three—taxes, subsidies, and regulations—and review the specific forms common to each. Forestry's unusual characteristic of four land use margins complicates the selection of an appropriate instrument because any particular policy instrument may affect some but not all margins and some policies have simultaneous but contrasting effects on those different margins. For example, price incentives encourage the expansion of lands under forest management, but they also encourage increases in the harvest levels at all margins and a contraction of the land area remaining in natural forest.

The discussion in previous chapters has established that forests generally occur at the geographic margin of other economic activity, that many forest resources and forest-based environmental services have relatively low unit values, and that these values tend to be geographically dispersed. These three forest characteristics play an important role in the selection of effective instruments for direct policy.

The same characteristics also mean that forests are easily affected by inadvertent spillovers from activities in higher-valued adjacent sectors such as agriculture or by modifications in the local infrastructure and institutions. These spillovers can be important for the forest, although they too vary in their impacts on both managed and natural forests. The second section of the chapter reviews these spillover effects on the forest from policies whose primary targets are activities in either of two economically adjacent sectors, agriculture and the wood and fiber processing industries. A third section examines the changes in infrastructure (primarily roads) and institutions (especially property rights) that can affect the action of the market or the enforcement of policy. While neither infrastructure nor institutions are commonly considered as components of policy, government decisions affect the design of both and both are important determinants of forest development. It will become apparent that policy changes in the two adjacent sectors and modifications in the local infrastructure or institutions, in some occasions, can be more effective in addressing forest policy objectives than the taxes, subsidies, or regulations that specifically target the forests themselves.

An appendix to the chapter turns from the instruments of policy to specific policy objectives. It will become apparent in the first section of the chapter that many forest regulations are ineffective in achieving their desired objectives and, as suggested above, some even produce results that conflict with those objectives. Therefore, it will be useful to identify the objectives that are central to contemporary discussions of forest policy and to consider the most effective means of accomplishing them. The appendix considers five of these: (1) carbon sequestration to protect against global climate change, (2) the protection of critical habitat and biodiversity, (3) the protection of natural environments and the aesthetic values that attract tourists and other recreational users of forest environments, and (4) erosion control and general watershed management. These first four objectives contain elements of non-market valuation, but all are affected by market activity and the policies that modify market-valued activity as well. The fifth objective is sustainable forestry and the control of deforestation. Sustainable forestry is often discussed as an objective unto itself, although its underlying purpose is the continuous long-term satisfaction of all values originating from the forest. Effective policies for protecting sustainability depend on the definition of sustainability. Satisfying the narrowest definition, perpetual forest management in all of the current locations of the forest, is futile. A better objective might be to improve the general level of sustainability and to guarantee the continued existence of those specific forest resources and resource services for which society has unique interest. In this case, while some policy designs are more effective than others, a case can be made that sustainability may be more dependent on regional economic development in general than on any particular policy that specifically targets the forests themselves.

We will delay for future chapters the discussions of one central element of modern forest policy and one critical external influence on the forest. This chapter focuses on government intervention in *private* sector forestry. Yet outright *public* ownership, particularly of natural forests, is another instrument of forest policy and the sale of products from the public forests is another important issue of global debate. These sales affect large blocks of forests, large flows of forest resources, both legal and illegal, and large flows of revenues to some government treasuries. Contracts for the exchange of these publicly-owned forest products and environmental services will be the topic of chapter 5. Still later, chapter 10 will return to a more thorough examination of public ownership and all of the objectives of the public forestry agencies.

Finally, macroeconomic planning and policy and the overall level of macroeconomic activity can also have significant effects on forests and the forest sector of the economy. Chapter 6 will discuss forestry's role in the aggregate economy and in international trade. We will delay a discussion of spillovers to forestry from macroeconomic policy and from the general macroeconomic environment until then.

The Instruments of Direct Forest Policy

The general economics literature separates government intervention in the market into two categories, standards and charges. Standards refer to physical limitations such as regulations that restrict the harvest and shipment of logs or environmental regulations that limit timber harvest activity in riparian zones. Charges refer to economic instruments that permit managerial discretion in achieving publicly acceptable levels of resource allocation. Taxes or fees for licenses or permits are common examples of charges. Financial incentives and other government assistance that reduce the costs of production are a second class of economic instruments that occur widely in forest policy. This section of the chapter reviews the effects of forest taxes, financial incentives, and standards or regulations in that order.

Taxes

Governments impose three categories of taxes on forests in one place or another: income taxes, property taxes or their substitutes, and severance taxes. This chapter reviews income and property taxes. Severance taxes, also known as royalties, are charged on non-renewable resources like oil and minerals and also on mature timber. They are central to the discussion of contracts for the use of mature forests in chapter 5.

Some taxes are primary sources of public revenues. For these, the objective is to collect revenues without altering the use of inputs or the level of output. Taxes that alter neither the use of inputs nor the output level are called "neutral." Other taxes are designed as disincentives for undesirable activities. These taxes are never neutral, nor are they intended to be. Non-neutral taxes are intended to limit undesirable activities such as those that cause environmental pollution. They are less common in forestry where physical standards tend to be the preferred instruments for controlling undesirable activity—as discussed later in this chapter.

Income taxes: Income taxes are levied on personal income or corporate profits. They are a primary source of revenues for many central governments. When applied equitably across productive activities, they tax equal proportions of income or profits from all activities. Therefore, the tax burden is the same on all activities. They are neutral with respect to the allocation of land and other inputs between competing productive activities.

In terms of the figures and the three-stage model, income taxes shift the forest value function V_f downward an equal proportional amount throughout its profitable range—from the broken line to the solid line in Figure 4.1. An equal proportional shift means a greater absolute shift at the left extreme of the function where net value is greater, but no change whatsoever after point B'' where the net return on land is zero. Income taxes also affect agriculture, shifting the agricultural value function V_a downward by the same proportional distance as for forestry. Therefore, equal proportional income taxes do not alter the critical land use margins at B', B'', and D.

However, the tax codes of some countries include provisions that do not apply equally to all productive activities. For example, they may tax income derived from capital gains at a lower rate than other income. Capital gains are the appreciation in asset value that occurs during the period an individual holds the asset. Most timber is held for long periods and, as it grows, it also appreciates. Therefore, a lower tax rate for capital gains favors investments in timber (and some

112 Forest Policy

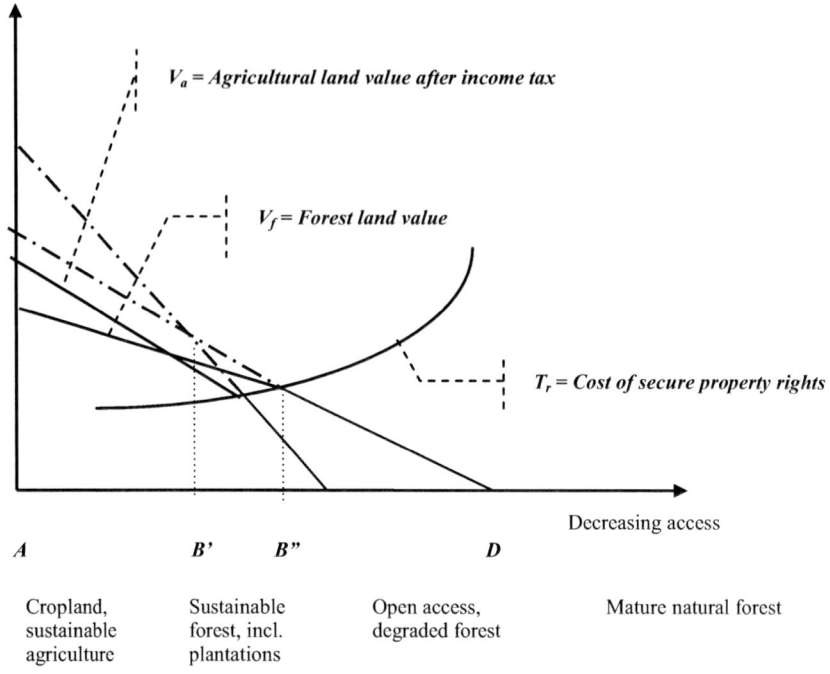

FIGURE 4.1 Income taxes

other assets like real estate, minerals, and oil that grow in value if not in volume) in preference to activities like agriculture, retail sales, and manufacturing whose production periods are shorter. The effect is not trivial. The favorable treatment of corporate capital gains in the United States—effective until 1986—may have been responsible for as much as 20 percent of the forest industry's after-tax profits (Russakoff 1985).[1]

The favorable treatment of capital gains, with its positive impact on profitability, is an incentive to shift investments in land, labor, and capital away from alternative activities like agriculture and into favored longer term activities like forestry. This shifts the competitive agriculture-forestry land use margin (point B') to the left in our figures and increases the total land area in managed forests. The increase in managed forest area means that total production from the managed forest also increases.

---

1   The favorable treatment of capital gains was reinstated in the United States for *personal* income in the early 1990s. The combined effects of the favorable treatment of capital gains in forestry for private individuals, plus a provision in the US tax code that permits the expensing of timber management costs, together accounted for reductions in corporate and individual income taxes amounting to $220 million annually in the late-1990s (Joint Committee on Taxation of the U.S. Congress, cited in Kripke and Dunkiel 1998).

Favorable treatment of capital gains has no effect on the extensive margin of managed forest at B'' because income taxes are proportional to profits and the profits from management are zero at this point. In general, favorable capital gains also have no direct effect on land beyond B'' that is unprofitable for forest management.[2] However, the additional managed production demonstrated by the leftward shift of the intensive margin at B' may substitute for some production from the natural forest frontier at D. Therefore, the favorable treatment of capital gains may indirectly conserve some amount of natural forest. The magnitude of this effect depends on the magnitude of the capital gains effect on harvest volume at the intensive margin and the relative costs of production at both locations. There have been no empirical assessments of this conserving effect on the natural forest but it can only have been positive in countries with regions in the third stage of forest development and which have a provision for preferential tax treatment of capital gains.

Estate and inheritance taxes, collectively known as death taxes, are another component of the income tax codes of some countries. They may have an impact on the allocation of land and other inputs to forestry. Corporate land owners are unaffected by death taxes because corporate life spans are unlimited. Individual land owners, however, may be placed at disadvantage if the death that precipitates the inherited wealth is unexpected. In this case, the sudden large tax bill may cause some individuals to sell large inherited properties like forests in order to obtain the funds to pay the death taxes. This is conjecture, however, and the effect on forestry is unclear because it is unclear how much, if any, of the land that changes ownership as a result of death taxes is removed from forestry. The empirical evidence is lacking here as well.

The tax codes of the UK and Chile provide the contrasting example. Both countries exclude inherited forests from death taxes. With this exclusion, the ownership of forestland may become a hedge against death taxes in general, and this favorable treatment may induce an expansion in the ownership of forests in preference to ownership of the many other assets that are subject to death taxes. The magnitude of the effect on land use and timber production is also uncertain, however, and for the same reason as before: careful estimates taken from the empirical evidence from the UK and Chile are lacking.

Property taxes: Property taxes are taxes on the value of real and personal property: land, its capital improvements, equipment, and intangible assets such as stocks and bonds. Property taxes are a primary source of revenues for many local governments. Like income taxes, their collection is annual (or sometimes biennial).

Property taxes are applied to the assessed value of each property. This is generally some proportion of the property's market value in its "highest and best"

---

2 A qualification in this case is necessary where those who purchase timber at the frontier for one price, but who delay logging until the value of their purchase increases, are eligible for favorable capital gains taxation on the increase in value.

use. If all lands within a locality are assessed at the same proportion of their true market values and taxed at the same rate, then the property tax on land shifts the agriculture and forest land value functions downward the same proportional amount throughout their profitable ranges. A diagram of the effect of the tax would be identical to Figure 4.1 for income taxes. The effect of the tax on the land would be neutral with respect to input and output decisions.

However, standing timber and various other assets, as well as land, are also subject to the property tax and their taxation introduces a bias into resource allocation. Since timber grows in place for long periods of time before it is ready for harvest, each annual collection of the property tax on standing timber repeats the taxation of unharvested growth from all prior years. The final accumulation of annual property taxes paid on the timber up to the time of harvest is much greater than the accumulated taxes would be if timber production were an annual activity and each period's growth were taxed only once. This effect of the property tax is known as the "time bias" against forest management. It encourages landowners to harvest their timber at an earlier age—in order to avoid some of the repeated and accumulating taxation.[3]

The accumulated property tax on timber makes some land at the extensive margin B'' unprofitable for forestry. The combined effects of the time bias on managed forests and the conversion of the extensive margin into unprofitable forestland were a major reason that many firms in the southern U.S. forest industry "cut and ran" to the American West in the 1920s. Southern firms harvested and then abandoned tens of millions of hectares rather than reforest and pay the property taxes.[4] The forest industry in the Lake States of the U.S. shared this same experience to such an extent that, as the abandoned cutover lands reverted to state ownership between 1910 and 1940, they became identified with their own name, "the new public domain" (to contrast with the original public domain of unsettled frontier not previously under state or private ownership.)

The extent of the effect of the property tax on the intensive margin of forest management is less certain. The time bias against timber works in favor of competing land uses like agriculture that do not accumulate a single standing product that is taxed and re-taxed year after year until one final harvest. However, the more developed land uses that compete with forestry at this margin tend to require more buildings and equipment per land unit and these buildings and

---

3   Time, or the cost of growing capital, is only one input to forest management. We could also trace the effect on the labor and manufactured capital inputs for forestry. However, these are lesser inputs for most forest management and the effect of property taxation on the harvest timing decision has easily been more important than its effect on the use of either labor or capital inputs. Since these latter effects are only relevant for managed forests, the simplest ways to show any of them would be to enter a term for an accumulating tax t_x on production $Q(.)$ in the Faustmann formula, eq. (3a.3), and to determine the resulting changes in the optimality conditions for the harvest period and for all labor and manufactured capital inputs, eqs. (3a.8) and (3a.9).
4   Recall the discussion of the South's first forest in chapter 3.

equipment are also subject to property taxes. They raise the tax bill for competing land uses relative to the tax bill for forest management. Do they raise the agricultural tax bill, for example, more than the affect of repeated taxes on accumulated timber growth raises property taxes on forestry? The net effect on competition between alternative land uses is not clear—and, once more, the question has not been examined empirically.

The effect of the property tax on timber management in places like the South and the Lake States of the U.S. caused policymakers to re-examine the local tax systems. In some cases, they reduced the appraised value of forestlands. In others, they introduced yield taxes as an alternative to property taxes. Yield taxes are assessed only once, at the time of harvest and they are generally lower than the accumulated annual property taxes would have been. Since a yield tax is assessed only once, it has a smaller effect on the optimal harvest age. Since it is generally lower than the accumulated property taxes would have been, or serves as a competitive alternative that limits higher property tax assessments, it has a smaller negative effect on the intensive and extensive margins of land use than an unfettered property tax would have.

Finally, as environmental protection became a more important policy objective in the last quarter of the twentieth century, some local governments searched for ways to protect their forests and open spaces. A few began to assess forest and agricultural land according to its current use, rather than its highest and best use. This reduces the tax bill on those forests and agricultural lands that would have greater value in more developed uses and preserves managed forestry's competitive advantage at the intensive margin. The preservation effect is temporary, however, if the value in alternative developed land uses continues to rise. The preservation effect lasts only until the developed land value exceeds the land's current value in forestry plus the difference in tax bills between the two land uses. The total impact of these "green space" taxes on forestland has probably been small (Boyd and Turnbull 1989).

Incentives

Forest incentives come in the various forms of direct financial aid for forest management, free seedlings, or advice and technical assistance for forest managers. The primary objective of all three generally has been to increase wood supply, and all three generally target supply from non-industrial private landowners. Financial incentives are more common in developed countries. They can afford them. The United States, some Canadian provinces, the UK, and the Nordic countries all provide financial assistance for small landowners. Even Chile's economy, well-known for its free market orientation, has been modified with a financial assistance program for forest management and the success of Chile's forest sector has led policy makers in some other countries to recommend these subsidies as important ingredients in the development of successful forestry sectors. Free or discounted seedlings and technical assistance are also common to forest policy, and rural development programs around the world, regardless of a country's development status, often include a seedling distribution component.

All three—direct financial incentives, seedlings, and technical assistance—provide cost savings in forest management. Therefore, their impacts are largely restricted to regions in the third stage of forest development. Regions in the second stage but at the cusp of entry to the third stage can also benefit if the financial incentive of cost saving is sufficient to make a difference in their profitability for long-term forest management. Policymakers and program managers often overlook this fundamental point and waste public resources as they attempt to offer management incentives in regions in the first stage or earlier in the second stage of development where managed forestry is not viable. Landowners in these regions will not be receptive to the incentives because the products of frontier forests are still less expensive than the products of managed forests. The economically productive lands in these regions yield greater returns in other, non-forest, activities.

Financial assistance: Direct financial assistance, or forest incentive payments (FIP), are monetary inducements to participate in forest management. The United States, for example, has administered a financial assistance program in one form or another since the 1920s. The current program began in 1974. It provides government cost sharing for up to 65 percent of all reforestation and management expenses for landowners with fewer than 210 hectares. China's program has paid farmers 4.5 yuan per mu (15 mu = 1 ha.) to plant trees since at least 2001 (Can et al. 2007). Chile, as a third example, also began a FIP program in 1974—with the objective of improving the international competitiveness of its forest sector. Chile's program returns 75 percent of costs to the landowner one year after successful afforestation. Reforestation without compensation is compulsory for subsequent timber rotations. Chile restricted this program to landowners with fewer than 500 hectares in 1992.[5]

Figure 4.2 traces the impact of FIP programs on land use for a region in the third stage of development. Financial assistance replaces some private management costs and, therefore, allows the forest value function V_f to shift upward by the per hectare value of the assistance. Land at both the intensive and extensive margins, B' and B", shifts away from competitive uses and into forest management and total production from the managed forest increases as a result. Regions in the second stage of development (not shown) could move to the third stage if the financial assistance is sufficient. Then they too would have land newly converted from open access to forest management as a result of the financial assistance.

The cost share received by private landowners potentially affects the use of other, non-land, inputs as well, and it may have an indirect effect on extraction from the natural forest at point D. It may induce increases in the optimal levels

---

5 Personal communication, E. Morales, November 8, 2002.

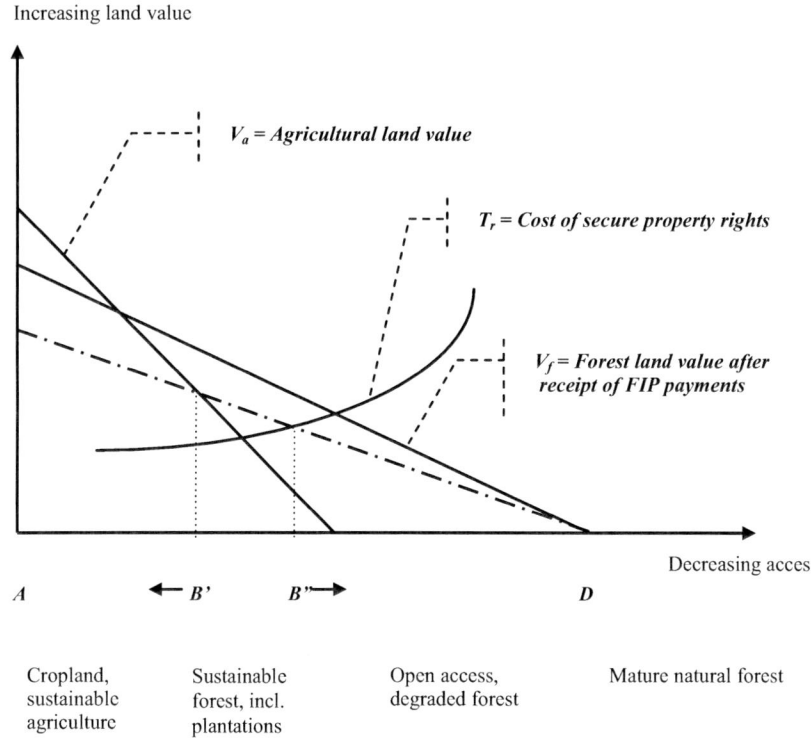

FIGURE 4.2 Forest incentive programs

of labor and manufactured capital inputs and decrease the optimal harvest age for managed forests.[6] Its potential for an indirect effect on the natural forest is comparable to the conserving effect discussed for favorable capital gains treatment in income taxes. That is, some of the increased production from managed forests may substitute for harvests from the natural forest.

Each of these potential effects is likely to be small for the U.S. FIP program because many landowners fail to take advantage of the government assistance, and because other factors are more important for those landowners who do obtain the government assistance. Some landowners accept the assistance without ever intending to harvest their forests (Boyd and Hyde 1989).

The effects of the programs in Chile and the UK may be greater. These programs apparently influenced numerous landowners for whom forest management was on the cusp of being profitable. The financial assistance for Chile's non-industrial private landowners is a much larger share of their total costs than it is for

---

6 Once more, the effects on the use labor and manufactured capital in managed forests are easily demonstrated with a manipulation of the Faustmann equation and its optimality conditions, conditions (3a.8) and (3a.9). The cost terms in these conditions decrease by the per hectare amount of the government financial incentive and the optimal input levels change accordingly.

their counterparts in the United States and, arguably, many Chilean landowners would not have afforested their lands without the initial promise of government assistance. The Chilean landowners who receive assistance account for nearly 40 percent of the timber supplied to that country's large wood processing industry, the fifth largest source of Chile's GDP and the third largest source of its export earnings.[7] Therefore, government financial assistance may be a source of significant increases in managed forestry and timber production in Chile.

The UK experience is related but still somewhat different. Prior to 1988, the UK program was part of an overall tax and incentive program that favored investment in commercial timber species to the extent that some felt that trees were replacing too much open space. The overall program was revised and the general investment climate is no longer as favorable for tree planting. The UK's current forest incentive program (the Woodland Grant Scheme) requires compliance with environmental and landscape guidelines, and it provides larger payments for reforestation with native broadleaf species. Therefore, the program now has a smaller increasing effect on forest management that is restricted to native broadleaf species.

Distributive arguments sometimes supplement the timber supply arguments generally used to justify FIP programs. These should be accepted with caution. In developed countries, even small forest landowners are not among the poorest or most disadvantaged for whom we generally design redistributive policies. For example, Boyd and Hyde's (1989) assessment of North Carolina landowners observes that those who do take advantage of the FIP program are not among either the poorer or the smaller of all private landowners. Furthermore, it is an open question whether FIP programs are of greater benefit to the landowners or to the wood processing industry that obtains its timber for lower prices because of the government assistance for forest management, and surely this industry is not the intended target of income redistribution.

Free seedlings and financial assistance for afforestation and reforestation: Various public programs around the world donate seedlings to smallholders and communities or build nurseries that provide seedlings at discounted prices. These have the same general effect as FIP programs. They decrease the costs of forest management.

Like FIP programs, the effects of free or discounted seedlings are restricted to managed forests within regions in the third stage of forest development, or to landowners at the cusp of profitable entry to the third stage. Those development assistance and forest conservation programs that distribute seedlings in regions in the first and most of the second stages of development overlook the clear facts that any management, even with free seedlings, has its opportunity costs and that resources from the natural forest remain less costly for those regions in the earlier stages of economic development.

---

[7] http://economist.com/countries/Chile/profile.cfm?folder= Profile-Forecast (accessed July 22, 2007).

Also like FIP programs, the effect of seedling distribution programs can be traced from the private financial saving for the seedlings through its increasing effect on the net forest value function V_f. The final effect on land use and forest production, even in the third stage of forest development, is probably small in most cases, although there are numerous examples from countries of various levels of economic development where farmers willingly accept free or discounted seedlings and tend them to maturity (Godoy 1992; Amacher, Hyde, and Rafiq 1993; Molnar, Scherr, and Khare 2003).

Technical assistance: Technical assistance programs are also known around the world, generally in the form of forest extension programs. These programs are designed to bring the latest information on modern technologies to local forest operations. Like FIP programs, forest extension generally targets smaller landowners. Occasionally, it includes advice on improved logging techniques. In this latter case, it can have an impact on timber harvests in all three stages of forest development—leading to additional recovery of usable material from the residual open access forests as well as more complete recovery of woody material from the area of current harvest while, in some cases, reducing the impacts on the natural environment.

More generally forest extension, like agricultural extension, is designed to assist landowners improve their land management. Therefore, like FIP programs and free or discounted seedlings, the effects of most technical assistance programs in forestry can be traced through their impacts on the forest value function over the range of managed forests in the third stage of forest development or lands adjacent to it.

The greater questions for many forest extension programs have to do with obtaining rapid and widespread adoption of the preferred new technologies. In fact, these are important questions for FIP and seedling programs as well. The experience of agricultural extension is relevant. In agriculture, better-informed landowners and landowners who can afford the uncertainty inherent in trying a new variety of seed or a new production technology are the obvious targets for initial assistance. Other landowners follow rapidly once they observe the successes of their neighbors.[8] Of course, the greater the management cost saving due to a new technology (the larger the positive impact on the forest value function V_f), the more rapid and widespread the rate and final level of adoption.

Regulations

Forestry has a long history of government regulation. The earliest regulations protected the forest as a hunting ground for kings, nobles, and other large estate

---

8 Feder et al. (1985) review the agriculture experience. The agroforestry evidence follows the agricultural experience, although the analytical evidence is not as extensive. Pattanayak et al. (2003) and Mercer (2004) review this literature.

owners. (Remember Robin Hood and his illegal hunting in Sherwood Forest?) Subsequent regulations protected the best trees for eventual use by the kings' navies. The emphasis on the rights of kings and large landlords changed in the latter part of the nineteenth century. First a sustained flow of timber and then, later in the twentieth century, a sustained flow of other environmental services from the forest, became new justifications for government regulation.

Table 4.1 traces the general pattern of forest regulation through the twentieth century. It lists the most common regulations in the order of their introduction in most countries. It also identifies the reason generally given for each regulation. In each case the more fundamental reason is the protection of some perceived broader social value.

Some North American and North European countries began, as early as the 1930s, to require reforestation following timber harvests. Today, some developing countries require a permit before allowing harvests even on private lands. Others restrict the shipment of timber across provincial boundaries. The objective for these regulations has always been to control unlimited harvesting and to insure a resource supply for local mills.

As the laws and policies of the developed countries began to focus on the environment, those countries introduced new restrictions on harvest systems. In North America, restrictions on clearcuts or regulations assuring selective forest management became widespread. Restrictions on other specialized forest practices followed as the public became aware of the detrimental effects of those practices. Herbicide and pesticide usage in forests were controlled, for example, once the public recognized the human health effects of substances like DDT.

TABLE 4.1 Common regulations of private forestlands

Regulation	Public Objective
Reforestation requirements and required silvicultural practices	First, long-run timber supply. Later, environmental protection
Restrictions on specific activities:	
On log harvests and shipments	Long-run timber supply. Protection of jobs, mills, and communities
On clearcutting	Aesthetic appeal and environmental protection
On herbicide and pesticide use	Public health
Water quality guidelines	Public health, environmental quality
On streamside management	Habitat protection
More recent concerns—often discussed but less frequently imposed:	
Protection of endangered habitats	Environmental quality, future public welfare
Forest certification	Sustainable forestry, environmental protection

Water quality restrictions are another example. As water quality in general became a public concern, the laws and policies protecting water quality first focused on point sources like effluent from pulp and papermills. By the mid-1980s concerns for water quality shifted to non-point sources like agriculture and forestry. Pulpmills are point sources because we can identify the exact point of their detrimental discharge, the pipe from which it spills. Agriculture and forestry are non-point sources because their contributions to water pollution cannot be traced to a single discharge point. Rather, their water quality effects originate from activities like herbicide applications or general soil disturbance that are dispersed over large areas.

Finally, the most recent regulatory issues for private forestry in North America refer to endangered species and forest certification.

The remainder of this section of this chapter reviews the design of forest regulation, and then examines the impacts of four illustrative categories of regulations: (1) restrictions on harvests and shipment, (2) reforestation requirements, (3) environmental requirements such as restrictions on herbicide use, streamside management, and clearcuts, and (4) forest certification. The appendix to the chapter reviews five additional categories of public interest—but from the perspective of their policy objectives rather than from the impacts of the regulation itself.

The design of forest regulation—and its merit: Regulations are generally set as absolute physical standards such as no clearcuts larger than some limited area, no harvests within a limited distance from a watercourse, diameter limits on felled trees and distance between felled trees, or zero disturbance within the habitat of an endangered species.

Economists often argue that it would be more efficient to identify the desired behavior; scenic improvement, erosion control, natural forest regeneration, and protection of endangered habitat, respectively, in our examples; and then charge landowners who fail to achieve the desired objective. This approach allows distinctions in the levels of charges between landowners who are more, or less, successful in achieving the objective. Greater charges increase the incentive for the desired behavior on the part of landowners who deviate farther from the objective.

A system of charges also encourages landowners to tailor their management plans to the unique characteristics of their own lands. For example, where erosion control is the policy objective, landowners might be assessed charges based on sediment flow at the outlet of watercourses or at the boundaries of each ownership. Each landowner would make his or her own decision, according to the characteristics of his or her own lands, regarding the most efficient management practice. One landowner might minimize the charge and, therefore, the sediment flow, by harvesting up to five meters of a stream course over gently sloping land but only within thirty meters on steeper and more erosive lands. Another landowner would make different decisions according to the different erosive characteristics of that second owner's land.

In fact, environmental charges have become an acceptable alternative to regulatory standards for some forms of pollution. There is good evidence that they induce the desired behavior in papermills in countries as diverse as the Netherlands and China (Bressers and Lulofs 2004; Xu, Hyde, and Amacher 2003). The worst polluters pay higher fees and the higher fees improve their incentive to control future discharges.

However, assessing charges at the pipe where a mill discharges effluent is one thing. Assessing charges for a non-point source like forestry is much more difficult. It may be easy to identify clearcuts in the forest but identifying the sources of stream sediment or the degrees of modification of endangered habitat would impose a large burden on the public agencies responsible for monitoring and enforcing the regulations. Environmental standards are generally simpler to monitor and enforce for low-valued and dispersed resources like forests, and this is one reason why forest policy often relies on environmental standards rather than on systems of graduated charges.

Restrictions on harvests and shipment: Restrictions on harvests are intended to insure a long-run supply of the resource or, in several more recent cases, to protect the natural forest from degradation and deforestation. Restrictions on shipments are intended to control external demand and, thereby, to insure a supply for local mills and protect the welfare of local communities. Some countries confine shipments within state or provincial boundaries. Some impose log export restrictions—which have the same effect, except that they restrict shipments from leaving the country rather than from leaving the forest or a region within the country.

Harvest restrictions occur in two basic forms. Some restrictions, known as logging bans, are intended to be absolute or nearly absolute, halting all harvests on certain classes of forests, such as remaining natural forests or forests managed by the state forestry agency. Numerous Asian countries have imposed logging bans in recent years (Durst et al. 2001). Other restrictions, often identified as logging quotas or limits to the official allowable cut, are intended to control certain categories of harvests or to limit harvests to approved levels. Logging bans are difficult to enforce because governments cannot hire enough guards to protect all of any class of forest and because it is difficult to prove that delivered timber came from the protected source—rather than from some farm woodlot. Discussions of logging bans quickly raise the related question of illegal timber harvests, a problem we will return to in the next chapter on the sale of products from publicly owned forests.

More limited restrictions such as quotas for timber harvests suffer these same problems and, in addition, create uncertainty in the minds of landowners and mill operators regarding their right to conduct future harvests. Will the quotas be more or less limiting in the future? Will future administrators enforce the policy more or less stringently? Will permits be available when the landowner is ready to harvest or when the mill manager needs wood? Uncertainty of this sort encourages landowners to harvest earlier than they would without the restriction—while they can still be certain of an economic return. Therefore, uncertainty induces short-run increases in the harvest level as landowners hurry to obtain some level of certain return. The long-term effect of uncertainty runs the opposite direction as timber harvested now is unavailable to grow to larger size for later harvest and as some landowners even transfer some forest lands into alternative non-forest activities where the economic returns are less subject to administrative whim and, therefore, more certain. The long-run timber supply decreases as a result.

The costs of timber harvest permits, when available, also diminish the returns to landowners. Sometimes these costs are small and their impact may be inconsequential. In other cases, obtaining the permit consumes significant amounts of time and financial resources and, even then, the availability of permits may be uncertain. In these cases, the high cost makes lands at the intensive and extensive margins of managed forestry less competitive as managed forests and it causes landowners to manage their remaining inframarginal forests and trees less intensively.[9] Once more, the standing timber resource and the long-run supply decline as a result—in direct contrast with the policy objective to insure long-run supply.

Indonesia, India, and Sri Lanka provide examples. Indonesia began regulating harvests of its high-valued sandalwood resource in colonial times and the modern governments of Indonesia and India have increased their restrictions on this aromatic wood more recently. India bans harvests of sandalwood, while Indonesia's government sets sandalwood prices at a small fraction of the market price. Both governments intend their regulations to preserve the remaining sandalwood resource. However, illegal harvesting thrives and the standing sandalwood inventory has declined in both countries. Garcia Garcia (1997) tells a similar story for Indonesia's harvest restrictions on jululing trees as well.

Sri Lanka restricts all harvests from its remaining natural forest and requires permits and inspections before allowing harvests on private lands as one means of assuring that harvested logs do not come from the protected natural forest. Permits have become such a serious constraint on supply that delivered log prices are as much as seven times the price received by producing landowners. The various costs associated with obtaining the permit absorb the difference. Of course, as landowners lose the incentive of the higher price they decrease their management and the levels of managed forest inventory and timber production decline. Moreover, the higher market price has been an incentive for illegal logging of the natural forest—also in direct contrast with the original policy objective. Timber production is now about one-fourth the level it was prior to the permit system and perhaps as much as one-fourth of all wood delivered to Sri Lanka's mills is the result of illegal logging (Gunatilake and Gunaratne 2001).

---

9 Once more, the changes in the optimality conditions from the Faustmann equation, conditions (3a.8) and (3a.9), demonstrate the effects of the permit on the remaining economically viable forestlands. The price term in these conditions decreases by an amount equal to the landowner's expenditures obtain the permit.

Even where the primary effects of harvest restrictions are desirable, the indirect effects may not be so beneficial. In Cambodia, the government established a logging ban in 2002 in response to strong external pressure from international NGOs and international donors. The objective was to preserve the remaining depleted natural forest and that objective is being served—at least to a limited extent. Illegal logging continues, but the local supply of dark hardwoods has been limited. Lumber prices doubled between 2002 and mid-year 2004 as a result, and this had at least two unanticipated results. First, the urban poor no longer have access to their preferred material for house construction and there has been an increase in shantytown construction from tin and cardboard. The second unanticipated result comes from the substitution of poorer grade material from palm trees for some prior local uses of tropical hardwoods. Palms are a source of multiple products (oil, sugar, alcohol, fruit) and of income diversity for agricultural households. The harvest of palms as a substitute construction material has cut into these sources of rural income, at least in the short run until a new crop of various palms can replace those harvested for substitution in Cambodia's market for tropical hardwoods. The obvious policy question should be whether the logging ban has caused an environmental saving in Cambodia's natural forests that is sufficient to offset the negative effects on housing for the urban poor and on income from palm derivatives. This question has not been part of the policy dialogue.

Finally, restrictions on shipments have effects on the forest that are similar to those of logging restrictions. The first effect of shipment restrictions is to reduce the number of loggers and mills eligible to compete for a resource. The United States, the Philippines, Nepal, and China, for example, each restrict at least a share of some timber sales for local mills (Hyde et al. 1997; Dangi and Hyde 2001; Hyde et al. 2003). These countries may succeed in protecting some less-competitive local mills in the short run. However, it is not clear that local community welfare is significantly enhanced because wood processing is generally a small share of local economic activity in the regions being protected and because loggers and millworkers have other employment opportunities (Stevens 1978; Daniels, Hyde, and Wear 1991; Ruiz Perez et al. 2003). The long-run effect of shipment restrictions is identical to the long-run effect of harvest restrictions. Where forest management is a viable option, the restriction decreases demand and, therefore, landowners obtain lower prices for their timber and their incentive to manage and produce a long-run timber supply declines. Both margins of managed forestland shrink as less profitable, non-forest, uses of the land replace managed forestry. Certainly, this is not in the interest of local welfare.

Reforestation requirements: Required reforestation following a timber harvest was originally introduced as another means of insuring long-run timber supply. More recently, the objective has shifted to general environmental sustainability. Sixteen U.S. states introduced laws requiring reforestation to satisfy the first objective between 1903 and 1950. Nine states revised or introduced new laws after 1968 to address the broader second objective. All the Nordic countries, several other Western European countries, Brazil, Chile, and Ghana, for example, have similar laws, all with similar objectives.

Forests regenerate rapidly and naturally on some lands. On these lands, the reforestation requirement imposes no real cost. On others, since reforestation is a contingency on harvest activity, required reforestation adds the reforestation cost as an expense of timber harvesting. This shifts the forest value function V_f downward. Where the requirement is enforced, harvests at the frontier decline and many would consider this an environmental improvement.

Required reforestation has no impact on the managed forests of the third stage of forest development because reforestation is already economically viable for these forests. Boyd and Hyde's (1989) empirical examination of the reforestation component of Virginia's state forest practice act demonstrates this point. Boyd and Hyde examined private forest management in the adjacent states of Virginia and North Carolina, states that are similar in their forest lands. Virginia has a reforestation requirement for its private landowners, while North Carolina does not. If Virginia's requirement were effective, that state should experience a greater level of reforestation and, over time, Virginia's standing forest inventories should also be greater. However, after controlling for differences in site quality and regional prices, Boyd and Hyde observed no measurable differences in standing forest volumes between the two states. As expected, Virginia's reforestation law has no effect on the private lands that are already viable for economic forest management.

Silvicultural prescriptions: Several developed countries and a few developing countries impose a variety of additional silvicultural and other environmental prescriptions on timber harvests and continuing forest management activities. These include limitations on clearcuts, harvest spacing restrictions designed to ensure the regeneration of mixed forests, restrictions on the use of herbicides and pesticides, and standards for streamside management and the construction and use of logging roads.

The added costs implicit in these environmental regulations shift the forest value function downward in either of two characteristic ways. Added management costs, such as those incurred for restrictions on herbicide use, pivot the function downward, affecting only the range of managed forestry. Added harvest costs, such as those incurred for restrictions on clearcuts, shift the function downward over its entire range, reducing the land area in managed forests at both the intensive and extensive margins and also decreasing management intensity on the remaining managed forests. Harvests at the frontier decline as well in this second case.

The net effect can be substantial. Sedjo (1999) estimates that silvicultural prescriptions add an average of 5 to 18 percent to the costs of forestry in the U.S. South, British Columbia, and Finland—although the impacts on individual landowners vary with local land quality and the level of enforcement. If these added costs are roughly comparable to a similar decrease in the stumpage price received by the landowner, and the supply price elasticity is in the range of 0.36 to 0.50 (as many estimates suggest), then they imply a 2–9 percent decrease in timber supply in each of the three regions. Finland, where the timber supply elasticity may be as great as 2.25, could suffer as much as a 35 percent decline in timber supply.[10]

Clearly, these silvicultural prescriptions alter forest production in important ways, decreasing production while improving the local forest environment. The magnitude of these costs and their effect on production raise two new issues, one for private land managers and the other for public environmental values.

Private landowners challenge the "taking" of their land use rights that occurs when governments impose new regulations that restrict them from making some choices and benefiting from some opportunities that were available before the new regulations were imposed. This concern is probably greatest in the U.S. South where non-industrial private landowners manage two-thirds of the forestland and account for 60 percent of annual timber harvests and where reforestation, water quality, and endangered species restrictions on private lands have become more confining in recent years (Flick, Tufts, and Zhang 1996; Zhang and Flick 2001; Zhang 2002). Taking is also an issue of contention, however, in parts of Canada and in the Nordic countries. In fact, it is an issue anywhere that increasing public environmental awareness threatens to affect private forest landowners—but promises them no restitution.

As with uncertainty about harvest restrictions, uncertainty about increasing environmental regulation causes some landowners to act preemptively. They harvest early, before the policy is formally implemented, thereby insuring themselves of some level of return on their timber investment, but forgoing any opportunity for greater return on the future growth of their timber and also permanently damaging the forest environment that the new regulation intends to protect.

The new issue for public environmental values has to do with shifts in the geography of production, and of the environmental losses that must occur when the environmental regulations are imposed in selective regions only. Consumer demand is unaffected by these regulations. Therefore, the significant decreases in production that occur in the U.S. South, British Columbia, and Finland, for example, will be largely compensated by increases in production from bordering regions that are not affected by the regulation—other parts of the United States, inland Canada, and Russian Korelia, respectively—as well as with additional imports from developing countries. In each of these cases, the production shifts are largely from managed forests of regions in the third stage of forest development (where the restriction was imposed) to the natural forest frontiers of regions in the second stage of forest development. Environmental standards are generally lower or non-existent in these regions of alternative production. Therefore, while the higher environmental standards may improve environmental quality in the U.S. South, British Columbia, and Finland, the same regulations effectively export environmental damage and deforestation to other forests in other parts of the country or the world.

---

10 See Buongiorno et al. (2002) for a summary of estimates of supply elasticity. Kuuluvainen and Salo (1991) provide the estimate for Finland.

Those who are interested in improving the forest environment generally overlook this transfer of effects. The same interests that demand environmental improvement at home and desire to reduce deforestation globally are forcing a share of all production to shift away from sustainably managed operations at home to unsustainable harvest operations at the frontier of other regions. While the tradeoff between environmental improvement in one region and environmental decline elsewhere has not been assessed empirically, the magnitude of additional environmental costs in the U.S. South, British Columbia, and Finland in our examples suggests that this tradeoff must be substantial. Decreasing production by 9 percent in only these three regions is equivalent to shifting more than 20 percent of the global production of industrial roundwood to other regions with generally lower environmental standards.[11]

Forest certification: Certification refers to the formal assurance that forest products originate from sustainably managed forests. This assurance generally originates with an independent authority that inspects the product or the process of production and provides its own stamp of approval. The procedure is similar to a licensing procedure where, in this case, the license guarantees the standards of sustainability.

Although certification is not the official policy of any government at this time, it is widely encouraged, especially among Western Europe's environmentally conscious consumers, and it could become policy in some countries in the future. Until a requirement that some or all forest products are certified becomes a formal government policy we can anticipate that landowners will seek certification when they perceive that it imposes little cost while bringing them advantage in the markets for their products, either a price advantage, an improved market share, more certain future access to an existing market, or an opportunity for penetration into new markets.[12] This means that the anticipated benefits of certification, the higher prices or increased share of an environmentally aware market, must offset the costs of obtaining the certification.

Certification itself is evidence of the land manager's long-term commitment to forest management and it is an indication that the manager is likely to be able to satisfy future environmental standards (Forsyth 1998; Ozanne and Smith 1998; Bass et al. 2001). Some land managers in Brazil use certification to assure potential creditors of their long-term commitment to forest management and the assurance has improved these managers' access to credit. Some Brazilian wood processors recognize this commitment and the advantage it brings to them as well in the form of a more certain long-run source of raw material for their mills (Sobral et al. 2002).

---

11 Calculated as [0.09 (200,000 + 80,000 + 50,000)/1,590,000] x 100, where the latter four values are estimates of 1000s of cubic meters of industrial roundwood produced in the three regions and globally in 2002 (USDA Forest Service 2005; Ministry of Forests 2004; FAO 2002).
12 See, for example, Karna and colleagues (Karna et al. 2001; Karna, Hansen, and Juslin 2003).

In terms of the figures and the three-stage model, we can anticipate that the owners of managed forests from regions in the third stage of forest development can provide legitimate evidence and make successful requests for certification. Whether certification causes a change in land use at either margin of managed land depends on whether the landowners' revenue gains from certification exceed the costs of certification. Certification will be more problematic, and we can doubt its validity, for landowners and loggers at the forest frontier.

Therefore, it is not surprising that Northern Europe's wood products industries seem to accept the idea of certification.[13] Northern Europe is largely within the third stage of forest development and many firms in its industries rely on wood supplies from managed forests. Certification does not impose a significant cost for them and more than 25 million hectares had been certified in the Nordic countries alone by year 2000 (Bass et al. 2001; Hansen and Juslin 1999). This was more than 40 percent of the forest in the Nordic countries, and it accounted for about half the global total of certified forest at that time. The experience is similar in the Republic of South Africa where all the commercial wood comes from plantations that were originally established in the 1890s by a government that anticipated Europe's demand for a reliable timber supply. Certification imposes no significant additional management cost on South African plantations and almost all of them are certified.

The story is similar in Brazil. Secure sources of raw material are more important to those wood processing enterprises along Brazil's coastal plain, and less important to those at the Amazon frontier where the resource is more plentiful. Therefore, it is not surprising that the 63 percent of enterprises participating in a Brazilian survey and indicating their perceived advantage to certification were located in the coastal plain (Sobral et al. 2002).

It is also not surprising that consumers express greater concern for the certification of production from tropical and developing countries. A much greater share of the forest products from these countries originates from unmanaged natural forests in regions within the first two stages of forest development.[14] Examples from Indonesia, Mexico, and Guatemala illustrate some of the problems for certification in these latter cases. Indonesia is the largest exporter of forest products in Asia and forest products are the third largest source of Indonesia's export earnings. As of 2003, Indonesia's large producers were eager to obtain certification and, thereby, gain potential European market share. The small new Indonesian certification institute could not keep up with the demands for its services.

Nevertheless, it is difficult to imagine successful certification for a significant

---

13 Similarly, it is not surprising that a simulation by Schwartzbauer and Rametsteiner (2001) concludes that certification would have only a modest effect on wood supply in Western Europe.
14 In fact, the International Tropical Timber Organization estimates that only 25 million hectares, or approximately one percent of all tropical forests, are managed (ITTO 2006). Some of these 25 million, no doubt, are managed in name only. That is, they appear as numbers in a management plan covering some larger area, but receive no other physical management inputs themselves.

share of Indonesia's forest products. Less than 10 percent of Indonesia's annual harvests originate from managed forest plantations and most of its timber markets are characterized by the first two stages of forest development where sustainable management is not yet financially viable. Certification faces an additional problem where the unsustainably managed share of all forest products is so large. The products of managed forests and of natural forests are similar and one is generally an easy substitute for the other at the mill. In this case, we can anticipate that either the monitoring costs associated with certification will be great or that certification itself will not be a reliable guarantee of production from genuinely managed forests.[15] Finally, there are scale economies in the process of certification and, partly for this reason, smaller producers in Indonesia as of 2003 had difficulty obtaining certification. In some cases these scale economies could drive small plantation owners out of the market for certified wood—and surely this is not an objective of the environmentally aware consumers that support the movement toward certification.[16]

Mexico and Guatemala share a still different experience. The forests of these two countries are not generally in the third stage of forest development. Yet approximately 50 communities in these two countries managing 1.1 million hectares of forest have been certified under the Forest Stewardship Council system of certification. Certification has helped communities and their smallholders obtain more secure land rights from the central governments and better technical support from foreign donors. It may have improved market access in some cases, but the communities and smallholders have not benefited from higher prices (Molnar 2003).

The certification process itself has been expensive for these communities—about US$60,000 for five years' certification. So—why would these small and generally poor communities in the first and second stage of forest development absorb this steep cost? They didn't. Environmentally motivated donors from Western Europe and North America paid for the certification. Therefore, the questions have to be "how long can donors afford this expense and for how many communities and forests?" And "how long will these communities manage their forests sustainably when they have no apparent incentive to do so beyond the period of their receipt of donor assistance?"[17]

---

15 Guariguata (2011) also questions the reliability of certification as the different participants in the certification process have different objectives which may lead to agreement only on weak standards for certification.
16 Some smallholders may be able to solve this problem by obtaining collective certification but even this imposes additional costs.
17 Hunt (2001) provides another illustration. He calculates that, for Papua New Guinea in the late 1990s, the premium on sawtimber necessary to justify certification would have to have been greater than US$47 per m^3. This compares with an average sawtimber market value of only US$132 per m^3 at that time. Landowners would have had to increase their stumpage to an unlikely price of $179 per m^3 to cover the additional cost of certification.

In sum, the concept of certification is young and the procedures for its administration are still developing. It faces its most serious challenge in producing regions that are characterized by the first two stages of forest development and reliance on the natural forest, yet regions that are also significant suppliers for consumers who desire the assurance of certification. Indeed, the measure of certification's success should be whether the benefits of certification become a justification for the conversion of any logging operation at the forest frontier to a *bona fide* program of sustainable forest management.

Certification and other programs seeking to insure sustainable forest management are a central focus of contemporary global forest policy.[18] We can anticipate that they will remain important policy issues for some time. Therefore, we might take this discussion a step farther and consider a description of their effect if governments do sort out the difficulties of successfully implementing these policies and, thereby, secure a larger degree of sustainable forestry than we currently observe.

Successful implementation of certification, and other policies and programs promoting sustainable forestry, can only be measured by a decrease in extractive activity at the forest frontier. Extractive activity at the frontier could be replaced with increased production from managed forests from regions in the third stage of forest development, as well as with a degree of wood-saving technological change in the mills and some amount of substitution away from wood products. The first of these means that, to the extent that certification is successful, and in regions where it is successful, those other policies and programs intended to affect extraction at the frontier (e.g., harvest restrictions, reforestation requirements) will become relatively less important and, simultaneously, those policies and programs (e.g., FIP, technical assistance, silvicultural prescriptions, most forest taxes) that affect managed forests will become relatively more effective tools of forest policy. The challenge is to make certification sufficiently successful for this to happen.

Direct Forest Policy—Conclusions

Governments have experimented with a range of direct forest policies: taxes, incentives, and regulations. Income taxes and property taxes, with various modifications are intended as sources of government revenues. They have accomplished that objective and modern tax laws do not generally impose unusual burdens on forestry relative to alternative land uses. In fact, some provisions of individual and corporate income tax codes have been highly beneficial to forest management.

Most forest incentives and regulations have a different justification and a different history. Most incentives and regulations intend either to encourage timber supply or to protect certain classes of trees or forests or the broader environment. Financial incentives are common in developed countries. Technical assistance and the distribution of free or discounted seedlings are also common in developed and

---

18 See the section on sustainable forestry in the appendix to this chapter.

in many developing countries as well. For most of these programs to have any effect whatsoever, they must carefully target either the managed lands in regions within the third stage of forest development or those lands at the two margins of managed forestry, lands which could readily support forest management in the presence of some level of financial assistance or cost saving.

The effects of most forest regulations are more problematic. Some (harvest and shipment restrictions) actually discourage long-run timber supply. Others (e.g., required reforestation) often have little effect. More recent silvicultural prescriptions may have substantial impacts, both decreasing timber supply and improving the environment in selective regions. However, it is unclear to what extent the decrease in supply in these regions is offset by expanding timber harvests and a deteriorating environment in other regions and countries where the environmental prescriptions are less stringent.

Finally, certification is a newer regulatory concept for forestry. Its meaningful implementation will be difficult. The fair test of its effect is not the share of all forests that are certifiable because those forests that are already managed are sustainable, and certifiable, by definition. Certification requires no change for them. The fair test of certification is its inducement to convert away from extractive activity on marginal forestland into sustainable forestry either at the frontier or through the substitution of expanding production from managed forestlands.

Spillover Effects from Policy Designed for Adjacent Sectors

As policy changes induce expansion or contraction in the sectors that compete with forestry for inputs, they also affect forestry itself. Similarly, as policy changes induce expansion or contraction in sectors that consume forest products, they too affect the forest. The competition for land, most generally between agriculture and forestry, is an important example of the first category.[19] For the second category, a variety of industries use the products of the forest in their own manufacturing processes: rubber; processors of exotic fruits, nuts, and natural herbs; those who use the forest for outdoor recreation, etc. Of these, the wood products industries have the greatest effect on forests. The next sections of the chapter focus on the effects of policy change in these two sectors, agriculture and wood products.

Agricultural Policy Spillovers

Agricultural policies affect agricultural land use, including those lands where agriculture competes with forestry. Therefore, agricultural policies may affect forests as well. The discussion in chapters 2 and 3 showed, however, that the

---

19 Labor is another crucial input. Agricultural policies (discussed in this section) and policies affecting migration, particularly roads, and regional development have the greatest effects on forest labor. The next section of the chapter considers the effects of roads and discussions of sustainability in the appendix and of macroeconomic policy effects later in chapter 6 include consideration of the impacts of regional development.

relationship between agricultural and forest land use is not consistent across the three stages of forest development. Furthermore, it showed that not all agricultural improvements use more land. The impacts of agricultural policy on the forest are complex for the same reasons.

Many countries have introduced policies that encourage agriculture. For example, the United States and Canada encouraged agricultural settlement on their frontiers in the nineteenth century, Indonesia financed the movement of new settlers to the forest frontier in a policy called "transmigration" in the 1980s, and Finland compensated farmers for clearing new land for agricultural use in the early 1990s (e.g., Hibbard 1965; Heydir 1999). For a period in the 1990s, Brazil gave title to Amazon forestland to anyone who cleared it for agricultural or livestock use (Binswanger 1989; Serôa da Motta 1993; Schneider 1994; Young 2002). In each case, agriculture (or population) policy shifted the agricultural value function V_a upward, at least at its far extremity, and encouraged agricultural expansion within a region in the first stage of forest development.

With progress into the second and third stages of forest development, policies designed to affect local agriculture no longer have an impact on the forest frontier. By this time, the more common agricultural policies are subsidies for agricultural inputs and price supports for agricultural outputs.

Input subsidies tend to encourage the use of fertilizers and water, in particular. They decrease costs per unit of capital and increase the marginal revenue product of these private capital inputs. They cause producers to shift input proportions, increasing the relative use of the subsidized capital and decreasing the relative use of alternative labor and land inputs. In terms of our figures, these capital-using and land-saving policies shift the agricultural value function upward along its vertical axis and increase its slope—as from the broken to the solid lines in Figure 4.3A. Depending on the magnitude of the subsidy and the substitution between capital and land, input subsidies can either expand or contract agriculture's use of the degraded open access lands (between points B and D) in the second stage of forest development and they can either increase or decrease agriculture's competitive position at the intensive margin of forest management (B') in the third stage of development.[20]

The effects of agricultural input subsidies are compounded by the effects of government research programs. These are not among the programs usually considered when in discussions of government interventions in the market. However, within agriculture, government research programs absorb significant amounts of public funding and some of the research they produce has been a source of phenomenal increases in agricultural productivity and decreases in agricultural

---

20 Where agriculture expands into the open access frontier (stage II) or converts land away from managed forestry (stage III), the policy induces *absolutely* greater use of both capital and land, but it remains *relatively* capital using and land saving.

Increasing land value

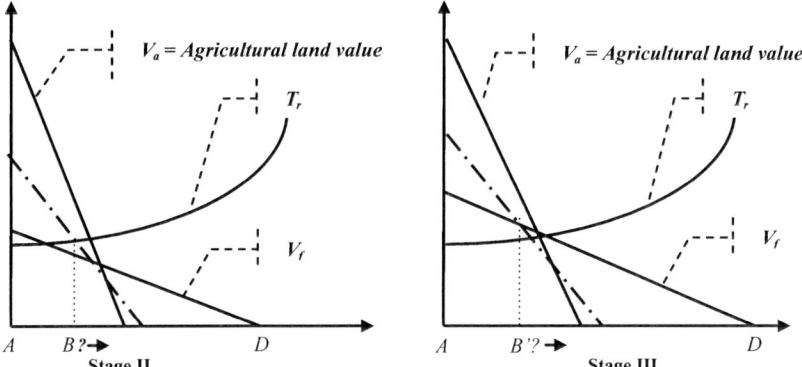

A. Input subsidies tend to favor capital inputs relative to land inputs. They increase the slope of the agricultural value function. Whether they shift the function outward over its entire range or only at its vertical axis (as shown in the figures) depends on the magnitude of the subsidy. Therefore, they can either expand agricultural use into the degraded open access forest in the second stage of forest development or contract it (as shown), and either expand agricultural use into previously managed forest in the third stage (as shown) or contract it and allow the managed forest to expand at its intensive margin.

Increasing land value

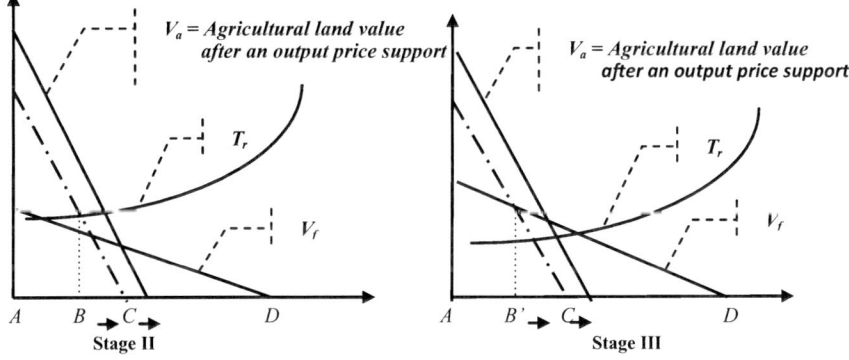

B. Output price supports raise the agricultural value function over its full range and unambiguously extend agricultural land use into the degraded open access forest in the second stage of forest development and into previously managed forest in the third stage.

FIGURE 4.3 Agricultural policies

costs.[21] It tends to make capital inputs less expensive and more productive. Therefore, publicly funded agricultural research, like other subsidized inputs, tends to be relatively capital-using and land-saving, and its effect on the forest is comparable to that of agricultural input subsidies.

---

21 See Ruttan (1982) for a survey of the agricultural research experience.

The effects of price supports on agricultural outputs contrast with the effects of input subsidies and government research. Agricultural price supports encourage output and the use of all inputs. They shift the agricultural value function upward throughout its range—as in Figure 4.3B—and induce agricultural expansion onto some of the degraded open access lands for regions in the second stage of forest development. For regions in the third stage of forest development, they improve agriculture's ability to compete with the intensive margin of managed forestry and, therefore, induce the conversion of some land away from forest management.[22]

Some countries have also introduced incentives to remove land from agricultural production. Some of this land, no doubt, reverts to forest. In fact, the U.S. agricultural Conservation Reserve Program specifically subsidizes the reversion of cropland to forest. Ireland, in another example, has received subsidies from the European Union to encourage the removal of some land from agricultural production (Forest Service, Government of Ireland [GOI] 2000). China's sloping land program offers tree seedlings, cash, and grain as incentives to remove steep lands from crop and livestock use. More than seven million hectares were enrolled in the first five years of China's program (Hyde 2003; Xu et al. 2004). Therefore, in the United States and some other countries, while some public programs support the conversion of forest to cropland in some parts of the country, other programs support the opposite, cropland reversion to forest, in other parts of the country.

The mixture of effects of agriculture policy is further complicated by the variability of programs between agricultural products. For example, in the United States wheat may be a greater beneficiary of output price supports while cotton and livestock may be greater beneficiaries of input (water and grazing) subsidies. The total affects of U.S. agricultural programs cause some products, such as milk, cotton, and peanuts to be produced in regions where they would never appear without the government assistance. In all cases, the pattern of effects traces through the entire agriculture sector, with government assistance to higher-valued crops affecting the land use margins those crops share with lower-valued crops and their margins with still lower-valued corps until eventually the shifts affecting some lower-valued agricultural product alter its ability to compete for land with either the degraded open access forest (stage II) or the managed forest (stage III). The net effect of agricultural policies on the forest of any one country

---

22 A few developing country governments have introduced contrasting policies, importing agricultural products and holding food prices artificially low in an effort to support urban populations and encourage industrialization (Timmer 1986). These food price controls decrease the value function for local commercial agriculture and, with it, decrease commercial agriculture's demand for labor. In these cases, for regions in stage I and stage II, some unemployed farm workers return to more rudimentary subsistence agricultural activities that are more land-using and that compete for land with open access degraded areas and the natural forest.

remains an unanswered empirical question. In many cases, it would be difficult even to speculate on the direction of the net effect without detailed analysis. Nevertheless, we can anticipate that, for many countries, agricultural policy does alter both the land area in forest and the standing forest volume.

Spillovers from Policies Designed for their Impacts on the Wood Products Industry

Policies that target the wood products industries affect the demands of those industries for their raw materials, and through these demands, they also affect forestry and the forest itself.

The common objectives of these policies are the protection of the domestic industry from international competition and the promotion of its development—although controls on its contribution to environmental pollution can also be important, especially for the pulp and paper industry. Log export bans are the common means of protecting the processing industries. Their effects are similar to the effects of restrictions on domestic timber shipments discussed previously. A policy to restrict exports removes international competition from the domestic market. With only domestic processors competing for timber, demand declines and domestic prices shift downward. This is a benefit for the domestic wood processing industry but it is a disincentive for forest management. The net forest value function V_f shifts downward throughout its range—as in Figure 4.4. Harvests at the natural forest frontier decline—for all three stages of forest development—and both the area in managed forestry and the level of managed forest production decline for regions in the third stage of forest development.

Several Asian countries have imposed log export bans and both the United States and Canada have banned the export of certain classes of logs from the

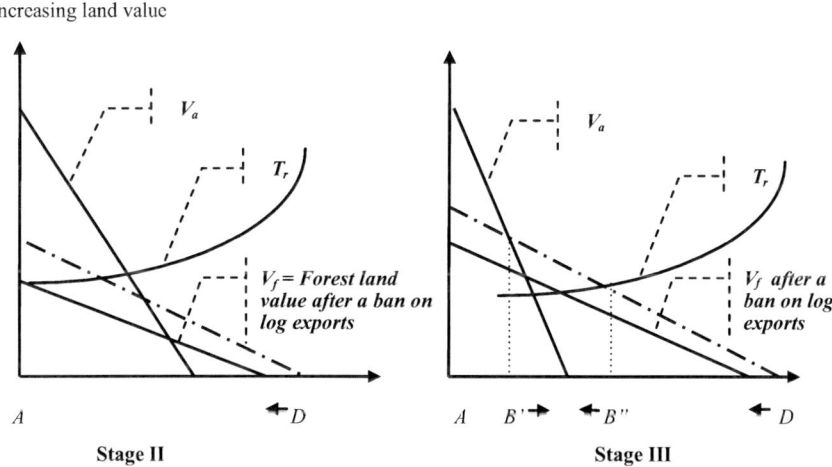

FIGURE 4.4 Policies targeting the wood processing industries

Pacific Northwest. Indonesia's experience provides a good example. Indonesia's log export ban was intended to promote domestic forest-based industrialization and employment. It did help accomplish this objective, especially for the plywood industry. Indonesia was the world's largest exporter of tropical hardwood logs in 1980, exporting 60 percent of its total harvests. Gradually tighter restrictions on log exports, beginning from the early 1980s, extended to a total ban in 1985.[23] The domestic wood processing industry expanded immediately. Both plywood and sawnwood exports increased from a very low level in the mid-1970s until their exports accounted for more than half of Indonesia's total production of processed wood by the mid-1980s. Indonesia became the world's largest exporter of hardwood plywood and wood products became the third largest source of Indonesia's export earnings prior to the East Asian financial crisis in 1997 (ITTO 2004; BPS Statistics Indonesia 2005).

However, growth in the plywood industry came at a cost. Domestic log production tapered off and domestic log prices fell relative to world prices throughout the 1980s and 1990s. Fitzgerald (1986) estimates that for every dollar Indonesia gained in the value of plywood exports, it lost four dollars in foregone log exports. The employment gains in the wood processing industry were less than the losses in employment in the logging industry (GOI 1985, 1997). The declines in timber demand and price coincided with a reduction in harvests by as much as 50 percent. Most of the reduction in harvests must have occurred at the frontier of natural stands as plantations account for only 3 percent of Indonesia's total forest.

Government sponsored research in wood processing, as in agriculture, has been another source of industrial development—and a source of reduction in the demand for wood as a raw material. For example, U.S. government sponsored research for the southern pine plywood industry produced rates of return in the neighborhood of 300 percent per annum throughout the 1960s and 1970s (Seldon 1987; Seldon and Newman 1987). One research improvement alone, the powered back-up roller, improved wood utilization and reduced the plywood industry's demand for logs by 17 percent. Overall research-induced cost reductions in southern pine plywood were a crucial component in that industry's expansion from three to 66 establishments between 1964 and 1981. This cost containment was the source of a sharp increase in demand for southern pine logs. However, southern pine plywood is a close substitute for western pine plywood and also a substitute for lumber in some of their common final use as construction material. The processing of southern pine plywood typically utilizes more of the log than either of the former products. Furthermore, the southern pine region is in the third stage of forest development with a significant share of its timber originating from managed forests, while forests in the western United States are more generally characterized by the second stage of forest development, and a much larger share of western harvests originate from mature natural forests. For all of these

---

23 The ban on log exports was removed in 1992. A prohibitive log export tax replaced the ban—with the same net effect.

reasons, we can be fairly certain that, while government sponsored southern pine research decreased costs and precipitated an increase in the demand for southern pine timber, it also had a conserving effect on the total U.S. demand for logs and, in particular, on harvests at the (mostly western) natural forest frontier.

Finally, pollution control policy may have an effect on the demand for logs and, therefore, on the net forest value function and the forest itself. Pollution control policy shifts marginal productive effort away from conventional capital, labor, and raw material inputs and toward the means of pollution abatement. This means that pollution control policy decreases the demand for logs and decreases the impact on the forest—for both managed and natural forests in all stages of forest development. However, the effect is probably small. The raw material of pulp and paper production in developed countries is composed of many wood substitutes (old corrugated board, rags, chip by-products of lumber production, etc.) as well as solidwood. In many developing countries, agricultural residues are yet another substitute for wood as a raw material for pulp and paper. China, for example, aggressively controls pollution in the pulp and paper industry with both regulations and economic instruments, but agricultural residues are three-fourths of total raw material inputs for China's industry (Xu et al. 2003). The effect of pollution control policy on the forests that supply China's pulpmills must be small when wood is such a small share of total raw material for the industry and when the opportunity for substituting other raw materials is so great.

Spillover Effects from Adjacent Sectors—a Conclusion

In sum, spillover effects on the forest from policies intended for their primary effects on agriculture and the wood products industries vary in impact. Agriculture and population policies have had important effects on forests within regions in the first stage of forest development—inducing substantial movement into the forest frontier in some cases. Agricultural policies have had mixed effects on regions in the third stage of development. Many individual agricultural policies have resulted in intensified production on fewer hectares, thereby reducing agriculture's competition with managed forestry. However, some (output price supports in particular) have also induced agriculture's expansion and competition for land with forest management. Therefore, it is difficult to speculate on the combined effects of all agricultural policies on forest management during the latter stages of forest development. Nevertheless, the very great magnitude of many countries' public expenditures on agricultural programs makes their impact a truly important question. Chapter 6 will revisit this question once more during its consideration of macroeconomic and trade policy effects on the forest.

On the other hand, we can speculate that policies directed at the forest products industries, while also mixed, have generally had forest conserving effects, increasing the utilization of each unit of wood consumed, decreasing the total use of wood as a raw material, and decreasing the demand for wood from both managed forests and the forest frontier. This observation has a very interesting environmental implication. That is, policies intended to promote domestic wood products industries may induce net decreases in the demand for timber and in the level of timber harvests. In this event, these policies would have favorable environmental effects. Among these, the government research investments that have led to wood-saving improvements in the manufacture of wood products may offer an inviting alternative to the many environmental regulations of the forest itself that are more difficult to monitor and enforce.

Infrastructure and Institutions

Most extensions of the local infrastructure and most modifications to the local and national institutions are results of public agency activity or public policy decisions. Neither is considered to be a component of forest policy but both can be important determinants of forest development.

Infrastructure

The term "infrastructure" includes public utilities and public services like education, hospitals, and transportation and communication networks. Antle (1983) convincingly shows that the full collection of items identified with infrastructure has a positive effect on rural economic development in general. Specifically, he shows that investments in infrastructure have a statistically significant output elasticity in the range of 0.20 to 0.25 for agricultural production in a sample of 66 countries. For forestry, roads may be the most important component of infrastructure. We can anticipate that improved roads have an important effect on the general condition of natural forests. They have a smaller effect on managed forests, one that is comparable to their effect on agriculture because, like agriculture, managed forests occur in regions that already have at least limited roads.

In the earliest stage of forest development, the community and the forest are in the closest proximity. All roads affecting the community also affect both agricultural development and the natural forest. As roads improve accessibility, they increase local agricultural and forest values, shifting the right-hand extremity of functions for both values to the right. As a result of the increase in value, local farmers convert some forest into permanent agricultural land and they or others in the local community also degrade and deforest additional land—as in the first diagram of Figure 4.5A.

In the second and third stages of development, roads that extend to the forest continue to have a direct effect on the forest. The improved access they provide makes the region's land more valuable in all uses. Once more, the extremities of the agricultural and forest value functions shift right, moving the important land value margins (*B* and *D* in stage II; *B', B"* and *D* in stage III) with them. In stage II these roads extend the claims of permanent agriculture and shift the entire degraded open access area farther into the geographic interior. In stage III, they extend agriculture into the area of previously managed forest and extend managed forestry into the area of previously degraded forest. The natural forest frontier recedes farther into the interior and the deforested area expands in both the second and third stages of forest development.

Increasing land value

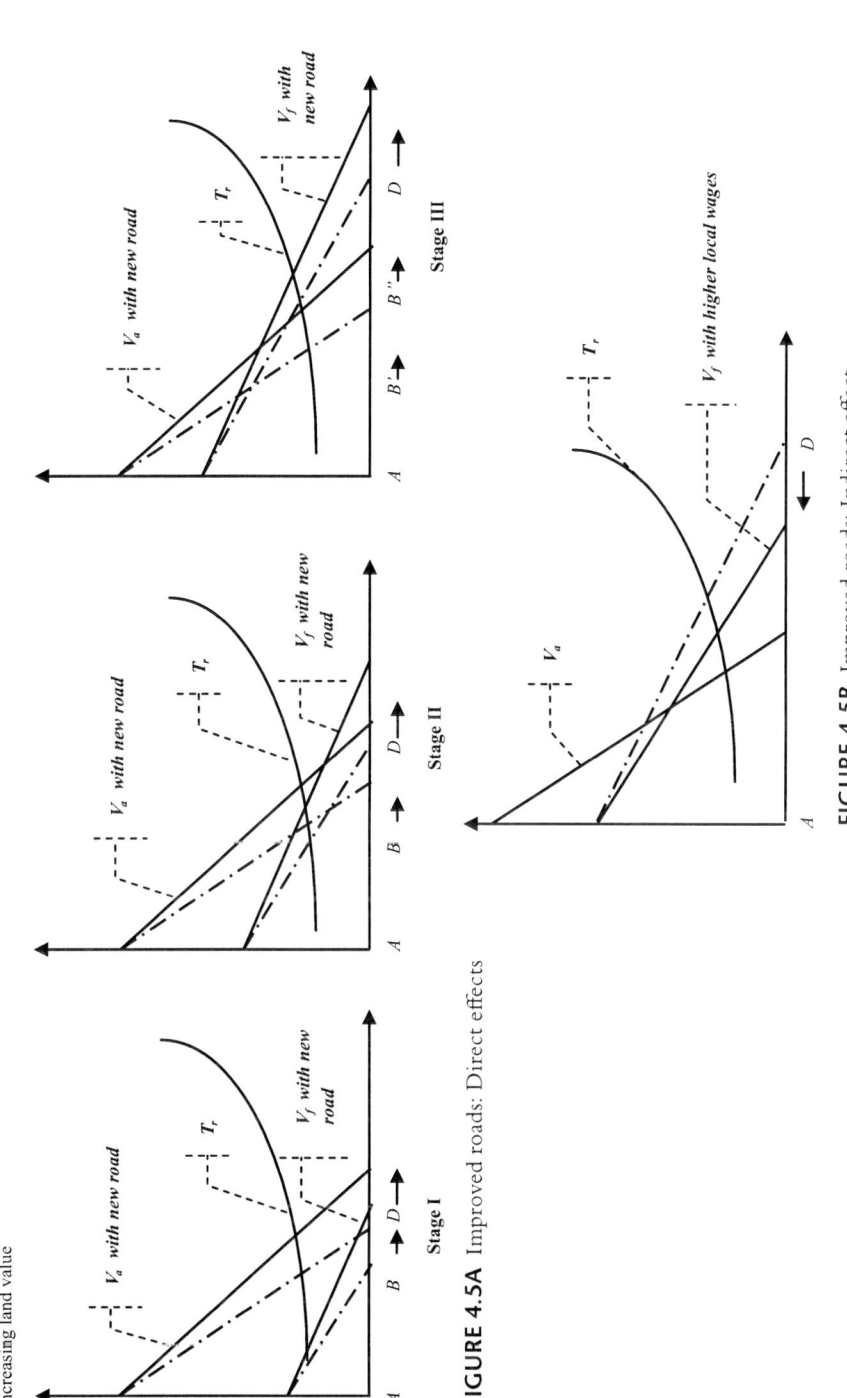

FIGURE 4.5A Improved roads: Direct effects

FIGURE 4.5B Improved roads: Indirect effect

The examples are numerous. In the United States, timber consumption expanded simultaneously with the extension of narrow gauge railroads into the Rocky Mountains in the late nineteenth century. Rail extension also made vast areas of timber accessible for logging in the South in the early twentieth century—as we discussed in chapter 3. Thailand provides a more recent example. Thailand built roads into its sparsely inhabited northeast in the 1960s. The policy objective was security—military access and encouragement for Thai settlers to secure the region against encroachment from Laos and Cambodia during the Vietnam War. Of course, timber harvests followed.

In fact, the rights to adjacent lands and timber are often part of a government's payment to private roadbuilding contractors. In the nineteenth century the U.S. government transferred alternate square-mile patches of public domain for ten miles either side of some railroads as inducement for their construction. The government of Laos made a similar transfer of timber rights in the late 1990s in exchange for an improved road built through its northeastern forests to provide access to the rapidly developing markets of southern China (Hyde and Kuuluvainen 1995).

Roads can also have an indirect effect on the forest through their impact on overall regional development. By the second and third stages of development, some new or improved roads may extend only to the rural community and not to the forest itself. These roads improve access to external markets for local products and local labor. Local agricultural and forest values may increase as a result but labor is the critical factor. Some labor finds new employment in the locally higher-valued agriculture. Some finds greater employment opportunities out of the region and, with better roads, these opportunities becomes more accessible. As the local labor supply declines, local wages must increase. With higher wages, the forest value function's intersection with the horizontal axis shifts inward and the boundary of extraction from the remaining natural forest shifts inward as well—as we explained in chapter 2. The areas of degraded forest and deforestation decline and some natural forest recovers—as in the stage III example in Figure 4.5B.

Bluffstone (1995) first illustrated the indirect effect of external labor opportunity on deforestation in Nepal. Others (Gunatilake 1998, Foster et al. 1997; Escobal and Aldana 2003; Tachibana, Nguyen, and K. Otsuka 2001; Rudel Perez Lugo, and Zichal 2000; Lamb, Erskine, and Perotta 2005; Fisher and Shively 2005) have confirmed it for Sri Lanka, India, Peru, Vietnam, Puerto Rico, and Malawi. The U.S. experience in the twentieth century supports it. At the same time the United States industrialized and urbanized, its highway system improved, its rural population declined, and its forest stock increased. (We discussed the effect on the southern forest industry in chapter 3.) The Philippine experience in the 1980s and 1990s supports the indirect effect of labor opportunity from the opposite perspective of a declining economy. As the Philippine economy suffered, wages declined, the urban share of the Philippine population declined, the rural population increased, and subsistence agriculture extended its competition with the remaining natural forest farther into the uplands (Amacher et al. 1998).

In sum, the design of rural roads is an important question for forestry. Roads that extend to the forest frontier, and even beyond, extend the range of forest extraction in all three stages of forest development. In the second and third stages of forest development, road improvements that extend only to the local human community, and not beyond it to the forest, can actually reduce the pressure of harvest activities at the forest frontier.

Institutions

Like infrastructure, the institutional arrangements for local exchange, and particularly the arrangements for local property rights, have an effect on forest development. In developed countries, we tend to think of property rights in terms of formal titles to land and other resources. However, Feder et al. (1988) use an example from Thailand to demonstrate that formal titles are insufficient protection unless the titles are enforced and also transferred easily when higher-valued uses of the property present themselves. On the other hand, Migot-Adholla et al. (1991) use examples from sub-Saharan Africa to show that formal titles are unnecessary where customary rights permit transfers and the local community respects the rights and the transfers.

Both points are important in forestry today, as many countries find that formal rights to those large areas of natural forest that tend to be the official responsibility of the central government and its forest ministry are not easily enforced. On the other hand, they also observe that some local communities protect their forests even when they have few formal rights to these resources. In fact, these two considerations provide a justification for the widespread discussion of transfers of public forestland to some form of local private or local community-based forest management. The familiar terms "community forestry," "joint forest management," "public participation," and "devolution" all refer to transfers to local users of some or all of the rights to forests that were previously the unambiguous responsibility of the central government's forestry agencies.

The economic argument for these transfers is that local users of the land and forest resources know the resources and the demands on them better than the officials of the forest ministry. Local users live closer to the forest and their daily activities put them in closer contact with it. They also know the other users of the forest and their habits and objectives. They can manage the resources more efficiently than the ministry and they have greater success than the ministry in enforcing their management objectives.[24] In terms of our figures, the shift to local

---

24 In Mexico, for example, 80 percent of the forests are managed by approximately 8,000 local agrarian communities. The deforestation rate on these community forests is comparable to the very low rate in nationally protected areas. This contrasts with a nation-wide deforestation rate of 1.1 percent annually occurring almost entirely from the remaining 20 percent of Mexico's forest managed by the national forestry agency (Bray, Merino-Perez, and Barry 2005, FAO/UN 2001). In partial contrast, Palmer and Engel's (2007) survey of 60 communities in Indonesia's East Kalimantan, found that "many communities engaged in self-enforcement both before and after decentralization" and that "little evidence exists" of either financial or environmental gains from decentralization.

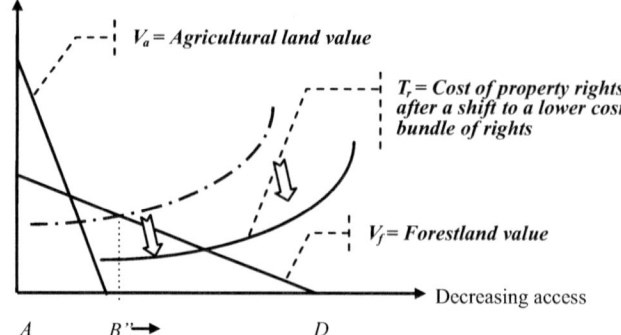

FIGURE 4.6 Improved property rights

management reduces the level of the transactions cost function T_r as in Figure 4.6, more land becomes sustainably managed as a result (points B or B'' shift to the right), and the area of degraded open access forest declines. The value function V_f may also rise (not shown) shifting points B or B'' even farther to the right.

These arguments and the transfer to local management work well where the forest values at stake are local and shared, and where these local values are relatively large enough to justify management of the resource. Where the values are shared by most of the community, local management can improve long-term land management for agriculture, timber and other extractive products of the forest, and also for local non-market values such as some erosion control activities and some aesthetic forest use. The list of successful examples of community management is almost endless and it comes from all parts of the world.[25] Nevertheless, local property rights and local management do have their limits.[26]

Successful transfers to local management: The most successful transfers to local management are those that establish substantial net benefits for local stakeholders while avoiding any serious externalities due to the local management. The examples of successful arrangements cover the spectrum from individually-held private property to group management by collectives of local individuals or households to more widely held community management, and from short-term

---

25 For example 25 percent of Nepal's forest area (1.19 million hectares) has been transferred to 14,227 community forest user groups, joint forest management committees manage 17 million hectares of forest in India, and the community forest program in Thailand covers a smaller 200,000 hectares for the benefit of 5,331 communities (Chhetri 2006; Bahuguna 2004; Wichawutipong 2005). White and Martin (2002) estimate that 22 percent of developing country forests operate under community ownership or management. Also see Landell-Mills and Ford (1999).
26 Stroup (2000) summarizes the conceptual arguments for institutional arrangements at any level of collective decision, whether central government or local committee. He also warns that less-than-perfect success at one level of government does not necessarily imply greater success at another level, and that it certainly does not rule out the possibility the even imperfect private market activity can be better than even less perfect collective action.

use rights for certain characteristics of the resource to permanent and transferable rights to the land and all of its resources.

Collective action seems to be a preferred solution of some development practitioners and there are numerous good examples to support their preference. However, while it is clear that examples of successful transfers from central to local responsibility exist for a spectrum of local arrangements for the property rights, it is also clear that the most successful transfers do not always involve widely-held and administered community activity.[27]

The obvious summary factors for successful management are contained in the net value function V_f and the cost of property rights T_f. Net values are greater near the existing limits of effective management (B or B'' in our figures), and we can anticipate that transfers of lands near these limits will be managed more successfully than transfers of less accessible lands nearer the frontier and beyond. Furthermore, larger net values will be assured over time if the transfers include the right for local managers to make subsequent transfers of indefinite term. This right assures that the current user obtains the highest value from the choice between either using or exchanging the resource. Therefore, it assures that the resource itself is used in its highest-valued activity.

The costs associated with the property rights depend on the institutional arrangements and different institutional arrangements are associated with different costs. The least-cost arrangement is generally the most effective. It takes greatest advantage of the scale economies of management. Since forestry is a relatively low input activity, its greatest scale economies often have to do with monitoring and enforcing boundaries. For this reason individuals with numerous smaller or less accessible plots often form associations to manage a single combined plot—with a smaller perimeter per unit of total area and smaller unit costs for fencing and monitoring. One example is recreational home associations in forested areas of developed countries that form associations for protection against vandalism or for fire protection. Another example comes from China's experience following its rural reforms in the early 1980s. When lands in China's forest collectives were redistributed for use by individual households, some received as many as six separate forested plots totaling an average of only 0.5 hectare. Households in some regions responded by forming associations to manage their combined forested plots more efficiently (Liu and Edmonds 2003). Various examples of successful collective management of less accessible grazing lands in southern Africa are another example (e.g., Runge 1981).

Collective management also succeeds for those more accessible plots from which individuals or households each obtain relatively smaller, occasional, or irregular benefits but from which the full community receives regular and significant benefits. Many centrally located community parks in almost every country in the world

---

27 For example, Bluffstone, Boscolo, and Molina (2007), with evidence from 32 communities in Bolivia, find "difficulty" in identifying any generalizable factors explaining successful common property management.

are good examples. Management by a single authority that acts for the members of the community is less costly than either individual management of smaller plots or joint management by a committee of all benefiting members.

We would be ill-advised, however, to assume that an entire local population is always the best association for local management. We would also be ill-advised to presume that community management is always better than individual private management. Certainly there are examples where individual households successfully manage even single trees of relatively higher local value (Fortmann and Bruce 1988). In fact, individual private management may be more common than collective ownership for market-valued forest resources such as timber, horticultural crops, bamboo, and fuelwood. Private managers may be more likely to form collective associations for only a subset of the activities related to their management. They may cooperate in the purchase and use of expensive manufactured capital like tractors, or in the conduct of specialized activities like fire control or the marketing of their products (e.g., Hammett 1994 for Nepal; Hultkranz n.d. for Sweden; RECOFTC 2008 for Laos). Indeed, it should not surprise us that this is also the experience of agriculture—where land ownership tends to be individual and private, but specialized activities like harvesting and marketing are often conducted collectively. Harvesting often requires brief periods of use of expensive equipment and additional labor which either demand the collective participation of many local farmers or contracts with external specialists. Marketing is a specialized skill that is not the comparative advantage of most farmers. Therefore, there can be advantages to a region's farmers for joining cooperatives that jointly market their very similar products. Sunkist is just such a cooperative for Florida's orange growers. However, farmers most everywhere tend to prefer to manage their own land independently throughout the planting and growing process.

In sum, in the presence of the wide diversity of possible local institutional arrangements, policymakers seeking successful local management are well-advised to restrict land transfers from public forestry agencies to locations where local values dominate and to take advantage of local preferences for the arrangement of the transferred property rights. Furthermore, wise policy permits local managers to make subsequent changes in land uses and subsequent transfers of lands and use rights, and to accept the spontaneous development of institutions that respond to local needs.[28]

---

28 Heltberg (2001) and Campbell (1994) support this summary contention with both developing and developed country examples. Heltberg, with evidence from 37 villages in India's Rajasthan, finds that required dependence on prior institutions actually has a negative effect on the likelihood of successful collective action. He argues that local households must be allowed to develop their own preferred institutional arrangements, arrangements that fit the specific issues at hand. Campbell, referring to the experience of Australia's very successful Landcare program, draws a similar conclusion—that local autonomy with regard to both institutional arrangements and management prescriptions is an important prescriptor of successful collective action. Larson (2002) continues the consideration of local institutions. She examines 21 local governments in Nicaragua and concludes that differences in their management capacity, along with other economic and political characteristics, are important predictors of the success of decentralization.

Limitations: Transfers of rights to local communities are less successful where the local values for forest products are low relative to the community's values for other uses of their land and their own time. This is the general case in the first stage of forest development. The forest is plentiful and members of the local community have little interest in using their own scarce resources to protect it. This is also the case for the lands that remain in an open access condition under the best institutional arrangements (the lands between B or B'' and D) in the second and third stages of development).

Local management is also less successful in several other cases within the second and third stages of development. Transfers to local community interest are less successful where important groups within the community compete in their demands on the forest. For example, community management has not been successful for the eroding hillsides around the village of Basantapur in Nepal. The higher income agricultural households in this community prefer to exclude all activity from the forests on these hillsides in order to restrict erosion and downstream damage to their own agricultural lands. The poorer landless households in the same community rely on the same hillside forests for fuelwood. Higher income households in Basantapur formed a forest protection committee and hired forest guards, but the guards have been unable to prevent fuelwood collection and further degradation of the forest by their trespassing poorer neighbors (Dangi and Hyde 2000).[29]

Some system that allows sharing of the forest might work for Basantapur, but the arrangements for successful sharing are not always simple, particularly when the sharing is between a local and a more distant authority. An example from Honduras illustrates. A central government agency (COHDEFOR) maintains the rights to Honduras' timber but it allows associations of local rubber tappers to collect latex and resins. Tapping retards tree growth, thereby reducing the volume of mature timber at harvest time. COHDEFOR established guidelines that limit the tapping of younger trees in an attempt to minimize the effect on harvest

---

29 Heltberg's (2001) experience in Rajasthan is similar but more specific. He argues that "efforts to improve collective management cannot be confined to the poorest households." Agrawal and Gupta (2005) restate this problem in a positive form. They show that successful decentralized governance of common pool resources in Nepal's terai depends on broad participation. However, they emphasize that the likelihood of participation in common activity is greater for those who are economically and socially better off—and that marginal households tend to become disenfranchised. Kumar (2002) makes the related points that joint forest management (India's version of community forestry) reflects the preferences of the rural *non*-poor, and the poor have been net losers over a forty-year period. Bwalya (2008) adds support for these arguments with evidence from Zambia. He shows that participation in community forest activities is greater for better-off households and for households with greater dependence on the forest. On the other hand, he urges caution in accepting community diversity as a negative factor in participation. Some forms of diversity (ethnic diversity in Bwalya's sample) may add breadth to local experience and actually increase participation and successful management, while others (religious diversity in his sample) may decrease participation. Sikor and Nguyen (2007) provide further evidence, from Vietnam's central highlands, that decentralization and community management may not benefit a region's poorer households.

volume, but the guidelines are difficult to enforce and the tappers have no incentive to follow them. COHDEFOR is unwilling to allow the tappers a share in the timber harvest revenues that could provide that incentive. As a result, the tappers extract latex and resins from younger trees and for longer than optimal periods, tree growth is deterred, and the forest yields less than the maximum combination of timber and latex (Johnson 1998).

When sharing between two very different groups does work, as it may in some examples of joint forest management in India, Kant (1996) shows that the shared arrangement depends on the levels of shared values at risk. There is no simple formula and a uniform formula for sharing between the government authority and all individual communities is not optimal. Successful arrangements must vary from community to community just as the shared values at risk also vary.

In addition to these limitations, local management may be no more successful than management by the forest ministry where either the local property rights are incomplete or the policy environment is uncertain. Incomplete property rights are common in forest transfers as the central government's ministries often prefer to maintain some degree of oversight—but succeed only in removing the incentives for local management. The Philippines provides an example of the former, incomplete rights. The Philippine Bureau of Forest Development (BFD) established a policy of land transfers to communities in 1994. The policy was supported by a US$40 million loan from the Asian Development Bank. However, various public interest groups (NGOs) were concerned about the potential distribution of benefits from the transfers and the BFD was concerned that the communities would fail to follow its own perception of good forest management. Therefore, the BFD required each community to hire a forester, to report management plans to regional and central advisory committees for approval, and to return 44 percent of gross revenues from the community forests to a central account used to repay the loan. These requirements were too severe and not one community in the entire country applied for the transfer of community forests until these requirements were withdrawn in 1997 (Hyde et al. 1997).

Yin and Newman (1997) illustrate the latter problem, an uncertain policy environment, by contrasting experience from two regions in China. Following the introduction of agricultural reforms in 1978, China's authorities gradually extended to individual farm households the rights to manage and harvest trees on the lands of China's former farm and forest collectives. In the north central plains, the authorities paid little attention to forestry and farmers initially harvested the few remaining trees as they obtained the new rights, but they also planted and both the standing forest and the level of harvests increased (by eight percent and almost 20 percent annually, respectively) over a period of only 10 years. In a second region just south of the first, the authorities gave land use rights to farmers and then, subsequently, rescinded these rights for some farmers. In fact, the authorities had changed the system of farmers' rights an additional three times in the 20 years preceding reforms. Farmers in this southern region responded to the uncertainty about the permanence of their rights by harvesting just as soon as they received the right—and not subsequently reforesting. Both the level of the standing forest and the level of successive years' harvests declined over several years despite the fact that the official rights to the land and trees were similar for both regions by 1985.[30] In sum, uncertainty was the difference between the two regions. The negative effect of uncertainty on sustainable production in the southern region overwhelmed the apparent advantage of land use transfers from farm and forest collectives to household management.

Finally, local management is never a successful substitute for regional or national management, or even global assistance, when the values at stake are public values that are shared by the broader regional, national, or global community. Carbon sequestration, biodiversity, and some classes of tourism are all in the global interest. In these cases, broader institutional management may be necessary to assure that broader interest, but local involvement is also necessary—in order to insure local cooperation instead of local trespass in search of locally valued extractive products. Numerous national parks in both developed and developing countries successfully include local participation in the benefits of the park (e.g., providing various tourist services and also employment in park management itself). Even in these cases, however, some amount of trespass is inevitable. We can anticipate that creating a local interest in sequestering carbon or protecting biodiversity will be even more difficult.[31]

What should be clear, but often is not, is that the institutional arrangement that works best is the arrangement that internalizes to its own operation the management incentives for the largest portion of all values at risk. For global treasures, like unique habitats or high profile national parks, national or even global institutions are appropriate, but either local values must also be included in the management prescriptions or the costs of monitoring and enforcing management prescriptions will be high. For local values, like erosion control on a small watershed, local institutions are more effective.

Conclusion

This chapter has summarized the policy instruments and the policies themselves that commonly affect forestry. It began with a discussion of direct forest policy

---

30 In a second example of uncertainty originating with the local forestry institutions, a Philippine Minister of Natural Resources and the Environment once explained that the forest industry was not to be trusted on the environment. In his opinion, the best way to manage the industry was to negotiate one arrangement today and then, after a short period of time, to re-open the negotiations and to insist on a more limited settlement. Of course, this process creates uncertainty for the industry. As a result loggers will extract what they can whenever they have the opportunity and with little expectation of future opportunity. Reinvestment in the industry will lag, and older and less efficient equipment will eventually dominate. The minister's own policy insured short-sighted behavior and poor environmental performance.

31 Songorwa (1999), referring to wildlife management in Tanzania, provides an example of the difficulties inherent in organizing effective local participation in global resource, and Fischer, Muchapondwa, and Sterner (2005) also referring to the experience of park and wildlife management in Tanzania, express the general caution that resource sharing does not always result in either improved management or greater net benefit for the local community.

instruments—taxes, financial incentives, and regulations—and continued with a discussion of policies and other public actions external to the forest sector that often spillover to affect forests and forestry.

The diversity of policies and policy impacts is great. This chapter provides a catalogue of them and there is no need to repeat it here. It is clear that many modern forest policies either have minimal impacts or even impacts that are opposite to the original policy objective, just as many others have unintended and unexamined effects on lands that were not forests before the policy or on lands that the policy causes to be withdrawn from forestry. Some have unexamined effects on forests and environments away from the targeted forests and even in other countries, on other forests, and even on non-forest environments that provide the products that substitute for the products of the forest that was targeted by policy. It is also clear that policies and programs designed for other sectors and other purposes can have large impacts on forestry. These are perplexing observations.[32] They must leave us confused about just what we can do about the objectives of modern forest policy. The appendix to this chapter speculates on five of these objectives.

Literature Cited

Agrawal, A., and K. Gupta. 2005. Decentralization and participation: the governance of common pool resources in Nepal's terai. *World Development* 33(7): 1101–1114.

Amacher, G., W. Cruz, D. Grebner, and W. Hyde. 1998. Environmental motivations for migration: population pressure, poverty and deforestation in the Philippines. *Land Economics* 74 (1): 92–101.

Amacher, G., W. Hyde, and M. Rafiq. 1993. Local adoption of new forestry technologies: with an example from Pakistan's Northwest Frontier Province. *World Development* 21(3): 445–454.

Antle, J. 1983. Infrastructure and aggregate agricultural productivity: international averages. *Economic Development and Cultural Change* 31(3): 609–619.

Bahuguna, V. 2004. Root to canopy: An overview. In V. Bahuguna, K. Mitra, D. Capistrano, and S. Saigal, eds., Root to canopy—Regenerating forests through community state partnerships. New Delhi, India: Commonwealth Forestry Association and Winrock International, pp. 15–24.

Bass, S., K., Thornber, M. Markopoulos, S. Roberts, and M. Grieg-Gran. 2001. *Certification's impacts on forests, stakeholders and supply chains.* London: International Institute for Environment and Development.

Binswanger, H. 1991. Brazilian policies that encourage deforestation in the Amazon. *World Development* 19: 821–829.

Bluffstone, R. 1995. The effect of labor market performance on deforestation in developing countries under open access: an example from rural Nepal. *Journal of Environmental Economics and Management* 29(1): 42–63.

Bluffstone, R., M. Boscolo, and R. Molina. 2007. Does better common property forest management promote behavioral change? On-farm tree planting in the Bolivian Andes. Unpublished Manuscript, Economics Department, Portland State University, Oregon.

Boyd, R. G., and W. F. Hyde. 1989. *Forestry sector intervention: The impacts of public regulation on social welfare.* Ames: Iowa State University Press.

Boyd, R. G., and J. Turnbull. 1989. The impact of greenspace laws on urban development. *The Review of Regional Studies* 19(2): 20–39.

---

32 Indeed, Persson (2003, p. x), in a review of his own extensive experience with international programs concludes that the history of forestry assistance is "not impressive."

BPS Statistics Indonesia. 2005. Foreign trade sector webpage: http://www.bps.go.id/sector/ftrade/tables.shtml (accessed July 17, 2008).

Bray, D., L. Merino-Perez, and D. Barry (Eds.).2005. The community forests of Mexico: Managing for sustainable landscapes. Austin: University of Texas Press.

Bressers, H., and K. Lulofs. 2004. *Industrial water pollution in the Netherlands: a fee-based approach.* In W. Harrington, R. Morgenstern, and T. Sterner, eds., Choosing environmental policy: Comparing instruments and outcomes in the United States and Europe. Washington, DC: Resources for the Future Press, pp. 91–116.

Buongiorno, J. S. Zhu, D. Zhang, J. Turner, and D. Tomberlin. 2002. *The global forest products model.* Rome: Food and Agriculture Organization of the UN.

Bwalya, S. 2008. Forest dependence, socio-cultural heterogeneity and participation in joint forest management in Zambia. Working paper, Economics Department, University of Zambia.

Campbell, A. 1994. Landcare: *Communities shaping the land and the future.* St. Leonards, Australia: Allen & Unwin.

Can, L., W. Sen, Z. Wei, and L. Dan. 2007. Compensation for forest ecological services in China. *Forestry Studies China* 9(1): 68–79.

Chhetri, R. 2006. From protection to poverty reduction: a review of forest policies and practices in Nepal. *Journal of Forests and Livelihoods* 5(1): 66–77.

Dangi, R., and W. Hyde. 2001. When does community forestry improve forest management? *Nepal Journal of Forestry* 12(1): 1–19.

Daniels, S., W. Hyde, and D. Wear. 1991. The distributive effects of Forest Service attempts to maintain community stability. *Forest Science* 37(1): 245–260.

Durst, P., T. Waggener, T. Enters, and L. Tan. 2001. *Forest out of bounds: Impacts and effectiveness of logging bans in natural forests in Asia-Pacific.* RAP publication 2001/08. Bangkok, Thailand: Food and Agriculture Organization of the UN,

Escobal, J, and U. Aldana. 2003. Are nontimber forest products the antidotes to rainforest degradation in Madre de Dios, Peru. *World Development* 31(11): 1873–1877.

Feder, G., T. Onchan, Y. Chalamwong, and C. Hongladarom. 1988. *Land ownership security, farm productivity, and land policies in Thailand.* Baltimore: Johns Hopkins University Press.

Feder, G. R. Just, and D. Zilberman. 1985. Adoption of agricultural innovations in developing countries: a survey. *Economic Development and Cultural Change* 33: 255–297.

Fischer, C., E. Muchapondwa, and T. Sterner. 2005. Shall we gather around the campfire? Zimbabwe's approach to conserving indigenous wildlife. *Resources* 158 (Summer):12–15.

Fisher, M., and G. Shively. 2005. Can income programs reduce tropical forest pressure? Income shocks and forest use in Malawi. *World Development* 33(7): 1115–1128.

Fitzgerald, B. 1986. An analysis of Indonesian trade policies: countertrade, downstream processing, import restrictions and the deletion program. Discussion paper SPD 1986-22. Washington, DC: World Bank.

Flick, W.A., R. Tufts, and D. Zhang. 1996. Sweet home as forest policy. *Journal of Forestry* 94 (4): 4–8.

FAO/UN (Food and Agriculture Organization of the United Nations). 2001. *Global forest resources assessment 2000.* FAO forestry paper 140. Rome: FAO.

FAO/UN. 2004. *Forest products: 1998–2002.* FAO forestry series no. 37. Rome: FAO.

Forest Service, Government of Ireland (GOI). 2000. *Afforestation grant and premium schemes.* County Wexford, Ireland: Department of the Marine and Natural Resources.

Forsyth, K. 1998. *Certified wood products: The potential for price premiums.* Edinburgh, UK: LTS International.

Foster, A., M. Rosenzweig, and J. Behrman. 1997. *Population and deforestation: Management of village common land in India.* Draft manuscript, Department of Economics, University of Pennsylvania, Philadelphia.

Fortmann, L., and J. Bruce (Eds.). 1988. *Whose trees? Proprietary dimensions of forestry.* Boulder, CO: Westview Press.

Garcia Garcia, J. 1997. *Rural markets and local institutions in Indonesia: the case of the provinces of Jambi and East Nusa Tanggara.* Jakarta: report to the World Bank. Washington, DC: World Bank.

Godoy, R. 1992. Determinants of smallholder tree cultivation. *World Development* 20(5): 713–725.

Government of Indonesia. 1985, 1997. *Statistic industri: Besar dan Sedang.* Jakarta, Indonesia: Budan Pusat Statistic.

Gunatilake, H. 1998. The role of rural development in protecting tropical rainforests: evidence from Sri Lanka. *Journal of Environmental Management* 53: 273–292

Gunatilake, H., and L. Gunaratne. 2001. *An assessment of alternative policies for conservation of natural forests in Sri Lanka.* Unpublished report (submitted to the Environmental Economic Program for South-East Asia) Department of Agricultural Economics, University of Peradeniya, Sri Lanka.

Hammett, A. 1994. Developing community-based market information systems. In J. Raintree, and H. Francisco, eds., *Marketing of multipurpose tree products in Asia.* Proceedings of an international workshop held at Bagio City, the Philippines, Winrock International. Bangkok, Thailand. pp. 289–300.

Hansen, E., and H. Juslin. 1999. The status of forest certification in the ECE region. Geneva timber and forest discussion papers. UN Economic Commission for Europe discussion paper 14. Geneva: ECE

Heltberg. R. 2001. Determinants and impact of local institutions for common resource management. *Environment and Development Economics* 6: 183–208.

Heydir, L. 1999. Population-environment dynamics in Lahat: Deforestation in a regency of South Sumatra province, Indonesia. In B. Baudot and W. Moomaw, eds., *People and their planet: Searching for balance.* New York: St. Martin's Press, pp. 91–107.

Hibbard, B. 1965. *A history of the public land policies.* Madison: University of Wisconsin Press.

Hultkranz, L. Commitment, irreversible investment & the utilization of forest resources: The role of forest owners associations in the development of paper pulp production in Sweden 1959–1985. Arbetsrapport 103. Umea, Sweden: Sveriges Lantbruksuniversitet Institutionen for Skogsekonomi.

Hunt, C. 2001. *Production, privatization, and preservation: PNG forestry.* Unpublished report for the International Institute for Environment and Development.

Hyde, W., B. Belcher, and J. Xu (Eds.). 2003. *China's forests: Global lessons from market reforms.* Washington, DC: Resources for the Future.

Hyde, W., M. Dalmacio, E. Guiang, and B. Harker. 1997. Forest charges and trusts: shared benefits with a clear definition of responsibilities. *Journal of Philippine Development* XXIV(2): 223–256.

Hyde, W., and Y. Kuuluvainen. 1995. *Timber price policy in Lao PDR.* Unpublished report prepared by Helsinki University Knowledge Services for GOL and World Bank.

Hyde, W. 2003. *Natural forest protection program.* Unpublished report prepared for the Chinese Academy of Forestry.

International Tropical Timber Organization (ITTO). 2006. Status of tropical forest management 2005. Yokohama, Japan: ITTO.

Karna, J., E. Hansen, and H. Juslin. 2003. Environmental activity and forest certification in marketing of forest products—a case study in Europe. *Silva Fennica* 37(2): 253–267.

Karna, J, H. Juslin, V. Ahonen, and E. Hansen. 2001. Green advertising: Greenwash or a true reflection of marketing strategy. *Greener Management International* 33: 59–70

Landell-Mills, N., and J. Ford. 1999. Privatising sustainable forestry—a global review of trends and challenges. London: International Institute for Environment and Development.

Johnson, R. 1998. Multiple products, community forestry and contract design: the case of timber harvesting and resin tapping in Honduras. *Journal of Forest Economics* 4(2): 127–145.

Kant, S. 1996. The economic welfare of local communities and optimal resource regimes for sustainable forest management. Unpublished PhD thesis, University of Toronto, Canada.

Kripke, G., and B. Dunkiel. 1998. Taxing the environment. *Multinational Monitor* (September): 9–15.

Kumar, S. 2002. Does "participation" in common pool resource management help the poor? A social cost-benefit analysis of joint forest management in Jharkhand, India. *World Development* 30(5): 763–782.

Kuuluvainen, J., and J. Salo. 1991. Timber supply cycle harvest of non-industrial private forest owners: An empirical analysis of the Finnish case. *Forest Science* 37(4): 1011–1029.

Lamb, D., P. Erskine, and J. Perotta. 2005. Restoration of degraded tropical forest landscapes. *Science* 310: 1628–1632.

Larson, A. 2002. Natural resources and decentralization in Nicaragua: Are local governments up to the job? *World Development* 30(1): 17–32.
Liu, D., and D. Edmunds. 2003. Devolution as a means of expanding local forest management. In W. Hyde, B. Belcher, and J. Xu (Eds.). China's forests: Global lessons from market reforms. Washington, DC: Resources for the Future, pp. 27–44.
Mercer, D. E. 2004. Adoption of agroforestry innovations in the tropics: A review. *Agroforestry Systems* 61(1): 311–328.
Migot-Adholla, S., P. Hazell, B. Barel, and F. Place. 1991. Indigenous land rights systems in sub-Saharan Africa: a constraint on productivity? *World Bank Economic Review* 5(1): 155–175.
Ministry of Forests, British Columbia. 2004. The state of British Columbia's forests. http://www.for.gov.bc.ca/hfp/sof/2004/pdf/sof.pdf (accessed July 7, 2006).
Molnar, A. 2003. *Forest certification and communities: Looking forward to the next decade.* Washington, DC: Forest Trends.
Molnar, A., S. Scherr, and A. Khare. 2003. *Who conserves the world's forests?* Washington, DC: Ecoagriculture Partners.
Ozanne, L., and P. Smith. 1998. Segmenting the market for environmentally certifies wood products. *Forest Science* 44(3): 379–389.
Palmer, C., and S. Engel. 2007. For better or for worse? Local impacts of decentralization of Indonesia's forest sector. *World Development* 25(12): 2131–2149.
Pattanayak, S., D. E. Mercer, E. Sills, J. Yang, and K. Cassingham. 2003. Taking stock of agroforestry adoption studies. *Agroforestry Systems* 57: 173–186.
Persson, R. 2003. *Assistance to forestry: Experiences and potential for improvement.* Bogor, Indonesia: Center for International Forestry Research.
RECOFTC (Regional Community Forestry Training Center). 2008. Is there a future for forests and forestry in reducing poverty? Unpublished working paper. Bangkok, Thailand: RECOFTC.
Rudel, T., M. Perez Lugo, and H. Zichal. 2000. When fields revert to forest: development and spontaneous reforestation in post-war Puerto Rico. *The Professional Geographer* 52(3): 386–397.
Ruiz-Perez, M., B. Belcher, M. Fu, and Xiaosheng Yang. 2003. Forestry, poverty, and rural development: perspectives from the bamboo sector. In W. Hyde, B. Belcher, and J. Xu, eds., *China's forests: Global lessons from market reforms.* Washington, DC: Resources for the Future, pp. 151–176.
Runge, C. 1986. Common property externalities: isolation, assurance and depletion in a traditional grazing context. *American Journal of Agricultural Economics* 63(4): 595–606.
Russakoff, D. 1985. Timber industry is rooted in tax breaks. *Washington Post* (March 24) pp. A2 ff.
Ruttan, V. 1982. *Agricultural research policy.* Minneapolis: University of Minnesota Press.
Schneider, R. 1994. *Government and the economy on the Amazon frontier.* LAC Regional Studies Program Report no. 34. Washington, D.C.: The World Bank.
Schwartzbauer, P., and E. Rametsteiner. 2001. The impact of SFM-certification on forest products in western Europe—an analysis using a forest sector simulation model. *Forest Policy and Economics* 2: 241–256.
Sedjo, R. 1999. Land use change and innovation in US forestry. In R. D. Simpson, ed., *Productivity in natural resource industries.* Washington, DC: Resources for the Future.
Seldon, B. 1987. A nonresidual estimation of welfare gains from research: the case of public R&D in a forest product industry. *Southern Economic Journal* 54 (1): 64–80.
Seldon, B., and D. Newman. 1987. Marginal productivity of public research in the softwood plywood industry: A dual approach. *Forest Science* 33(4): 872–888.
Serôa da Motta, R. 1993. *Policy issues concerning tropical deforestation in Brazil.* Rio de Janeiro, Brazil: IPEA.
Sikor, T., and T. Nguyen. 2006. Why may forest devolution not benefit the rural poor? Forest entitlements in Vietnam's central highlands. *World Development* 35(11): 2010–2025.
Sobral, L., A. Verissimo, E. Lima, T. Azevedo, and R. Smeralsi. 2002. Acertando o alvo 2: consume do madiera e certificacao floresta; no Estado do Sao Paulo. Belem, Brazil: Imazon.
Songorwa, A. 1999. Community-based wildlife management (CWM) in Tanzania: Are the communities interested? *World Development* 27(12): 2061–2080.
Stevens, J. 1978. *The Oregon wood products labor force.* Unpublished manuscript Agricultural and Resource Economics Department, Oregon State University, Corvallis.

Stroup, R. 2000. Free riders and collective action revisited. *The Independent Review* 4: 485–500.
Tachibana, T., T. Nguyen, and K. Otsuka. 2001. Agricultural intensification versus extensification: A case study of deforestation in the northern-hill region of Vietnam. *Journal of Environmental Economics and Management* 41(1): 44-69.
Timmer, P. 1986. Getting prices right. Ithaca, NY: Cornell University Press.
White, A., and A. Martin. 2002. *Who owns the world's forests? Forest tenure and public forests in transition.* Washington, DC: Forest Trends.
Wichawutong, J. 2005. Thailand community forestry 2005. In N. O'Brien, S. Matthews, and M. Nurse, eds., *First Regional Forestry Forum—Regulatory frameworks for community forestry in Asia.* Bangkok: Regional Community Forestry Training Center, pp. 101–129.
Xu, J., W. Hyde, and G. Amacher. 2003. China's paper industry: Growth and environmental policy during economic reform. *Journal of Economic Development* 28(1): 49–79.
Xu, Z., M. Bennett, R. Tao, and J. Xu. 2004. China's sloping land conversion program four years on: Current situation, pending issues. *International Forestry Review* 6(3-4): 317–326.
Yin, R., and D. Newman. 1997. The impact of rural reforms on China's forestry development. *Environment and Development Economics* 2(3): 289–303.
Young, C. 2002. Land tenure, poverty and deforestation in the Brazilian Amazon. Unpublished draft manuscript available from the author at young@ie.ufrj.br
Zhang, D., and W. Flick. 2001. Sticks, carrots, and reforestation investment. *Land Economics* 77 (3): 443–456.
Zhang, D. 2002. *Endangered species act and timber harvesting: The case of red-cockaded woodpeckers.* Unpublished manuscript, College of Forestry, Auburn University, Auburn, Alabama.

Appendix 4A: Contemporary Policy Objectives

The focus of the policy discussion in forestry has shifted over the last forty years. Timber supply to support a domestic industry and domestic economic growth, and its European counterpart, the creation of a strategic timber reserve, were central to the discussions of forest policy in Europe and North America for at least 100 years. The colonial powers on these two continents exported their concerns to government forestry agencies and forestry schools in the developing world until timber supply became the dominant forest policy issue in almost all countries. Tighter budget constraints for the government forestry agencies, the expansion of international trade in wood and wood products, and an increasingly well-off and environmentally aware public have made a difference and the focus of the policy discussion in forestry has changed. The policy discussion at the beginning of the twenty-first century focuses on several issues related to the forest ecosystem services: carbon sequestration to mitigate the conditions of global climate change; the protection of biodiversity and critical habitat; management of the resources that support environmental tourism, forest-based recreation, and the broader aesthetic appeal of the forest; erosion control and general watershed protection; and sustainable forestry and the limitation of global deforestation.

The first two are newer environmental issues while the latter three have always been with us in some form. A wealthier and more mobile public has been the source of exploding demand for forest-based recreation in almost every part of the globe. The roles of forested watersheds as sources of clean and dependable supplies of water and as protection against the erosive forces of nature have always been clear to affected local populations. However, better information about the role of water in public health and the impacts of public works on our watercourses and seashores has improved the awareness of their importance. The fifth, sustainable forestry and the limitation of global deforestation, contains elements of the other four and of the old concern with timber supply as well. This appendix reviews each of these five issues and speculates on the effective means for providing them.

The markets for each of these five are either thin or incomplete in important cases. This fact alone makes each an appropriate concern for public policy. Fortunately, each has its own combination of economic and physical characteristics and, in locations where more than one occurs, the value of one generally tends to dominate. Therefore, on particular sites where joint production is important, it tends to involve only a small number of key products—often one market-valued product like timber or an agricultural crop, and one non-market valued product.[33] Of course, the particular combination of products with higher values varies from site to site. For these reasons, we can consider the typical physical locations of each of these five non-market values in turn and inquire as to the likely joint production or spillover effects involving each independently.

Carbon Sequestration and Global Climate Change

The topic of carbon sequestration is a part of the discussion that arises from the modern concern that our global climate is warming (Watson et al. 2000, Stern 2007).[34] It is warming because we are burning and discharging carbon into the atmosphere. The discussion of the role of forestry in global climate change has two components: (1) the effect of climate change on forests, which has to do with changes in the extent of forest cover in association with anticipated global climate change (Sedjo and Solomon 1990; Lewis and Clough 2009); and (2) the mitigating effects of the forest on climate change (Sedjo and Amano 2006). The latter is probably of greater interest and various researchers have shown the cost effectiveness of forests in sequestering carbon (Richards et al. 1993; Sohngen and Mendelsohn

---

33 Indeed, as Bowes and Krutilla (1989) prepared their research on the economics of multiple use forestry, they began with the complex theory that could address problems of numerous beneficial and detrimental outputs produced jointly on the same land unit (Bowes 1983). However, they found that reality is simpler. The most complex cases for any local forested site tend to involve only two products of high and competitive value. Therefore, the book that is the culmination of their research features five empirical examples of timber produced jointly with only one other single product: (1) water, (2) grazing, (3) hardrock minerals, (4) hunting, and (5) non-consumptive forest recreation.

34 Also see Nordhaus (2007), the symposium on "The economics of climate change: the Stern review and its critics" in the *Review of Environmental Economics and Policy* (Winter 2008), and further comments in the next issue of the same journal (Summer 2008). Aldy et al. (2010) summarize the discussion of policy choices to mitigate climate change. Also see the papers in the section on Critical Issues in National Climate Policy Design of the 2011 Papers and Proceedings issue of the *American Economic Review*, 101.

2003; Tavoni, Bosetti, and Sohngen 2007; *CIFOR News* 2007b; Langford 2007). This latter topic receives even greater attention in current discussions of REDD+ (Reducing Emissions for Deforestation and forest Degradation).[35]

Increases in atmospheric levels of CO_2 cause global warming. On this much, most are in agreement. The level and importance of global warming remain debatable issues for a few, although most scientists (including many Nobel laureates and even the George W. Bush administration) agree that the global climate is warming and that the source of change is increasing discharges of CO_2 (Mendelsohn and Neumann 1998; Nordhaus and Boyer 1999; Millenium Ecosystem Assessment 2005; Landner 2007). The importance of the forest in mitigating global warming remains a debatable issue to some. However, it is clear that fossil fuel combustion and land use change (mostly land conversion from forest to agricultural use) do release CO_2 into the atmosphere. The acts of harvesting timber and processing wood products also release CO_2, although not nearly at the rates of fossil fuel combustion and land use change. Indeed, wood processing releases only a portion of the carbon originally contained in a tree because so much is stored in the soil (roots, soil carbon, detritus on the forest floor), in the product (lumber), and in landfills (paper).

Reducing the rate of forest conversion to agriculture, particularly where conversion involves burning the forest cover, would decrease the rate of atmospheric buildup of CO_2. The resulting "avoided deforestation" has become a key international policy issue of its own as almost one-fifth of total annual greenhouse gas emissions are now the result of forest conversion (Stern 2007; *CIFOR News* 2007a).[36] Furthermore, the global social costs of carbon emission are such that additional forest conversion for agriculture and commercial development seldom generates much net economic benefit—on average, less than US$5 of benefit for each metric ton of carbon released. In Indonesia, for example, less than two percent of new forest conversion generates clear economic benefit (Swallow et al. 2007).

On the other hand, the process of growing trees in new plantations, harvesting, and then storing the final biomass retains additional CO_2. In fact, about two-thirds of the globe's terrestrial carbon, exclusive of that sequestered in rocks and sediments, is sequestered in standing forests, forest understory plants, leaf and forest debris, and forest soils (Sedjo and Amano 2006). However, growing trees and storing them on the stump produces only a short-term increment to stored carbon because little additional storage occurs once the trees approach maturity and their rate of growth declines. Nevertheless, additional tree planting and tree growth would be a means of buying time until world markets substitute other energy sources for fossil fuels. Sedjo and Amano (2006, p. 20) argue that that "up to 20 percent of excessive emissions can be captured in forests and biological sinks over the next 50 years"

---

35 See Angelsen (2011) for a brief summary.
36 Some estimates are even higher. Kindermann et al. (2008) and IPCC (2007) suggest that losses of tropical forests alone cause between one-fifth and one-quarter of all annual carbon emissions.

The problem confronting any attempt to use trees to mitigate global climate change, even for a limited period, is the mismatch between the non-exclusive global public impact and the more identifiable but still general location of the forest activities to control it. Protection from climate change is a public good in the broadest sense of the concept. Literally everyone everywhere benefits, and the exclusion of anyone from the benefit is impossible. On the other hand, the depletion of mature natural forests and the loss of carbon stored in forests is most rapid at the forest frontier of the developing tropical countries where trees tend to be removed by any number of local people acting independently for their very individual benefit under conditions that are difficult to monitor and limit.

Various import regulations, taxes, and subsidies have been suggested for addressing the problem. Import restrictions such as required certification are in favor today, but enforcing the forest management requirement for certification (and with it, sustainable management) must be difficult while less expensive timber from the natural forest frontier remains available to local loggers. The natural solution for many economists is to tax the negative externalities (land conversion and fossil fuel consumption), while using the tax revenues to subsidize the positive externality (forest management). However, taxes on land conversion are difficult to impose in the very many countries where conversion is largely a response to population growth, insecure land tenure, or domestic development policy.

Taxes on fossil fuel consumption have attracted more attention. Weimar (1990) shows that even a small tax on fossil fuel emissions comparable to 0.5 percent increase in the price of a barrel of petroleum would extract US$9 billion annually from the developed countries.[37] Such a tax would deter some combustion of carbon and the deterrence might be greater in the developed countries that tend to be the larger consumers of carbon. However, it might be more difficult to enforce the deterring tax on those who convert forests at developing forest frontiers—even if the revenues from the carbon taxes were used to compensate those at the frontier for refraining from the forest conversion. In this latter case, even a large sum of tax revenues would rapidly dissipate in (1) the transfers necessary to induce each of the very many marginal users of the tropical forest to discontinue their extractive activities and (2) the costs of monitoring and enforcing (M&E) the promised behavior by so many individuals spread over a vast forest landscape, as well as (3) the administrative costs of the entire transfer.

More recently, voluntary carbon markets have developed some appeal. Five annual international conferences on voluntary carbon markets had been held by October 2009 and this discussion has taken on its own acronym, REDD (for reducing emissions from deforestation and degradation) (Ecosystem Marketplace 2009). The concept is similar to the previous tax and transfer idea, except that the carbon market is voluntary. Carbon emitters (largely in developed countries) would charge themselves a fee for their emissions and use this fee to pay others

---

37 Parry and Williams (2010) suggest that such a tax, because of its magnitude, would have the additional benefit of helping reduce central government budgetary deficits.

(often, not always, in developing countries) to refrain from forest land conversion and to grow more trees to the extent that their efforts to sequester carbon would offset the carbon emissions of the payers. A few such arrangements are now active. However, the general concerns with this approach are similar to those with the tax and transfer approach and it remains a fact that past efforts to curb forest losses have not had great success. In Costa Rica, for example, payments have prevented deforestation on less than one percent of enrolled forestlands—because the other 99 percent are ill-suited for conversion to higher-valued use (Robalino et al. 2005). That is, they are beyond the frontier of commercial value, beyond point D in the figures of the three-stage model.[38]

The figures from the three stages of forest development illustrate the difficulty—even for lands at the forest frontier. They also illustrate two potential but challenging solutions.

For regions in all three stages of development, the costs of establishing and maintaining property rights at the natural forest frontier (D) are greater than the market value of the resource. In the first two less developed stages, these costs also exceed the value of marginal agricultural land use in the vicinity of the frontier (from point B to point C). This means that the costs of M&E, for carbon sequestration or any other purpose, are greater than the market value of the protected resource. Few developing country resource management agencies have either the budgetary means or the personnel to accomplish the M&E task under these conditions, while the local population continues to have competing private incentives to convert land and extract forest resources.

However, this same description also suggests an institutional solution: improved property rights. Any policy improvement that reduces the costs of establishing and maintaining property rights (reducing the level of the function C_j) will extend the area of sustainable management (shifting points B or B'' to the right). This will decrease the level of forest conversion for regions in stage I and reduce the area of open access forest degradation for regions in stages II and III—thereby protecting threatened forests and stored carbon for regions in stage I, increasing the forest stock in the degraded area for regions in stages II and III, and adding to the managed forest area and sequestering additional carbon for regions in stage III. Two changes that could satisfy this description are the institutional modifications that would permit local farmers to register their land use claims more easily, and the transfers of natural forests from public agencies to local land management. Community forestry, where it is successful, is an example of the latter. General improvements in the overall macroeconomic policy environment are a third example, one which we will discuss in chapter 6. These are the kinds of improvements that increase the confidence of local land managers in their future prospects in general, and cause them to extend their planning horizons. As a result, long-term investments such as trees and other conservation improvements on their lands would become more attractive.

---

38 Andam et al. (2008) and Blackman et al. (2009) add examples from Mexico and draw similar conclusions.

The second set of potential solutions is related to unplanned spillovers from agricultural activity. Reductions in the price supports for agricultural outputs would reduce agriculture's impact on the natural forest for regions in stage I (shifting point C to the left), decrease agriculture's infringement on the degraded open access forest for regions in stage II (also shifting point C to the left), and permit a relative improvement in the ability of forest management to compete with agriculture and expand into some former agricultural land for regions in stage III (shifting point B' to the left). Agricultural price supports in North America, Japan, and the European Union are immense, and the favorable effect of their reduction on forests and, therefore, on carbon sequestration could be substantial. Of course, altering these agricultural incentives will not be an easy task as their political support is very strong in almost all developed countries.

The effects of other agricultural incentives are not so clear. Some, such as those that encourage shifts to higher yield varieties, can induce agricultural intensification. As they do, they can have a beneficial effect opposite to that of the current large agricultural price supports. That is, maintaining price supports for higher yield varieties has the same beneficial effect on the forest and on carbon sequestration as that achieved by diminishing the world's numerous price incentives for agricultural expansion. Furthermore, agricultural intensification often increases the demands for labor, raises rural wages, and attracts workers away from the forest. This too saves on forest exploitation and diminishes the destruction of carbon already stored in mature trees.

Therefore, any proposal to modify agricultural incentives for the purpose of improving carbon sequestration must be selective regarding the incentives in question. Many agricultural incentives extend agriculture's competition with the forest and decrease the sequestration of carbon. Others reduce agriculture's competition with forests, allow the forest to expand, and improve carbon sequestration.[39]

Finally, and as with each of the other policy objectives discussed in this appendix, most would argue that the world needs to find a way to overcome the difficulties of providing for this global value. Even many of those who disagree on the current relative importance of global change and carbon sequestration, would agree that this is a problem that will not go away easily. The world must eventually, if not immediately, find a means for reducing carbon emissions and mitigating climate change.

Biodiversity and Critical Habitat

The protection of biodiversity is based on the idea that currently undeveloped, and even unknown, species possess characteristics of unknown but potential

---

39 Subsequent to drawing these conclusions, I learned that Humphreys and Palo (n.d.) share the assessment that the best instruments for combating deforestation and mitigating climate change are (1) to remove biased subsidies or taxes, and decrease the rate of road construction, and to (2) improve systems for monitoring forest resources. Their third recommendation is similar to the recommendation for forest sustainability (later in this appendix); that is, to promote economic growth and diversification.

future value. Preserving their habitats helps preserve the species and the option to obtain future advantage from them as we learn more about them. Various examples demonstrate the potential social and economic gain from preserving select species for which we had no prior scientific knowledge (e.g., Putz et al. 2001; Norton 1988; Ehrenfeld 1988)—although some doubt remains regarding the significance of the very probabilistic values to be gained from preserving unknown species in general (Simpson, Sedjo, and Reid 1996).[40]

Most of us do accept the importance of preserving species, and conventional wisdom posits that the remaining natural forest, the forest beyond the frontier at point D in our figures, contains most of the critical habitat. That is, natural forests (rather than forest plantations, other managed forests and trees, and the degraded open access lands) will be the focus of most of the global preservation effort because only these forests contain undisturbed habitat. Furthermore, among natural forests, the tropical moist forest contains more than one-half of all species of flora and fauna, including many that are still unknown. This particular forest is all the more important because its marginal habitats are more threatened than those of either the tropical dry forests or the temperate and boreal forests. Therefore, the remaining mature tropical moist forest is an appropriate focus for a large share of the global interest in protecting biodiversity.

The economic problem, once more, is the mismatch between the location of the most concerned public and the location of critical habitats. It contains elements of similarity with the carbon sequestration problem. That is, the greater wealth of the developed countries puts them in a better position to act on the potential future value of biodiversity and to pay the price necessary to protect endangered habitats, while the poorer local farmers and extractive users of tropical forests bear most of the costs of foregone development and habitat protection.

These elements of similarity diminish in importance, however, once we examine the physical sources of carbon and biodiversity. Carbon sequestration improves with the protection and expansion of all forests and trees. Most lands and trees anywhere (e.g., in backyards, along roadsides, or in natural forests) are of similar value for carbon sequestration, but all lands and trees are definitely not similar for biodiversity. Biodiversity requires the protection of selective "islands" of specialized habitat. In terms of the three stages of forest development, the greater numbers of remaining islands of unique but unprotected habitat are generally at the natural forest frontier or beyond it. The foregone future development opportunities associated with them are often smaller than even the current opportunities foregone when protecting forests for carbon sequestration at the margins of economic activity (points B or B' and B'', and D). This means that many critical habitats remain unthreatened today simply because they are beyond the limits of economic access for either agricultural land conversion or extractive forest activities. Their inaccessibility protects them—for now.

---

[40] Meilleur and Hodgkin (2004) review current efforts around the world to protect habitats of the wild relatives of commercially valuable agricultural crops.

In other cases, some habitats can be protected in the normal course of management for other forest products and forest-based environmental services. This is the case, for example, for many red-cockaded woodpecker habitats in the southern pine region of the United States. Careful planning of the sequence of timber management activities, with no change in the overall activities themselves, is probably sufficient to protect this endangered species (Hyde 1991).

This leaves two fundamental problems for policy resolution, protecting that currently threatened habitat that is in competition with other consumptive uses of the forest (or monitoring to ensure that regions of consumptive uses are not in competition with endangered habitat), and identifying additional specialized habitat that is inaccessible now and protecting it before it becomes threatened in the future.

Protecting that share of currently endangered habitat that is in the neighborhood of the forest frontier requires the establishment of boundaries around the habitat and either the permanent exclusion of incompatible land uses or carefully monitored management for compatible joint uses. It means either public ownership of the habitat within the boundaries or public regulation to insure that private landowners exclude incompatible land uses. Preventing land conversion and excluding agricultural use of the forest will be a problem for habitats at new frontiers in regions in the first stage of forest development. Therefore, reducing the incentives for agricultural expansion will be important for the preservation of habitat in newly settled regions.

Otherwise, for regions characterized by the more developed forestry described by the second and third stages of forest development, the threats to diverse forest habitats originate from either the development of new infrastructure or from extractive activities in the remaining natural forest. The first means that designing roads to avoid critical habitat is an important element of its protection. The second, extractive activities in these natural forests, are typically undertaken in a transitory manner by loggers or other operators who are here today and gone tomorrow. These operators have all the private economic incentives to harvest resources, often timber, and controlling their activities is a difficult task. Protecting against these extractive activities requires continuous monitoring of the boundaries of the protected habitat as long as the habitat is to be protected. Therein lies another problem. The resource management agencies of most developing countries have neither the human nor the financial resources to adequately ensure the protection of many critical habitats. International donors could assist, and many are willing to assist for a brief period, but most donors prefer not to assist with long-term continuing maintenance costs. Therefore, the long-term protection of currently threatened habitat remains a largely unsolved policy problem. Furthermore, in the presence of very limited resources, the question of which of many threatened habitats are most worthy of the long-term demands on these very limited resources is an additional problem.

The second fundamental problem has to do with those critical habitats that are currently inaccessible and unthreatened. The identification of these critical habitats is a task for field research and it is one justification for the interest in developing indicators of forest quality (WCFSD 1999). As current opportunities for biotechnology expand, the values of some endangered habitats may become clearer to private investors and those private interests may undertake the effort to save some of them.[41] Otherwise, this too is a public problem and one that the donors and public research institutions of the developed countries may be particularly well-suited to address. Once public researchers do identify critical habitat, however, policy makers are still left with the previously discussed and largely unsolved task of finding a publicly-funded means for permanently excluding extractive users of the forest from the critical habitat.

Finally, *ex situ* preservation can be a partial solution for long-term preservation. *Ex situ* preservation is the off-site storage of the endangered biological material. Once the threatened flora or fauna is identified, its germplasm can be catalogued and stored, preferably in a public institution that maintains the biological material for the benefit of all of the world's people in this generation and future generations as well. Zoological parks are one example. The institutions of the Consultative Group on International Agricultural Research that maintain cold storage banks of germplasm are another. Gene banks, however, are static institutions. They fail to capture and preserve dynamic genetic evolution but they do contribute to the preservation of existing biodiversity while the search continues for means to identify and permanently protect critical forest habitat.

Environmental Tourism and Forest-based Recreation

A broad range of people participate in environmental tourism and a broad range of forested sites provide for it. Those who benefit range from wealthier globe-trotting tourists to local picnickers or others just looking for a pleasant moment in a natural setting. The sites they visit range from unique global resources like Yellowstone, the Serengeti, or Sagarmantha to pleasant local forested groves and village parks. The unique sites are often focal points for substantial demands for tourist support services like restaurants, motels, guide services, and outdoor equipment shops; and these can be important sources of employment for the local economy.

Once more, the economic problem is either one of protecting islands of specialized forested sites within lands that are valuable for other, extractive uses (out to point *D* in our figures), or identifying forestlands that are inaccessible for extractive land uses (beyond *D*) but uniquely attractive for nature tourism, and then protecting these sites before they become accessible for those extractive uses. The latter has been more common in the past (e.g., Yellowstone National Park, established in 1872 as a unit within the midst of other public lands). Examples of

---

41 A few examples already exist. Merck & Co., the world's largest pharmaceutical firm, made a one-time payment of US$1 million (plus a share of any eventual profits) to the National Biodiversity Institute of Costa Rica (INBio). INBio collects plant, insect and soil samples for Merck which screens and, potentially, develops marketable products from them (Richards 1999).

the former are becoming more common (e.g., the Serengeti in Tanzania or Redwood National Park in the United States, established in 1968 and expanded in 1978 from original purchase of private lands), but they are more expensive at the time of their establishment and special arrangements for affected local economic interests are often a major part of the expense.

For the most unusual sites, fees can be charged at points of limited access and the collected revenues can be used to establish boundaries and systems for monitoring and enforcing the exclusion of undesirable uses. For example, Kenya charges a larger national entry fee for global tourists than for other visitors, Nepal charges international tourists for trekking permits which must be obtained at the offices of guide services, and various countries charge gate fees for those national parks that do have natural boundaries.

Two problems remain, however, even in the case of these unique resources. First, the fee is general. It makes no distinction for particular resources or for specialized services within the site covered by the fee. For example, fees collected at national borders do not distinguish between different parks within a country or between particular resources within a park. Therefore, there is no market signal to assist managers in the allocation of revenues among numerous resources and services within the site of the natural attraction. Sometimes this problem can be addressed by placing the general management of all resources under one integrated operation; a national park service, for example; and competitively allocating concessions for specialized services like hotels and guide services within each park boundary. Each concession then discriminates by charging for its own specialized service.

The second problem exists even when this first problem is solved. While tourists who arrive from long distances can be charged or excluded at the national border or the park boundary, the exclusion of local users is more difficult. Therefore, monitoring and enforcing restrictions against competing local uses of the unique resource is more difficult. Poaching within East Africa's game parks and illegal timber harvesting within the boundaries of Southeast Asia's natural reserves are examples.

A partial solution to this problem can often be obtained by establishing an interest in the park's tourist services within the local population, or even some sharing of revenues collected for the park with local interests. When some of the local population gain, as from employment in park concessions or by providing guide services, then local citizens have their own incentives to assist in the M&E of park boundaries, to help maintain or increase the park's quality, and to discourage their neighbors' violations of the objectives of park management. However, the determination of effective revenue sharing mechanisms is not an easy task and, even with them, some amount of local trespass is still likely.[42]

---

42 Naidoo and Adamowicz (2005) discuss these problems with respect to the specific example of nature-based tourism in Uganda.

For those forest resources that are not unique and do not attract global tourists the most common non-consumptive users are members of the local community. Local institutions are generally better suited to manage these resources and we observe many successfully protected village parks and forest sanctuaries around the world. The employees of local institutions have a better understanding of both the pattern of demands for these resources and the arrangement of their most appealing physical characteristics. Nevertheless, the cost of management is still positive and, since the exclusion of local users is difficult, the local community must bear this cost either as part of its community budget or as part of a commonly respected decision not to exploit the extractable resources within the protected area.

Erosion Control and General Watershed Protection

Erosion control and watershed protection incorporate all the services of trees and watersheds in managing wind, water, and soil movement; for example, provision of a regular flow of water of acceptable quality, storm protection—especially in coastal areas, and control of the upstream loss of soil nutrients and downstream effects of sediment deposition. Like carbon sequestration, watershed protection can be divided into two broad classes of activities; those that require new conservation interventions such as tree planting in shelterbelts to deter wind erosion and along streams and gullies to deter water erosion, and those that maintain the services of existing forested watersheds and limit their deterioration.

The difference in economic effect is that watershed values tend to be local or regional, and those who benefit are individually or as a group easier to identify, while carbon sequestration and the protection of critical habitat are global public values. Depending on the watershed management activity, it can be of greater benefit to an individual landowner who makes the conservation investment and improves the productivity of his or her own land, or it can yield greater benefit to a range of downstream or other off-site land users in the same watershed. In the latter case, watershed management is a public good but it involves a smaller and more local public than the global society affected by carbon sequestration or the protection of biodiversity. The benefits of the activity are clearer and more immediate for the local community than for carbon sequestration or the protection of biodiversity, and using local administration and to achieve support for and compliance with management objectives may be somewhat easier for erosion control than for carbon sequestration or the protection of critical habitat.

Many of the first class of watershed management activities, those requiring new conservation investments, are responses to human development. They are a means of improving the productivity of existing (often agricultural) land uses. These investments typically occur on the private lands—to the left of the intersection of the agriculture and forest value functions with the rising property rights cost function in our figures (points B or B''). Therefore, increases in the manager's private long-run productivity are often sufficient to induce private conservation investments. In fact, this observation conforms to the established evidence for developed countries—where the land use rights are clearer than in some developing countries (e.g., Crosson 1985; Crosson and Stout 1984). More recently, Yin (2003) and Alemu (1999) demonstrate its reliability for developing countries as well, once farmers in those countries obtain longer-term land use rights. Farmers in China in the 1970s and in Ethiopia in the 1980s responded almost immediately to rural reforms that gave them more secure and longer term land use rights. They increased all inputs, including conservation inputs like trees to protect against wind and water erosion. Yin and Alemu trace significant shares of the overall increases in private agricultural productivity directly to these conservation investments.[43]

The second class of watershed management activities protects the upland watersheds or the coastal wetlands for the benefit of off-site residents of the same local area. Grazing livestock are a common source of upland erosion and downstream sedimentation (e.g., the Lake Victoria watershed in central Africa and the central uplands of Luzon in the Philippines (Ikiara, Kazoora, and Kulindwa 2003; Cruz, Francisco, and Conroy 1988). The upland collection of fuelwood and fodder is another source in some severely degraded forest areas (e.g., Nepal's hills; Dangi and Hyde 2001). Even less intrusive activities like picnicking can also affect local water supplies to a sufficient extent that the city of Portland, Oregon, for example, restricts all human activity on a critical section of its adjacent Mt. Hood watershed. Finally, logging and land conversion in many coastal areas have damaged local fisheries and diminished the protection the local forest provides against the effects of tropical storms and tidal hazards (Ruitenbeek 1992; WWF 2005).

These examples all characteristically occur within either the open access degraded forest or the neighborhood of the mature natural forest frontier in the second and third stages of forest development (between points C and D). The costs of protecting watersheds in these areas exceed the open access private (grazing, fuelwood collection, recreation, timber harvesting, or land conversion) values of these lands. Therefore, unfettered private management will be unsuccessful and regulation or public ownership along with a degree of monitoring and enforcement is the usual means for insuring the common watershed benefits for the local community.

Where most or all members of the local community share in the common watershed benefits, local residents also share a common incentive to protect the watershed, and monitoring and enforcement may be a relatively simple matter. In other cases, where the local incentives are dissimilar and members of the community compete for different uses of the land, M&E consumes more resources and protecting the public benefits is a more difficult task.

For example, we previously considered the village of Basantapur in Nepal's hills where some poorer landless households rely on the watershed for fuelwood.

---

43 Landell-Mills and Portas (2002) provide further evidence of the private market nature of some classes of watershed management. They identify over 180 cases of markets for watershed services from countries around the globe and in a multitude of local institutional arrangements. The website of the Katoomba Groups identifies many more (www.ecosystem marketplace.com).

Their fuelwood collection degrades the watershed and increases the off-site flow of soil to the detriment of the agricultural productivity of the better-off households at the base of the watershed. This community's two interests in the watershed conflict with each other, and the community's forest guards have been unable to fully restrict access. The degradation of the watershed continues (Dangi and Hyde 2001).

Finally, some cases require broader regional or national oversight. Private or local collective action will be insufficient in these cases. They are the cases of vast watersheds and upstream actions that effect human communities great distances downstream. The Chinese authorities feel this was the case with the flooding of the Yellow and Songhua Rivers in 1998. Upstream deforestation and construction damaged agricultural lands more than 1,500 kilometers downstream. China's solution was to begin a massive long-term national program for improved management of the upper portions of these watersheds in 2000.

The city of New York provides another example that requires broader regional oversight. The city realized, in 1997, that changing agricultural practices meant that it needed to act to preserve the quality of its drinking water. It could have installed new water filtration plants at a cost of $4–6 billion. The city chose, as a much less costly alternative, to preserve the rural nature of its Catskill Mountain watershed. It purchased some land on which it restricts further development and it pays farmers in the watershed to refrain from farming practices that pollute (Kenny 2009).

Sustainable Forestry and the Control of Deforestation

The concern for sustainable forestry grew out of a much older concern for resource depletion. It originally focused on market-valued resources in general and, in the case of forests, timber in particular. The evidence of concern may begin with Solomon's use of the cedars of Lebanon to build his temples. Formal public records of the concern are at least as old as 1546 when the Viceroy of the City of Mexico wrote home to the King of Spain alerting him that North America was running out of timber. In 1876, F. B. Hough, in his address to the American Association for the Advancement of Science, described the environmental damage suffered following deforestation, and that organization formed a committee to encourage the U.S. Congress to address its perception of a coming timber shortfall. The U.S. National Forest System was eventually created in 1891 to address this problem—and also to address watershed management issues in the East (Clepper 1977). Western Europeans have been concerned about a timber shortfall at least since Jevons (1865) wrote about the limited sources of mine props in England in the mid-nineteenth century. In addition, Europe's periodic wars over the last several hundred years have regularly depleted the existing mature timber—and, thereby, demonstrated the need for a supply of timber as a strategic material.

Of course, North America has not run out of timber. In fact, U.S. timber stocks are greater today than they were 50 years ago and greater yet than they were 100 years ago. Fortunately, Western Europe seems to have entered a period with fewer large wars, and the usefulness of wood as a strategic resource has declined. Furthermore, the stocks of most market-valued forest-based goods are not declining in economic terms. That is, their costs of production are not increasing over time.

However, residual doubts exist in the minds of many regarding the potential for timber shortages. Others, while not so concerned about the depletion of market-based forest products, are concerned that we may be depleting our stock of global means to provide the non-market environmental services of forests. These arguments are the basis for modern discussions of forest sustainability and controls on deforestation and forest degradation.

The modern discussions take a number of perspectives.[44] Perhaps a useful perspective for our purposes would be "sustainable options." That is, a definition of sustainability, restated as a useful objective, could be "maintain, in perpetuity, options for all the various uses of forest resources, market and non-market, consumptive and non-consumptive, known and unknown." This would mean controlling environmental destruction. It would mean maintaining for the future the potential for all the different uses of the land and other forest resources. It would also mean using the forest to help maintain other future options.[45] It would include using the forest to help control erosion, to protect critical habitat and important aesthetic resources, and to limit global climate change. Such a statement of sustainability would allow some shifts of forestland to agriculture, others from agriculture back to forest, and still others from natural forest to managed forest so long as both the land's productive base and also the genetic material of the forest biota remain undamaged. Relative values will change with time and preferred patterns of land use will change with them, but we can insure that changes in relative values and changes in land use do not destroy opportunities for new and different land or forest resource uses in the future.

In the context of the three-stage model of forest development, this perspective of sustainability is consistent with minimizing the area of degraded open access forest while locally regulating specific eroding watersheds, critical habitats, and important aesthetic resources, both within and outside the degraded area. Minimizing the degraded area is an objective because its elimination is impossible as long as secure property rights impose a cost and so long as the public agencies responsible for managing the degraded open access area have limited budgets.

---

44 See Toman and Ashton (1996) for a summary.
45 This perspective is consistent with the idea of "safe minimum standards" first proposed by Ciriacy-Wantrup (1968). It is also consistent with the idea of "sustainable livelihoods," and with the definitions selected by the Ministerial Conference on the Protection of European Forests and the "Helsinki Process" (1993), and also by the CSCE Seminar and the "Montreal Process" (also 1993). The definitions of sustainable forest management emerging from these meetings focused on maintaining the diversity and productivity of forests while ensuring future opportunities from the forests (FAO 2002). The perspective of "sustainable options" does not depend on a "permanent forest estate." In fact, a forest estate with permanent boundaries is a futile objective, as we have seen that forest boundaries must change as local relative prices adjust and forest development proceeds through its three stages.

166 Forest Policy

The fundamental means for minimizing the degraded area involve (1) reducing the cost of establishing and maintaining property rights and (2) attracting extractive human activity away from the forest. The first requires finding the least cost bundle of property rights and the institutions that can provide it, thereby insuring the lowest cost function T_r' in Figure 4A.1. Of course, the most effective bundle of property rights and the institutions that can provide this bundle will vary with local values—as we discussed in the body of this chapter. Various arrangements of private rights, local community rights, or state ownership will be appropriate in different local situations but none of these will be a universal solution.

We discussed the effect of the opportunity cost of labor on the natural forest in chapter 2. Forest users with lower opportunity costs can afford the time to travel farther into the natural forest to extract its products. Because their opportunity costs are low, they can also justify removing material in the degraded area down to a low level. Providing these low wage or low opportunity cost forest users with improved employment opportunity outside the forest will cause some of them to change their employment from extractive activities in the forest to the more lucrative activities away from the forest. In terms of the mature forest economy depicted in Figure 4A.1, the net forest value remains unchanged at the market, but the forest value function shifts inward along the horizontal axis and becomes steeper as some users leave the forest for higher wage alternatives, frontier labor becomes scarcer, frontier wages increase, and the foregone labor opportunity costs of removing additional resources from the frontier forest also increase.

The combined effects of these two fundamental improvements are a decline in the extent of the degraded open access area from $B_1''D$ to $B_2''D$ and an increase in the density and volume of the forest in the remaining open access area.

Stating the argument a different way, poverty is a crucial source of forest degradation and forest depletion. Economic development induces improvements in

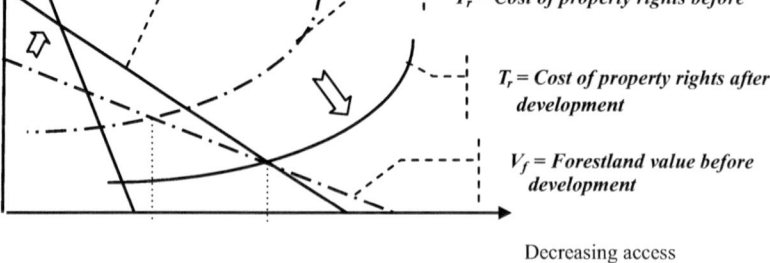

FIGURE 4A.1 Sustainable forestry and the control of deforestation

the forest environment as it shifts land into sustainable activities. In fact, economic development is likely to have a second round of beneficial effects as well. Improved wages and better labor opportunities create the first round. Then, along with improvements in overall welfare, the local institutions also tend to become more effective over the course of general economic development. Their budgets probably increase, and they improve in their ability to insure property rights and in their ability to manage economic transitions and provide for economic stability. Both improved institutions and a more stable economy lower the transactions cost function and cause a second round of reductions in the degraded area.

Figure 4A.2 is further illustrative of this argument. It contrasts two regions, one more developed than the other in terms of overall economic welfare. In this illustration, both are in the third stage of forest development, although similar comparisons can be drawn for regions in the first or the second stages as well. Agricultural land values are comparable in both regions. The transaction cost function for property rights is lower and alternative wage opportunities for forest users are greater in the more developed region. Therefore, the degraded area between points B'' and D is smaller in this region.

The depictions of the open access degraded area and the mature natural forest under the figure show the contrasts in degradation between regions of greater and less overall economic development. In the more developed region on the right, only the small open access forest is degraded and even this forest is not heavily degraded. It is smaller in area and also better stocked because the rewards of open access trespass onto lands with formal title are small compared with the risks incurred for local populations whose incomes, while modest, are well above those of many forest users in the poorer and more degraded region at the left in the illustration.

In fact, we know that an area of open access forest exists, even in the most economically developed countries. In some cases, open access exploitation may be almost unnoticeable to a casual observer. In chapter 2, we mentioned open access harvesting of ginseng and Christmas greenery in southwestern Virginia and illegal logging in the western United States and Canada. The effects of these activities are often so minimal as to be difficult to detect in developed countries and that difficulty is one reason they continue as open access activities, even in forests with identifiable (but imperfectly enforceable) formal property rights.

The amount of illegal activity is probably greater in less developed regions and countries. Countries undergoing rapid change in economic welfare provide good examples, and the most notable of these in recent years have been the countries of the former Soviet Union. Many of these countries suffered serious economic decline and instability as they as they adjusted to new arrangements independent of the former Soviet Union. The effectiveness of their formal institutions declined as well. They moved from the greater economic welfare characterized by the illustration on the right of Figure 4A.2 to the lesser level of welfare characterized by the illustration on the left. Indeed, illegal logging increased dramatically in many of these countries and the increase occurred simultaneously with the decline in overall economic welfare. In Estonia, for example, up to three-fourths

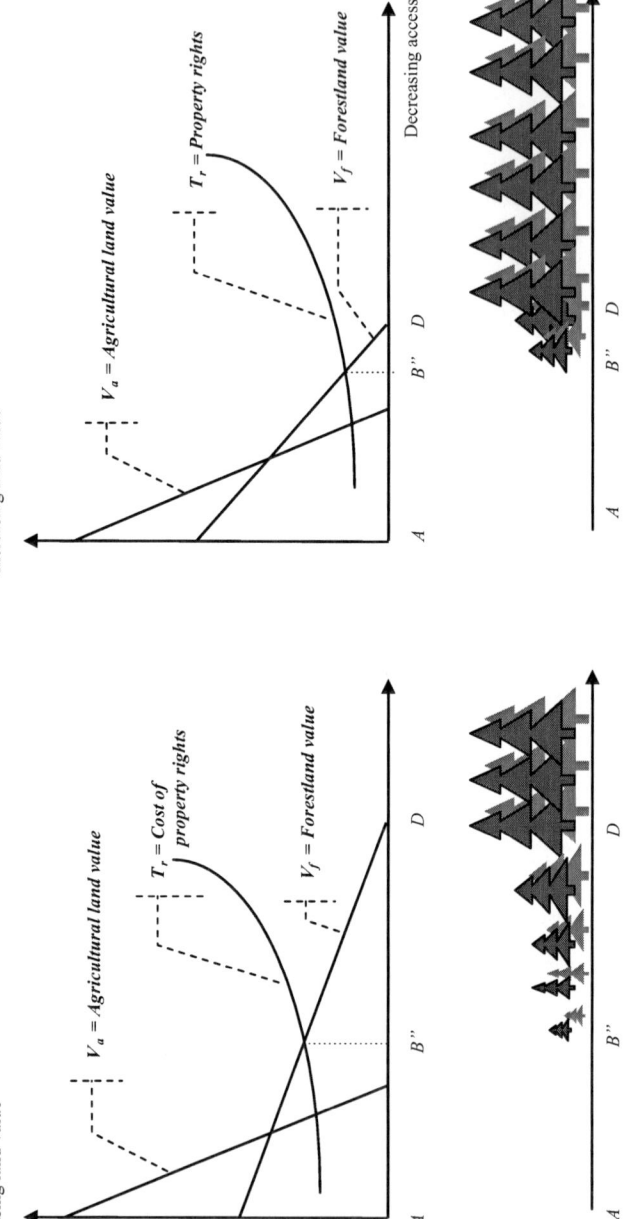

FIGURE 4A.2 Sustainability and deforestation: The effects of development

of the timber harvested between 1998 and 2002 may have been in violation of legal regulations (Hain and Ahas 2004).

The broader historical evidence is also consistent with these arguments. Countries draw down their stocks of natural resources like trees and forests as they enter periods of initial economic development. However, they also build back their stocks of forests after some point in the development process. For example, natural forest cover has doubled in Switzerland and France and tripled in Denmark since the early or mid-nineteenth century (Mather 200; Kuechli 1997). In the developed northeastern United States, forest cover has grown from around 15 percent in the early twentieth century to as great as 90 percent in some states in the early twenty-first century (USDA Forest Service 2005). Forest cover has increased by more than 30 percent over the 25 years since the first agricultural reforms marked the beginning of China's double-digit annual economic growth and six-fold increase in rural income (Hyde et al. 2003). In each of these regions and countries, agricultural land use has remained relatively constant or even declined over the period of economic growth and the increase in forest cover has been greater than any decrease in agricultural land area. The only explanation must be that, in a period of economic development, forest cover has expanded into areas of previously degraded land.

India's Punjab provides a more specific example. The Punjab is India's most productive agricultural region. The region began a period of rapid and sustained development in 1960. Crop yields per hectare had tripled by 1990 and real income per capita doubled. The land area in agricultural crops more than doubled while the principle agricultural prices remained relatively constant or declined—depending on the crop. Meanwhile, the rural share of the region's population remained steady. Forest cover in the Punjab increased six-fold and horticultural tree cover increased more than 250 percent. Before 1960 a large area had been cleared of its forest cover. It existed only as an open access wasteland. Since 1960 both the land area in agriculture and the forest stock have increased. Furthermore, the land area in wasteland has declined as a large share of the previously open access land has been converted into cropland and an additional large share has been reforested (Singh 1994).[46]

In sum, sustainability in the sense of maintaining options for land and resource use over time is important everywhere, but it may be most difficult to insure in the less developed regions of the world. These tend to be characterized by the first two stages of forest development where loss of some non-market value from the natural forest is inevitable. The institutions responsible for managing the natural forests of these regions may be neither well-established nor well-funded. Therefore, establishing priority locations and resources for protection is especially important for these less-developed regions. However, rural economic

---

46 The examples of the last two paragraphs suggest a turning point and forest recovery above some level of regional welfare. The discussion of an environmental Kuznets curve for forests in chapter 6 returns to this point.

development and the alleviation of rural poverty must be central to long-run improvements in forest sustainability, and also to any attempt to decrease the rate of global deforestation.[47] Rural economic development is everyone's objective, not just the objective of forest policy. Accomplishing it is not an easy task, but it is certainly no more difficult than trying to accomplish sustainable forestry and decrease the rate of deforestation through the imposition of government regulations on the uses of relatively low valued and widely dispersed forest resources by a scattered and poor rural population.

Literature Cited

Aldy, J., A. Krupnick, R. Newell, I. Parry, and W. Pitzer. 2010. Designing climate mitigation policy. *Journal of Economic Perspectives* XLVIII(4): 903–934.

Alemu, T. 1999. *Land tenure and soil conservation: Evidence from Ethiopia.* Unpublished PhD thesis, Economics Department, Goteborg University, Sweden.

Andam, K., P. Ferraro, A. Pfaff, J. Robalino, and A. Sanchez. 2008. Measuring the effectiveness of protected area networks in reducing deforestation. *Proceedings of the National Academy of Sciences* 105(42): 16089–16094.

Angelsen, A. 2011. What does REDD+ really cost? POLEX (July) [Blog].

Blackman, A., A Pfaff, J. Robalino, and Y. Zepeda. 2009. Mexico's natural protected areas: enhancing effectiveness and equity. *Interim narrative report to the Tinker Foundation.* Washington, DC: Resources for the Future.

Bowes, M. 1983. *Economic foundations of public forestland management.* Resources for the Future discussion paper D-104. Washington, DC: Resources for the Future.

Bowes, M., and J. Krutilla. 1989. *Multiple use management: The economics of public forestlands.* Washington DC: Resources for the Future.

CIFOR News. 2007a. *Forests and climate change: Tough but fair decisions needed.* 43: 2–4.

CIFOR News. 2007b. *CIFOR and CPF launch first "forest day" at UN global climate change talks in Bali.* 44: 1, 4.

Ciriacy-Wantrup, S. 1968. *Resource conservation: Economics and policies.* Berkeley: University of California Press.

Clepper, H. 1977. *Professional forestry in the United States.* Baltimore, MD: Johns Hopkins University Press for Resources for the Future.

Crosson, P. 1985. Impact of erosion on land productivity in the United States. In S. el Swaify, W. Moldenhauer, and A. Lo, eds., *Soil erosion and conservation.* Ankeny, IA: Soil Conservation Society of America, pp. 217–236.

Crosson, P., and T. Stout. 1983. *Productivity effects of cropland erosion in the United States.* Unpublished manuscript, Washington, DC: Resources for the Future.

Cruz, W., H. Francisco, and Z. Conroy. 1988. The onsite and downstream costs of soil erosion in the Magat and Pantabangan watersheds. *Journal of Philippine Development* 15(1): 48–85.

Dangi, R., and W. Hyde. 2001. When does community forestry improve forest management? *Nepal Journal of Forestry* 12(1): 1–19.

Ecosystem Marketplace. 2009. Voluntary carbon markets. http://www.ecosystemmarketplace.com/documents/cms_documents/StateOfTheVoluntaryCarbonMarkets_2009.pdf (accessed April 9, 2010).

Ehrenfeld, D., 1988. Why put a value on biodiversity. In E. Wilson and F. Peter (eds.), *Biodiversity.* Washington, DC: National Academy Press, pp. 212–216.

---

47 In fact, Heath and Binswanger (1996) argue that the resource degradation effects of poverty are largely policy induced. Therefore, once more, the task is not just one of designing good policy but also of correcting ill-advised existing policy.

FAO/UN (Food and Agriculture Organization of the United Nations). 2002. Proceedings: Second expert meeting on harmonizing forest-related definitions for use by various stakeholders. Rome: **FAO**.

Guariguata, M. The timber may be certified: But is it sustainable? CIFOR News Update (May): pp. 1, 2–5.

Health, J., and H. Binswanger. 1996. Natural resource degradation effects of poverty and population growth are largely policy-induced: the case of Colombia. *Environment and Development Economics* 1: 65–83.

Hein, H., and R. Ahas. 2004. The structure and estimated extent of illegal forestry in Estonia. Unpublished manuscript, Institute of Geography, University of Tartu, Estonia.

Humphreys, P., and M. Palu. n.d. *Forests in global warming*. Tokyo: UNU World Institute for Development Economic Research.

Hyde, W. 1991. The marginal costs of endangered species management: The case of the red-cockaded woodpecker. *Journal of Agricultural Economics Research* 41(2): 12–19.

Hyde, W., B. Belcher, and J. Xu (eds.). *Introduction to China's forests: global lessons from market reforms*. Washington, DC: Resources for the Future.

Ikiara, M., C. Kazoora, and K. Kulindwa. 2003. *Environmental sustainability in the Lake Victoria Basin: A proposal for economic policy analysis and capacity building*. Unpublished manuscript, Environmental Economics Unit, Gothenburg University, Sweden.

Intergovernmental Panel on Climate Change (IPCC). 2007. Climate change 2007: The physical science basis. In S. Solomon, D. Qin, M. Manning, Z. Chen, M. Marquis, K. Averyt, M. Tignor, and H. Miller, eds., *Contribution of working group I to the fourth assessment report on climate change*. Cambridge, UK: Cambridge University Press. Available at http://www.ipcc.ch/publications_and_data/ar4/wg1/en/contents.html

Jevons, W. S. 1865. *The coal question: An inquiry concerning the progress of the nation and the probable exhaustion of our coal-mines*. London: Macmillan.

Kenny, A. 2009. Ecosystem services in the New York City watershed. Available at http://www.ecosystemmarketplace.com (accessed June 27, 2009).

Kindermann, G., M Obersteiner, B. Sohngren, J. Sathaye, K. Andrasko, E. Rametsteiner, B. Schlamadinger, S. Wunder, and R. Beach. 2008. Global cost estimates of reducing carbon emissions through avoided deforestation. *Proceedings of the National Academy of Sciences* 105(30): 10302–10307.

Kuechli, D. 1997. *Forests of hope: Stories of regeneration*. London: Earthscan.

Landell Mills, N., and I. Portas. 2002. *Silver bullet or fool's gold: A global view of markets for forest environmental services and their impacts on the poor*. London: International Institute for Environment and Development.

Landner, M. 2007. Nobelists feel validation on climate. *International Herald Tribune* (October 11), pp. 1, 4.

Langford, K. 2007. Less than $1.00 per ton of CO_2: Research suggests Indonesia can reduce emissions with sustainable benefits. *CIFOR News* 44: 4.

Lewis. S., and G. Clough. 2009. Average tree size increasing as trees absorb more carbon. *CIFOR News* (June), pp. 3–4.

Mather, A. 2001. The transition from deforestation to reforestation in Europe. In A. Angelson and D. Kaimowitz, eds., *Agricultural technologies and tropical deforestation*. Wallingford, UK: CAB International, pp. 35–52.

Meilleur, B., and T. Hodgkin. 2004. *In situ* conservation of crop wild relatives: status and trends. *Biodiversity and Conservation* 13: 663–684.

Mendelsohn, R., and J. Neumann (Eds.). 1998. *The impacts of climate change on the American economy*. Cambridge, UK: Cambridge University Press.

Millenium Ecosystem Assessment. 2005. *Ecosystems and human well-being: Synthesis*. Washington, DC: Island Press.

Naidoo, R., and Wictor Adamowicz. 2005. Biodiversity and nature-based tourism at forest reserves in Uganda. *Environment and Development* 10:159–178.

Nordhaus, W. 2007. A review of the Stern review on the economics of climate change. *Journal of Economic Literature* 45(3): 686–702.

Nordhaus, W., and J. Boyer. 1999. *Warming the world: Economics models of global warming.* Cambridge, MA: MIT Press.
Norton, B. 1988. Commodity, amenity, and morality: the limits of quantification in valuing biodiversity. In E. Wilson and F. Peter, eds., *Biodiversity.* Washington, DC: National Academy Press. pp., 200–211.
Parry, I., and R. Williams. 2010. Is a carbon tax the only good climate policy? *Resources* 176: 38–41.
Putz, F., G. Blate, K. Redford, R. Fimbal, and J. Robinson. 2001. Tropical forest management and conservation of biodiversity: An overview. *Conservation Biology* 15(1): 7–20.
Richards, K., D. Rosenthal, J. Edmonds and M. Wise. 1993. *The carbon dioxide emissions game: playing the net.* Unpublished manuscript Available from K Richards, Duke University, Durham, NC.
Richards, M. 1999. *'Internalising the externalities' of tropical forestry: a review of innovative financing and incentive mechanisms.* European Union Tropical Forestry Paper 1. London: Overseas Development Institute and European Commission.
Robalino, J. A. Pfaff, G. Sanchez-Azofiefa, F. Alpizar, C. Leon, and C. Rodtiguez. 2008. *Deforestation impacts on environmental services payments: Costa Rica's PSA program 2000–2005.* Environment for Development discussion paper 08-24. Washington, DC: Resources for the Future.
Ruitenbeek H. 1992. *Mangrove management: An economic analysis of management options with a focus on Bintuni Bay, Irian Jaya.* Unpublished manuscript, Dalhousie University, Halifax, Nova Scotia.
Sedjo, R., and M. Amano. 2006. The role of forest sinks in a post-Kyoto world. *Resources* 162: 19–22. Also see the more detailed version, Forest sequestration: Performance in selected countries in the Kyoto period and the potential role of sequestration in post-Kyoto agreements. 2006. Available at http://www.rff.org/rff/Documents/RFF-Rpt-ForestSequestratinKyoto.pdf
Sedjo, R., and A. Solomon. 1990. Climate and forests. In N. Rosenburg, W. Easterling, P. Crosson, and J. Darmstadter, eds., Greenhouse warming: abatement and adaptation. Washington, DC: Resources for the Future, pp.. 105–119.
Simpson, R., R. Sedjo, and J. Reid. 1996. Valuing biodiversity use in pharmaceutical research. *Journal of Political Economy* 104(1): 163–185.
Singh, H. 1994. *The green revolution in Punjab: The multiple dividend, prosperity, reforestation and the lack of rural out-migration.* Unpublished student paper, JFK School of Public Policy, Harvard University, Cambridge, MA.
Sohngen, B. and R. Mendelsohn. 2003. An optimal control model of forest carbon sequestration. *American Journal of Agricultural Economics* 85(2): 448–457.
Stern, N. 2007. *The economics of climate change.* London: Cambridge University Press.
Swallow, B., M. van Noordwijk, S. Dewi, D. Murdiyarso, D. White, J. Gockowski, G. Hyman, et al. 2007. *Opportunities for avoided deforestation with sustainable benefits.* Bogor, Indonesia: World Agroforestry Centre.
Toman, M., and M. Ashton. 1996. Sustainable forest ecosystems and management: a review. *Forest Science* 42: 366–377.
Tavoni, M., V. Bosetti, and B. Sohngen. 2007. *Forestry and the carbon market response to stabilize climate.* Working paper 15.07. Rome: Fondazione Eni Enrico Mattei.
USDA Forest Service. 2005. USDA Forest Service forest inventory and analysis webpage: http://fia.fs.fed.us (accessed July 17, 2008).
Watson, R., I. Noble, B. Bolin, N. Ravindranath, D. Verardo, and D. Dokken (Eds.). 2000. *Land use, land-use change, and forestry.* Special Report of the Intergovernmental Panel on Climate Change. Cambridge, UK: Cambridge University Press.
WCFSD (World Commission on Forests and Sustainable Development). 1999. *Final report.* Cambridge, UK: Cambridge University Press.
Weimar, D. 1990. An earmarked fossil fuels tax to save the rainforest. *Journal of Policy Analysis and Management* 9(2): 254–259.
Weitzman, M. 2007. A review of the Stern review on the economics of climate change. *Journal of Economic Literature* 45(3): 703–724.
WWF (World Wildlife Fund). 2005. Tsunami issues paper. http://www.iema.net/news/envnews?aid=4974 (accessed November 1, 2005).
Yin, R. 2003. Measures of the effects of improved property rights, a stable policy environment, and environmental protection. In W. Hyde, B. Belcher, and J. Xu, (eds.), *China's forests: Global lessons from market reforms.* Washington, DC: Resources for the Future. pp. 59–84.

5

FOREST CONCESSIONS

A Specialized Topic in Forest Policy and Management

This chapter continues the discussion of policy begun in chapter 4 with a specialized topic of modern forestry, the temporary transfer of some of the rights to forest properties. Landowners (and government resource management agencies) often contract with others to perform specialized services for a period of time while retaining the right to determine the fundamental objectives of land management for themselves. Tourist concessions in forest parks are one example. Contractors obtain the right to sell the tourism services (food, lodging, guide services, etc.), but they compensate the landowner or park management for this right and they operate within contractual limitations that are consistent with the overall management objectives set by the landowner.

Timber sales or timber concessions are a second example and they are an issue of greater controversy in modern forest policy. These are arrangements between landowners and second parties who obtain the right to harvest mature timber over a fixed period of time in return for financial compensation to the landowner and subject to certain constraints on the harvest practices and land management responsibilities of the contractor.

The economic principles for transfers of both kinds of resource responsibilities are similar, but transfers of timber rights have developed from a culture of their own, originating with the management of forested estates in eighteenth-century Germany. This chapter focuses on these timber transfers. Common usage distinguishes between two classes of transfers, timber sales and forest concessions. Timber sales are short-term arrangements ranging from the right to harvest a few thousand board feet from a farmer's woodlot over the brief period of a few weeks or months to the right to harvest several hundred thousand board feet over a three- to five-year period from a well-identified unit of land within the U.S. National Forest System. Forest concessions refer to the right to harvest and manage tens and even hundreds of thousands of hectares of public land for longer periods like 25 or 50 years in Canada and some developing countries. Despite the distinction, the economic principles behind timber sales and forest concessions are similar. We can discuss them jointly in this chapter and we will use the terms "contractor" or "logger" or "concessionaire," and "sales" or "concessions," interchangeably.

The general issues for negotiation between the landowner and the contractor—or concessionaire or logger—are the financial compensation for the timber rights, the duration of the contract, and the environmental obligations of the contractor. These are the central issues for contracts between private landowners and private contractors and they are also the focal points in the policy discussion regarding timber concessions on public forestlands. Illegal logging and the landowner's effectiveness in monitoring and enforcing the terms of the contract is an additional issue of broader policy discussion. Finally, it will become apparent that the arrangement for the transfer of rights itself, the structure of the auction or other mechanism for the timber sale, includes options that could be fruitful issues of policy discussion for the administrators of public lands.

The chapter begins by establishing the place of concession management within the three-stage spatial model of forest development, and continues with an assessment of the common alternative contractual arrangements for the transfer of rights from the landowner to the contractor. (An appendix develops the formal mathematical model.) The final section of the chapter reviews applications in Western Europe, within the U.S. National Forest System, on private lands in the U.S. South, on public lands in Ontario in Canada, and in Indonesia. These applications illustrate key issues of the previous conceptual discussion, as well as some of the issues of contention and opportunities for improvement in the contracts for forest concessions.

Timber Concessions within the Spatial Model

The economics of harvest rights, whether for managed or natural timber, are similar. They can be described in terms of Figure 5.1—which is closely related to the three-stage model of forest development. The fundamental product for exchange is some form of mature timber—known as "logs" in the delivered form. The vertical axis of Figure 5.1 refers to the delivered price for a fixed volume of logs. The delivered price is the same for all logs of similar quality regardless of their original location in the forest. The horizontal axis refers to the relative accessibility of the unharvested product, standing timber, to the processing facility or other central market point.

Loggers and concessionaires are interested in mature timber or timber that will be mature when they are ready to harvest it. For managed stands, the landowners have already incurred the expenditures necessary to grow the timber to maturity. Natural stands at the frontier are already mature and there were no costs associated with their growth. Therefore, in either case, for managed or for natural stands, the relevant costs at the time of the contract are only the costs of logging mature timber and hauling the product to the mill.

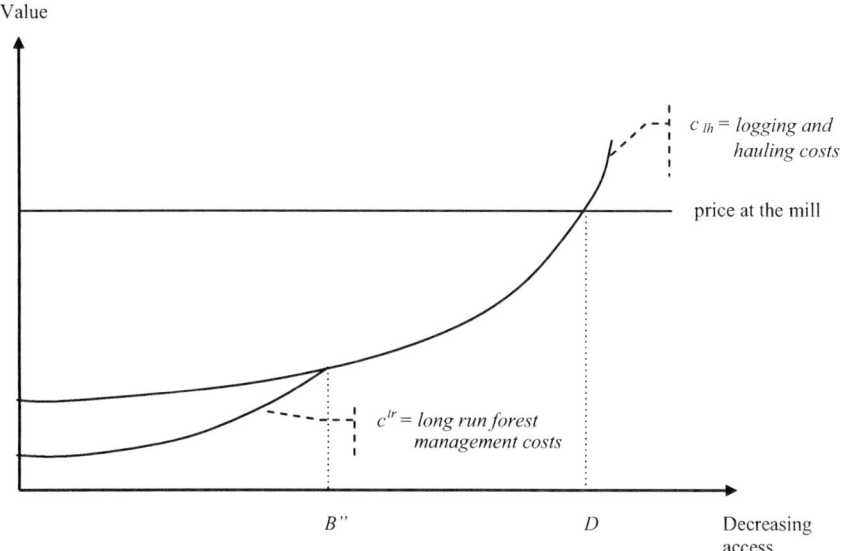

FIGURE 5.1 Contracting for mature timber

Logging and hauling costs c_{lh} rise with decreasing ease of access. As logging continues at locations of decreasing access or ever increasing distance from the mill, these costs continue to rise until, at some distant location, they eventually equal the delivered price of logs at the mill. At this point, point D in Figure 5.1, the delivered log price exactly covers all the costs of removing a unit of timber from the forest. These are the full costs for removing timber from natural forests but only the short-run costs of harvesting timber from managed forests.

Logging to this point leaves the owner of managed timber growing at point D with no excess revenue to compensate for the prior activities associated with timber management and growth. Owners of managed stands calculate their full long-run costs, c^{lr} in Figure 5.1, as the costs of timber management incurred over the full timber rotation plus the short-run logging and hauling costs. The delivered log price is sufficient to compensate these full long-run costs at point B''. Beyond B'' the costs of harvesting at the natural frontier are less than the combined costs of growing and harvesting managed stands. While landowners with mature timber on lands beyond B'' may be willing to sell that timber for any price in excess of c_{lh}, they will refrain from subsequent management on these lands and they may even abandon them. Therefore, point B'' in Figure 5.1 corresponds to point B'' in the three-stage model of forest development, and the vertical distance between the horizontal price line and c^{lr} corresponds to the vertical distance between the net forest value function V_f and the higher of either the net agricultural value function V_a or the transactions cost function T_r in the three-stage model.

Point D in Figure 5.1 also corresponds to point D in the three-stage model, and the distance between price and c_{lh} in Figure 5.1 is comparable to the distance between the net forest value function V_f and the horizontal axis in the model. At any point in Figure 5.1 this distance is the *in situ* value of the standing unharvested timber (the stumpage value) at that point. As in the three-stage model, previous logging will have degraded the natural forest between B'' and D, and loggers will not extend their harvest activities into the mature forest beyond D until such time as either delivered log prices increase or logging costs decline.

Landowner Objectives

Some landowners do their own logging but most contract their logging to others who possess specialized equipment and skills. The objective of most landowners in their contracts with loggers is to obtain the maximum financial return consistent with protecting certain characteristics of the surrounding forest environment.

The constraint on the forest environment is necessary because most landowners desire continued productivity from their lands after the logging is complete and the loggers move on. Loggers do not share these longer-term environmental concerns because they retain no continued interest in either the land or the residual forest after the logging operation is complete. Because logging operations do have the potential to damage the land's continuing productivity, landowners insure their longer term interests with environmental restrictions written in the terms of the contract and they obtain the logger's agreement to these terms before they allow the logging to begin.

The conditions for environmental protection vary with the characteristics of the land and trees and with the specific longer term objectives of the landowner. For example, where forest regeneration and continued timber production are a landowner objective, the contract may specify the species, location, and minimum size of the residual standing trees that form the basis for forest regeneration and future growth after the logging has been completed. Where erosion is an issue, the contract may specify restrictions on logging technologies, on logging in riparian zones, and on the location and maintenance of skid trails and logging roads. Where fish and wildlife are a concern, the contract may restrict some logging activities and specify the condition of the residual habitat after the logging has been completed. Where scenic value is important, the contract may specify restrictions on the location of the logging itself as well as restrictions on the disposal of the slash that remains after logging. Where the risks of losses to fire, insects, and disease are critical, the contract may impose additional restrictions on the disposal of logging slash and on the physical design of harvest boundaries. Of course, many landowners have multiple environmental objectives and their contracts specify multiple environmental restrictions.

These environmental concerns are especially important for natural forests at the economic frontier. Public agencies are responsible for much of the frontier forest because it had little financial value at the time private rights to this land were originally assigned. The public agencies often obtained this land and the stewardship responsibility for it by default. Environmental concerns can be greater on these lands, in part because public agencies have a greater responsibility for non-market values. However, the environmental concerns for these lands are also greater because the lands themselves tend to be more topographically, geologically, and environmentally diverse than the gentler terrain and lower elevation lands which tend to support most managed forests. Their diversity makes these frontier lands aesthetically appealing, but it also increases the risk of damage to them during logging operations. Finally, the lower *in situ* value of the frontier timber relative to the *in situ* value of timber from managed forests means that even comparable environmental values and environmental risks take on greater importance *relative* to timber harvest values at the frontier.

Indeed, the analysis in chapters 2 and 3 predicts that market timber values decline to zero at the frontier. This proposition demands further examination. In contrast with the argument that timber prices decline to zero, it is clear that loggers pay very large sums for some timber sales at the frontier; for example, for some competitive timber sales of the U.S. Forest Service. In economic terminology, these values are rents—values that are not due to investments in timber itself. Observations of significant rents beg explanation, and descriptions of the characteristic conditions under which significant rents have occurred should help us anticipate those situations under which other examples of significant prices and significant opportunities for timber concessions at the frontier will occur in the future.

Normal gradual upward adjustments in log prices and downward adjustments in logging costs create only small rents, and even these disappear where those who bid for the timber anticipate the adjustments. Where larger rents are apparent, they must have been created by something more than these modest adjustments. The obvious candidate is the relaxation of previous restrictions on the timber harvest activity. Restrictions could have occurred in the forms of either limited access to the forest or public policies that made prior harvest activity uneconomic. In both cases, the rents are truly returns to the activity or the decision that changed the prior restriction rather than returns to the land or the timber harvest operation itself.

As examples of the former, many forests in northeastern Thailand, the Amazon, and the interior of the United States were unavailable for harvesting until roads were built into these territories, and forest resource values alone generally could not justify the cost of these roads. Thailand's military built roads into the northeast for domestic security purposes in the 1960s. These roads opened the region to human settlement, but also to logging that was uneconomic until the roads were built. A 360-mile pipeline linking gas fields in Bolivia to a distribution center in Brazil is one of several Amazon examples. The pipeline and its adjacent road, completed in 2001, opened fifteen million hectares of the Chiquitano forest to logging (Grimaldi 2002). For the United States, Barlow and his colleagues (Barlow and Helfand 1980; Barlow et al. 1980) and Wolfe (1989) show that the National Forest System has lost money on timber sales in every year of its existence despite receiving substantial revenues from the sale of its timber. The losses were often due to the large cost of building roads into timber harvest sites (Zimmermann and Collier 2004). In fact, external financial support for roads into heavily forested areas has created and continues to create substantial rents in numerous heavily forested parts of the world.

The allowable cut policy is an example of a second general source of rents at the forest frontier. Appendix 3b discussed this policy, and its impact on the forest. The allowable cut policy is the basic model for long-term forest management and sustainable forestry taught in most forestry colleges around the world. It is common to western European and North American forest ministries, and also to most government ministries in the developing world. Where the allowable cut policy is enforced successfully, it restricts some harvests of financially viable timber and allows timber rents to build up in some highly valued public forests—such as those in western Oregon and Washington in the United States (Walker 1974; Kutay 1977; Hyde 1980; Clawson 1976). At the same time, it has justified building roads and creating new rents in some otherwise uneconomic areas—such as some forests in the Intermountain West (e.g., Hyde 1981).

Enforcement is the final element in the landowner's interest. Wherever the rents are substantial and the landowner desires to recover them as revenues, or wherever the timber rents are positive but the local environmental values are also great, or both, then monitoring and enforcement (M&E) to ensure the terms of the contract becomes an important additional cost for the landowner. We can anticipate that M&E is more difficult at the frontier than for managed forests because it is part of the cost of establishing and maintaining property rights, a cost that increases with decreasing access.

The Complete Model: Landowner Objectives and Logger Objectives

Each of these elements, revenue recovery, environmental restrictions, and contract violations and enforcement, can be incorporated in Figure 5.2—which is only a variation on Figure 5.1. Figure 5.2 can represent either the entire frontier, as in the right-hand extent of Figure 5.1, or it can describe the neighborhood of one characteristic timberstand or forest concession located at the frontier. The axes of Figure 5.2 are identical to those of Figure 5.1. The same functions continue to describe the delivered log price and the cost of logging and hauling to the mill, and point *D* continues to identify the financial frontier.

Lump sum contracts: The area *pdc* between the price line and the logging cost function is the rent. A landowner who is determined to collect a maximum share of this rent sells the timber within this land area for a single lump sum, and competitive bidders offer up to the full net value of the standing timber described by the area *pdc*. The contract is known as a lump sum contract because the sale arrangement is for the entire timber resource within the land area in question and the bid value is one single sum. Since each logger's estimate of *pdc* will differ (as their alternative sources of timber, their own processing facilities, the markets for their products, and their own abilities to appraise timber all vary), the landowner obtains the maximum revenue by asking each potential logger to submit a best bid

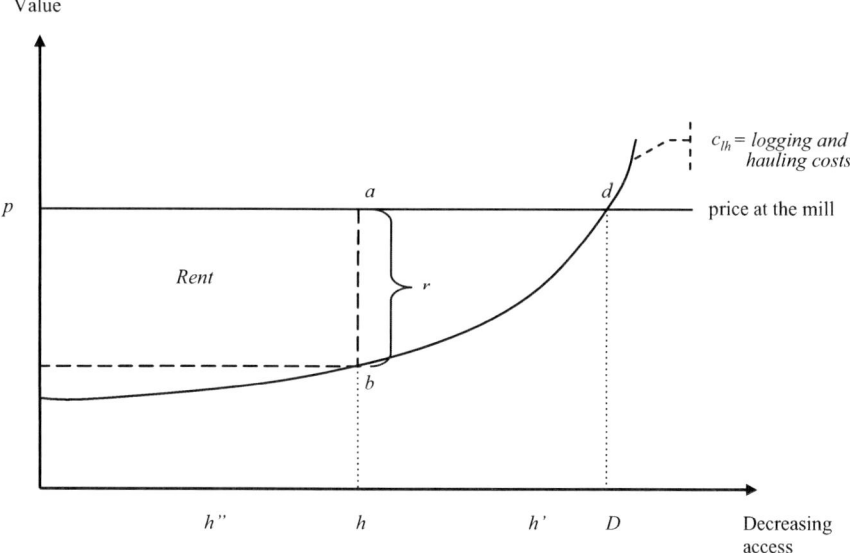

FIGURE 5.2 Rent under a uniform royalty

for the full harvest region out to point D. The revenue maximizing landowner then selects the highest bid from among those submitted.

The logger who submits the successful bid then begins harvesting and continues the harvest operation all the way to point D. At any location such as h within a timberstand or forest concession, the logger harvests large and high quality trees of net value greater than or equal to bh. Smaller and lower quality trees do not yield a return sufficient to cover their logging and hauling costs. The logger leaves these smaller and lower quality trees, and some of them may be damaged in the effort to remove the more valuable trees in the least-cost manner. Therefore, at less accessible locations, such as h', the logger harvests fewer trees than at h and leaves a larger residual stand.

The logger's decision to take the most valuable trees and leave the others is known to foresters as "high grading". Foresters with long-term interests in the quality of the remaining forest detest high grading as it leaves trees of lower quality as the basis for the growth in the successive future forest and, therefore, the next forest that grows from these trees will also be of lower quality. Regardless of this longer term disadvantage and regardless of the structure of the timber contract, the logger's incentives always favor taking the most valuable trees. Therefore, the logger's incentives always favor high grading, and high grading always occurs unless the contract between the landowner and the logger specifies different preferred and enforceable characteristics for the residual forest after the logging operation is complete.

Stumpage price contracts: Unfortunately, the tradition for timber sales differs from this model. Foresters generally assess timber value by log volume or by volume on the stump, and not by the land unit. Therefore, they tend to sell timber at a uniform price per standing unit known as a royalty to economists or a stumpage fee to foresters. For a royalty of r, for example, loggers harvest out to point h were the royalty equals the delivered value of a log minus its logging and hauling costs. Alternatively, where landowners desire to sell the timber within an area—out to point h for example—but ask for bids in terms of log value, then loggers bid up to a royalty of r. As with lump sum sales, successful loggers harvest large and high quality trees of standing net value greater than or equal to *r* and leave a residual of smaller and lower quality trees standing when their harvest operations are complete.

The landowner who sells timber for a royalty of r receives revenues equal to r times the quantity harvested out to point h, or area *pab(p-r)*. The landowner who desires to maximize revenue under this contractual arrangement sets the royalty at a level that induces the winning logger to harvest all the way to the point where the elasticity of the royalty with respect to recovered revenues is equal to one. The elasticity of the logger's cost function is also equal to one at this point. Total revenues decline for greater royalties (and smaller harvest levels) to the left of this point and also for smaller royalties (and greater harvest levels) to the right of this point.

The logger claims any log value in excess of r. Therefore, the logger captures a portion of the rents equal to the triangular area *(p-r)bc* for timber harvested out to point h. The residual area *adb* in Figure 5.2 represents a stand of trees of positive net value that remains in the forest after the contracted logging is complete. This positive value is a harvest incentive for any logger who can obtain access to these lands and trees. It is an incentive for the logger to violate the contract and harvest logs from the farther extent of the financially viable timberstand beyond point h out to point D. Thus, the form of the contract itself is one incentive for illegal logging.[1]

Monitoring and Enforcement

Landowners oversee two categories of contractual obligation on their loggers, obligations to satisfy environmental standards and, for contracts written in terms of stumpage prices, obligations to pay the royalty. These responsibilities impose the additional costs of monitoring and enforcing (M&E) the contracted performance.[2] Therefore, they impose new complexity on the system. These costs are least when the landowner imposes no environmental restrictions whatsoever and

---

1 Some arguments in this section and following sections may be clearer in a mathematical exposition. See appendix 5a.
2 Foreign governments and international agencies have become involved in attempts to regulate against the international flow of illegally harvested logs. M&E becomes very complex when a chain of agencies and individuals—from foreign governments and international agencies to central government agencies and officials through layers of regional agents and on to various local officials—become involved. For those cases, M&E in our discussion includes the sum of all

permits logging all the way to the financial frontier at D in Figure 5.2. They increase as the landowner adds environmental restrictions and increases the royalty and, therefore, the incentive to log illegally. For any specified set of environmental standards and any set level of royalty, the M&E costs per unit of land or unit of stumpage may rise as additional land comes under the environmental restriction, or as we move to the right in the figure because less accessible land is often on more difficult terrain and, therefore, the environmental risks associated with harvesting it can be greater.

Lump sum contracts: With lump sum contracts, the logger agrees to a single total contracted fee before the harvesting begins. The logger pays the contracted fee and the landowner's M&E activities can focus on the logger's compliance with environmental restrictions. Both the landowner and the logger recognize new costs, the M&E costs for the landowner and the costs of environmental compliance for the logger. The incremental M&E costs for the landowner now rise to the level of the delivered price at a lower harvest level—to the left of point D in Figure 5.2. Similarly, the combined incremental harvest and compliance costs for the logger also rise to the level of the delivered price at a lower harvest level. However, there is no a priori reason that the landowner's marginal M&E costs should equal the logger's marginal compliance costs. Therefore, there is no reason to expect that the new optimal harvest levels will be the same for both parties.

The landowner has two options for enticing convergent behavior from the logger: The landowner can design the pattern of M&E costs such that the last increment of landowner M&E costs ensures an equal increment of compliance costs by the logger. This is difficult because both costs vary with local environmental conditions and, furthermore, because the logger has no incentive to share reliable information on compliance costs with the landowner.[3]

---

domestic landowner and international activities designed to regulate timber harvests. Of course, the more layers involved, the greater the expense and the greater the opportunity for malfeasance. Gunatilake (2007) suggests that these layers of malfeasance may account for 40 percent of total log value in Sri Lanka. Paudel, Keeling, and Khanal (2006) estimate that the cost of malfeasance can be six times the royalty received by the government in Nepal.

3 Becker (1968) first formulated a general economic framework for optimal enforcement. He includes the opportunity for bribe taking as well as M&E costs and, of course, bribe taking is a complication in forestry. Becker also points out that it might not be optimal to prevent all illegal activity. Mookherjee and Png (1995), Burlando and Motta (2007), Klerman and Garoupa (2002), and Macho-Stadler and Perez-Castrillo (2006) each develop these ideas further for general and for environmental cases. Burlando and Motta show that where enforcement costs are high and monitoring is difficult, then governments may optimally permit some level of illegal activity. Macho-Stadler and Perez-Castrillo show that firms are likely to participate in some level of illegal activity unless the government M&E budget is very large and that, in the presence of limited budgets, governments optimally focus first on those situations that are easiest to monitor. Of course, each of these general cases is relevant in the forestry case, and they introduce complexity beyond our simpler example. Robinson and Lokina (2008) apply these ideas for forestry with special reference to examples in Tanzania.

Alternatively, the landowner can assess a lump sum for environmental costs at the outset. This removes incremental environmental considerations from the daily decisions of the logger, and the landowner only has to satisfy his or her own concern that the pattern of M&E is worth the effort in terms of the logger's improved environmental performance.

Uniform fixed royalties: The prospect of illegal logging creates an entirely new problem for uniform fixed royalties, a two-step problem. The landowner and logger accept an agreed-upon royalty as one of the terms of the contract. This is the first step. The royalty identifies the level of contracted harvests. The logger's actual harvest decision is the second step. As the logger retains the incentive to harvest additional, uncontracted or illegal, logs under contracts of this form, the logger's actual harvest level may include additional and illegal harvests beyond the limits of the contract.

Adding environmental standards complicates the problem still further. The effect of environmental standards on these contracts is comparable to its effect on lump sum contracts in that the costs imposed by the environmental standards cause both landowners and loggers to decrease their own perceptions of the optimal harvest level. As in that prior case, there is no reason to expect that the reductions in the final optimal harvest level for landowners and loggers are similar.

Furthermore, both the logger and the landowner recognize another round of costs associated with the potential for illegal logging, a second round of M&E costs for the landowner and, for the logger, either the costs of anticipating the landowner's monitoring and avoiding detection or else the cost of the penalty itself when the landowner does detect the logger in illegal activity. The landowner's monitoring activities are easier when access from the harvest site to the mill is limited and monitoring can be conducted at the mill or, similarly, when transshipment is limited to a few easily monitored points. Nevertheless, the monitoring responsibility always imposes at least small costs and the logger can almost always find a way to avoid some of the monitors.

These M&E costs are an incentive for the landowner to increase the contracted harvest level and, thereby, to decrease the new M&E costs—as M&E is less costly when the landowner imposes fewer restrictions on the logger. The logger's avoidance costs or the cost of the penalty itself are an incentive to decrease the level of illegal activity and, thereby, to decrease the avoidance costs or penalty. However, since the costs of M&E as well as the costs of avoidance vary from site to site, and because neither the landowner nor the logger knows the other's cost schedule, there is still no reason to anticipate that the optimal harvest preferences of the two parties will converge.

Since government agencies manage the overwhelming majority of land at the frontier, most contracts for timber on this land tend to be between government agencies and loggers, or between government agencies and concessionaires or millowners who arrange for their own logging. Government agencies, as large operations responsible for many forest lands, tend to rely on standardized rules and procedures for monitoring and standardized penalties for contractual violations. Under such arrangements, broad categories of penalties are generally the same for all sites managed by the same agency. More specialized, site specific, procedures, and penalties would impose an additional administrative burden. Therefore, while the logging activity is site specific, these standardized arrangements are not site specific and they create all the more reason for divergence between the preferred harvest levels of the landowner and the logger, and all the more reason for us to anticipate that some level of illegal logging will be common.

This is an important point for forest policy. It means that the structure of the logging contract is a source of illegal activity in the forest. Poor governance and a lack of political will are often blamed for illegal activity. Surely they are the sources of some illegal activity, but it is incorrect to blame them as the only sources while overlooking the characteristics of forest value functions, the role of the contract itself, and the arrangements for its enforcement.

Specialized Considerations within the Contract

Four additional broad issues arise in many discussions of timber sales and forest concessions: the time period or duration of the contract, the use of environmental bonds to insure the satisfactory performance of loggers or concessionaires, the effect of a limited number of bidders, and the preferred distribution of timber rents.

Time period or duration of contract: Timber contracts range from brief periods of a few months to 25 or even 50 years. Shorter contracts are associated with harvest operations alone, and often with harvests from managed stands of private timber, although shorter-term contracts are also used by some landowners with frontier stands of natural timber. Longer periods are associated with larger timber concessions at the frontier. Of course, loggers or concessionaires participating in longer-term contracts have a greater opportunity to return to the same land unit at some future time to harvest timber from the new growth that follows their prior harvest operations. Therefore, concessionaires with longer-term contracts have a greater incentive to conduct logging operations in a manner that leaves the residual forest in good condition for valuable future growth, and these concessionaires also have an incentive to participate in improved management on the land following their logging activities and before they return to log a second time.

There are limits, however, to improved concessionaire performance because even the long-term objectives of concessionaires are not necessarily consistent with the overall objectives of the landowner. For example, the concessionaire may not even share long-term timber stewardship objectives with the landowner if the concessionaire expects that alternative future sources of mature timber will be available elsewhere and at less than the costs of careful logging and continued management at the site of the current contract. Moreover, even if the concessionaire and the landowner do share long-term timber management objectives, the landowner may have additional environmental objectives that are not shared by the concessionaire.

Therefore, we can conclude that longer-term contracts may improve concessionaire performance—and that is their advantage—but even longer term contracts require some form of monitoring to insure that the concessionaire satisfies the environmental standards of the contract and the landowner.

Environmental performance bonds and performance awards: Fines, the confiscation of illegal timber, and the cancellation of timber contracts are common penalties for logging operations that are in violation of the terms of the contract. More recently, some have suggested environmental bonds as an alternative means for ensuring concessionaire performance in accordance with the environmental standards of the contract (e.g., Paris, Ruzicka, and Speechly 1994; Magrath et al. 2007).

Environmental bonds are financial deposits made as guarantees of satisfactory concessionaire performance. Concessionaires make these deposits with the landowner or with an independent authority prior to the start of the logging operation. The landowner either returns the bond upon satisfactory completion of the contract or keeps the bond or a portion of it as a penalty for less than satisfactory performance. The potential return of the bond is the logger's incentive to satisfy the terms of the contract.

For penalties of any sort to be effective the penalty itself must match the magnitude of the potential damage. Therefore, the size of an environmental bond must match the magnitude of potential environmental risk due to the logging operation. This means that an environmental assessment of the site and of the cost of mitigating potential environmental damage is necessary before the landowner or manager writes the contract and sets the level of the bond. Since environmental risks and the costs of their mitigation vary from site to site, the size of the bond must also vary from site to site and, therefore, from contract to contract. There is no easy solution to this problem. For example, there is no reason to anticipate that effective bonds will bear a regular proportional relationship to contracted timber revenues because sites with lower-valued timber are often associated with steeper terrain and shallower soils and, therefore, with the riskier environments that would require larger environmental bonds.

Environmental performance awards are similar to environmental bonds. They reward satisfactory performance at the completion of the timber contract with a financial supplement or with favored recognition in the negotiation of future contracts. These too, to be effective, must vary from site to site, just as environmental bonds vary from site to site.

Limited numbers of bidders: Some timber sales are competitive among many bidders. Many others attract only one or a very few bids. In fact, the latter is common because timber markets are contained within the distance that loggers can profitably transport their logs, a distance that seldom exceeds 350 km. anywhere in the world and is often much less. The number of independent mills (and loggers working for them) within this distance from any timber concession is seldom more than a handful. The cost of transporting greater distances is too great and the volume of mature timber within a region of this radius seldom permits more than a few mills of efficient scale. Moreover, not all the mills within a timbershed will choose to bid on every available timber sale. The design of some mills restricts the size and species that comprise their preferred input of logs. Other mills will have sufficient inventories of logs or standing timber from previous contracts.[4]

Where the number of bidders is small, those who do bid may have market power over the many local landowners, and the bidders' offer prices for timber may be less than the competitive market price. In this case, landowners looking to increase their revenues may estimate the competitive market price themselves and attempt to use this estimate, rather than the mills' or loggers' bid prices, as a minimum sale price. Landowners can compare the characteristics of their timber, the accessibility of their site, and the difficulties of logging it with other sites and other markets in the same general region, and then estimate a fair price for their own timber based on the price experience at these other sites adjusted for the differences between those sites and their own. This method, relying on market comparisons, accurately reflects competitive market prices as long as the sources of comparison are themselves competitive markets with numerous buyers and numerous sellers.

Alternatively, landowners can derive an estimated competitive market price by first observing the price in a competitive market for a processed wood product, the price in a competitive lumber market for example. From this lumber price, they can subtract the estimated cost of mill processing as well as the hauling and logging costs to arrive at an appraised competitive price for the landowner's own standing timber.[5] This appraised price becomes the landowners' minimum offer price (a "reservation price" in economic terminology).

In some cases, the appraised price has been only a starting point for competitive bidding and the final sale price is much higher. A much higher sale price is evidence that the appraisal was faulty and faulty appraisals call into question the reliability of the estimated costs in the calculation. They should also cause large landowners, such as public land management agencies, to inquire into the reliability of their similar appraisals for other timber sales. Fortunately, mill processing, hauling, and logging costs from other, competitive, markets provide a basis for accurate estimates of these costs in non-competitive markets and, therefore, a basis for accurate revisions of inaccurate appraisals in those non-competitive markets.

---

4 Mead's (1966) examination of bidding for National Forest timber in the U.S. Pacific Northwest provides detailed examples. Mead considers oligopsony (limited buyers) to be the result of collusion. Of course, collusion may exist but oligopsony is also a natural result of geographically contained markets and limited numbers of buyers who are aware of each others' logging inventories and, therefore, each others' willingness to bid on the next timber concession regardless of any attempt to collude. Baldwin, Marshall, and Richard (1997) also discuss bidder collusion in U.S. Forest Service timber sales.

5 Weintraub (1959) reviews the economic rationale of this second approach as it is applied by the U.S. Forest Service.

The distribution of timber revenues: The distribution of timber revenues from state-owned forests has become an issue of contention for observers of many developing country forest ministries. Some observers feel that the state forest ministries or public treasuries should receive larger shares of these revenues (FAO 2001; Vincent and Gillis 1998; Gautam et al. 2000). On the other hand, some of the same observers (FAO 2001; Gautam et al. 2000; World Bank 1995), and others as well (Jakarta Post 2000; Cerutti, Lescuyer, and Mvondo 2010), observe that these government agencies dissipate the revenues they do receive in unwise or at least dubious investments and in personal payments to favored citizens; often this is simply graft and corruption. What is the best distribution of these timber revenues? The answer depends on source of the revenues in some cases and their disposition in others.

The distinction between managed and frontier forests is crucial to this discussion. The payments for timber from managed forests are returns for investments that landowners or land managers previously made in their lands and forests. The prospect of these payments was the incentive that originally induced the owners to make the investments. Therefore, the payments belong to the landowners. Without these payments, landowners lose the return on their investments and lose the incentive to continue investing in their land and standing timber. Where the state is the owner of these managed forests, these payments are appropriately allocated to the forest ministry or the government treasury.

Timber revenues from frontier forests, however, are not returns on forest investments. They may be returns to activities like roadbuilding or to a change in forest policy. Returns from investments in roadbuilding might become the property of the roadbuilder as part of the incentive for building the road. Returns to a change in forest policy that frees up previously unavailable timber could become the property of the policymaking government agency. However, the existence of these policy rents raises questions about the original policy and the judgment of the agency that established it. If the agency made an unwise decision that restricted harvests and created these rents, a decision that it now decides to revise, will the same agency make related unwise decisions in the future and, perhaps, use its new timber revenues unwisely as part of these decisions? Is such an agency the best consumer of rents that could produce greater benefit in other public programs? The answers to these questions must vary from agency to agency and, even then, with individual opinion.

In any event, revenues from timber harvests that occur only because of a revision in harvest policy are not a return for timber investments. In economic terminology, they are an "unearned increment" and their allocation has no effect on the optimality of future forest investment decisions. Future investment decisions should be made on their own merit and not because some unearned accumulation of these prior rents already exists.[6]

Therefore, these frontier rents might reasonably be allocated wherever they

---

6 Hirschleifer (1974) discusses this point with respect to the allowable cut policy.

create the greatest benefit, wherever they can be invested for the greatest public gain. They might be deposited in the public treasury from which the government allocates funds for a wide variety of socially rewarding activities. However, we recall that some observers feel that some government agencies dissipate their revenues in ways that fail to benefit the public and the country. Furthermore, we also know that private investors often make investments in their own self-interest that require the purchase of local goods and the employment of local workers. These private investments often are more beneficial to the country than many misdirected public expenditures. The only possible conclusion is that the best allocation of these frontier rents is an empirical question. Its answer will vary from case to case depending on the public or private recipient's expected use of the funds and the broader benefits for the public that may be anticipated from this expected use.

Empirical Applications

The final section of the chapter reviews the arrangement of temporary timber rights in a number of different situations—by public and private landowners, over shorter and longer terms, and in developed and developing countries. These examples illustrate most of the issues discussed in the last few pages.

Nineteenth-century Germany

Professional forestry traces its roots to the management of forested estates in nineteenth-century Germany. The early German foresters managed the forest, including the logging operation. Each estate's own crew, under the direction of the estate's forester, harvested the timber, hauled the harvested logs to the roadside, stacked them ready for sale, and supervised the sale itself. The estates sold logs, not the right to harvest forested land. Sales that were priced according to log volume and value were the best mechanism for extracting the full value of the product for the seller. This organization of forest operations continues in much of Germany today. A forester manages the land, supervises the logging crew, and arranges sales of harvested logs. In this case, the log is the true product, and sales by log units remain the most transparent focus of market exchange.

The organization of timber sales is different, however, in most of the Americas, Africa, and Asia. Most landowners do not sell harvested logs. Rather, they sell the right to harvest timber as it stands on the stump. They may monitor the logging operations, but what they are selling is the right to harvest marked units of their forests. A crew that is responsible to the purchaser of this right, the logger or concessionaire, does the logging. The seller's interest is in obtaining the greatest value for a different product, in this case the unit of mature standing forest. We showed that a lump sum sale accomplishes this objective and that sales that are priced by the log or by a unit of standing timber do not. Nevertheless, tradition is strong, and many modern timber sales follow the procedure that was appropriate in nineteenth-century Germany.

The U.S. Forest Service

Professional forestry in North America, and the government forest management agencies themselves, trace their early history through the careers of a few distinguished public servants who were trained in Germany or by German-trained foresters who taught in the United States and Canada.[7] The government agencies continue to follow the German tradition, modified for sales of standing timber in the forest rather than sales of logs at a roadside landing. For example, the system followed by the U.S. Forest Service begins with an appraisal of stumpage value. A Forest Service appraiser records the local lumber price and subtracts mill processing, hauling, and logging costs, plus a premium for the risk and uncertainty faced by the logger and the millowner. The appraiser converts the residual from this calculation from lumber units into log units and the Forest Service advertises the standing timber within a well-marked forest area, generally less than fifty hectares, for sale at this price per unit of stumpage volume. Potential buyers participate in a sealed bid auction for which the advertised price is the lowest admissible bid. The bidder who offers the highest value per unit of stumpage wins the right to harvest the stumpage at that winning price and within a period that generally does not exceed three years.

Some potential timber sales appraise at negative prices. Some receive only a limited number of bids—as Mead (1966) documents. Others are more competitive, and bid prices occasionally exceed 300 percent of the advertised price. In the latter event, the appraisal clearly was not an accurate reflection of the market. Furthermore, the analysis explained along with Figure 5.2 anticipates that, under this contractual arrangement, loggers have a substantial incentive to add illegal logs to those that are legally contracted and we know that illegal logging does occur. Consideration of the lump sum alternative to the U.S. Forest Service system of appraisals and contracts might be in order.[8]

Private Landowners in the U.S. South

The timber sales of some private landowners in the U.S. South, and some in Western Europe as well, provide a contrasting example. The non-industrial landowners of the South are individually small but, as a group, they are important, producing more than 60 percent of annual southern timber harvests (USDA Forest Service 2005). The primary occupations of these landowners are often different from forestry and the landowners themselves are not steeped in forestry's traditions. They may even have other sales experience in their primary professions. Many manage their forests as an avocation and as a source of supplementary income. Once they are ready to harvest their timber, some hire the services of

---

7 See Clepper (1971) for extensive discussion of this history.
8 The annual volume of timber sales by the U.S. Forest Service has declined substantially since the spotted owl court decision in 1991. However, that timber the Forest Service does sell still follows the same procedure as before.

a consulting forester to advise them, but they do not do the logging themselves. They sell their timber on the stump, just as the U.S. Forest Service does, but these non-industrial landowners often sell in lump sum sales. Their success with lump sum sales makes a recommendation that the U.S. Forest Service examine this contractual alternative all the more compelling.

Ontario

A different arrangement, common in much of Canada and many developing countries, incorporates elements of both short- and long-term obligations between the landowner and the concessionaire. The Province of Ontario in Canada provides an example.

The province has negotiated concessions ranging up to 1.8 million hectares for periods of twenty years, with extensions following each five years of satisfactory performance. The original contracts, in the 1920s, obligated concessionaires to build a mill. Some original concessionaires also constructed roads and other infrastructure. All concessionaires today must own mills. Concessionaires do not contract for a fixed fee or a fixed volume of stumpage at the outset. Instead, they make annual payments according to the volume they harvest and an appraised stumpage price (OMNR 2001). Therefore, concessionaires can adjust their harvest levels (within limits established by the Ontario Ministry of Natural Resources) in response to market demand and their own changing mill requirements. This system suffers the disadvantage of uniform fixed royalties (for each year), just as the system of the U.S. Forest Service suffers the same disadvantage. However, the longer term of the Ontario contracts creates the opportunity for some level of annual harvest flexibility and, from the concessionaire's perspective, this is a very desirable feature of the Ontario contracts.

The allowance for variation in harvest levels shifts back to the landowner some of the long-term uncertainty the concessionaire absorbs when contracting for a single lump sum sale at the outset (Leffler and Rucker 1991; Niquidet and van Kooten 2006). Long-term lump sum sales contain greater uncertainty for the concessionaire than sales based on actual harvest volumes, and loggers or concessionaires lower their bids for such sales accordingly. Therefore, for the long-term concessions of the OMNR, there is a tradeoff between efficient long-term lump sum auction pricing on the one hand, and higher competitive market bid prices based on periodic harvest volumes on the other. The preferred alternative must depend on total revenues collected, the volume of illegal harvests, and the costs of monitoring and enforcement. It is not apparent which alternative is preferable in any general way for long-term concessions.

Wood processing in Ontario is strongly oriented toward the very capital-intensive pulp and paper industry. This industry's capital intensity makes it immobile and it makes a reliable timber supply more important than for a less capital-intensive, more mobile, industry like sawmilling. However, the very large scale of Ontario's pulp and paper operations means that most forested regions cannot support more than one mill. This means that long-term concessions provide the necessary security of raw material supply to induce industrial development, but it also means that most local markets are non-competitive and that competitive market pricing is unlikely. In this situation, some form of appraisal by the OMNR becomes necessary regardless of the form of the contract.

The OMNR conducts short-run appraisals for the purpose of pricing each concession's annual harvests. It conducts these appraisals according to a formula similar to that for U.S. Forest Service timber sales. Each concession's own mill is the basis for the estimate of processing costs within the appraisal for that concession. Other mills with variant technologies elsewhere in the province may have lower costs but the OMNR makes no attempt to estimate costs under competitive technologies. By using the costs of only the one local mill for each concession, the OMNR appraisal system allows the deduction of higher processing costs for older and less efficient mills. This removes the raw material cost incentive for modernization and it must result in an underestimate of competitive stumpage prices for some concessions.

Ontario has a modern forest industry, and its capital intensive pulp and paper mills have other market-based incentives to remain efficient by the most modern standards. Therefore, it is difficult to speculate on the amount that estimates of competitive stumpage prices would vary from the current estimates. Nevertheless, a revision of the appraisal system to rely on costs of current best technologies would create a better estimate of competitive stumpage prices and restore the raw material price incentive for mill modernization. It would probably increase the collection of stumpage revenues by the OMNR as well because Ontario's capital intensive mills are unlikely to either modify their capacity or to move their operations in response to moderate adjustments in their stumpage prices.

Indonesia

The Ontario experience is similar to that of many developing countries—although most developing countries do not have a concentration of pulp and paper facilities as a justification for long–term concession management. Indonesia provides an especially interesting example because Indonesia's forests and forest industry are so large and because many international environmental and economic advisors express concern over Indonesia's apparently low stumpage fees and large losses of both illegal timber and government timber revenues (Gautam et al. 2000; Brown 1999; Casson and Obidzinski 2002; World Bank 1995). Sizer (2005) estimates that half of all Indonesian logging is illegal and, further, that unreported log exports were worth US$1.4 billion in 2003.

Indonesia's Ministry of Forestry has granted large concessions of up to 400,000 hectares for 35 years to support new mills. It charges all concessionaires a virtually uniform royalty per unit of log volume for timber removed each year.[9] The

---

9 The actual system is somewhat more complex in Indonesia but the principle remains the same as for Ontario. Indonesia's royalty is actually a combination of two fees, the royalty itself (*IHH/ PSDH*, allocated to general revenues) and the reforestation fee (*Dana Rebosasi*, originally intended

International Monetary Fund (IMF; with advice from the World Bank) required the Ministry to increase the royalty as one of the conditions for assistance to Indonesia following the East Asian financial crisis of 1997. IMF and World Bank advisors anticipated that the increase in royalties would increase government revenue recovery and also decrease harvest levels (IMF 1998).

Some long-term concessions may be appropriate for Indonesia for reasons that are different from those in Ontario. Indonesia, like Ontario, requires that its forest concessions are tied to a wood processing facility. However, many of Indonesia's wood processors are plywood operations that are less capital intensive than Canada's pulp and paper mills. The argument for long-term arrangements based on the capital intensity of the processing operations is not as valid for plymills as for pulpmills. Therefore, the justification for longer term concessions in Indonesia must be based on the contention that Indonesia's concessionaires can manage better than the Ministry of Forestry. Perhaps they have resources for forest management that are unavailable to the Ministry and, therefore, they can manage better. And perhaps not. However, if they do, and if the concessionaires do manage their forests rather than simply operate as loggers of the frontier natural forest, then long-term concessions may have some basis, and the risk-shifting advantage of periodic revenue collection based on the volume of timber removed in each period could be an argument for Indonesia's system of uniform fixed royalties.[10]

Nevertheless, the merit of an approximately uniform national royalty is doubtful. Indonesia is a large and diverse country, ranging more than 4,800 km from east to west. Private log prices vary widely across this expanse, and the revenue maximizing royalties for government concessions must vary as well. Calculating the revenue maximizing royalty for each concession would be a difficult task—just as estimating the competitive stumpage price for each concession is a difficult task in Ontario. The data are more readily available for estimating the elasticity of the regional logging and hauling cost functions. (The appendix to this chapter shows that the two elasticities are identical.) We can anticipate that these too vary across broad geographic regions of the country. The Ministry could use the estimated regional cost elasticities to indicate where regional adjustments in the royalty would increase its collection of revenues. Of course, increasing the royalty will only increase revenues in regions where the cost elasticity is greater than one. There is no current evidence as to which, if any, of the Indonesia's 39 forested provinces fall into this category. Adjusting the royalty in the direction indicated by the regional elasticity would not ensure maximum revenue collection. However, periodic reassessments of the elasticities, and subsequent adjustments of the

---

for reforestation), both charged per unit of volume and both collected by the Ministry of Forestry until 1998 and the Ministry of Finance thereafter. Minor adjustments in the two basic fees are made for species differences and for price differences in outlying islands, and all concessionaires pay an additional small licensing fee. By far the greatest revenue collection is from the two essentially flat fees, the IHH/PSDH and the Dana Rebosasi.

10 The alternative of lump sum payments for each block of harvested land, and flexibility regarding the number of annually harvested blocks might have merit in Indonesia as well as Ontario.

regional royalties in the directions indicated by the elasticities, would insure that the royalty always moves in the direction of revenue maximization, and it would also accommodate periodic adjustments in logging and hauling costs and market prices.

Adjustments in the royalty would also affect the incentive for illegal logging. Therefore, it would be advantageous for the Ministry to estimate the elasticities associated with its observations on illegal activity and its experience with enforcement as well. These would be instructive of the relative effectiveness of enforcement in different regions. Adding this information to the elasticity for hauling and logging costs—as described in the appendix—would provide even better instruction for regional adjustments in the royalty.

Conclusions

This chapter reviews the specialized characteristics associated with the temporary transfer of the use rights to trees and forestland. The problems associated with such transfers, particularly the problems of pricing and illegal logging, are especially important for modern timber policy and timber management.

The two principle contractual forms for these transfers are lump sum transfers and transfers payable in uniform fixed fees per unit of product. Lump sum transfers are payable as one sum agreed upon by the landowner and the concessionaire or logger for the contracted use of the land or for harvestable trees and within a fixed period of time. If the contract is for standing timber, then the logger harvests all the timber of value greater than its logging and hauling costs. The landowner collects the maximum revenue from transfers made according to this contractual structure.

Unfortunately, the tradition in forestry is different and most transfers are conducted in terms of a price per unit, a stumpage price or royalty per unit of harvested logs. In this event, the landowner obtains maximum revenue from the timber sale by setting the royalty such that the elasticity of the royalty relative to total revenue collection is equal to one, and by setting the contract to allow harvesting out to the point consistent with this elasticity. However, some timber standing beyond this point will have a value greater than its logging and hauling costs. Its positive net value is an incentive for additional logging in violation of the contract, and this incentive is one explanation for observations of illegal logging around the world.

Both contractual forms can be modified to account for differences between the landowner and the logger in their long-term environmental objectives and for the potential for illegal logging. These modifications create further differences between the landowner's and the logger's incentives and, therefore, differences between their perspectives of the optimal level of logging. These differences reinforce the potential for illegal logging.

Various other adjustments occur in practice, notably for concessions of public timber at the natural forest frontier. The U.S. National Forest System sells such timber in 3- to 5-year contracts. The forest agencies of many other countries sell their frontier timber as large concessions of thousands of hectares over periods as long as 50 years. The latter arrangements introduce new problems of long-term risk management and noncompetitive markets. We can anticipate some of the problems and we can understand some of the preferred solutions, but many of the best solutions vary with the local markets and local environmental conditions.

The most important lesson from this discussion is that landowners and land managers must be clear about their own objectives and they must design their contracts to emphasize those objectives. Furthermore, their contracts must also provide for the landowners' understanding of the loggers' or concessionaires' objectives. Finally, the landowners must recognize that they cannot satisfy their own multiple objectives without some financial tradeoffs. (The tradeoffs should be clearer with reference to the mathematical appendix.) The ultimate success of any contract lies in understanding the conflicting objectives of the two parties involved and in each party understanding how much of one objective it is willing to forego to insure more of another objective or to insure its preferred behavior from the other party.

Literature Cited

Baldwin, L., R. Marshall, and J-F. Richard. 1997. Bidder collusion at forest service timber sales. *Journal of Political Economy* 105(4): 657–699.

Barlow, T., and G. Helfand. 1980. Timber giveaway — a dialogue. *The Living Wilderness* 44: 38–39.

Barlow, T, G. Helfand, T. Orr, and T. Stoel. 1980. *Giving away the national forests: An analysis of U.S. Forest Service timber sales below cost.* Washington, DC: Natural Resources Defense Council.

Becker, G. 1968. Crime and punishment: an economic approach. *Journal of Political Economy* 76: 169–217.

Brown, D. 1999. *Addicted to rent: corporate and spatial distribution of forest resources in Indonesia.* Jakarta: DFID/ITFMP.

Burlando, A., and A. Motta. 2007. Self reporting reduces corruption in law enforcement. Working paper 0063, Dipartmento di Scienze Economiche "Marco Fanno," University of Padua, Italy

Casson, A., and K. Obidzinski. 2001. From new order to regional autonomy: shifting dynamics of 'illegal' logging in Kalimantan, Indonesia. *World Development* 30(12): 2133–2151.

Cerutti, P., G. Lescuyer, and S. Mvondo. 2010. The challenges of redistributing forest-related monetary benefits: a decade of logging area fees in Cameroon. *International Forestry Review* 12(2): 130–138.

Clawson, M. 1976. The national forests. *Science* 191(4227): 762–767.

Clepper, H. 1971. *Professional forestry in the United States.* Baltimore, MD: Johns Hopkins University Press for Resources for the Future.

Food and Agriculture Organization of the United Nations (FAO/UN). 2001. *Governance principles for concessions and contracts in public forests.* FAO/UN Forestry Paper 139. Rome: Food and Agriculture Organization of the United Nations.

Gautam, M, U. Lele, H. Kartodihardjo, A. Khan, I. Erwinsyah, and S. Rana. 2000. *Indonesia: The challenges of World Bank involvement in forests.* Washington, DC: The World Bank.

Gray, J. 1983. *Forest revenue systems in developing countries.* FAO/UN Forestry Paper 43. Rome: Food and Agriculture Organization of the United Nations.

Grimaldi, J. 2002. Enron pipeline leaves scar on South America. *Washington Post* (May 5), A01.

Gunatilake, H. 2007. *Efficient technology and the conservation of natural forests: evidence from Sri Lanka.* ERD Working Paper no. 105. Manila, Philippines: Asian Development Bank.

Hirschleifer, J. 1974. Sustained yield versus capital theory. In B. Dowdle, ed., *Economics of sustained yield forestry.* Unpublished conference proceedings. Seattle: College of Forest Resources, University of Washington.

Hyde, W. 1981. Timber economics in the Rockies: efficiency and management options. *Land Economics* 57(4): 630–639.

Hyde, W. 1980. *Timber supply, land allocation, and economic efficiency*. Baltimore, MD: Johns Hopkins University Press for Resources for the Future.

IMF (International Monetary Fund). 1998. *Memorandum on economic and fiscal policy*. Washington, DC: IMF.

Jakarta Post. 2000. 'Illegal' logging involves central government officials. August 19.

Klerman, D., and N Garoupa. Optimal law enforcement with a rent-seeking government. *American Law and Economics Review* 4(1): 116–140.

Kutay, K. 1977. Oregon economic impact assessment of proposed wilderness legislation. In Oregon Omnibus Wilderness Act. Publ. 95-42, part 2:29-63. Hearings before the Subcommittee on Parks and Recreation of the Committee on Energy and Natural Resources, United States Senate, 95 Cong, 1 sess., April 21. Washington, DC: Government Printing Office.

Leffler, K., and R. Rucker. 1991. Transaction costs and the efficient organization of production: a study of timber-harvesting contracts. *Journal of Political Economy* 99(5): 1060–1087.

Magrath, W., R. Grandalski, G. Stuckey, G. Vikanes, and G. Wilkinson. 2007. *Timber theft prevention: Introduction to security for forest managers*. Washington, DC: The World Bank.

Mead, W. 1966. *Competition and oligopsony in the Douglas fir lumber industry*. Berkeley: University of California Press.

Mookherjee, D., and I. Png. 1995. Corruptible law enforcers: How should they be compensated? *Economic Journal* 105(1): 145–159.

Niquidet, K., and C. van Kooten. 2006. Transaction evidence appraisal: Competition in British Columbia's stumpage markets. *Forest Science* 52: 451–459.

Ontario Ministry of Natural Resources. 2002. *State of the forest report, 2001*. OMNR Forest Information series. Toronto, Canada: The Queen's Printer for Ontario.

Paris, R., I. Ruzicka, and H. Speechly. 1994. Performance guarantee bonds for commercial management of natural forests—early experience from the Philippines. *Commonwealth Forestry Review* 73(2): 106–112.

Paudel, D., S. Keeling, and D. Khanal. 2006. *Forest products verification in Nepal and the work of the commission to investigate the abuse of authority*. Country case study 10, VERIFOR. Forest Policy and Environment Programme. London: Overseas Development Institute.

Robinson, E., and R. Lokina. 2008. *To bribe or not to bribe*. Working paper, Department of Economics, University of Dar es Salaam, Tanzania.

Sizer, N. 2005. Halting the theft of Asia's forests. *Far Eastern Economic Review* 168(5): 50–53.

USDA Forest Service. 2005. USDA Forest Service Forest inventory and analysis webpage: http://fia.fs.fed.us (accessed July 18, 2008).

Vincent, J., and M. Gillis. 1998. Deforestation and forest land use: A comment. *World Bank Research Observer* 13(1): 133–140.

Walker, J. 1974. *Timber management planning*. San Francisco: Western Timber Association.

Weintraub, S. 1959. Price-making in Forest Service timber sales. *American Economic Review* 49(4): 628–637.

Wolfe, R. 1989. Managing a forest and making it pay. *University of Colorado Law Review*. 60(4): 1037–1078

World Bank. 1995. *The economics of long-term management of Indonesia's natural forests*. Draft forestry report. Washington, DC: World Bank.

Zimmermann, E., and S. Collier. 2004. *Road wrecked: Why the $10 billion Forest Service road maintenance backlog is bad for taxpayers*. Washington, DC: Taxpayers for Common Sense.

Appendix 5A:
Contracts, Rents, and Royalties

This appendix provides a formal mathematical description of contractual behavior between landowners and loggers or concessionaires for the rights to a stand of mature timber. It follows the same sequence as the chapter. The first section of the appendix examines unconstrained net revenue maximization for the landowner, an objective that is satisfied with lump sum contracts. The second section examines revenue maximization when timber sales are conducted in terms of uniform fixed stumpage fees or royalties. Both sections begin with the simplest analytical case in the absence of landowner concerns for environmental standards and noncompliant behavior on the part of the logger. However, environmental standards and noncompliant behavior are crucial elements of the contracts and of the modern policy discussion. Therefore, subsequent sections introduce new terms into the revenue maximizing equation to explain these standards and behaviors, and these terms cause the optimal behaviors of the landowner and the logger or concessionaire to diverge.

Revenue Maximization: Lump Sum Contracts

The net revenue available from a stand of timber is equal to the gross revenue from its sale minus the costs of obtaining this revenue. Gross revenue is equal to the price per unit of timber delivered to the mill p times the delivered volume V. The costs are the costs of logging the timberstand and hauling its logs to the mill c_{lh}. They vary with the area and volume harvested ($dc_{lh}/dV>0$). For stands of mature natural timber on which there was no prior management, the net revenue is also known as the economic rent R.

$$R^1 = pV - c_{lh}(V) \tag{5a.1}$$

The harvest volume associated with maximum recoverable net revenue can be determined by differentiating eq. (5a.1) with respect to volume and setting the derivative equal to zero.

$$\frac{dR^1}{dV} = p - \frac{dc_{lh}}{dV} = 0 \tag{5a.2}$$

Therefore, for a representative timberstand depicted by Figure 5.2, the landowner maximizes net revenues when the logger harvests to the point where the delivered price for the last unit equals the unit harvest cost at that point. In the figure this occurs at point D. The maximum net revenue is the area pdc, and competitive loggers bid up to this amount for a lump sum contract for the rights to harvest timber out to point D. Loggers have no interest in harvesting beyond this point because the unit harvest costs exceed the revenues gained from logging increments beyond this point.

Environmental Restrictions

The problem becomes more complex when the landowner introduces restrictions on environmental performance into the contract. The landowner must now absorb the additional costs of monitoring and enforcing environmental compliance c_{me1} and the logger must accept the additional costs of either complying with the environmental restrictions or paying a penalty for failure to comply c_e ($dc_{me1}/dV>0$ and $dc_e/dV>0$).

The objective functions for the landowner R_{lo} and the logger R_{lg} are no longer identical,

$$R_{lo}^2 = p\ V - c_{lh}(V) - c_{me1}(V) \tag{5a.3a}$$

$$R_{lg}^2 = p\ V - c_{lh}(V) - c_e(V), \tag{5a.4a}$$

and the rules for optimal behavior diverge.

The landowner anticipates receiving maximum net revenues if the logger harvests out to the point where

$$p = \frac{dc_{lh}}{dV} + \frac{dc_{me1}}{dV} \tag{5a.3b}$$

This point is short of the previously determined unrestricted optimum, or to the left of point D in Figure 5.2.

The logger's optimum bid is for the timber out to the point where

$$p = \frac{dc_{lh}}{dV} + \frac{dc_e}{dV} \tag{5a.4b}$$

This point is also short of the unrestricted optimum at D, but there is no reason it should be identical with the point of the landowner's optimum.

The landowner's and the logger's optima diverge, unless $dc_{me1}/dV = dc_e/dV$. The landowner's M&E costs and logger's compliance costs both vary with local environmental conditions. Therefore, they vary from site to site and from contract to contract, and the conditions for convergence of the landowner's and the logger's objectives must vary similarly.

Solution: An Environmental Bond

The landowner can control the extent of the divergence by assessing an environmental bond at the outset of the contractual agreement. The landowner sets a value on the bond such that an increment in bond value, minus the landowner's expected monitoring and enforcement (M&E) costs necessary to guarantee compliance, equals the value of the environmental protection it ensures. If the landowner assesses this bond as a lump sum to be deposited at the outset of the contract and only returned to the logger upon satisfactory completion of all terms of the contract, then the bond has no effect on the logger's incremental logging decision. On the other hand, bonds that are returned in part for partial satisfaction of the environmental terms of the contract do have an effect on the logger's incremental logging decision and designing a system of M&E that satisfies the landowner's optimal level in this case is a more complex problem.

Uniform Fixed Royalties

Uniform fixed royalties are an alternative to lump sum contracts. For landowners who charge a fixed fee (a royalty) r per unit of harvest, the total revenue recovered is equal to this royalty multiplied by the harvest volume V.

$$R_{lo}^3 = r \bullet V(r) \tag{5a.5a}$$

In this case, harvest volume itself is a function of the royalty since the harvest area and harvest volume both decrease as larger uniform fixed royalties absorb ever larger shares of the delivered price per unit of volume ($dV/dr<0$).

The royalty that maximizes landowner revenues can be determined by differentiating eq. (5a.5a) with respect to the royalty and setting the derivative equal to zero:

$$\frac{dR^3}{dr} = V(r) + r\frac{dV(r)}{dr} = 0. \tag{5a.5b}$$

Rearranging terms:

$$\frac{dV(r)}{dr} \bullet \frac{r}{V(r)} = -1. \tag{5a.5c}$$

The left-hand-side of eq. (5a.5c) is the elasticity of harvest volume with respect to the royalty e_r. An inspection of Figure 5.2 shows that the absolute value of this elasticity is identical to the absolute value of the elasticity of harvest volume with respect to harvest costs e_{lh}. Therefore, the landowner obtains maximum net revenues under a uniform fixed royalty by setting the royalty at the level corresponding to $e_r = e_{lh} = |1|$. Empirical estimates of these elasticities require data on the observed harvest volumes across several harvest sites and also data on either the individual royalties or the harvest costs associated with each of those sites. Uniform royalties are invariant by definition, but harvest costs vary from site to site. Therefore, the elasticity of harvest costs will generally be easier to estimate.

Reference to Figure 5.2 shows that e_{lh} is less than one to the right of h at points like h' and greater than one to the left of h at points like h''. Increasing the royalty in the range of $e_{lh}<1$ causes the logger to back down the cost curve and bid on a smaller total harvest volume, but it increases the landowner's collection of revenues. Increasing the royalty in the range of $e_{lh}>1$ also causes the logger to back down the cost curve and bid on a reduced total harvest volume, but it decreases the landowner's total revenue collection. In either case, total revenue collection is less than the full area pdc that would have been collected under a lump sum contract.

The logger's decision involves a different, two-step, process. In the first step, the logger agrees with the landowner on the royalty. Perhaps the landowner specifies the royalty and the logger agrees, or perhaps the logger wins an auction by placing a bid for a given volume or area of harvest that is greater than the bid of any competing logger. The second step follows, as the logger attempts to maximize his or her collection of net revenues.

The logger's objective function is identical with eq. (5a.1) and the logger's optimal harvest level occurs where $p=dc_{lh}/dV$. This is the same condition (5a.2) that identified the optimal harvest level under lump sum contracts. Reflection on Figure 5.2 reminds us that this condition is satisfied at point D, which is far to the right of the harvest area and volume associated with the landowner's optimum under uniform fixed royalties (identified by $e_r = e_{lh} = |1|$). Clearly, the logger can satisfy the terms of the contract by harvesting to the landowner's optimum and paying the contracted royalty per unit of contracted volume. However, the potential additional financial return available to the logger encourages harvesting beyond this level whenever the logger finds it possible to under-report the level of total harvests and avoid payment of the royalty due on a share of all harvests. That is, the landowner's and the logger's optima diverge, and the financial incentive encourages the logger or others who have access to the timberstand in question to harvest some timber in excess of the contracted amount.

Environmental Restrictions

Environmental restrictions on harvest practices introduce new costs for both the landowner and the logger—just as they did in the previous case of lump sum contracts. For the landowner, they add the cost of environmental enforcement c_{me1} to eq. (5a.5a) ($dc_{mc1}/dV > 0$).

$$R_{lo}^4 = r \bullet V(r) - c_{me1}[V(r)] \tag{5a.6a}$$

Taking the derivative with respect to the royalty, as in eq. (5a.5b), and rearranging terms

$$\frac{dV}{dr}\frac{r}{V} = -1 + \frac{\frac{dc_{me1}}{dV}\frac{dV}{dr}}{V}. \tag{5a.6b}$$

A comparison with eq. (5a.5c) and Figure 5.2 shows that, in the added presence of environmental restrictions, the landowner's optimum is even more restrictive of harvesting. The landowner's optimal harvest level is somewhat less than that associated with the point where $e_r = e_{lh} = |1|$ because the numerator of the second term on the right-hand-side of condition (5a.6b) is negative. The larger the royalty, the smaller the harvest volume; the smaller the harvest volume, the lower the environmental risk and the lower the landowner's environmental M&E costs.

Of course, the logger still prefers to maximize net revenues. This means that the logger's objective function, in the presence of costs for environmental compliance, remains as in eq. (5a.4a) and the logger's optimal harvest level is still determined by condition (5a.4b), $p = dc_{lh}/dV + dc_e/dV$. Therefore, in the presence of environmental restrictions, the logger harvests less than if the restrictions had not existed (to the left of point D in Figure 5.2 but probably to the right of $e_r = e_{lh} = |1|$).

In sum, the rules for the landowner's and the logger's harvest optima are not similar and there is no reason to anticipate convergence in logger and landowner preferences or behavior. They might approach convergence if the logger's compliance costs are large while M&E is easy and the landowner's M&E costs are small. This is unlikely. It is more likely that there is a considerable difference between the two optima and that this difference is further evidence of the incentive for illegal logging under contracts based on uniform fixed royalties.

Illegal Logging

The presence of illegal logging introduces a second M&E responsibility for the landowner and, with it, a second M&E cost, c_{me2}.[11] This cost increases with higher royalties and smaller harvest volumes because these mean that a larger unharvested volume is available for illegal harvesting and, therefore, the likelihood of illegal activity and the cost of limiting it is greater ($dc_{me2}/dV<0$).

$$R_{lo}^s = r \bullet V(r) - c_{me2}[V(r)] \tag{5a.7a}$$

The condition for the landowner's optimum in the presence of a uniform fixed royalty is

$$\frac{dV}{dr}\frac{r}{V} = -1 + \frac{\frac{dc_{me2}}{dV}\frac{dV}{dr}}{V}. \tag{5a.7b}$$

The numerator of the second term on the RHS of condition (5a.7b) is positive in this case. Therefore, the landowner's optimal harvest level is somewhat greater that that associated with $e_r = e_{lh} = |1|$.

As before, the logger's interest remains in maximizing net revenues. The logger's objective function remains similar to eq. (5a.4a)—with a new term for the added costs of illegal harvesting. These are the costs of assessing the landowner's pattern of M&E and then avoiding detection or else paying the penalty for being caught c_a. These costs increase as the logger's illegal harvests increase—or as the level of legal harvests decreases, thereby leaving a greater residual stand available for illegal harvests ($dc_a/dV<0$).

$$R_{lg}^s = p \bullet V - c_{lh}(V) - c_a(V) \tag{5a.8a}$$

The condition for the logger's optimum is

$$p = \frac{dc_{lh}}{dV} + \frac{dc_a}{dV} \tag{5a.8b}$$

Therefore, the logger has a price incentive to illegally harvest beyond the point where the landowner maximizes gross revenue ($e_r = e_{lh} = |1|$), but the additional costs of avoiding detection prevent the logger from harvesting all the way to the logger's own gross revenue maximizing point at the frontier (D).

The landowner's and the logger's optima both fall between the economic frontier at D and the point of revenue maximizing royalty. However, convergence of the two optima, once more, would be unique because conditions (5a.7b) and (5a.8b) are so different. Convergence requires the landowner to structure M&E activities to fit both the environmental conditions and the logger's opportunities for avoidance. Both are unique to each site. Such a careful design for M&E is unlikely since most illegal activity occurs in frontier forests that tend to be publicly owned and agency policy tends to restrict public land managers to broadly uniform criteria applicable for M&E and to standard penalties for illegal activity

---

11 We will separate the assessment of environmental standards and illegal logging for clarity of exposition.

both applicable to the many diverse lands managed by the agency. Therefore, uniform fixed royalties themselves must be a source of some level of illegal logging.

Conclusions

Lump sum contracts and contracts that specify prices in terms of units of products are the two basic arrangements for exchanging the rights to mature timber. For lump sum contracts in their simplest form, the objectives of both landowners and loggers are identical and their optimal harvest levels are also identical. Contracts that specify prices in terms of units of production tend to use uniform prices (royalties or stumpage fees) for broad categories of timber. Landowners who rely on such contracts maximize their collection of revenues at lower harvest levels than those associated with lump sum contracts and these lower harvest levels leave a residual of profitable standing timber. The profitability of this residual is an incentive for loggers and others who have access to the harvest site to extract additional timber in violation of the contract.

Landowners with environmental objectives in addition to their short-term timber revenue objectives and loggers with opportunities for profitable illegal logging must introduce new terms into both forms of contracts. The landowner's cost of monitoring environmental restrictions and the logger's cost of environmental compliance reduce the optimal harvest levels for each of these participants to the timber contract—regardless of the lump sum or uniform royalty design of the contract. The reductions will vary from site to site, as forested sites themselves are diverse in their geography and their timberstands, and the new optimal harvest levels for landowners and loggers are unlikely to be similar because the additional costs of these principal actors are unlikely to be similar.

The potential for illegal logging introduces a second monitoring and enforcement responsibility for landowners, and a new cost (for either complying with or avoiding the monitors) as well as potential new revenues for loggers. These potentials exist wherever the landowners' and loggers' optimal harvest levels (before illegal activity) are dissimilar, but the potential is greater with uniform fixed royalties. Recognition of the potential increases each landowner's optimal harvest level and decreases each logger's optimum—but convergence of the two remains unlikely and some incentive for illegal logging remains. Therefore, it is clear that some level of illegal logging is a result of the structure of the contracts themselves.

This discussion suggests that one real challenge with respect to logging contracts is an institutional challenge. This is the challenge of improving the institutions that work with logging contracts. And this would require improving the general understanding logging contracts and improving the selection and design of those contracts themselves.

6

THE EFFECTS OF TRADE, MACROECONOMICS, GROWTH AND DEVELOPMENT

Trade, by definition, involves at least two markets. Macroeconomics includes trade between the various sectors and markets in an economy, as well as trade with other economies. Therefore, the natural progression for our analysis is from the microeconomics of forestry in one market—the topic of previous chapters—to trade in forest products. Furthermore, as previous chapters added the impacts of markets and policies in adjacent sectors on the forest, the progression to trade between markets must include the impacts of macroeconomic activity on the forest. These latter topics, trade and macroeconomics, are the themes of this chapter. Over time, entire economies grow and develop. This dynamic aspect of macroeconomics is comparable to the intertemporal aspects of the microeconomics of forestry discussed in chapter 3. This chapter also inquires into these dynamic macroeconomic effects on forests.

Macroeconomics includes both the effects of the sector in question on the broader economy and the reverse, the effects of adjustments in the broader economy on the sector in question. For forestry, the latter is often more important. Globally, the forest sector employs about 0.4 percent of the labor force, contributes about 1.2 percent to gross domestic product (GDP) and accounts for about 2.3 percent of global merchandise trade. Other than the small island countries of Vanuatu and the Solomons, Finland is the only country in the world for which the full forest sector contributes more than 6 percent of gross domestic product (Lebedys 2004). This means that forestry is generally a small sector in the national economy. It is not uncommon for market and policy adjustments that are exogenous to small sectors to have larger effects on a small sector than policies that are specifically designed for their direct effects on that sector and this, by itself, is a good reason for those interested in forestry to examine the relationships between forestry, the aggregate economy, and macroeconomic policy.

The effect of international trade has become a more important topic for forestry in recent years. While forestry contributes only a small share to global trade and global GDP, recognition of the role of global trade in forest products is increasing. Global exports increased from 4 percent of all wood products production in 1961 to about 8 percent in the early twenty-first century. Trade in forest products increased 400 percent in the decade following the mid-1990s until it now exceeds 140 million m^3 of industrial roundwood equivalent. Canada and the United States account for one-third of the global export value of forest products, while Japan and the United States account for about 30 percent of all imports. However, the developing country share of all imports is increasing, particularly for the developing countries of Asia. For exports, developing countries in the tropics are responsible for slightly more than 10 percent of the total global value. They dominate the export of plywood and their share of wood-based furniture is increasing rapidly (Rythonen 2003; FAO 2005).[1]

Questions about trade are questions about the exchange of production away from higher cost markets and toward lower cost markets. This exchange improves economic welfare in both markets—otherwise neither would participate in the exchange. Nevertheless, the displaced producers in the importing market may suffer economic loss and, while one of the markets involved in any exchange generally experiences environmental gains, the other can suffer environmental losses. These potential losses are one source of public doubt about the merit of trade. The mass demonstrations against the Seattle meetings of the World Trade Organization in 1999, for example, were an expression of this doubt.[2]

In order to understand the different effects of trade, it is necessary to distinguish between initial or short-term effects and longer-term and more permanent effects. The short-term effects of large shocks on the patterns of trade are the evidence that disturbs the opponents of greater freedom in trade. The East Asian financial crisis of 1997 and its impact on Indonesian forests and the effect of the economic malaise in Russia on shipments of Siberian timber to China are two recent examples. However, this focus on short-term effects may not serve the public interest well. Many short-term effects are not permanent and the longer-term effects of trade generally have greater economic and environmental impacts.[3]

The analysis of this chapter explains and then provides examples for most of these points. The first section of the chapter builds from the analytical structure of previous chapters, this time with two markets rather than one, to reflect on

---

1   Buongiorno et al. (2003) and also Hashiramoto, Castano, and Johnson (2005) summarize recent trends in global trade in forest products.
2   Buongiorno et al. (2003) and Mersmann (2005) survey current issues and institutions focusing on trade in forest products.
3   The distinction between short-term and long-term effects is likely to be more important for trade effects and for macroeconomic policy effects on the forest than the effects of domestic forest policies discussed in chapter 4. Domestic forest policy adjustments are likely to occur in small increments with only small marginal effects, while even incremental adjustments in macroeconomic policy can create large impacts on the forestry sector of some regional economies.

the impacts of trade on the forest economies and forest environments of both markets. The economic effects are clear but the net effect on the combined forest environments of the two markets is not so clear. The analysis is straightforward but its conclusions are uncertain, and they remain uncertain after reflection on the empirical evidence. The best we can do is to speculate on reasonable hypotheses regarding net forest environmental impacts for trading economies.

If the effects of trade itself are mixed, then the effects of trade policy are also mixed. The effects of direct policy restrictions and inducements for trade in forest products tend to be small for most countries and most products. Furthermore, adjustments in trade policy generally occur after considerable negotiation and they tend to develop gradually and in small increments. Therefore, the impacts of many single adjustments in trade policy are likely to be limited. We will review the direct instruments of trade policy briefly, and then turn to external influences on trade and on the forest itself in the second part of the chapter. These external influences have greater impacts in some important cases.

As trade leads to general improvements in economic welfare, the improvements, and other exogenous factors as well, have their own impacts back on the forest. Any number of factors may be important in one economy or another, but five broad categories stand out:

- external shocks such as the East Asian financial crisis of 1997;
- shocks from within the domestic economy but external to the forestry sector, shocks such as those created by substantial sectoral policy adjustments or by severe civil disturbance;
- specific domestic fiscal and monetary policies, including management of the terms of trade (exchange rate policy);
- substantial and long-lasting external policies, such as European Union and North American agricultural policies that may have large impacts on the forests of other countries;
- substantial and long-lasting domestic policies such as a pattern of interventions to support regional development, or broad biases in the system of general taxation.

The first three categories of factors in this list have notable immediate and short-term effects that may deviate from their impacts on the longer-term equilibrium. The latter two categories typically develop more gradually and may become more permanent. Their impacts are often overlooked. The second part of the chapter examines all five categories. All five have effects on the production of forest products as well as on land use. The occurrence of these effects across multiple regional or national economies and the opposing effects that often occur in two or more trading economies make these trade and macroeconomic effects more complex than the effects of the policies discussed in chapter 4.

The final section of the chapter reflects on the overall effects of macroeconomic growth and development on the forest. Observations that developing countries tend to draw down their forest stocks as they enter the path to development contrast with observations that mature economies tend to add to their forest stocks as they continue to grow. A general observation known as the "environmental Kuznets curve" (EKC) reconciles these two divergent observations. The final section of the chapter examines the concept and the support for it in forestry. Finally, an appendix introduces the related issue of the contribution of forests to the national income accounts, and the extensions of these accounts to incorporate non-market environmental values, extensions known as environmental or green accounting.

Trade

For trade in extractive forest products, we can focus on regions in the second and third stages of forest development. Regions in stage I are lesser participants in external trade of any sort, and the forest products they do trade are not generally transported very far, or at least they are not transported very far before their initial processing. On the other hand, many regions in stages II and III do participate actively in trade in primary products like industrial wood and even in trade for many non-timber forest products.[4] This means that all four margins of forest land use will be relevant for trade: the intensive and extensive margins of managed forestry for regions in stage III, the boundary of active harvests from the natural forest for regions in stage II and stage III, and marginal quality species and size classes of timber and other products from the open access components of regions in both stages. As we previously suggested, in many cases this open access region has a significant standing forest inventory. It can be an important source of industrial wood in regions of those developing countries where the population is still low and the institutions that define property rights are not as well developed.

We can begin by assuming that most commercially active regions look like either stage II or stage III before they engage in trade. Two regions engage in trade only when some participants in both regions gain from the exchange. Consumers in one region purchase from a second region when the imported goods cost less than identical goods produced in the local market. Producers in the second region export when consumers in the first region are willing to pay the transportation costs plus a price that is at least as great as the local market price in the producing region. As a result of trade, prices decline in the consuming region and rise in the producing region until a new market equilibrium develops for both regions.[5]

Economic Welfare and Land Use Impacts: Forest Management and the Natural Forest Frontier

The net effects of trade on forest land use will be uncertain—almost regardless of the initial condition of the forests in the two trading regions. These effects become even less certain as we extend the analysis from two regions to multiple

---

4   Of course, regions in all three stages participate in trade in non-consumptive products such as forest recreation.
5   Samuelson (1948) provides the classic economic description.

trading regions. However it is clear that not all trade creates environmental losses. In fact, all trade affects forests in more than one region, and trade often creates net global environmental gains. Continued reference to the figures will help make these points clear.

Consider the case of trade between two regions that are both in the mature third stage of forest development. Trade in sawnwood from Finland to Germany, from southeastern Canada to the U.S. Northeast, and even from Finland and Sweden to Vietnam are modern examples. After their markets open for trade, consumers in the importing region recognize that they can decrease costs by purchasing from the exporting region. These consumers are now willing to pay only the new lower price for forest products and some of their demand shifts from the domestic market to the exporting region. As the demand for local products declines in the importing region, the forest value function V_f declines—following the broad arrows in the left-hand diagram of Figure 6.1. The value of forestland declines in this region and, along with it, the region's own output of forest products also declines. Its employment of forest workers and its demand for the other inputs to forest production declines along with the decline in forest production.

The exporting region absorbs an increase in demand from its new customers in the importing region, and its prices rise to a new higher level. As prices increase in this region, its forest value function shifts upward and outward, following the broad arrows in the right-hand diagram in Figure 6.1. The land area under forest management expands. The region's output of forest products also increases, with some of the increase coming from the larger area of managed forests, some from additional extraction at the natural forest frontier, and perhaps some from additional removals from the region's open access lands. Employment in the forest sector also expands with the increase in production.

Some loggers in the importing region may continue to extract timber from their old harvest sites as long as the existing capital and infrastructure of this region are in good condition and as long as the revenues from their harvest activities cover the variable costs of their logging operations. However, with time, the lower price in the importing region must cause its loggers to decrease the level of their harvest activity, allowing some land at both margins of forest management to revert to other land uses, allowing some forest recovery in the open access lands, and delaying some harvesting at the region's mature forest frontier.[6]

The net effect of trade on land use in the combination of the two regions is uncertain. It depends on the elasticities of consumer demand in the two regions and the relative productivities of lands at all margins in both regions. Harvests

---

6 Prestemon (2000) provides a tidy empirical example for this importing country case. He examined the effect of trade liberalization on Mexico's forests. He separated Mexico's forests into private and public components and further separated the public component into managed (or protected) and open access categories, and then estimated different supply functions for each forest component. (Prestemon's organization is similar to the managed, open access and frontier, and remote arrangement of the three-stage model.) Under most scenarios, Prestemon finds that the lower prices induced by trade liberalization induce a net increase Mexico's total forest cover, although forest cover on managed private lands may decrease in the short run.

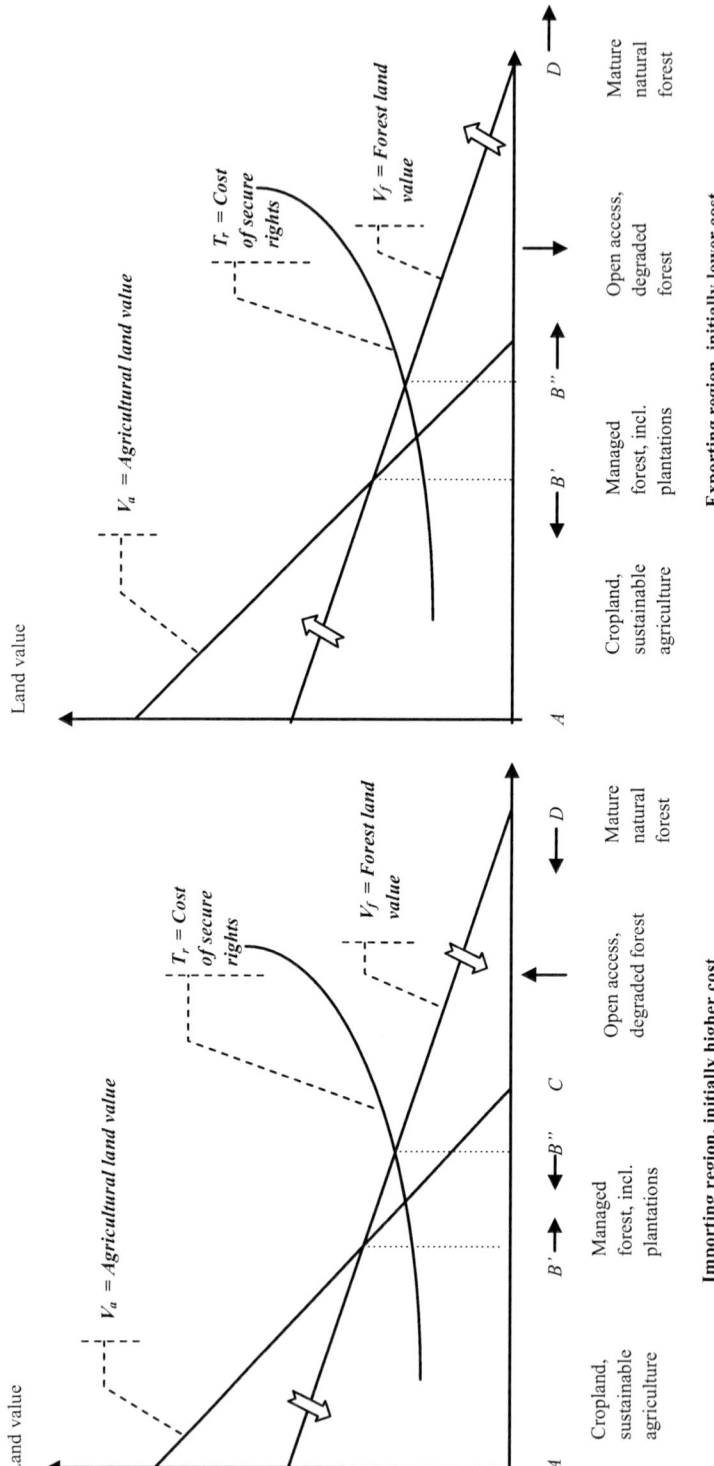

FIGURE 6.1 The land use impacts of trade between two regions in the mature stage of forest development

from low cost, highly productive, managed forests in the exporting region may replace harvests from the natural forest frontier of the importing region. In this case, investment in the exporting region's managed forests will increase. Many would see this as an environmental improvement and most would also see the saving of natural forest in the importing region as an environmental improvement.

On the other hand, managed forests generally tend to be among the more costly producers while harvests per land unit tend to be greater from mature natural forests. Therefore, trade may cause a net shift that favors relatively greater reliance on the exporting region's natural forests. Most would see this as an environmental loss.[7]

Either case is possible. The only certainties regarding land use are that harvest activity declines in the importing region and some harvest activity is likely to continue in the natural forest of the exporting region—because its costs of production are low and the complete prevention of all illegal extractive activity from this land is unlikely.

This characterization of trade between two regions in the mature stage of forest development is only one of four possible cases. Two regions both in stage II can engage in trade. Shipments of logs from Laos and Cambodia to Thailand are examples. Regions in stage II can export to regions in stage III. The many tropical countries that ship logs to developed countries are examples, as is shipment from Russian Korelia to Finland and from Siberia to northeastern China. Finally, regions in stage III may ship to regions in stage II—although it is difficult to think of substantial examples of this fourth case.

The three sets of diagrams within Figure 6.2 demonstrate these remaining possibilities. The broad arrows show the shifts in the forestland value functions associated with each possibility. The bold arrows below the figures show the effects on the margins of forest land use in each region. Table 6.1 summarizes the effects on forest management, on the degraded open access lands, and on the unmanaged natural forest for all four possible cases.

Conclusions: The first general conclusion to draw from Table 6.1 is that there are always effects on consumer welfare in the importing region, and also on

---

[7] Sedjo reasons similarly that restrictions on domestic production are an inducement to increase trade. For example, increased forest regulation in the United States leads to a decrease in domestic U.S. production and an increase in the U.S. demand for imported wood. Therefore, by promoting domestic forest conservation the United States encourages environmental loss among its trading partners (Sedjo et al. 1994; Sohngen Mendelsohn, and Sedjo 1999). Somewhat similarly, Bolkesjo, Tromborg, and Solberg (2005) show that increased policy-induced forest conservation in Norway results in higher local prices (and gains to Norway's forest landowners, but not to its sawmill operators) and also significant harvest increases in regions and countries unaffected by the new environmental policy. Uusivuori and Kuuluvainen (2001) find that the substitution effect between domestic, and imported wood is particularly strong for softwood and it is their softwood forests that North America and Western Europe most actively protect. Buongiorno et al. (2003), in another example, demonstrate that China's more recent natural forest harvest restrictions are compensated by the substitution of roundwood imports from Southeast Asia, although with only a negligible net effect on the global forest.

208 The Effects of Trade, Macroeconomics, Growth and Development

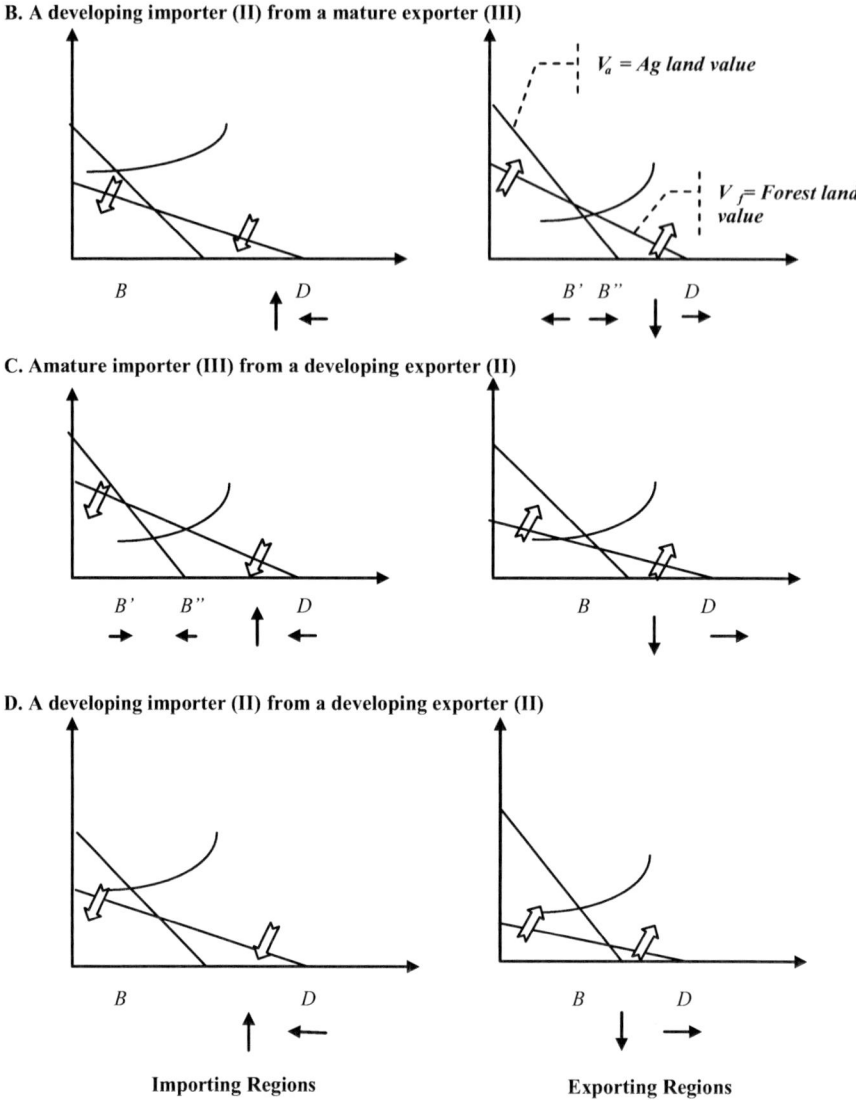

FIGURE 6.2 Trade and land use—the remaining cases

employment and on the forest itself in both regions. The public discussion seems to find it easy to overlook the consumer gains, while focusing on employment losses in the importing region and on the possibility of environmental loss in the exporting region. In fact, the economic value of employment losses in the importing region are smaller than the value of consumer gains in this region and the employment losses are temporary, lasting only as long as it takes disemployed loggers and other woodsworkers to find alternative employment in a local economy which is now better off due to the consumer gains. Employment and

TABLE 6.1 Effects of trade on consumers, employment and production, forest management and on the unmanaged natural forest, by importing and exporting region

Case	Effect on	Importing region	Exporting region	Net effect
A. Stage III region imports from stage III region	Consumers Production and employment	gain decline	uncertain expand	gain gain
	Managed forest Mature natural forest Degraded forest	contract recover recover	expand contract degrade further	uncertain uncertain uncertain
B. Stage II region imports from stage III region	Consumers Production and employment	gain decline	uncertain expand	gain gain
	Managed forest Mature natural forest Degraded forest	n.a. recover recover	expand contract degrade further	expand uncertain uncertain
C. Stage III region imports from stage II region	Consumers Production and employment	gain decline	uncertain expand	gain gain
	Managed forest Mature natural forest Degraded forest	contract recover recover	n.a. contract degrade further	contract uncertain uncertain
D. Stage II region imports from stage II region	Consumers Production and employment	gain decline	uncertain expand	gain gain
	Managed forest Mature natural forest Degraded forest	n.a. recover recover	n.a. contract degrade further	n.a. uncertain uncertain

n.a.: not applicable

production always increase in the exporting region and the increase is generally greater than the counterpart decline in the importing region because the overall lower prices in the importing region mean that the aggregate quantity of wood products demanded in both regions increases and, with it, aggregate production and employment must increase as well. Finally, trade affects the environment of both regions. The environment of the importing region must improve as a result of the decline in this region's harvest level.

China's recent experience is illustrative. China's natural forest declined rapidly during its period of rapid economic growth from 1978 through the late 1990s. More than nine million hectares were removed from the natural forest base (Rozelle, Huang, and Benzinger 2003). International environmentalists expressed alarm. China's own leaders perceived the negative downstream effects from deforested uplands and, in 1998, they banned logging on many of China's natural forests and reduced it on others. The planned harvest levels in northern and western China alone declined by more than 18 million m^3 annually in four years (Hyde, Belcher, and Xu 2003). However, China's demand for forest products was largely unaffected and its demand for imports increased to offset the declining domestic harvest levels. Imports of forest products, largely from Siberia and from Indonesia and other Southeast Asian countries, increased from 45.8 to 98.1 m^3 of roundwood equivalent over the same four years, and they continue to increase to this day (Sun, Katsigris, and White 2005). Of course, many of these imports originated from natural forests and China's draw on the natural forests of other countries creates environmental losses in the exporting countries. Many of the same international environmental interests that previously expressed alarm over China's declining natural forests, interests that were encouraged by the government's ban on logging in the uplands, now say little about China's recovering forest cover, but they express new alarm over the volume of China's imports (Lague 2003; *Economist* 2005a). The obvious conclusion continues to be that trade imposes opposite effects on forest environments in each of the trading partners and we must not overlook either one.

The second general conclusion is one of uncertainty regarding these net affects of trade on the combined forestlands and forest environments of both regions. The important global environmental question is whether the reduction in forest activity in the importing region is more beneficial than any environmental loss caused by the expansion in forest activity in the exporting region. For the example above, is the recovery of China's forests of greater or lesser global importance than the accompanying decline in Siberian and Southeast Asian forests?

Returning to the discussion of Table 6.1, the land area under forest management increases unambiguously only for the case of imports to a region in stage II from a region in stage III (case B in Table 6.1). The net effect of trade on the natural forest is ambiguous in all cases: Harvests at the frontier decline in the importing region and expand in the exporting region, but what is the net effect and is it possible that forest recovery in one region is more important than forest decline in another region? Furthermore, the ambiguity about the net effect across the two regions only increases as we extend the analysis from two regions or countries to global trade. Global trade—for almost any commercial forest product—undoubtedly includes regions and countries that fit the descriptions of each of the four cases in Table 6.1. The sum of many uncertain cases is surely only more reason for caution when considering generalized views of the net effect of trade on the world's forestlands.

To this point the discussion has been conceptual, although we have referred to valid examples for each point of discussion. We might inquire whether the environmental uncertainty could be resolved with empirical estimates of the shifts in land use associated with each example of international trade. Perhaps. Prestemon (2000) provides a good example for a single importing country (see footnote 6.) However, the calculations necessary when including all trading partners would be difficult and the answer could be subject to change over relatively short periods of time. As the appendix to chapter 2 explains, the available data are physical, not economic, and the economic measures of managed forests, degraded forests, and natural forests vary from market to market. Many countries contain multiple wood markets—or "regions" from the perspective of our three-stage analysis. Therefore, the physical data within each nation's official forest inventory can only provide a loose impression of what is happening at each economic margin.

Even that impression would be subject to change as logging technologies, wood utilization technologies, and rural infrastructure (especially roads) change. This is a crucial point because the rates of technical and institutional change can exceed the growth rates of commercial forest plantations and they certainly exceed the growth rates of mature natural forests—as chapter 3 explains. Technological and institutional change can be the most important determinants of trends in the preferred species and size classes of industrial wood over the long run. They explain why loggers repeatedly return to harvest previously degraded forests, and why the dominant varieties of technological and institutional change for forestry—roads, mill utilization, or logging utilization—are the common sources of the most important shifts in long-run land use margins for most regions and countries.

Further speculation on trade and the environment: While neither our figures nor most available forest survey data provide a satisfying basis for generalized conclusions regarding the relationship between trade and the forest environment, we can speculate further based on knowledge of the relationship between logging costs and potential environmental damage. We can also reflect on the conditions of those countries that are least open to trade. What are the conditions of their environments in general and their forests in particular, and what does their experience suggest for the effect of trade on the environment.

Greater logging costs are associated with less accessible lands and reduced accessibility is associated with more difficult terrain—either steeper slopes and more erodable soils in one case or poorly drained lands in more gentle terrain in the other. Lands in these categories tend to be more environmentally diverse than well-drained flatter and gentler terrain. Therefore, we can speculate that greater logging costs are associated, in a broad general way, with greater environmental risk and that trade that permits a shift from higher to lower cost regions of forest production bears a general relationship to net improvement of the combined forest environment of two trading regions.

This is neither an observation nor a conclusion. Rather, it is a general speculation that merits further examination. There will be exceptions. Highly productive and lower cost coastal redwood forests in California and, more generally, most riparian lands are two examples. Both possess great environmental value. Trade that induces additional forest extraction from these lands while reducing production in a second region might not yield net environmental improvement. Nevertheless, the speculation that trade improves the aggregate forest environment of two trading partners may still be accurate as a broad general rule.

Observations from the People's Republic of (North) Korea (PRK) and from Myanmar provide additional perspective and further basis for speculation that trade is generally good for the forest environment. PRK and Myanmar may be the two most closed economies in the world today. They also suffer some of the worst environmental problems. Although the data from both countries must be accepted with unusual caution, the official FAO measures of forest cover, 0.3 and 0.8 hectares per capita, respectively, are low by global standards. Myanmar, which still has significant forest cover, suffers a rate of deforestation (1.4 percent) that is among the highest in the world and illegal logging in that country is a subject of serious international concern (FAO/UN 1999). (There are no recent official data that allow a calculation of annual deforestation for PRK.)

Moreover, these two countries suffer in particular contrast with their neighbors. Those neighbors, the Republic of (South) Korea (ROK) and Thailand, are countries with similar fundamental ecologies but relatively open economies. The ROK and Thailand experience similar low measures of forest cover per capita (0.1–0.2) but much smaller rates of deforestation (0.1 and 0.7, respectively) than PRK and Myanmar (FAO/UN 1999).

The comparisons of PRK with ROK and Myanmar with Thailand are not conclusive. Closed economies and rapidly deteriorating forest environments may not be related. Both may be functions of some other greater problem in both PRK and Myanmar. Nevertheless, when the most closed economies suffer some of the greatest environmental problems, their experience should be a caution to advocates of reduced trade and less open economies. Indeed, their experience seems to support the opposite. It supports speculation that trade is good for the forest environment.[8]

The Instruments of Trade Policy

This next section of the chapter turns from the discussion of patterns of trade on the forest environment to the instruments of trade policy. The three fundamental instruments are tariffs, non-tariff barriers, and export restrictions. The first two limit imports while the third limits exports.

---

8   Antweiler, Copeland, and Taylor (2001) confirm the general positive effect of trade on the environment in a statistical survey of forty countries for which they had full data. Unfortunately for our interests, they, like most others who have investigated this issue, focus on environmental pollutants and not on natural resources like forests. Copeland and Taylor. (2004) survey the literature on trade and the environment and conclude that "most available studies suggest that the effect (of trade on the environment) is small." Esty (1994) summarizes the forestry assessments prior to the mid-1990s and concludes that trade is not an important source of forest-based environmental problems. Buongiorno et al. (2003) speculate that the immediate effects of tariff liberalization might be neutral with regard to the global natural forest environment. On the other hand, they speculate that as trade liberalization in forestry is part of broader global reductions in trade barriers in general, the ensuing general economic growth would lead to increases in demand for non-consumptive forest amenities and, therefore, improvement in the forest environment.

Tariffs: Tariffs are fees levied on imports. They raise the price of imports to consumers. This limits consumer demand for imports, shifts some of that demand to domestic production, and induces an increase in domestic prices. In terms of our figures, tariffs in the importing country reduce the demand and the forest value function in the exporting country, thereby decreasing production in that country. The increase in demand for domestic production in the importing country raises the forest value function in that country and increases its production above the level that would exist in the absence of the tariff.

The U.S. tariff on Canadian softwood lumber imports is illustrative.[9] The United States imposed a 27 percent duty on Canada's imports in 2002. The objective was to protect American lumber producers, particularly those in the U.S. Pacific Northwest. The duties caused approximately fifty Canadian mills to close, laying off thousands of workers (*Economist* 2004). Canada continues to supply approximately one-third of the U.S. market but this is less than before the tariff. Boyd and Krutilla (1987, 1992) estimate that even a 10 percent duty on Canadian timber probably decreases Canadian softwood lumber exports to the United States by a minimum of 4.5 percent, decreasing U.S. consumer welfare by 5 percent of the total expenditure on lumber and decreasing Canadian producer welfare by 7 percent of Canada's revenues from lumber. The current 27 percent tariff surely has even greater negative effects—while assisting U.S. producers at a much smaller level.

Despite this exception, tariffs on forest products in the early twenty-first century generally tend to be low in comparison with tariffs on many non-forest products. Therefore, their general effect is usually small. Tariffs of less than 5 percent are common for most forest products imported to developed countries, although tariffs on wood-based panels and some paper products are as great as 20 percent in a few countries and the effects of these higher tariffs can be significant (Bourke and Leitch 1998). The general trend, following the Kennedy Round of the General Agreement on Trade and Tariffs (GATT) in 1967, the Tokyo Round in 1979, and the Uruguay Round in 1994, has been for lower tariffs and particularly for limitations on tariff escalation for more processed products. Furthermore, many countries are members of regional trade associations (e.g., North American Free Trade Association, Association of South-East Asian Nations) and some countries have unique arrangements with close trading partners or former colonies

---

9 The United States argues that the manner in which Canada's provincial governments sell public timber to their mills creates a market advantage for Canadian mills relative to their U.S. competitors and that the tariff is necessary to remove the unfair advantage. The U.S. government does not acknowledge the counter-argument that the U.S. Forest Service similarly sells much of its timber below cost, thereby providing an advantage to U.S. producers that has been, in many ways, comparable to the advantage offered to Canadian mills by the Canadian provincial governments. (See the discussions of timber sales procedures for the Canadian province of Ontario and for the U.S. Forest Service at the end of chapter 5.) Zhang (2007) provides a thorough review of the political economy of the Canada-U.S. softwood lumber trade debate. A special issue of Forest Science (2006, v. 52, n. 3) also reviews the U.S.-Canada experience.

(e.g., European Union-ACP, Commonwealth States).[10] These associations generally negotiate lower rates for their members. Finally, many developing countries obtain lower rates from developed countries through the Generalized System of Preferences. In sum, tariffs on forest products, while variable across countries and products, have not generally been serious barriers to trade and they probably do not seriously distort most markets for forest products (Bourke 2003).[11]

Non-tariff barriers: Non-tariff barriers (or NTBs) are minimum physical standards on imported goods. Some NTBs are designed to protect forest health in the importing country—as in restrictions on imports of exotic species—or to protect tropical forests in the exporting country. Others are designed to protect engineering standards or public safety—as in minimum standards for grades of imported construction lumber.[12] Still others are simply substitutes for tariffs as restrictions on trade for the protection of the domestic industry. All have the same effect as a tariff, decreasing the demand for imports, decreasing trade, and shifting some production from a lower cost exporting country to a higher cost importing country and raising prices in the country that would import more without the NTB.

The more important question for current debate is whether NTBs can have much success in protecting tropical forests. NTBs are often suggested as a means for developed countries to reduce their impact on tropical deforestation in general and to help control illegal logging in tropical countries in particular (e.g., Brack Marijnissen, and Ozinga 2003). The success of NTBs in controlling deforestation is subject to all the uncertainties of certification discussed in chapter 4. In fact, when certification is a requirement of the importing country, it is an NTB.[13]

Furthermore, as one impact of successful NTBs is a geographic shift in production, the geographic shift itself must be to alternative sources of similar products. In practice, this often means a shift in consumption, away from the hardwood products of tropical countries to the products of temperate hardwood forests in the developed countries. Temperate hardwoods are not generally managed. Therefore, NTBs on imports of tropical forest products contain an element of negative environmental impact on unmanaged temperate forests and often these are in that same developed country that considers imposing the NTB with

---

10 Barbier (1996) summarizes the Uruguay Round effects on global forest products.
11 Furthermore, Buongiorno et al. (2003) project that further tariff reductions in forest products following the Accelerated Tariff Liberalization proposal to the World Trade Organization (WTO) would have only a slight effect, increasing the value of global consumption by 2.3 percent and decreasing the global cost of production by 1.4 percent.
12 WTO sets rules for its member nations regarding NTBs. Its Agreements on Sanitary and Phytosanitary Standards sets rules for the establishment of standards to protect human and plant health. Its Technical Barriers to Trade governs the use of technical regulations such as building codes.
13 Taylor, Tomaselli, and Hing (2005) present the view that NTBs are an increasing problem for exporters of tropical timber and timber products. They list various NTBs and the countries that tend to use each as a restriction and give particular attention to ecolabelling as it is required for government procurement in the European Union and North America.

the objective of protecting global forests. The policy objective cannot be met in this case. (Developed country producers welcome this kind of environmental "improvement" but the proponents of global environmental improvement should be more thorough in assessing their own preferences.)

Export restrictions: Export restrictions control shipment from the producing country. They have had two objectives. The usual objective has been to assist the development of the producing country's processing sector, but recent logging bans by several countries in Asia and the Pacific were imposed in response to natural disasters like flooding that were blamed on deforestation in the uplands. The local effectiveness of this group of logging bans has been mixed. Moreover, even where these bans may have had some diminishing effect on local deforestation, they have also exported the problem—to an uncertain net global effect. For example, the logging ban imposed in Thailand in 1989 precipitated increases in logging in the neighboring uplands of Myanmar, Cambodia, and Laos, and the logging ban imposed in China in 1998 accompanied an immediate 35 percent increase China's demand for imported wood and wood products (Durst et al. 2001; Sun et al. 2005).

Where the objective has been protection and development of a domestic processing sector in a country that previously exported logs, the common restrictions have taken the form of either export duties or limitations (or bans) on the shipment of logs. In terms of our figures, these restrictions decrease the local demand for logs in the exporting country by diminishing or even eliminating the demand for log exports. This shifts the forest value function downward, decreases forest production, and decreases the land area subject to timber harvests. As demand decreases, the market price for domestic production must also decrease and the domestic wood processing industry can now obtain its raw material inputs at lower cost. Kishor and Constantino (1993), for example, calculate that Costa Rica's log export ban caused domestic log prices in that country to fall to between 20 and 60 percent of their previous level. This means, however, that the log input costs for the domestic wood processing industry must decline, and the competitive position of that processing industry in the world market must improve. That is, the domestic timber producing industry declines but the domestic wood processing industry expands.

Many developing countries, and the United States and Canada as well, have imposed log export bans of some form at one time or another. Chapter 4 reviewed Indonesia's dramatic experience. In brief, Indonesia was the world's largest exporter of tropical hardwood logs before the imposition of a log export ban in the mid-1980s. Indonesia's annual harvest level declined by approximately 50 percent in response to the ban but its sawnwood and plywood industries benefited, expanding from a small base until Indonesia became the world's largest exporter of hardwood plywood by the mid-1990s. The general expectation is that the losses in the timber production industry exceeded the gains in the processing sector and Indonesia's net economic welfare probably declined as a result. (The real gains were to a select group of businessmen with close ties to the political leadership, businessmen who had invested in the plywood and the shipping industries [Barr 1998].)

Margolick and Uhler (1992), in another example, show that British Columbia's log export ban caused domestic prices in the 1980s to decline to a level 28 percent less than prices for the same species and grade in other Pacific Rim producers. Removing the ban (under reasonable assumptions of demand and supply elasticities) would allow BC's log prices to adjust upward, thereby decreasing quantity demanded from the BC wood processing industry by 25 percent but increasing quantity demanded in the market for BC log exports by more than 300 percent. Employment in BC's logging industry would increase less than employment in the wood processing industry would decline—and therein lies the policy concern. However, the net effect on the full BC economy in 1983 would have been more than a $100 million gain.

Recent trends: In general, there may be a global trend toward trade liberalization in forest products. However, as many countries have reduced export taxes and import tariffs, they have replaced them with other export restrictions (such as NTBs) and also with producer subsidies (discussed in chapter 4). Therefore, the overall net effect of recent policy adjustments must remain speculative.

Feedback to the Forest from Macroeconomic Activity

Several categories of broader macroeconomic influences have impacts back onto the forest of any one region or country. These influences may have external sources, such as general malaise in the economy of a major trading partner, and in such cases they are related to the trade and trade policy topics of the previous section of the chapter. The broader influences can also originate from within the region or the domestic economy, from sources such as domestic macroeconomic policy or overall macroeconomic adjustment and growth. Typically, these influences are more difficult to trace than the direct effects of forest markets or policies that we examined in previous chapters. However, when these broader influences are sufficiently strong, their impacts on the forest can be greater than the direct effects of forest markets or forest policies.

The next section of the chapter reviews the effects of five categories of broader influences on the forest before the final section turns to the question of even more general overall economic growth and development and its impact back onto the forest.

Exogenous Shocks

If trade with another country or region consumes a large share of domestic production in general, and forest production in particular, then a decline in the economy of the trading partner can have a measurable impact on the forests of the producing and exporting country. The impact is likely to occur in two stages.

The importing region's aggregate demand declines in the first stage. That region decreases its demand for all imports including imports of forest products from the exporting producer country. Prices must fall. Initially, producers in the exporting country may try to maintain their former production levels despite the lower prices, and the impact on the forest may continue relatively unchanged in the short run. (Indeed, the high fixed costs of roads and processing facilities make the short-run supply response to decreasing prices inelastic.) Producers will maintain production levels in order to recover what financial return is available for their capital equipment and from their managed forests. They can maintain production as long as their product prices cover the variable costs of forest operation.

Eventually, manufactured capital (e.g., logging equipment, plant and machinery in the processing industries), infrastructure, and managed forests all require major repair or replacement. This signals the second stage. Major repair and replacement is a fixed cost of operation. There are no means to pay fixed costs until the export market recovers and prices return to a higher level. Therefore, some forest operations will continue with deteriorating, less efficient, equipment. Others will discontinue operation. The less efficient equipment will be more destructive of the forest than better maintained, more modern, equipment. That is, it contributes more damage to remaining timberstands and recovers a smaller volume of useful wood from any given tree or forest stand. The net effect must be a long-run decline in both managed forests and in harvesting operations at the natural forest frontier, a decline that continues until the manufactured capital deteriorates beyond capability for any further use. After this, both forest management and harvests from the natural forest can continue only at still lesser levels than before the decline in the importing country's economy, and some natural forest may even begin to recover.[14]

Indonesia's experience with the East Asian financial crisis illustrates this sequence. Indonesia was the world's largest exporter of hardwood plywood, shipping $8.6 million or 90 percent of its total plywood production in 1996, the year before the crisis (GOI/BPS 2001). Most of these exports went to other Asian countries—to Japan, Korea, and Taiwan in particular. Many East Asian countries suffered financial decline and severe devaluation of their currencies in July 1997. Indonesia's own currency declined to approximately one-quarter of its former value. Aggregate demand declined throughout East Asia and, with it, the demand for Indonesia's forest product exports also declined. North America and

---

14 Sudden sharp increases in exogenous demand are less common than sharp decreases. Sharp increases in demand create a similar two-stage adjustment process. Initially, producers in the exporting country increase output by adding variable capital and labor to the existing fixed capital and fixed area of mature forest. They may add new fixed capital and increase their forest management activities as quickly as possible, but there will be delays until these new facilities and improved forests are ready for production. During the first stage, operations will be neither as economically efficient nor as environmentally sound as they had been, but production will increase. Eventually, greater economic efficiency and better environmental performance will return as the new greater demand attracts new fixed capital and as the new managed forests mature.

Europe did not provide alternative sources of demand for Indonesia's suddenly-less-expensive exports because the economies in those regions were entering their own period of stagnant growth.

Nevertheless, Indonesia's production continued at its previous level for several months. The officially reported timber harvest level even increased in 1997 and again in early 1998. The longer term trend in sawnwood and plywood production only began to slacken in 1998. Sawnwood production declined from 7.2 million m^3 in 1997 to 2.5 million m^3 in 1998, while production in the somewhat more capital intensive plywood industry declined only from 7.8 in 1997 to 7.4 million m^3 in 1998. The pulp and paper industry, the forest products industry with the highest fixed costs and relatively lowest variable costs, maintained its production level through 1998, and then declined from 2.0 to 1.2 thousand tons of pulp, a decline of 40 percent, in 1999. Production levels could not continue even at these lower levels. Production continued to erode in all three wood products industries. The high fixed cost, low variable cost, pulp and paper industry declined the least and showed the earliest signs of recovery in 2001. Sawnwood and plywood production declined to less than one-quarter their former levels and only began to recover slightly in 2003 (GOI/DF 2004).

In sum, Indonesia's experience demonstrates the anticipated two-stage adjustment to substantial external stimuli. As expected, the initial adjustment was delayed longest and recovery began to occur first in the pulp and paper industry, the wood product industry that is characterized by highest fixed costs and lowest variable costs. Nevertheless, the entire forest products sector did decline in response to the general economic malaise and the decrease in demand from Indonesia's trading partners. Subsequently, timber harvests, and both sawnwood and plywood production, varied greatly from year to year during the first years of the new century, indications that these industries had not yet completed their second-stage of adjustment to the changed economic conditions even by 2008—shortly before a new round of global economic and financial difficulty made its presence.

Shocks from within the Domestic Economy

We might identify four varieties of broad and general shocks from within the domestic economy: significant social and institutional change and adjustment, a sharp increase in another (non-forest) sector of the economy, comprehensive macroeconomic policy adjustment, and the effects of more specialized fiscal and monetary instruments of macroeconomic policy. Civil disorder is an example of the first and civil war is an extreme example. "Dutch disease," named for the economic impact on the Dutch economy of the discovery of natural gas in the North Sea, is an example of the second. What has become known as "structural adjustment," often imposed externally by international lending authorities, is an example of the third. Structural adjustment is generally composed of a collection of fiscal and monetary and other broad macroeconomic policy instruments. We will consider social and institutional change, Dutch disease, and structural adjustment in this section and reserve the specific effects of selective fiscal or monetary policy instruments for the next section of the chapter.

Social and institutional change—including civil disorder: Substantial social and institutional change breeds uncertainty. In general, managers postpone investment in the presence of uncertainty. For forestry, this means that unpredictable broader social adjustment causes loggers to delay major repairs in their equipment and forest managers to delay new investments in forestry. They draw down the existing forest stocks while covering their variable costs of operation—until the general social and institutional outlook becomes more settled and predictable.[15]

The initial behavior of loggers and forest managers, in this case, is comparable to behavior in first stage of the previous category of exogenous shocks. Longer term behavior, comparable to the second stage in the presence of exogenous shocks, depends on whether the social and institutional change and eventual new stability engenders growth or decline in the demand for forest products.

In the presence of extreme uncertainty, loggers and managers not only postpone investment but they also may become more aggressive in their harvesting of existing economic forest stocks. Military activity is an example and an unfettered military can become the source of severe forest extraction. The military can redirect soldiers, trucks, and other equipment that are not normally used in logging activity to harvest trees and forests that would not be within the range of economic activity for either private loggers or the government forestry agencies. At least some of the financial support for the military's labor and capital comes from other sources. Therefore, the variable costs it associates only with logging are lower than private or government agency logging costs. This means that the short-term financial frontier for the military can extend farther into the hinterlands and farther up the mountainsides beyond the logging frontier for normal market activity.

Nevertheless, there are limits to the extent of harvesting in the presence of great uncertainty and even in the presence of low cost logging operations such as those of an active military. The limits identify new financial frontiers comparable to point D in any of our figures and any of the three stages of forest development, although these limits may extend farther than the peacetime limits of an effective and stable private market.[16]

Consider the logging activities of the competing armies in Cambodia in the 1980s and 1990s. These armies redirected some of their men and equipment into

---

15 Page (2006, p. 1058) makes the summary observation for instability and all natural wealth: "political and economic instability allow for unchecked exploitation of resources and ecosystem mismanagement. Thus, political instability causes environmental degradation." Of course, exploitation is not totally unchecked. Short-run variable costs do set a limit on it.
16 Deacon (1994) demonstrates the short-term effect with data on deforestation and civil disorder from eighty-four countries. Countries with the weakest governance suffered the greatest rates of deforestation.

logging operations and then sold the logs across the border in Thailand to obtain financing for their continued military activities. They extended logging beyond the areas that civilian loggers could afford to log, increasing revenues from the country's annual timber exports more than ten-fold by 1992 (Lebedys 2004). Elements of various military groups were still active in Cambodia in 2005. It should not be surprising that the local political instability and economic uncertainty that goes hand-in-hand with their presence was also accompanied by continued deforestation.

It is also not surprising that, in year 2000, illegal logging was responsible for a greater share of gross national product in Cambodia than for any other country in the world. Indeed, we should anticipate that the ratio of the value of illegal logging to aggregate economic activity will be greater for countries whose institutions protecting property rights are less developed and, among those, greater for countries suffering serious social disorder.[17]

Liberia provides another example. Liberia suffered a similar experience from 1997 to 2003. Timber exports increased more than 300 percent to support Charles Taylor's military and political ambitions. The increase in timber harvests began as Taylor supported his rebel military with timber revenues and continued after he assumed political power. It only slackened when the UN imposed a sanction on all Liberian timber exports.

The uncertainty caused by large social change and even civil disorder also deters interest in long-term investment, including investment in forestry. The modern contrasts between four planned economies, China and Vietnam, Laos and Cambodia, are illustrative. China and Vietnam have both experienced extended periods of stable policy. Market reforms began in China with land rural land reforms in 1978. The household land contracts that were fundamental to these reforms have been renewed and, despite discussion of additional reforms, China's farmers seem to have the confidence that comes with more than a quarter century without significant reversals in land use policy. They have shown their confidence with their willingness to make long-term investments, including investments in forestry. Forest cover on the former collective lands, now often managed by individual households, has increased more than 60 percent and timber production from these lands has increased as well (Hyde et al. 2003). Vietnam's experience is similar, if more recent. Vietnam began its *doi moi* reforms in the mid-1990s. Since then, its economy has also grown without substantial policy reversal and its farmers have also begun to plant trees (Tachibana, Nguyen, and Otsuka 2001). The area in forest plantation was negligible in the early 1990s but it covered 2.089 million hectares by 2005.[18]

---

17 The ratio for Cambodia was 2.3:1. Indonesia, another country suffering from financial and political instability at this time, had the world's fifth highest ratio—0.7:1. The second-fourth highest ratios belonged to Papua New Guinea (1.2:1), Cameroons (0.9:1) and Brazil (0.7:1). The latter three did not suffer the same instability but property rights were not well-established at the forest frontiers of these three countries. Calculations based on estimates of illegal logging reported in various sources cited by Contreras-Hermosilla (2001).

18 Available in Vietnamese at http://www.kiemlam.org.vn. Translated by Le Cong Uan, May 27, 2006.

Laos and Cambodia provide the contrast. Both countries suffer uncertainty throughout their economies. Laos suffers excessive regulation and little government transparency. One furniture manufacturer told the story of fourteen signatures from fourteen different government offices required before he obtained permission to export. Trucks line up at the bridges waiting for yet more signatures before the official permission is complete. Cambodia was still recovering from fifteen years of civil turmoil in 2006. The current government is stable but competing military units still controlled some parts of the country in 2006 and even today more than forty international donors and international NGOs advise often conflicting policy reforms for this small country. Long-term forest management is minimal in Laos and virtually non-existent in Cambodia despite many land use characteristics and international market conditions that are similar to those for Vietnam and for China's southern provinces.[19]

Dutch disease: The impact on a country's forests from rapid expansion in a large and different sector of the economy is more difficult to predict. Two opposing effects create an uncertain result. Wunder (2003, 2005) provides an explanation and recent examples from the experience of new oil or mineral wealth impacting the forests of eight tropical countries. He observes that, while the inflow of foreign exchange due to new oil revenues increases the value of domestic currency and weakens the price competitiveness of the forest sector in competition with other countries, it also strengthens the domestic demand for forest products. The net effect on the forest sector and the forest itself depends on the relative strength of these opposing forces and this must vary from country to country depending on the level of currency adjustment and the offsetting domestic and export demands for forest products. In Wunder's examples, the currency effect tends to dominate in less developed countries and rates of deforestation decline. In middle income countries, the increase in domestic demand is more likely to offset the decline in wood product exports, and the effect on the natural forest is more variable.

Structural adjustment: The impact of broad and general macroeconomic policy adjustment on a region's or country's forests is also difficult to assess. The breadth of adjustments often creates a number of effects, some with contrasting impacts on a country's forests. The financial reforms imposed in the 1990s by international lending institutions on several developing countries with serious currency imbalances and international payment deficits are an example. These structural adjustments were designed to improve deficits in the balance of payment and to reduce public debt and inflation.

---

19 Laos has 12.561 million hectares of forest, 0.006 million or less than 0.5 percent of which is in plantations. Cambodia has 9.335 million hectares of forest, 0.082 million or less than 1 percent of which is in plantation. Vietnam and southern China, in contrast, have 12.094 and 642 million hectares of forest, respectively, of which 2.089 and 287 million hectares or 17 and 45 percent are in plantations (various sources, as cited in Hyde 2005; Hyde et al. 2008, FEDRC 2006).

Undoubtedly, their initial effects on the forest are similar to the first stage effects of the previous categories of broad influences. Their longer term effects are more complex—and they surely vary from country to country as the implementation of the structural adjustment and the range of other policies also varies.[20] Therefore, assessing the chain of causality from any fundamental structural adjustment program through the various government policies and programs and on through the various affected sectors of an economy is a difficult analytical task.

Brazil's experience between 1970 and 1995 provides an example. Brazil suffered a 22-fold increase in international debt, a 40-fold increase in long-term interest rates, and annual inflation rates as high as 2,560 percent following the oil shock of 1974 and twenty years of diverse domestic economic policies (Young 1996). The International Monetary Fund imposed a program of structural adjustments on Brazil in 1982 as a condition for the Fund's assistance with Brazil's international debt.

Young (1996) traces the effects of this program on the Amazon's natural forest frontier. He identifies two fundamental relationships between the broad macroeconomic policy of the time and forest cover.

- Both the decline in general government expenditures, including reductions (1) in subsidized agricultural credit and (2) in expenditures for roadbuilding discouraged deforestation and improved the conditions of the forest at the natural frontier.
- Export incentives designed to improve Brazil's balance of payments led to appreciating prices for more capital-intensive commercial agricultural products like soybeans. This (3) increased land values for those small farms that could convert to larger commercial agricultural operations, and (4) decreased the real agricultural wage, thereby causing agricultural workers and small farmers to lose employment. The latter effect was reinforced by other programs of the central government designed to decrease the minimum wage. As a result, both agricultural workers and small farmers migrated to the frontier where they converted Amazonian forest for new smallholding agricultural activity.[21]

Rural wages, roadbuilding, and agricultural credit had the most elastic effects on deforestation—all in the neighborhood of 0.4. That is, a 1 percent decrease in rural wages caused and approximate 0.4 percent increase in land deforested at the frontier, while a 1 percent decrease in roadbuilding had the opposite effect, decreasing the rate of deforestation by approximately the same amount.[22] The decrease in agricultural credit raised interest rates and favored shorter-term agri-

---

20 See Benhin and Barbier (2004), Sunderlin and Pokam (2002), or Anderson et al. (1994) for specific examples from Ghana, Cameroon, and Bolivia. Kaimowitz et al. (1997) compare the experiences of Cameroon, Bolivia, and Indonesia. Their only general conclusion is that those structural adjustments that succeed in creating new off-farm employment may reduce deforestation.
21 Others (Mahar 1988; Mahar and Schneider 1994; Schneider 1995; Binswanger 1991) draw similar conclusions for the Brazil example.
22 These effects are exactly those anticipated for rural wages and roads in chapter 3 and rural wages in chapter 4.

cultural activities that were more land-using. As such, they favored land conversion at the Amazon frontier into agriculture and, thereby, increased deforestation.

The notable conclusion is the contrast in effects. The downward adjustments in wages and agricultural credit increased deforestation, while the downward adjustment in roadbuilding decreased deforestation. In sum, Brazil's structural adjustment program had multiple and mixed effects on the forest frontier resulting in an indeterminate net long-run effect on deforestation. Undoubtedly, the effects on Brazil's managed forests were also mixed. Brazil is only one example, but we might speculate that other broad-based macroeconomic policy programs in other countries have also had mixed impacts on their forests. The net effects must depend on the combination of specific effects from each individual more general macroeconomic adjustment.

Domestic Fiscal and Monetary Policy

While we cannot easily assess the net effect of broad macroeconomic policy adjustments on forests, we can anticipate the direct effects of the two fundamental varieties of macroeconomic policy, fiscal and monetary policy. Fiscal policy refers to adjustments in government expenditures or taxation. Monetary policy commonly refers to adjustments in the interest rate or adjustments in the money supply which, in turn, affect the interest rate. Where the value of a country's currency relative to other national currencies is controlled by the central bank or the central government, then the exchange rate can be a another component of monetary policy. Young's example from Brazil provides some insight but none of these macroeconomic policy effects on forests have been widely examined and each would benefit from closer scrutiny.

Fiscal policy: Fiscal policy is often used as a short-term first step out of a period of economic stagnation or decline. Fiscal policy uses government expenditures to provide the new demand to employ previously unemployed resources. It has its most beneficial effect when the sectors receiving the influx of government funds are sectors that both respond and grow rapidly themselves and also link with numerous other sectors of the economy through their demands for the products of those other sectors or their supply of inputs to them. These links mean that growth in the first sector is also a source of growth in the linked sectors.[23]

The construction industry is a good example and it is often a prime candidate for fiscal intervention. Construction responds rapidly to new demand for its product, either domestic housing or an input to production in other industrial sectors. The use of wood as an input to construction makes the demand of the construction industry of prime importance for forestry and the wood products industry. (In the United States, for example, construction accounts for approximately 40 percent of all roundwood consumption.) Therefore, fiscal policy that

---

23 Hirschman (1958) is the classic reference.

targets construction can have an important impact on the wood products industries and on forestry itself—although the latter impact is not immediate because it must account for the time lags between the introduction of new government expenditure on construction, the increase in the construction industry's demand for wood products, and the increased demands of the wood products industries for their raw material. Nevertheless, one final impact of an increase in construction is an increase in activity at all margins of forest operation.

The effects on forests of injections of government expenditures into other sectors depend on the links between those sectors and the forestry sector. For many injections the links will be small and the effects on forestry will also be small, and delayed as well. Tax reductions have popularity among some Americans as an alternative fiscal tool (known as "supply side" economics) for encouraging a stagnant economy. The speculation is that as tax decreases allow consumers to retain more of their income, they will use the increment on consumption and their additional consumption will fuel economic expansion. Of course, only the smallest share of additional consumption will be of forest products and the effect on the forest sector and forests themselves is probably minimal.

The longer term effects of fiscal policy (either expenditure or taxation) depend on the effectiveness of the policy in generating economic growth and the effect of general macroeconomic economic growth on forestry. That is the topic of the final section of this chapter.

Monetary policy: Central banks control the interest rate, raising it to dampen growth and control inflation in times of full employment and decreasing it in times of economic stagnation in an effort to renew investment and reinvigorate an economy with underemployed resources.

The effect of monetary policy on forestry has been a source of confusion among forest economists. The Faustmann model (appendix 3a) predicts that lower interest rates cause forest managers to decrease harvest rates as they lengthen timber rotations and increase their forest stocks. This is a partial prediction, however, because the Faustmann model only refers to existing managed stands. It provides no indication of the impacts of interest rates on the margins of land use. The more likely short-run scenario is that lower interest rates increase the demand for construction in all other sectors of the economy. The increase in construction increases each sector's demand for wood products and, after accounting for time lags in each stage of this adjustment process, timber harvest levels increase at all margins of forestland.

Once more, the long-run scenario depends on the effectiveness of government policy in generating economic growth and the effect of general macroeconomic economic growth on the forest sector and the forests themselves.

Exchange rate policy: Since the 1970s the currencies of most countries float according to daily transactions in global markets. Those currencies that do not float are maintained at specified levels as official government policy or the official policy of a nation's central bank. However, no government and no central bank has a sufficient stock of foreign currency to protect its own national currency from all long-term adjustment.

When those governments that control the exchange of their currencies permit adjustment, the adjustment can be either upward or downward—a currency revaluation or currency devaluation—and these adjustments may have opposite effects. Revaluations are a primary result of Dutch disease—discussed previously. Even when the effect is not as substantial as may occur with Dutch disease, the direction of effects and their explanation is similar.

Devaluation means that all domestic goods become less expensive for other countries and exports of all domestic production, including forest products, increase. Therefore, the short-run effect of devaluation is to draw down the stocks of both managed and frontier natural forests in order to satisfy the increased foreign demand. Indeed, the low variable costs of logging operations can make timber harvests an attractive source of foreign exchange during a time of devaluation and financial difficulty. Over longer periods the additional foreign demand may also be a sufficient incentive to increase forest management and managed forest production for some countries that were in the third stage of forest development before the devaluation.

Of course, the full long-term effects of currency devaluations are much broader as exporters and producers of export goods experience gains, but importers and those dependent on imported goods lose. The net effect on the forestry sector can be mixed, as log exports and exports of finished wood products increase, logging at the frontier increases, but forest management also increases for economies in the third stage of forest development. However, the costs of the imported capital equipment used in logging and wood processing also increases and this dampens all potential increases in forest production.

Substantial and Long-lasting Agricultural Policy

Macroeconomic policy usually refers to short-term policy-induced shocks to the overall economy that have the objective of correcting some general economic imbalance. Government activity also includes many broad-based longer-term programs that, while not usually considered macroeconomic policy, can have substantial effects on the forest. The next two short sections of the chapter reflect on longer-term programs of foreign governments that spillover to affect broad regions and sectors of a domestic economy, and long-term programs of the domestic government that affect the pattern of domestic growth and, through it, the domestic forestry sector. These longer term programs are generally more gradual in their development than most macroeconomic policy adjustments and we are unlikely to observe significant short-run effects. However, their full effects can be long lasting and substantial.

The agricultural programs of the North American and Western European countries and Japan may be the best example of long-term programs that spillover to affect the forests of other countries. We discussed the impacts of agriculture and agricultural policy on the forest in previous chapters. In those chapters our interest was on agriculture's effects on forests in the same region. In this chapter our interest is on the agricultural policies of one region or country that affect the forests of other countries.

The United States, the European Union, and Japan together spend more than US$300 billion each year in farm subsidies and price supports. The United States alone spends in the neighborhood of $60 billion in various agriculture support programs (Watkins and von Braun 2002).[24] These expenditures compare, for example, with a U.S. public budget for forestry programs of less than $6 billion for 2005 (USDA Forest Service 2007). These immense public agricultural expenditures fund additional use of all agricultural inputs, including large shares of North American and European land that would otherwise be under forest cover. The use of additional inputs produces large agricultural surpluses, some of which are exported to developing countries where their artificially low prices drive out competition from local commercial agriculture. The resulting lower level of local agricultural production means a reduced level of employment in commercial agriculture in the developing countries. Some local farmers and agricultural workers, no longer employed in commercial agriculture, return to subsistence farming. Subsistence farming uses land more extensively than commercial agriculture. This means that one effect of the agricultural policies of the developed countries is to convert frontier forest to use by the new subsistence farmers. In sum, the agricultural programs of the developed countries cause a loss of forest both at home and in developing countries around the world.

The full global effect is only speculation, but it is conceivable that the North American, European Union, and Japanese agricultural support programs are more destructive of global forests than all commercial forest activities at the frontiers of the developing countries. The effect has not been examined quantitatively. However, if developed country agricultural support programs are the source of serious forest destruction in the developing countries, then revisions of these agricultural policies could be most beneficial to the global environment, as well as to the economic well-being of many developing countries. We can speculate that revisions in agricultural policy would have greater forest conserving effects than direct environmental policies such as forest certification, improved silvicultural standards, or increased enforcement against illegal logging. Furthermore, their administration would be simpler because the revision would only require the reduction or discontinuance of existing government subsidies. Reducing or

---

24 The perversions within the distribution of these funds are simply amazing: 93 percent of all U.S. agricultural subsidies are for corn, soybeans, rice, cotton, and wheat (46 percent for corn alone), effectively depressing world prices for each of these commodities. Ten percent of American farmers receive 72 percent of the subsidies, while another 60 percent receive no federal moneys. Moreover, more than half of all U.S. farm subsidies are concentrated in only 25 of the 435 U.S. congressional districts. For the European Union, its Common Agriculture Policy is approximately 40 percent of the annual budget—for the support of only 2 percent of the EU workforce which produces an even smaller share of the EU countries' GDP (*Economist* 2005, 2006a, 2006b).

removing current policies would not impose the monitoring difficulties inherent in forest certification, silvicultural standards, or effective controls on illegal logging.[25]

Other Substantial and Long-lasting Domestic Programs

Two varieties of more general and long-term domestic programs may have significant effects on forestry: regional development programs and the entire range of policies and programs that favor capital investment over labor and especially unskilled labor.

Regional development programs generally target poorer rural regions or entirely undeveloped regions. The many programs over the years designed for southern Appalachia in the United States or China's newer Western Regional Development Program (and the associated Natural Forest Protection Program) are examples of programs designed to improve welfare in poorer regions of an economy. Programs such as Indonesia's transmigration program in the 1990s or the Homestead Act in the United States in the nineteenth century were designed to develop previously undeveloped regions.

Both poorer regions and undeveloped regions tend to be sources of natural forest. Managed forests, where they exist, tend to occur in better developed regions closer to the major markets—in the coastal plains and the Piedmont of the U.S. South, for example. The better developed regions also contain some natural forest, but the focus of most regional development programs has been in more mountainous inland regions such as Appalachia (also largely in the U.S. South) or southwestern China where the forests are almost entirely natural. Therefore, we can anticipate that successful regional development tends to encourage exploitation of the forest frontier, shifting it farther into the hinterlands and up the mountainsides. Moreover, to the extent that forest products from regions benefiting from public development programs substitute for production from other regions, these regional development programs also delay the progress of sustainable forest management in more developed regions.

Labor and capital investment policies, the second more general program of our interest, are seldom discussed in terms of their effects on forests. Nevertheless, these effects may be important in some cases. Most countries have a range of tax policies designed to encourage investment throughout the country. Some countries also provide favorable import duties encouraging investment in imported capital equipment. These programs favor capital relative to labor. In addition, many countries establish a minimum wage which effectively favors skilled labor in preference to unskilled labor. Each of these policies is intended to improve the conditions for economic growth, and they may accomplish that, but they also decrease the relative position of labor and especially unskilled labor. If growth

---

25 Of course, the much greater difficulty lies in the political transactions costs involved in changing the agricultural support policies and institutions in North America, Western Europe, and Japan.

occurs, then employment opportunity may improve and workers may be drawn toward better labor opportunities and away from forested rural areas. The condition of the natural forest at the frontier may even improve as a result. However, the relative position of the lowest wage workers must decline as a result of these policies.[26] Increased unemployment may even be one result. Employment in low wage subsistence agriculture may increase and the rural poor often turn to the forest to find land for their subsistence activities in this event. As they do, the natural forest suffers degradation and even deforestation.

The net effect on the forest frontier of economic growth but declining labor opportunity for the rural poor is another unexamined empirical question. Its answer surely varies as the package of capital-favoring policies and the opportunities for the rural poor vary from country to country. Thoughtful speculation on the magnitude and significance of the answer should be a precondition to the introduction of programs designed to encourage rural economic growth and especially for programs designed to improve the welfare of the poorest members of any rural society.

Neither regional development policies nor the range of policies favoring capital investment and skilled labor are among those we regularly consider as important to forestry. In fact, we often consider the existence of natural forests as indicative of potential for regional development. Perhaps we should reconsider. Under what conditions do the potentially destructive effects of these policies on the forest frontier offset their intended favorable effects on economic growth?

Economic Growth and Development

Economic growth and development generally require the full participation of the entire economy or at least most of its sectors. Growth in multiple sectors implies growth in their various demands for the products of other sectors, including the extractive forest products of the forestry sector. Development implies improvement in general economic welfare and it often introduces new demands for environmental protection and the non-consumptive services of forests, as well as increasing demand for the consumption of extractive forest products. Therefore, both growth and development have important effects on the forest.

Many factors are at work in both growth and development. Therefore, identifying a small number of characteristic factors that have significant impacts on forests becomes a difficult task. We can speculate, however, on the importance of the relative size of an economy's or a country's forestry sector, the relative importance of trade in forest products to that economy, and the effect on the forest of the general pattern of its growth.

---

26 See Boyd and Hyde (1989) on the effects of the U.S. minimum wage on loggers and workers in the forest products industries. Their overall observation is that a 1 percent increase in the minimum wage induces only a 0.5 percent increase in total wages paid to workers in the U.S. forest industries, and it is accompanied by a net 0.3 percent decrease in employment in those industries, as the demand for labor decreases with the increase in its cost for the producers who provide the employment.

The Relative Size of the Forestry Sector

Where the forestry sector is a substantial component of the domestic economy but aggregate economic growth is largely a result of expansion in other sectors, then aggregate growth may fuel growth in the local forestry sector as well, but the impact of aggregate growth on the forest sector is likely to be small relative to its own size. That is, the increased demands in other sectors of the regional economy are unlikely to absorb sufficient increases in forest product inputs to those sectors to make a difference in the basic pattern of regional forests and the regional forestry sector.

More generally, where forestry is a large component of the domestic economy, the interesting effects are likely to run the other direction, from forestry to the general economy, and increasing demands on the forest sector will play a role in aggregate economic growth. This means that external demand and interregional and international trade must be a source of aggregate economic growth—since trade must be the largest source of demand on the large forestry sector. Demand for forest products by the rest of the local economy is relatively small. It cannot absorb the increase in production that is necessary to fuel growth in a forest sector that is already large relative to the other sectors of the domestic economy.

Table 6.2 provides some perspective. It shows the size of the forestry sector relative to the aggregate economy for a selection of 26 countries. These countries either have unusually large forest sectors or they are the world's the largest exporters or importers of forest products. The entire forest and wood products sector (including forestry itself and also the manufacture of forest products) accounted for 4 percent of the aggregate economy in the year 2000 in only six of these countries (Finland—7.6 percent, Gabon—5.0 percent, Latvia—4.9 percent, Malaysia—4.7 percent, Guyana—4.6 percent, and Brazil—4.1 percent). A comparison of columns 3 and 4 in the table shows that the contribution of the forest itself to gross domestic product is much smaller than the contribution of the full forestry sector in all countries. Therefore, we can conclude that in all national economies the direct contribution of the forest itself is a small share of the total economy and the contribution of the full forestry sector, including the manufacture of forest products, is hardly large.

Nevertheless, the forestry sector is important in a number of regional cases. It accounts for 40 percent of employment in the Malaysian state of Sarawak (Anon. 2007). Six U.S. states provide examples of large forestry sectors for regions in the third stage of their forest development (Table 6.3). The forestry sectors in these states, including forestry itself and manufactured forest products, comprise up to 5.6 percent of the gross state product. The local economies absorb less than one percentage point of this. They export the bulk of their forest production to external markets in other states and countries. Clearly, in these states, a small increase in external demand and trade in forest products has a much larger absolute impact on forests and the full forestry sector of the local economy than a comparable percentage increase in the local demand for forest products. As the impact of an increase in the external demand for forest products traces through local forest

TABLE 6.2 Forest sectors as components of national economies[1]

Country	Gross domestic product	Forestry contrib'nto GDP	Forestry manufacturing sector (except furniture) contribution to GDP (value and share)		Forestry sector trade			
					Exports		Imports	
					Value	Share of all exports	Value	Share of all imports
Australia	390,113	389	3,355	0.9%	807	1.3%	1,769	2.5%
Belarus	29,950	161	250	2.5%	173	2.4%	171	2.1%
Brazil	595,458	11,682	19,098	4.1%	3,024	5.5%	978	1.7%
Cameroon	8,879	163	260	2.9%	536	29.2%	18	1.2%
Canada	687,882	4,292	19,843	2.3%	27,714	10.0%	4,203	1.7%
Chile	70,545	276	1,872	2.9%	1,787	9.8%	185	1.0%
China	1,079,948	79.9	14,930	1.3%	3,640	1.5%	14,699	6.5%
Estonia	4,969	111	277	3.8%	381	12.0%	106	2.5%
France	1,294,643	2,185	8,249	0.7%	5,790	1.8%	7,897	2.4%
Finland	121,687	2,460	7,914	7.6%	10,974	23.8%	904	2.6%
Gabon	4,932	142	176	5.0%	333	10.6%	2	0.3%
Germany	1,873,568	1,145	15,252	0.9%	11,497	2.1%	12,520	2.5%
Guyana	712	12	27	4.6%	38	7.7	3	0.5%
Indonesia	152,255	1,765	3,977	2.5%	5,517	8.9%	1,376	4.1%
Latvia	7,150	104	306	4.9%	627	33.6%	93	2.9%

Malaysia	89,659	2,195	3,694	4.7%	2,793	2.8%	972	1.2%
New Zealand	49,903	672	1,837	3.9%	1,483	11.2%	265	1.9%
Norway	161,769	255	1,632	1.1%	1,344	2.2%	1,021	3.0%
Portugal	105,054	683	1,938	2.1%	1,311	5.4%	976	2.4%
Republic of Korea	457,219	1,022	6,352	1.6%	1,627	0.9%	3,708	2.3%
Russian Federation	251,106	249	1,750	0.8%	3,792	3.6%	388	0.9%
South Africa	125,887	432	1,856	1.6%	791	2.6%	486	1.6%
Surinam	846	19	29	3.5%	3	0.8%	2	0.5%
Sweden	227,319	1,759	6,912	3.4%	10,127	11.6%	1,577	2.2%
United Kingdom	1,414,557	254	9,696	0.8%	2,194	0.8%	9,006	2.6%
United States	9,837,406	13,717	116,014	1.3%	16,612	2.1%	25,706	2.0%

1 All financial values in millions of 2000 US$.
Sources: compiled from Lebedys (2005).

TABLE 6.3 Forest sectors in the United States as components of regional economies[1]

U.S. State	Gross state product(GSP)	Forestry sector		
		Forestry itself (incl fisheries & related activities)	Forest products (incl. furniture)	Combined forestry & forest products share of GSP
Alabama	114,576	734	4,387	4.46%
Georgia	290,887	817	6,320	2.47%
Mississippi	64,266	622	2,951	5.56%
North Carolina	273,698	664	6,072	2.46%
Oregon	112,438	1,526	3,318	4.31%
Washington	221,961	2,078	3,124	2.34%

1 All financial values in millions of 2000 US$.
Source: US Department of Commerce Bureau of Economic Analysis website http://www.bea.gov/bea/dn/home/gdp.htm. (accessed July 14, 2006)

production, it also has a larger impact on employment, workers' demands for services, and the entire domestic economy than a comparable increase in the smaller local market demand for forest products.

The forest sector can also be large enough to lead economic growth in some regions in the second stage of forest development. Gabon and Guyana, and also the historical development of western Washington in the United States and British Columbia in Canada, provide examples. Refer again to Table 6.2. In these cases, the natural forest serves as a source of capital. As loggers draw down the stock of natural forest, their exports of logs and wood products create an inflow of financial capital available for reinvestment. Some of this financial capital may be reinvested in the domestic economy—if good local investment opportunities exist and the investment climate is sound.

Investment opportunity is crucial. Where the investment opportunity is lacking, the local economy may experience a "boom" period of forest production followed by an indefinitely longer period of "bust". This was the early experience of western Washington and coastal British Columbia as loggers first harvested around young settlements in Seattle and Vancouver, then moved progressively farther down the Washington coast and up the British Columbia coast as they depleted the local natural forests. Each local area experienced a temporary logging boom with little expectation of reinvestment and permanent activity to follow, although the broader regions of coastal Washington and British Columbia continued to produce logs and the cities of Seattle and Vancouver grew. The general investment opportunity, either in the forestry sector or in a diversified local economy, took time to develop.

Where the investment opportunities are favorable, reinvestment does occur. The forests themselves are not initial recipients of new investment because forest management is unrewarding during the second stage of forest development. Therefore, as investment opportunity arises the domestic economy diversifies and becomes more resilient. Forest production may lead the way in early development but, as the local economy diversifies, other sectors expand in their relative importance to the overall welfare of the regional economy. The forestry sector may remain important, but the role of the forestry sector declines in terms of its relative contribution to the aggregate economy. The aggregate economy becomes more resilient and less subject to the traumas that can be introduced by external market shocks because it is not as dependant on activity in one single sector.

Meanwhile, the forestry sector may still grow in absolute terms, as it has in Finland, Washington, and British Columbia, and also in Chile, the Brazilian state of Para, and the island of Java in Indonesia. In each of these examples the forestry sector itself has diversified and the role of the industries involved in the manufacture of forest products has expanded. These regions have all moved from the second to the third stage of forest development with at least a small and increasing amount of new investment in managed forests. Of course, harvests from the natural forest continue in each case and the newer managed forests are often some distance removed from each region's or country's remaining natural forests.[27]

This is the pattern for regional economies with large forestry sectors, but a forestry sector that is small relative to the rest of the domestic economy is the more common case. The forest sector is never greater than 7.6 percent of the national economy, even in Finland, the largest case identified in Table 6.2. In the more common case of a relatively smaller forestry sector, the sector may have little effect on aggregate economic growth, but the aggregate economy can have important effects on the forestry sector and the forest itself.

The Effect of Aggregate Economic Growth on the Forest

We examined the specialized effects of forest markets and direct forest policies in chapters 4 and 5; the effects of adjacent sectors, also in chapter 4; and the effects of trade, trade policy, and international markets earlier in this chapter. The effects on a smaller forest sector of a broadly expanding aggregate economy and of overall improvements in economic welfare remain unexamined. They are the subject of this final section of the chapter.

The effects are complex because the aggregate economy is composed of many small and often contrasting shifts in markets and policies, and also because the different components of the forest do not always respond in a similar manner to these different markets and policies. We have explained, for example, that forest product prices have opposite effects on the lands in managed and natural forests, and that agricultural prices are more likely to affect the lands in managed forests; and we expect that improvements in general welfare may have relatively greater effect on the demands for the environmental services of natural forests than on the products of managed forests. These points were established in previous chapters,

---

[27] Chapter 11 will revisit this topic in its discussion of the interactions between forests and the local human population.

but the available evidence on the overall effects of aggregate economic growth and development on the forest remains scattered and incomplete.

We will approach the assessment of aggregate economic effects in two steps. The first relies on an example to demonstrate the effects in one case. This example also illustrates a basic approach for examining other cases. The example is Hainan Island, a province in the southeast of China. Hainan is a good example because its unusually rapid growth over the last three decades allows a clear statistical measurement of the effects of growth, because its managed and natural forests have both undergone substantial and measurable transitions during this period, and because, as an island, its market boundaries are clear. The second step is to reflect on the broad and general hypothesis of the three-stage model, the hypothesis that the development process begins with forest exploitation but eventually, after some level of development, the incentives for forest exploitation shift and forests and their associated natural environments begin to recover.

Hainan Island: Hainan was originally covered with tropical rainforest and rainforests still covered half of the island at the outset of the Japanese occupation in the 1930s. The Japanese extracted more than 10,000 m^3 of logs annually to support eighteen mills, and then removed an additional amount of timber during World War II for road construction. Fire destroyed still more of the native forest during the 1930s and 1940s.[28]

Forest cover declined to only 30 percent of the total land area by the early 1950s. The remaining natural forest was located mostly in the mountainous center of the island. Forest cover continued to decline in response to expanding demands for wood, both within the island and from external markets. By the end of the 1980s, 40,000 hectares of rainforest had been logged with very little silvicultural reinvestment. Repeated high-grading reduced much of the forest to degraded shrubland or even bare land and Hainan went from being a net exporter of timber in the 1950s to a net importer by the 1980s.

Isolated patches of tropical tree crops (rubber, tea, and tropical fruits) began appearing in the early twentieth century, but the land area in these crops did not expand significantly until the 1950s. Land use for other agricultural crops remained relatively constant as well, until the 1970s and 1980s when the shifting cultivation practices of the native minority population annually destroyed 2–3,000 hectares of natural forest. By the late 1970s, Hainan's forests had shrunk to 15 percent of the island's total land area and the area of degraded former forestland extended to one-quarter of the its total area.

Policy played a role as well. From the 1950s onward, the government reallocated agricultural lands from larger landlords, first to small farmers, then to collectives, then to people's communes. Agricultural productivity declined and this may have contributed to the number of households that turned to subsistence

---

28 This section on Hainan Island relies on Zhang et al. (2000).

agriculture and the practice of shifting cultivation—causing additional deforestation and forest degradation.

China began its gradual return to a market economy in 1978. The earliest reforms occurred in agriculture where a policy known as the Household Responsibility System returned some land use rights to individual households. Similar reforms in the use rights for forestland followed shortly thereafter. Rural household incomes throughout China rose dramatically more than six-fold between 1978 and 2000 as a result. Both forest area and standing forest volume increased on the former collective lands, although their expansion was not as dramatic as the increase in household income (Hyde et al. 2003). Hainan Island shared these national experiences. Indeed, the general economy of the province general grew even more rapidly than that of the rest of China.

Meanwhile, the Ministry of Forestry continued to manage some of Hainan's interior forests. It began establishing 130,000 hectares of fast-growing plantations of eucalyptus and a few other species in 1982 and, by 1995, state-owned plantations accounted for more than a third of all managed forests on Hainan. These forests added 4 percent to Hainan's forest cover, and they have been the annual source of 100,000 tons of chip exports since 1989. Finally, the government, in recognition of growing environmental concerns, prohibited all harvests on the remaining natural rainforest after 1985. It also closed many degraded areas and began a forest rehabilitation program. In total, 6–8 percent of the land area of Hainan had been rehabilitated or reforested by 1997 as a result of the reforestation activities of the Ministry of Forestry (renamed the State Forest Administration after 1998), forest closure, and contracts between the collectives and local households.

In sum, Hainan's experience through the twentieth century seems to follow the general pattern of a region moving from the second to the third stage of forest development. The exception was the period from the 1950s through the early 1980s when a decline in general economic welfare caused some households to return to shifting cultivation and forest exploitation described by the first stage of forest development.

Since the early 1980s, however, economic growth has been rapid, making Hainan an exceptional case for examining the sources of change in both managed and natural forests. Of course, our interest is in the shares of these changes that are directly correlated to the improvement in aggregate economic welfare.

Zhang, Uusivuori, and Kuuluvainen (2000) assembled the data to evaluate these sources of change for the years between 1957 and 1994, including the years of most rapid market-driven growth since 1978. Table 6.4 repeats their regression results. Their data on prices, population, and gross output value (GOV) are in natural logarithms. Therefore, the coefficients on these variables are also elasticities.

The crucial price terms performed as anticipated. Increasing timber prices were an incentive to extend the area in managed forest, but also an incentive to harvest and deplete the remaining natural forest. Increasing agricultural prices were correlated with an increase in the natural forest, but they had no significant effect on managed forests, perhaps because the greater competition of agriculture was with commercial tropical forest crops (orchards, rubber, and tea). The latter is consistent with the argument that improved agricultural opportunity draws workers away from the natural forest. The inclusion of tropical forest crops is an important contribution of this analysis—just as the inclusion of agroforestry and other trees and tree values is important for other analyses where the volumes and land areas of those trees are substantial or where the analytical objective focuses on issues like carbon sequestration that must include all trees and not just those included in the measures of standard forest inventories (see the appendix to chapter 2). The tropical forest crop price coefficients in Table 6.4 indicate that these crops competed for land with managed forests, but their production substituted for the extraction of the same products from the natural forest.

The policy and institutional terms in the regression also performed as anticipated, and this lends additional confidence to the analysis. Hainan's state-owned forest enterprises were established to exploit the natural forest and to afforest degraded lands. Their coefficients confirm the effectiveness of these policies.

TABLE 6.4 Regression results for the sources of change in managed and natural forest cover in Hainan, 1957–1994

Variable	Managed forest cover (%)	Natural forest cover (%)
Constant	−3.74★	2.98★
	(7.03)	(3.40)
Timber price	0.61★	−1.52★
	(5.81)	(4.80)
Agricultural products price	0.11	0.33★
	(1.16)	(6.55)
Tropical forest crop product price	−0.40★	1.29★
	(4.22)	(7.68)
Per capita gross output value	0.59★	−0.31★
	(22.48)	(18.33)
Population density	2.21★	−0.61★
	(42.57)	(21.51)
Ratio of state-owned forest land to total forestland	0.42★	−0.77★
	(7.17)	(7.80)
Percentage of land under HRS to total collective land	0.79★	−0.54★
	(8.56)	(16.06)
Dummy variable for uncertainty	−0.09★	−0.76
	(2.73)	(0.88)
Joint effect of HRS and uncertainty	−1.55★	0.66★
	(15.49)	(17.07)
ρ (autocorrelation coefficient)	0.64	0.53
Log-likelihood	134.7	144.6

Notes: These are GLS regressions. Asterisks identify statistical significance at the 99 percent level. The t-values are in parentheses. The regional dummy variables are excluded from the table.
Source: revised from Zhang et al. 2000

The decentralization of authority and improvement of household rights in the late 1970s and 1980s were accompanied by increases in household investment in managed forests—although investment was conditioned by the households' uncertainty over the long-run continuation of the decentralization policies. Decentralization also led to the conversion of some natural forests for other household land uses.

Our greater interest is in the effect of aggregate economic welfare. Zhang et al. rely on gross output value (GOV) as an indication of aggregate welfare. (Data on gross domestic product were unavailable for the early years of their analysis.) Per capita GOV increased approximately ten-fold over the thirty-seven-year period of their analysis. The coefficient on GOV indicates that general economic growth had a positive effect on the land area in managed forests but a negative effect on the amount of remaining natural forest. Both effects were inelastic. GOV growth explained 17 percent of the increase in managed forests and 13 percent of the decrease in natural forest between 1957 and 1994.

These results are also not surprising. With economic growth, aggregate demand expands and the regional infrastructure also generally improves. Improvements in roads and improvements in the institutions that provide for and protect property rights have particular importance for forestry. Both improved roads and increased demand support increasing extraction from the natural forest and a decline in the natural forest cover. However, improved property rights and increased demand are also inducements for forest management and the expansion of the area in managed forestland. In fact, the expansion of managed forests in Hainan accelerated in the early 1990s, just as the economy experienced its most rapid growth. Furthermore, a wealthier society may be able to afford a longer term appreciation of the non-consumptive *in situ* benefits of forests. The fact that approximately 20 percent of the plantation forests in Hainan are intended for environmental pur poses seems to confirm this point. The political decision to restrict harvests from the natural forest beginning in 1985 seems to be further confirmation.

In sum, aggregate economic growth had measurable and statistically significant effects on both managed and natural forests (elasticities of +0.59 and -0.31, respectively). These were important but they were not the largest sources of adjustment in the two classes of forest. Timber prices and all the various factors that modify prices were more important. Timber prices had an effect on the managed forest comparable to that of general economic growth (elasticity of 0.61) but a much stronger effect on the natural forest (elasticity of −1.52). In sum, 1 percent of economic growth had a positive effect on the managed forest comparable to that of all the taxes, subsidies, and other factors that combined to alter the timber price by 1 percent. A 1 percent increase in the timber price had an even greater negative or debilitating effect on the remaining natural forest. From the perspective of sustainability, the net effect of aggregate economic growth on the two classes of forest combined was much more beneficial than the negative net effect of a comparable percentage increase in the timber price, and the effect of aggregate economic growth was certainly much more important than any single market or policy factor affecting the timber price.

The elasticities in Table 6.4 also indicate that a 1 percent increase in aggregate growth had a stronger effect on managed forests than either agricultural prices (elasticity of 0.11) or the prices of cultivated tropical tree crops (−0.40), although increases in the latter two prices actually had restorative effects on the remaining natural forest.

These results raise interesting prospects for sustainability in forestry. If timber prices decline as an economy grows, then economic growth can be a source of net improvement in both managed and natural forests. Furthermore, if timber prices decline relative to either agricultural prices or the prices of cultivated tree crops, they are a second source of sustainable forestry. (In fact, timber prices in Hainan rose ever so slightly relative to agricultural prices but declined slightly relative to the price of cultivated tree crops.)

Clearly, aggregate economic growth had important effects, both relatively and absolutely, on Hainan's forests. The question remains, however, whether Hainan's experience between 1957 and 1994 was unique, and whether we can anticipate a similar experience as Hainan's economy continues to grow and, indeed, in other regions and economies around the world as their economies grow too. A more recent assessment, similar to that of Zhang et al. but for all of China, provides broader results that are consistent with the observations from Hainan Island (Hyde, Wei, and Xu 2008).

Other examples support the broadest contentions from Hainan's experience as well. Between 1880 and 1910, more than one million hectares of farmland in the six small U.S. states of New England reverted back for forest (Williams 1988). In two of those states, Vermont and New Hampshire, forest cover continued to grow from approximately 15 percent to 77 percent over the course of the twentieth century (USDA Forest Service 2005). In France and Switzerland, forest cover is twice what it was two centuries ago (Peyron and Colnard 2002; Mather 2001). More recently, forest cover in France grew from less than eleven million hectares before World War II to around sixteen million hectares by the end of the twentieth century. The public national forestry fund (Fonds forestier national) financed the initial growth in tree cover in France but tree cover remained unaffected after the program was discontinued in 2000. In the UK, forest cover had declined to as little as 10 percent by the fourteenth century and 5 percent at the beginning of the twentieth century, but it has recovered to 12 percent today (FAO 2001). In each of these cases, New England in the United States, France and Switzerland, and the UK, per capita incomes grew several fold, more than keeping pace with the growth in forest cover over the course of the twentieth century. In a briefer period of time, forest cover in India's Punjab grew more than six-fold between 1960 and 1990 while per capita income doubled (Singh 1994). These are a diverse set of examples, but could they be unusual and different from the norm? The next section of the chapter reflects on the broader evidence.[29]

---

29 See Lamb and Gilmour (2003) for further discussion and more examples.

An environmental Kuznets' curve for forestry: Environmental Kuznets' curves (EKCs) refer to the hypothesis that development begins in association with natural resource and environmental depletion but, as development proceeds, the rate of depletion gradually tapers off until eventually, as growth and economic development proceed further yet, the natural environment begins to recover. This creates an inverted U-shaped relationship between development and environmental depletion similar to Figure 6.3. Initially, the economy is too poor to afford pollution control and some of the stock of natural capital becomes a source of financial capital for reinvestment and growth. Eventually, above some critical income level, environmental quality improvements and non-consumptive public goods such as the aesthetic values received from natural forests become relatively more important than additional natural resource consumption, the economy invests in pollution control and other environmental improvements, and overall environmental health improves along with improvements in human health and welfare.

The original statement of the EKC (Grossman and Krueger 1993) referred to the condition of the full environment, but many subsequent empirical examinations refer to specialized environmental conditions, pollutants like sulphur dioxide (DeBruyn 1997) or resources like forests (Bhattarai and Hammig 2001, 2004; Said 2001; Mather, Needle, and Fairbairn 1999; Koop and Tole 1999).[30] In the latter case, the vertical axis of Figure 6.3 generally refers to a rate or level of deforestation.

Data are often a problem for the evaluation of EKCs and this is certainly true for forestry. The intertemporal evidence relies on scattered examples like those for New England, France and Switzerland, the UK, and India's Punjab in the final paragraph of the preceding section of this chapter, but this evidence is too anecdotal to be convincing of a general pattern. The other alternative, cross-sectional evidence, requires comparable measures of forest cover or deforestation across many regions and countries of different income levels. The appendix to chapter 2 describes the great variation in the basic measures used in different national forest inventories, a variation that hinders the use of these data for completely satisfying cross-sectional comparisons.

Without comparable cross-sectional data or numerous long-term intertemporal assessments, what can we say about an EKC for forestry? We can consider the pattern of demands on the forest as incomes rise. In keeping with the broad environmental perspective of the original statement of the EKC, we should consider both consumptive demands for wood and non-consumptive demands for forest-based environmental services, including those satisfied by those trees and forest cover that are not included in most forest inventories. The pattern of these two sets of demands, together with the prospects for competing agricultural uses of

---

30 Dasgupta et al. (2002) and Copeland and Taylor (2004) review the literature. Copeland and Taylor conclude that this evidence supports the view that rising incomes have a positive effect on environmental quality.

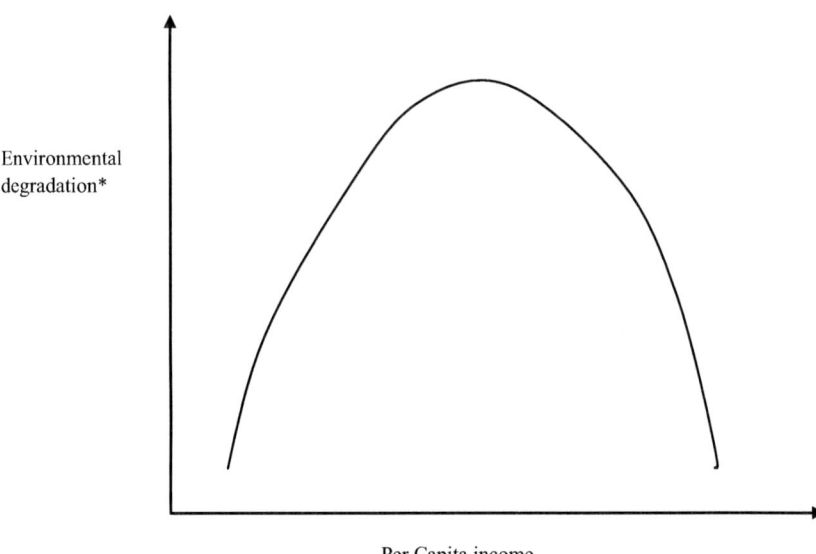

*For forestry, environmental degradation could be a rate of deforestation or it could be the inverse of a measure of forest cover per capita or per unit of area)

FIGURE 6.3 The Environmental Kuznets Curve

forestland, and along with reflection on the three-stage pattern of forest development, provide cautious support for the existence of an EKC for forestry.

The primary consumable wood products are fuelwood, sawnwood for lumber, other industrial roundwood, plywood and board, and various papers. Buongiorno et al. (2003) calculate the income elasticities for these products from a sample of 64 countries over the twenty-five-year period between 1973 and 1997. Not surprisingly, their income elasticities for fuelwood and other industrial roundwood are negative—implying that the consumption of these products actually declines with growth in income. The other elasticities, with the exception of paper products, are generally positive but less than one—implying that the consumption of these products grows, but less than proportionately with growth in income. Buongiorno et al. further separate their sample into 23 high and 41 low income countries and determine that, once more with the exception of paper products, income elasticities are generally smaller for the high income countries in their sample.[31] Taken altogether, these income elasticities suggest that the consumption of wood products other than paper declines, at least relatively, with improvements in income and these relative levels decline even further as income levels increase further yet.

---

31 In a related assessment, Zhu Tonberlin, and Buongiorno (1998) calculate the short- and long-term elasticities for the same products. Their long-run elasticities are all less than one.

The primary non-consumable forest-based activities are recreation and a number of ecosystem services: carbon sequestration to protect against global climate change, habitat for biodiversity, and erosion control. There is no comprehensive summary of income elasticities for these services. However, we can be confident that forest-based recreation grows with income, at least for developed countries. (Could the income elasticity for forest-based recreation be greater than one?) Furthermore, the various international negotiations on the environment make it clear that developed countries, with their higher incomes, place greater relative value on protection against climate change and protection of diverse habitats. Finally, the developed countries generally set aside more parks and forest preserves and generally impose higher environmental standards on the protection and use of their natural resources. This all suggests that these non-consumptive demands on the forest grow with development and, therefore, with income.

Erosion control is a different issue. Governments and local landowners of all income levels in countries in all stages of development do plant trees to protect against erosion. However, they only plant when they have relatively long-term security of land and tree tenure. The security of land tenure generally improves with overall economic development and welfare—just as the performance of all public institutions generally improves with economic development. Therefore, the correlation between improved institutions and overall economic welfare suggests that forest protection against erosion also probably improves with economic development.

Finally, competing agricultural uses of forestland apparently decline with economic development. The income elasticity for the consumption of agricultural products is positive but small, and agricultural productivity per hectare has been increasing. As a result, total agricultural land use has declined slightly for many countries over the course of the last several decades. Despite population growth and large agricultural support programs in several developed countries, agricultural land use apparently becomes gradually less competitive with forestland in higher income countries and, in one assessment, Waggoner and Ausubel (2001) argue that it has become less competitive with forestland over time for all countries.

In sum, the income elasticities for many consumable forest products decline with development and, for higher income countries, the income elasticities for recreation are greater than the income elasticities for most consumable forest products. The demands for other non-consumable forest services also seem to increase with development. Therefore, as incomes rise, the competing consumptive and non-consumptive demands on trees and forests both increase but the demands for non-consumables probably rise more rapidly. Meanwhile, the demands for agricultural land seem to decrease with rising income levels. Therefore, we can reasonably speculate that, with development, the demands for the management and protection of forests and trees eventually exceed both the demands to harvest them and the demands to convert the forestlands to agricultural use. This conclusion is consistent with an EKC for forestry.

The three-stage pattern of forest development tells little about the non-consumptive uses of the forest, but its sequence of consumptive forest use is also consistent with an EKC for forests. As we have seen, in the first stage, forests may even impede development, and new settlers remove trees and forests wherever they interfere with agricultural production. The forest continues to be a source of natural capital in the second stage of development and forest depletion continues. By the third stage, commercial values rise to the point where forest management is an economically viable activity. Some extraction from the natural forest continues, but the growth of newly managed trees and forests offsets it. The rate of forest depletion declines and it may even be reversed.

Thus, knowledge of the income elasticities for forest products and observations of the three-stage pattern of forest development are two reasons to accept the EKC hypothesis that economies in the early stages of development deplete their trees and forests but, as development proceeds and incomes rise, relative values shift, and the forest recovers. Better and more complete data would be more convincing but, for those interested in environment and development, our reasoning is a basis for cautious optimism about the future of the world's forests. Finally, it is clear, even from the less-than-perfect official evidence of forest inventories, that forest cover in the more developed countries of Western Europe and North America has increased over the last two centuries. In fact, none of the 50 most forested countries in the world, all with per capita incomes greater than US$4,600 per annum are experiencing deterioration in their forests—according to the most recent FAO data (as reported by Kauppi et al. 2006).

Several empirical estimates place the critical point for EKCs in general in the range of US$5,000 to US$8,000 of per capita annual income (Dasgupta et al. 2002). Is this the relevant income range for forestry, for both total forest cover and for natural forest? A recent empirical assessment of China's experience during its years of rapid economic growth between 1978 and 2003 provides additional confidence for the contention of an EKC for forestry, as well as evidence of a much lower turning point for forestry's EKC (Hyde et al. 2008). China's forest data has the advantage of consistent collection and reporting over an extended period and for a broad geographic range, both of which are advantages for econometric analysis. Hyde et al. observe an EKC for China's natural forests with a very low turning point of only 563 yuan (approximately US$64, although China's purchasing power parity may have been four times that) in annual per capita rural income, a level China attained in the mid-1980s. They also observe that rising incomes rapidly induced sufficient expansion in China's managed forests to create a net increase in the aggregate (managed and natural) forest area—even at very low income levels.[32]

---

32 Zhang et al. (2005) draw similar conclusions for the aggregate of China's managed and natural forests. They find that 21 of China's 31 provinces were in the latter stage of the EKC, past the turning point, by 2001.

Vietnam provides another interesting example. Vietnam's per capita income was only US$440 in 2004 and, until recently, its forest was in decline. However, Vietnam has begun reforesting, an indication that it has entered the third stage of forest development. One-sixth of the country's forest cover is now in plantation, and the area in plantations is increasing at a rate of 10 percent or 200,000 hectares annually (Hyde 2005). Does Vietnam's experience confirm the observation from China that the EKC for forestry has a very low turning point, or are China and Vietnam special cases? The former would be a satisfying conclusion, but it requires supporting evidence from many other regional and national examples. If the latter is correct and China and Vietnam are unusual special cases, then what makes them different and what can their differences teach us about the factors that may decrease the turning point of the EKC?

This becomes the truly important question. What factors shift the turning point of the EKC downward—reflecting an increase in total forest cover, and especially in the extent of the natural forest? If the critical income level at which forests seem to begin their recovery could be decreased, the natural environment would improve for billions of people.

Three certain factors and two more conditional factors are associated with this critical point. The certain factors are improved institutions, especially the institution of property rights; increased socioeconomic economic certainty and macroeconomic stability; and a more equitable distribution of income. The two conditional factors are technological improvements and freer trade.

All five are old acquaintances in our discussion of forest development. Improved property rights lower the cost function in all three stages of forest development and decrease the area of degraded open access forest.[33] Improved long-term macroeconomic stability has the same effect and it is particularly important for landowners who are considering long-term investments in activities such as forest management. A more equitable income distribution means fewer low wage workers and a reduction in their infringement on the forest. It probably also means a related increase in the demands for forest-related environmental services. Of course, each of these factors is a characteristic of development itself and improvements with respect to each of them are basic objectives of all development programs, not just forest development. The effects of technological improvements and freer trade are conditional—as chapter 3 showed that some, not all, categories of technological change improve the condition of the forest—and the first part of this chapter demonstrated that freer trade improves the condition of those natural forests in importing countries but we can only speculate on its net effect on global forests. Better understanding of all five factors would improve our understanding of what it will take to decrease the income level that defines the critical turning point for the EKC for forests.

---

33 Dasgupta et al. (2006) also highlight the independent effects of income and governance in their investigation of EKCs.

Conclusions

This chapter has reviewed the broader effects of general economic activity on forests and forestry, including the effects of trade with other regions and economies and the effects of general macroeconomic growth.

Some effects are unambiguous. Trade improves welfare in both the exporting and importing regions, otherwise it would not occur. It causes a loss of forest sector employment in the importing region but the loss is temporary if forest workers and those who depend on the forest for subsistence products find other opportunity in the now-wealthier local economy. The natural forest environment of the importing region improves, and the combined environment of both regions improves if the lower financial costs of logging and forest management in the producing and exporting region imply gentler forest terrain than in the importing regions and, therefore, relatively less environmental risk for forestry activity in the producing region. Of course, the latter is not always the case, and trade can have the effect of exporting environmental problems to the country that increases production. In this latter case, the question of whether the net environmental effect for the two trading partners is better or worse usually remains an unexamined empirical question.

Trade policies have had only limited effects on forestry because the tariff and non-tariff barriers for logs and forest products are generally low in most countries. There are, however, two notable exceptions. First, log export controls such as those in Indonesia, Canada, and previously in the U.S. Northwest, reduce the total demand for domestic logs and create a benefit in the form of lower prices for the domestic wood processing industry, but they have negative effects on forest production itself and on the aggregate economy. Second, many developed countries have regulated improved environmental standards for their domestic forests. The improved standards cause a shift of some production to other countries with lower costs and lower environmental standards. Therefore, the developed countries that impose higher standards on their own forests may not improve the global environment as they may only shift the environmental burden to those poorer developing countries that export primary forest products. Once more, the extent of these economic and environmental effects is conjecture—and worthy of further examination.

For most countries, forestry is a small sector relative to the rest of the economy and general adjustments in the full economy can have substantial effects on the domestic forest. We argued that the effects will be opposite for managed forests and frontier natural forests, and we demonstrated these opposite effects for the island province of Hainan in China. As Hainan's economy grew more than tenfold in 37 years, the elasticity of effects of general economic growth on managed and natural forests were +0.59 and −0.31, respectively, increasing the managed forest by an amount greater than any loss in natural forest. It would be unusual for Hainan or anywhere else to find a single forest policy adjustment that had effects nearly as great as these aggregate economic effects on either category of forest.

Some of the selective aspects of macroeconomic policy and growth are more difficult to trace. The effects of general structural adjustment programs on forestry are an example. There are simply too many potential channels of effect and each is weighted according to the specific conditions of the particular local economy. Other questions of macroeconomic policy remain largely unexamined despite available data and the importance of the questions. Nevertheless, we can hypothesize, for example, that

- Hypothesis 1: Instability and uncertainty are detrimental to long-term investment. Therefore, they are particularly injurious both to forest management and to forests at the natural frontier. They encourage expanding harvests and forest depletion in the short run, and they deter forest investment and investment in modern logging equipment. Unstable conditions are important sources of the poor financial and environmental performance of forestry activity in numerous developing countries.
- Hypothesis 1a: As general stability returns, local institutions recover, confidence in longer term investment returns, short-term harvest activity at the frontier declines, and investment in forest management and in the logging and wood processing industries also recovers.
- Hypothesis 1b. In fact, stability and effective governance may determine the success of other more directed policies—such as most specifically forest policies. Stiglitz (1998), for example, argues that governance is the key to effective trade liberalization in general and Richards (2005) argues that effective governance and (by implication) general stability are the key to effective trade liberalization in the forestry sector.
- Hypothesis 2: Market and policy shifts that induce rapid short-term harvesting have their own limits. Even aggressive harvest activities must cover the variable costs of their capital and these costs set a limit to the rate and extent of harvest activity. Furthermore, in the long run, as the manufactured capital used in harvest activities deteriorates and is not replaced, the rate of short-run harvest activity must decline. This is the case of, for example, timber harvesting by the Cambodian military in the 1980s and 1990s.
- Hypothesis 3: European Union, North American, and Japanese agricultural support programs expand agricultural land use and production in the developed countries of those regions, thereby reducing the land area in their own forests. These countries export some of their resulting agricultural surpluses to developing countries where those surpluses out-compete some local commercial agriculture production and cause some farm labor to return to subsistence agricultural activities to the detriment of the frontier forest in the developing countries. One testable hypothesis is that these agricultural support programs are a large source of global deforestation and forest degradation.
- Hypothesis 4: Capital-favoring investment and trade policies and minimum wage policies (where they are effective) decrease the relative position of labor and especially unskilled labor. Inexpensive manufactured capital facilitates commercial logging at the frontier. Displaced low-wage rural labor often finds its alternatives in subsistence agriculture within the natural forest. Therefore, these policies are also sources of deforestation and forest degradation.

- Hypothesis 5: There is a pattern to the role of forestry in economic development. Forest resources, where economic stocks are accessible to markets, may be a source of capital for development. (The simple existence of standing forests is insufficient.) Forest management becomes a participant in the development process once the local prices and level of development can justify forest investment. At this time, forestry may even be a leading sector for development in some regions, but economic diversification follows eventually, other sectors emerge and become more important to the economy, and the forest sector declines in relative importance even as it may continue to grow absolutely.
- Hypothesis 5a: The environmental Kuznets' curve is a valid concept for forestry. That is, the stock of trees and forests is drawn down in the early stages of development. But development eventually attains a level that supports higher relative values for forest environments. With more development yet, reinvestment occurs and the aggregate stock of trees and forests does recover, as does the specific stock of natural forest.

These are only hypotheses, but both theory and casual empirical evidence support each of them. The magnitudes of the respective impacts on forests and forestry may be great in some important cases. They deserve more critical inquiry.

Literature Cited

Anderson, R., L. Constantino, N. Kishor, G. Labadie, and J. Panzer. 1994. *Structural adjustment, regional development and forest degradation: Loggers and migrants in Bolivia*. LATEN dissemination note #11. Washington, DC: World Bank.

Anon. 2007. Sarawak goes for sustainable forest management. Malaysian National News Agency http://www.bernama.com.my/bernama/v3/news.php?id=267978 (accessed June 18, 2007).

Antweiler, W., B. Copeland, and M Taylor. 2001. Is free trade good for the environment? *American Economic Review* 91(4): 877–908.

Barbier, E. 1996. *Impact of the Uruguay round on international trade in forest products*. Discussion paper, FAO/UN. Rome: Food and Agriculture of the United Nations.

Barr, C. 1998. Bob Hasan, the rise of Apkindo, and the shifting dynamics of control in Indonesia's timber sector. *Indonesia* 65: 1–36.

Benhin, J., and E. Barbier. 2004. Structural adjustment programme, deforestation and biodiversity loss in Ghana. *Environmental and Resource Economics* 27: 337–366.

Bhattarai, M., and M. Hammig. 2001. Institutions and the environmental Kuznets curve for deforestation: A crosscountry analysis for Latin American, Africa and Asia. *World Development* 29(6): 95–1010.

Bhattarai, M., and M. Hammig. 2004. Governance, economic policy, and the environmental Kuznets curve for natural tropical forests. *Environment and Development Economics* 9: 367–382.

Binswanger, H. 1991. Brazilian policies that encourage deforestation in the Amazon. *World Development* 19: 821–829.

Bolkesjo, T., E. Tromborg, and B. Solberg. 2005. Increasing forest conservation in Norway: Consequences for timber and forest product markets. *Environmental and Resource Economics* 31: 95–115.

Bourke, I. 2003. *Trade in forest products, trade restrictions, and trade agreements affecting international*. Unpublished paper prepared for IIED and the FAO conference on Impact Assessment of Forest Products Trade in the Promotion of Sustainable Forest Management. Rome, February 2003.

Bourke, I., and J. Leitch. 1998. *Trade restrictions and their impact on international trade in forest products*. Rome: Food and Agriculture of the United Nations.

Boyd, R., and W. F. Hyde. 1989. *Forest sector intervention*. Ames: Iowa State University Press.

Boyd, R., and K. Krutilla. 1987. The welfare impacts of US trade restrictions against the Canadian

softwood lumber industry: A spatial equilibrium analysis. *Canadian Journal of Economics* 20(1): 17–35.
Boyd, R., and K. Krutilla. 1992. The impact of the free trade agreement on the U.S. forestry sector: A general equilibrium analysis. In P. Nemitz, ed., *Emerging issues in forest policy* Vancouver, Canada: University of British Columbia Press, pp. 236–253.
Brack, D., C. Marijnissen, and S. Ozinga. 2003. *Controlling imports of illegal timber, options for Europe.* London: Royal Institute of International Affairs.
Buongiorno, J. 2006. *Who will meet China's import demand for forest products?* Unpublished research paper, University of Wisconsin, Department of Forest and Wildlife Ecology.
Buongiorno, J. S. Zhu, D. Zhang, J. Turner, and D. Tomberlin. 2003. *The global forest products model: Structure, estimation, and applications.* Boston: Academic Press.
Contreras-Hermosilla, A. 2001. *Forest law compliance: An overview.* Rome: Food and Agricultural Organization of the UN.
Copeland, B., and M. Taylor. 2004. Trade, growth, and the environment. *Journal of Economic Literature* XVII(1): 7–71.
Dasgupta, S., B. Laplante, H. Wang, and D. Wheeler. 2002. Confronting the environmental Kuznets curve. *Journal of Economic Perspectives* 16(1): 147–168.
Dasgupta, S., K. Hamilton, K. Pandey, and D. Wheeler. 2006. Environment during growth: Accounting for governance and vulnerability. *World Development* 34(9): 1597–1611
Deacon, R. 1994. Deforestation and the rule of law in a cross section of countries. *Land Economics* 70(4): 414–430.
DeBruyn, S. 1997. Explaining the environmental Kuznets curve: Structural change and international agreements in reducing sulphur emissions. *Environment and Development Economics* 2(4): 485–503.
Durst, P., T. Waggener, T. Enters, and T. Cheng (Eds.). 2001. *Forests out of bounds: impacts and effectiveness of lagging bans in national forests in Asia-Pacific.* Bangkok: Food and Agriculture of the United Nations.
Economist. 2004. Canada and the United States: time to bury the hatchet. (September 18). 375:46.
Economist. 2005a. China: Green guise. (March 25) 376: 42.
Economist. 2005b. Europe's farm follies. (December 10): 25–27.
Economist. 2006a. World trade: under attack. (July 8): 65–66.
Economist. 2006b. Farm subsidies: Uncle Sam's teat. (September 7) 380(8494): 35.
Esty, D. 1994. *Greening the GATT: Trade, environment, and the future.* Washington, DC: Institute for International Economics.
FAO/UN (Food and Agriculture Organization of the United Nations). 2001. Global forest resources assessment 2000. Forestry paper 140. Rome: FAO
FAO/UN (Food and Agriculture Organization of the United Nations). 2005. Yearbook of forest products, FAOSTAT statistics database. http://apps.fao.org/) accessed August 23, 2008).
FEDRC (China National Forest Economics and Development Research Center). 2006. *Forest resource analysis for China's forestry supply.* Unpublished report. Washington, DC: World Bank.
GOI/DF (Government of Indonesia, Department of Forestry. 2004. *Statistic Kehutanan.* Jakarta: Department of Forestry.
GOI/BPS (Government of Indonesia, Badan Pusit Statistic). 2001. *Statistik Industri: Large and medium manufacturing statistics.* Jakarta: BPS.
Grossman, G., and A. Krueger. 1993. Environmental impacts of the North American Free Trade Agreement. In P. Garber, ed., *US-Mexico free trade agreement.* Cambridge, MA: MIT Press, pp 13–56.
Hashiramoto, O, J. Castano, and S. Johnson. 2005. Changing global picture of trade in wood products. *Unasylva* 219(55): 19–26.
Hirschman, A. 1958. The strategy of economic development. New Haven, CT: Yale University Press.
Hyde, W. 2005. Forest development and its impact on rural poverty. Unpublished report prepared for the Asian Development Bank, Manilla, Philippines.
Hyde, W., B. Belcher, and J. Xu, (Eds.). 2003. *China's Forests: Global lessons from market reforms.* Washington, DC: Resources for the Future.

Hyde, W., J. Wei, and J. Xu. 2008. *Economic growth and the natural environment: the example of China and its forests since 1978*. Unpublished research paper. Peking University and Economics Department, Gothenburg University.

Kaimowitz, D., Erwindo, O. Ndoye, P. Pacheco, and W. Sunderlin. 1997. *Forests under (structural) adjustment in Bolivia, Cameroon, and Indonesia*. Unpublished research paper. Bogor, Indonesia: Center for International Forestry Research.

Kauppi, P, J. Aussubel, J. Fang, R. Sedjo, and P. Waggoner. 2006. Returning forests analyzed with the forest identity. In *Proceedings of the National Academy of Science* 103(46): 17574–17579.

Kishor, N., and L. Constantino. 1993. *Forest management and competing land uses: an economic analysis for Costa Rica*. LATEN discussion note no. 7. Washington, DC: World Bank.

Koop, G., and L. Tole. 1999. Is there an environmental Kuznets curve for deforestation? *Journal of Development Economics* 58: 231–244.

Lague, D. 2003. Felling Asia's forests. *Far Eastern Economic Review* 166(51): 26–29.

Lamb, D., and D. Gilmour. 2003. *Rehabilitation and restoration of degraded forests*. Cambridge, UK: IUCN Publications.

Lebedys, A. 2004. *Trends and current status of the contribution of the forest sector to national economies*. Working paper FSM/ACC/07. Rome: Forests Products and Economics Division, FAO/UN.

Mahar, D. 1988. *Government policies and deforestation in Brazil's Amazon region*. Environment Department working paper no.7. Washington, DC: World Bank.

Mahar, D., and R. Schneider. 1994. Incentives for tropical deforestation: some examples from Latin America. In K. Brownand and D. Pearce, eds., *The causes of tropical deforestation*. London: University College London Press, pp. 159–171.

Margolick, M., and R. Uhler. 1992. The economic impact on British Columbia of removing log export restrictions. In P. Nemetz, ed., *Emerging Issues in forest policy*. Vancouver, Canada: University of British Columbia Press, pp. 273–296.

Mather, A. 2001. The transition from deforestation to reforestation in Europe. In A. Angelson and D. Kaimowitz, eds., *Agricultural technologies and tropical deforestation*. Wallingford, UK: CAB International, pp. 35–52.

Mather, A., C. Needle, and J. Fairbairn. 1999. Environmental Kuznets curves and forest trends. *Geography*. 55–65.

Mersmann, C. 2005. Links between trade and sustainable forest management: An overview. *Unasylva* 55: 3–9.

Page, S. 2006. Are we collapsing? A review of Jared Diamond's *Collapse: How societies choose to fail or succeed*. *Journal of Economic Literature* 43: 1049–1062.

Peyron, J., and O. Colnard. 2002. Vers des compts de las forest? In *Forest, Economie et Environnement: Rapport de la Commission des Compts et de l'Economie ole l'Environnement* [Forest Economy and Environment: Report of the Audit Board and the Economy of the Environment]. Orleans, France: Institut Francois de l'Environnement, pp. 169–190.

Prestemon, J. 2000. Public open access and private timber harvests: theory and application to the effects of trade liberalization in Mexico. *Environment and Resource Economics* 17: 311–334.

Richards, M. 2005. Forest trade policies: How do they affect forest governance. *Unasylva* 219(55) : 39–43.

Rozelle, S., J. Huang, and S. Benzinger. 2003. Forest exploitation and protection in reform China: assessing the impacts of policy and economic growth. In W. Hyde, B. Belcher, and J. Xu, eds., *China's forests: Global lessons from market reforms*. Washington, DC:. Resources for the Future, pp. 109–134.

Rytkonen, A. 2003. Market access of forest goods and services. *Background paper for the global project: Impact of forest products trade in the promotion of sustainable forest management*. Rome: Food and Agriculture of the United Nations.

Samuelson, P. 1948. International trade and the equalization of factor prices. *Economic Journal* 59(234): 163–184.

Schneider, R. 1995. *Government and the economy on the Amazon frontier*. LAC Regional Studies Program Report no.34. Washington, DC: World Bank.

Sedjo, R.A., C. Wiseman, D. Brooks, and K. Lyon. 1994. *Global forest products trade: The consequences of domestic forest land-use policy*. Discussion paper 94-13. Washington, DC: Resources for the Future.

Seidl, A. 2001. Intra-regional wealth-deforestation relationships in the Brazilian Pantanal: An examination of the environmental Kuznets curve hypothesis. Unpublished manuscript presented at the Western Economics meetings in Logan, Utah.

Singh, H. 1994. The green revolution in Punjab: The multiple dividend, prosperity, reforestation and the lack of rural out-migration. Unpublished student paper, JFK School of Public Policy, Harvard University, Boston.

Sohngen, B., R. Mendelsohn, and R. Sedjo. 1999. Forest management, conservation, and global timber markets. *American Journal of Agricultural Economics* 81: 1–13.

Stiglitz, J. 1998. *Toward a new paradigm for development: Strategies, policies, and processes*. Prebisch lecture. Geneva.

Sun, X., E. Katsigris, and A. White. 2005. *Meeting China's demand for forest products: An overview of import trends, ports of entry, and supplying countries, with emphasis on the Asia-Pacific region*. Washington, DC: Forest Trends.

Sunderlin, W., and J. Pokam. 2003. *Economic crisis and forest cover change in Cameroon; The roles of migration, crop diversification, and gender division of labor*. Unpublished manuscript. Bogor, Indonesia: Center for International Forestry Research.

Tachibana, T., T. Nguyen, and K. Otsuka. 2001. Agricultural intensification versus extensification: A case study of deforestation in the northern-hill region of Vietnam. *Journal of Environmental Economics and Management* 41(1): 44–69.

Taylor, R., I. Tomaselli, and L. Hing. 2005. How to hurdle the barriers. ITTO *Tropical Forest Update* 15(2): 18–20.

USDA Forest Service. 2005. Forest inventory and analysis webpage: http://fia.fs.fed.us (accessed April 16, 2010).

USDA Forest Service. 2007. *Highlights of the Forest Service performance and accountability report*. Washington, DC: USDA Forest Service.

Uusivuori, J., and Y. Kuuluvainen. 2001. Substitution in global wood imports in the 1990s. *Canadian Journal of Forest Research* 31: 1148–1155.

Watkins, K., and J. Von Braun. 2002. Time to stop dumping on the world's poor. 2002–2003 Annual Report: Trade policies and food security. *International Food Policy Research Institute*: 18. Washington, DC: International Food Policy Research Institute.

Waggoner, P., and J. Ausubel. 2001. How much will feeding more and wealthier people encroach on forests? *Population and Development Review* 27(2): 239–257.

Williams, M. 1988. The death and rebirth of the American forest: Clearing and reversion in the United States 1900–1980. In J. Richards and R. Tucker, eds., World deforestation in the twentieth century. Durham, NC: Duke University Press, pp. 211–229.

Wunder, S. 2003. *Oil wealth and the fate of the forest: A comparative study of eight tropical countries*. London: Routledge.

Wunder, S. 2005. Macroeconomic change, competitiveness and timber production: A five country comparison. *World Development* 33(1): 65–86.

Young, C. 1996. Economic adjustment policies and the environment: A case study of Brazil. Unpublished doctoral thesis, Department of Economics, University of London.

Zhang, D. 2007. *The softwood lumber war*. Washington, DC: Resources for the Future.

Zhang, Y., J. Uusivuori, and J. Kuuluvainen. 2000. Econometric analysis of the causes of forestland use/cover change in Hainan, China. *Canadian Journal of Forest Research* 30: 1913–1921.

Zhang, Y., S. Takibana, and S. Nagata. 2005. Impact of socioeconomic factors on the changes in forest areas in China. *Forest Policy and Economics* 9: 63–76.

Zhu, S., D. Tonberlin, and J. Buongiorno. 1998. *Global forest products consumption, production, trade and prices: Global forest products model projection to 2010*. FAO Forest Policy and Planning Division working paper GFPOS/WP/01. Rome: Food and Agriculture of the United Nations.

Appendix 6A:
Forestry in the National Accounts

National income accounts are measures of a region's or an economy's aggregate productivity, the accumulated productivity of all of its sectors. Forestry is one of the many sectors of economic activity that contribute to an economy and, therefore, to these accounts.

The national accounts typically accumulate monetary measures for all market-valued goods and services. Monetary units, as they are determined by the actions of consumers in an open economy, are measures of the social values that consumers place on goods and services. Therefore, national accounts based on these measures are one measure of the social welfare of an economy, and differences in an economy's accounts over periods of a few years may be fair indications of progress in that economy's level of welfare.

This first part of this appendix reviews the assembly of the national accounts and the incorporation of forestry activity within them. Discussions of the natural resource and environmental components of the accounts tend to focus on omissions from the accounts and on the depreciation in the value of natural assets such as standing forests. The second part of the appendix summarizes the modifications necessary to correct these errors and create an accurate measure of the contribution of the forest sector.

The National Income Accounts

National income accounts report accumulations of either income and product or expenditure for an entire region or economy. Table 6A.1 summarizes the distinctions in the several common measures of these accounts, along with the magnitudes of these measures for the U.S. economy for 2002. Gross domestic product or GDP is the market value of goods and services produced and not resold during a given accounting period, usually one year. Gross national product or GNP is GDP adjusted for income flows to and from the rest of the world. Net national product or NNP, the third common measure used in national income accounting, is GNP less allowances for capital consumption (or depreciation). National income or NI, the fourth common measure, is NNP less indirect business taxes (such as real property taxes) plus government subsidies less the surpluses of government enterprises.

The first column of Table 6A.2 is known as the income and product account. It summarizes the income flows to individuals and corporations that sum to GNP. The table summarizes the U.S. accounts for 2002. The full forestry sector (forestry itself and forest products) contributed approximately 1.3 percent to total U.S. GDP and a slightly larger share of approximately 2 percent of both U.S. exports and imports.[34] For comparison, tables 6.2 and 6.3 report the approximate

---

34 The value added by the forestry sector for 2002 was: forestry, fisheries and related activities US$24.6 billion; wood products $30.4 billion; furniture and related products $31.1 billion; and

TABLE 6A.1 Summary measures of national income—along with official estimates of these measures for the U.S. accounts for 2002 in billion US$

Symbol	Measure		2002 U.S. value
GNP	Gross domestic product		10,469.6
	plus: income receipts from the rest of the world		305.7
	less: income payments to the rest of the world		275.0
GDP	Gross national productof which:		
	forestry	?	
	wood products	30.2	
	furniture	31.1	
	paper manufacturing	50.3	10,500.2
	less: consumption of fixed capital		1,292.0
NNP	Net national product		9,208.3
	statistical discrepancy		−21.0
NI	National income		9,229.3

Source: U.S. Department of Commerce Bureau of Economic Analysis website http://www.bea.gov/bea/dn/home/gdp.htm accessed July 14, 2006

contributions of the forestry sectors of several other countries and regions. Those tables show clearly that, even where the forestry sector is important, its contribution to the total regional or national economy seldom exceeds 4 percent.

The second column of Table 6A.2 is the expenditure account. It identifies the contributions to GNP of expenditures in the four primary sectors of an economy: the personal sector, the business sector, the government sector, and the rest of the world (international trade).

The income and expenditure accounts for any one period should balance. However, since these two accounts are assembled independently, their sums are not identical. For the U.S. accounts in 2002, the discrepancy between the two was $76.8 billion or 0.7 percent of GDP.

Errors in the Accounts

While the national accounts accumulate measures for all market-valued goods and services, their reliance on market values implies the omission of at least three classes of goods and services that create value for an economy. Each of these include goods and services in which forests and forestry play an important role: non-market environmental services such as the carbon sequestration of trees

---

paper products $50.3 billion—for a total of $136.4 billion (less the contribution of fisheries, hunting and related activities). This is less than a 1.30 percent share of GDP for year 2002 (U.S. Department of Commerce n.d.).

TABLE 6A.2 2002 U.S. gross national product and gross national expenditure in billion US$

National income and product account		National expenditure account	
Wages, salaries, and supplements (of which:	6,091.2	Personal consumption expenditures	7,350.7
forestry, fisheries & related;	(15.7		
wood products;	21.5		
furniture & related prods;	22.4		
paper manufacture)	32.4)		
Proprietor's income of which:	768.4	Gross private domestic expenditure	1,582.1
forestry, fisheries & related)	(6.5)		
Rental income of persons	152.9	Net exports	−424.4
Corporate profits and inventory valuation adjustments (of which:	886.3	Government consumption expenditures and gross inventory	1,961.1
forestry, fisheries & related;	(1.0		
wood products;	2.0		
furniture & related prods;	3.1		
paper manufacture)	−0.8)		
Net interest	520.9		
of which: agriculture, fisheries, forestry, and hunting)	(10.4)		
Taxes on production and imports (of which:	762.8		
forestry, fisheries & related;	(0.04		
wood products;	0.3		
furniture & related prods;	0.4		
paper manufacture)	0.5)		
less: Subsidies	38.4		
Business current transfer payments	84.3		
Current surplus of government enterprises	0.9		
Net national income	9,229.3		
plus: depreciation, statistical discrepancy, and net payments to rest of world	1,240.3		
Charges against gross national product	10,469.6	Gross national expenditure	10,469.6

Source: U.S. Department of Commerce Bureau of Economic Analysis website http://www.bea.gov/bea/dn/home/gdp.htm. (accessed July 14, 2006)

and forests, illicit activities such as illegal logging, and household production for domestic use such as the largest share of the world's fuelwood. In addition, market valuation may erroneously account for the value of government activity and governments manage a large share of the world's forests. Therefore, the measure of government activity can be another source of error in accounting for the forestry sector's contribution to an economy. Finally, discussions of environmental accounting argue that the national accounts disregard the depletion of natural resources like forests, and that depletion is a critical source of undervaluation in the accounts. Since the national accounts are used to place aggregate economic activity in a policy perspective, undervaluation in one sector of the accounts could cause policy makers to diminish the importance of that sector or of a comprehensive issue of public policy involving that sector (such as long-term national or global deforestation) relative to other pressing national policy issues.[35]

Omissions from the accounts: At least three classes of omissions from market valuation contain a significant forestry component. The first is forest-based environmental services. We considered three of these non-market forest services in the appendix to chapter 4: carbon sequestration to protect against global climate change, the protection of biodiversity, and erosion control.

The first, carbon sequestration, may be the largest in value—and the importance of environmental accounting itself increases as the importance of global warming and carbon sequestration is increasingly recognized. Estimates of the total global value of carbon sequestration by forests range upward to $30 per forested hectare. The annual change in this value, the much smaller annual increase or decrease in carbon value per hectare, is the appropriate value for inclusion as an addition to a global income account—because the objective of these income accounts is to measure annual flows.

This is only one parcel of the carbon accounting associated with forests. A second is the loss in annual value due to the net conversion of forest land to other uses and, thereby, the release of additional carbon into the atmosphere. The gain due to reforestation occurs largely in developed countries, while the loss due to forest conversion occurs largely in less-developed tropical countries. The loss in value, however, tends to be more dearly felt in the wealthier developed countries. Therefore, while a global income account might measure a significant value for the annual loss in this non-market forest service, the mismatch between loss in value in one country and the decreasing physical resource in another leaves us uncertain about where to assign the lost value. Should we assign it to the accounts of the developed countries that feel the loss more dearly, or to the less-developed tropical countries where the physical resource loss tends to be greater?

---

35 In one extreme case, the government of Nepal has allocated public expenditures proportionately to shares in the national accounts Katila (1995) suggests that Nepal's official accounts undervalue forest sector activities by a factor of three, thereby causing a potentially serious underallocation of government resources to the forest sector.

Annual adjustments due to loss in biodiversity may be smaller than annual adjustments due to the loss in carbon sequestration. Their accounting also suffers from a mismatch similar to that of carbon sequestration as, once more, the perception of value loss is greater in the wealthier developed countries while the greater annual physical loss probably occurs as a result of tropical forest destruction and deforestation in developing countries.

Losses due to erosion, or to the losses in protection against erosion provided by forests, are easier to assign because they tend to be local or regional—and, therefore, both the physical resource adjustment and the adjustment in value tend to occur largely within the same country. The assigned values for each country's accounts are easier to identify. The appropriate value, once more, is only the annual adjustment, not the total value of all erosion or all erosion control. This annual value is likely to be small because local markets and local institutions often take the necessary steps to control erosion when and where the value of its effect is most important, and many (not all) of these costs are reflected in market exchange, for example, the costs of material and even labor contracted to build and maintain some physical controls on erosion (see appendix 4A.)

Therefore, we can hypothesize that, while the effects of forest-based carbon sequestration and biodiversity on the global accounts may be significant, the annual effect of omitted non-market forest-based environmental services that can be assigned to any single country's national accounts is very uncertain.

Illicit activities are a second class of activities that create value, but value that is not captured in formal markets of record and, therefore, is not included in national income accounts. Illegal logging is the most important example for forestry. Annual values of illegal logging as great as US$300 million have been estimated for British Columbia, for example (Smith 2002). Contreras-Hermosilla (2001) estimates that as much as 50 percent of the value of all log value in Brazil, Bolivia, Russia, Cambodia, Cameroons, Indonesia, Myanmar, and Papua New Guinea may be due to illegal activity. However, even these large amounts must be much less than 0.7, 0.4, 0.06, 2.3, 0.9, 0.7, 0.4, and 1.2 percent, respectively, of the gross domestic products of these economies.[36]

Household services are a third class of omissions from the national accounts. The standard example is the unpaid services of a spouse. That is, surely cooking, cleaning, childcare, etc., add value, and those households that must hire these services can testify to the effect of their costs on household expenditures and household welfare. However, unless they are hired, these services and their value escape record in the national accounts.

These household services are important in developed countries, but they can be a large share of all value created in subsistence economies and it is in these economies that forest activity becomes an important component of the omission. Perhaps more than three-fourths of all fuelwood consumed in the world is collected by subsistence households for their own domestic use (Arnold et al. 2003).

---

36 Calculated for year 2000 as one-half of gross value of forestry in the economy divided by GDP.

For example, de los Angeles and Peskin (IRG and Edgevale Associates 1994) observe that the adjustment to include the value of household-collected fuelwood is the largest of all adjustments necessary for an accurate account of natural resource and environmental value in the Philippines.

The value of government activity: Aggregate estimates of the net benefits of government activity are not easily calculated. Accurate estimates would require market observations of prices where they exist (e.g., public timber) and non-market estimates of willingness to pay where prices do not exist (e.g., some forest recreation). These measures, multiplied by the annual flow of resources (timber) or resource use (recreation), minus the cost of their government production, would yield improved estimates of the value of government activity. However, such estimates would be extremely expensive and, even then, they would be controversial. Therefore, the convention in national accounts is to value government activity at its cost. This implies the fond hope that all government production approximates its social value at the margin.

For many sectors, government activity is small and misestimates of its value are unlikely to have a significant effect on the estimate of overall activity for the sector. However, government activity is a large component of the forest sector in many countries, and a misestimated government account could substantially alter the forest sector's contribution to the national accounts.[37] Nevertheless, estimates of the net value of government forestry activity are nonexistent.

The common roles of government forestry agencies are regulation to insure private provision of public forestry services, technical assistance, research, and public ownership. We will examine each of these further in chapter 10. Doubts may arise about whether the marginal benefits of government activity equal the marginal costs in any of these roles. The costs of government forestry regulation seem to be excessive in some cases, some government research produces large positive net benefits while the costs of other government research are well in excess of its benefits and, finally, the net benefits of public ownership are surely debatable.[38]

Regarding the latter, the management of publicly-owned timber in the United States, for example, generally costs more than it returns in annual stumpage values (Barlow and Helfand 1980; Barlow et al. 1980). However, most would anticipate that the net values of public forest-based recreation are positive. One complete empirical economic analysis of both the benefits and costs of public recreation finds that the U.S. National Forest System set marginal costs very close to marginal benefits for campsite use in one valley in western Montana (Hyde and

---

37 Governments administer 34 percent of the forests in the United States, Japan, and the EU countries and 92 percent of the forests in the other developed countries (UN-ECE 2001). For those developing countries for which there are data, the public share often exceeds 80 percent of all forestland.
38 See Boyd and Hyde (1989) for a review of the economic effects of government regulations in U.S. forestry and also of the effects of government technical assistance, or Hyde, Newman, and Seldon (1992) for a review of the economic effects of U.S. government forestry research.

Daniels 1988). Therefore, the net benefits of public recreation were positive in this example. Is this example representative of the majority of publicly managed recreation? If so, are the net benefits of recreation on the public forest lands sufficiently positive to offset apparent losses from public timber management? What is the net economic effect across all government forestry programs? And would the conclusion for the United States be similar for other countries as well? These questions have not received even a modicum of speculation, and we do not have enough information to form even a reasonable hypothesis.

Environmental Accounting

The omissions from the national accounts are well known. The interest in correcting the accounts for them, and particularly for the effects of omissions in the environmental and natural resource sectors, dates from the 1970s when the greater concern was for errors in the depreciation of natural resource and environmental assets as those resources are depleted over time. The debate focuses on the appropriate conceptual approach for correcting this error, and occasional accounts for several countries now incorporate adjustments of one form or another for the forest sector.[39]

There is no consensus for a preferred approach for making the corrections. There is only agreement that most national accounts treat manufactured and natural wealth inconsistently, depreciating the former over its expected lifetime while disregarding changes in the stock of the latter.[40] This explains the focus on forest depletion in discussions of forest sector accounting—although depletion may not be the greatest source of error in the forest sector.

One approach to correcting the accounting discrepancy between manufactured and natural capital recommends separate physical accounts of natural resource activities—with links from these physical accounts to the income accounts. Norway and France use this approach. Alternatively, the United Nations Statistical Office recommends using government expenditures as proxies for willingness-to-pay for environmental services. Neoclassical economics teaches a third approach. It recommends treating all forms of capital uniformly, whether manufactured or natural. This implies revising the national income accounts according to measures of all non-market draws on natural capital in the current income or product and expenditure accounts (Table 6A.2), while incorporating previously unaccounted natural resource depletion or revaluation in a separate capital account.

These different approaches reflect different objectives and different key sectors in national economies. Norway and the United States, for example, use the national accounts as data bases for policy analysis, but Norway is a resource-

---

39 UN et al. (2003) provides more detail on alternative approaches. Binkley (1999) provides a summary list of completed national environmental accounts or green forestry accounts. A few others, notably China (Zhang 2007), have been added since 1999.
40 See Prince (1996) or Peskin and Lutz (1993) for reviews of the general literature, including both applications and debate about concepts.

oriented economy for which the physical data may be more important than for the United States. Therefore, a separate physical account may be more important for Norway than for the United States. For some (e.g., Repetto 1992; El Sarafy 1989), the objective is a reliable account that does not misestimate the rate of national economic performance. Those with this perspective focus on the contributions of physical changes in marketed natural resources, anticipating that these are the largest sources of misevaluation, particularly for many developing countries.

Our prior discussion of errors and omissions in the accounts and the means for their corrections in the current income and expenditure accounts follows neoclassical economics. Continuing that approach leads to the recommendation that each forest-based good or service in the current income or expenditure account should have a counterpart entry in a capital account that measures the impact of any change in the standing forest asset on the value of future production of the goods or services from that asset. The term for this change in value is "capital depreciation" (or capital re-valuation in the event of growth in value). Capital depreciation is a measure of the loss (or gain) in economic value of the asset over the accounting period. The environmental accounting literature identifies three approaches for its calculation, only one of which is conceptually reliable for forestry.

The change in the net present value of any asset service is equal to the difference between the expected discounted future net income stream from that asset at the beginning of the accounting period and that at the beginning of the next accounting period. This difference may be due to changes in the capital stock or to changes in the price of the asset, and both are common in forestry. Changes in the capital stock occur with harvests from natural forests and with reinvestment and growth in managed forests. Upward price adjustments are the natural and expected results of drawing down natural forest stocks and, in chapter 2, we explained the expectation that upward price adjustments will continue in any market or economy until the new price is sufficient to justify investments in forest management. Furthermore, we also know that, in many cases, the removal of decaying older forest stands is a necessary pre-condition to obtaining vigorous new growth and more productive forests. Harvests actually increase the capital account in this latter case.

These concepts do not evoke debate. Rather, the debate focuses on the difficult problem of calculating reliable estimates of the capital adjustment. This calculation requires projections to some distant future date of the annual changes in both the prices and the physical flows of all forest-based goods and services (Binkley 1999). This is a modeling and data intensive exercise that is relieved only partially by the technical good fortune that social discounting sets a reasonable terminal period for future projections of prices and capital stocks.

The modeling intensity encourages a search for simpler solutions. Two alternatives have been proposed. Neither reflects the neoclassical economic focus on flows of income or expenditure and neither is reliable for forestry, but the attention each receives mandates our reflection on them. The first focuses on the difference in the remaining life of the asset before and after the current period's draw on the stock (El Sarafy 1989). This approach provides reasonable estimates for those non-renewable resource assets with no close substitutes. It was not intended for, and is inappropriate for, both stocks of natural forest that must be drawn down before economic incentives and the available land base can support new managed forests, or for those currently managed forests that absorb regular reinvestment. Both are renewable and both have close substitutes. Forests, unlike deposits of non-renewable resources, do not have fixed remaining lifetimes.

The second alternative receives wider attention. It estimates capital depreciation as the full stumpage value of the current flow of timber harvests (Repetto 1992). Its appeal is its simplicity. The problem is that it incorporates neither expected price changes over time (in response to the anticipated shortage) nor the additions to the physical stock that those price changes would induce. Once more, both are important in forestry. Proponents of this alternative also tend to argue that current timber harvest rates are too great, often as a result of mismanagement by a public forestry agency (e.g., Repetto and Gillis 1988). This view is inconsistent with the accounting approach.[41] When public agencies allow their timber harvests to exceed the efficient rate, then harvest costs must exceed receipts, making net stumpage values negative. Therefore, where harvest levels are excessive, the observed positive stumpage price for marginal timber is entirely a subsidy created by the failure of the public agency to subtract its management costs. Stumpage prices formed in this manner have no basis as estimates of resource value in the national accounts. Finally, and regardless of this last point, the determining criterion for an approach that only intends to provide an estimate of the neoclassical economic value should be its performance. This second alternative performs poorly. It creates capital account estimates that are orders of magnitude greater than the preferred neoclassical economic estimate for assets like trees that have lifetimes greater than a couple years.

The summary Philippine account is instructive for this discussion. Of several applications of the neoclassical approach, the 1988 Philippine account has the advantage that it is fully detailed in its resource and environmental sectors. This Philippine account applies the detailed neoclassical approach to both current and capital accounts for the forest sector (Table 6A.3). The two columns reflect the standard income and expenditure accounts. The upper portion of the table is the conventional national account, while the lower portion adds the environmental and natural resource revisions necessary to create an account that accurately adjusts for environmental omissions and resource depletion. The items of our particular interest are non-market household production and natural resource depreciation. Fuelwood dominates the former and its value exceeds the value of both forest depreciation and total natural resource depreciation in the latter.[42] (Renewable resource depreciation exceeds nonrenewable resource depreciation

---

41 The appendix to chapter 2 discussed some of the difficulties with using stumpage prices as indicators of scarcity. Also see Toppinen (2002).
42 Consider the Philippine fuelwood and forest depreciation estimates further. It seems unlikely that the fuelwood value is actually greater than the estimate for all subsistence agriculture—as

because the forestry and fisheries sectors are much larger than the nonrenewable sectors of the Philippine economy. The results might differ in another economy that is more dependent on non-renewable resources.)

Two summary observations are striking. First, the net forest sector adjustment for the Philippine environmental account is positive. That is, the more complete accounting for the services of the forest environment adds to national wealth. This is the opposite of most common expectations, and it is due to the large positive household value for non-market fuelwood. Second, the revised environmental account varies less than 0.03 percent from the conventional national account. These unexpected findings reflect a) the opposing effects of household production and resource depletion and b) the fact that the Philippine economy is largely a modern market economy. Most important values exchange in well-defined markets and Philippine forests in particular are relatively low-valued resources that have only a small impact on the full economy. For countries that are more developed than the Philippines, we can anticipate that the variation between the revised environmental accounts and the conventional national income accounts will be even smaller.

In sum, we can anticipate that the approaches of El Sarafy and Repetto are justified in their focus on environmental accounts in those countries where markets are less well-developed and natural resource sectors are important components of the full economy. Nevertheless, the Philippine example shows that this description does not fit all developing countries. Therefore, we must select the particular economy with care before presuming that an improved environmental account will substantially alter the complete national accounts.[43]

Finally, the Philippine account is only one observation. The Philippine results in general and for the forestry sector in particular, are intuitively appealing, but they do beg confirmation from comparable environmental accounting exercises

---

it is in the 1988 Philippine account. Surely households consume a greater value of homegrown crops than of fuelwood. The large fuelwood estimate is probably due to an overestimate of the foregone opportunity of labor used in fuelwood collection (the proxy for unit fuelwood value in the Philippine income accounts). Other examinations suggest that labor used in fuelwood collection is actually a joint input to several products (Amacher et al. 1993). Therefore, the implicit wage for this time is an overestimate of unit fuelwood value. Furthermore, fuelwood collection is often seasonal, occurring when the labor opportunity costs are small (Cooke 1998). Nevertheless, further adjustments to account for these two points would probably not alter the conclusions for environmental accounting. The fuelwood value would still exceed losses due to forest depreciation even if the implicit value of fuelwood in this Philippine account were reduced by a factor of four.

The estimate of forest depreciation invites a different insight. It invites comparison with the alternative estimates arising from the El Sarafy and Repetto approaches. Separate calculations show that the El Sarafy approach yields an estimate that is approximately equal to that in Table 6A.3. An estimate following the Repetto approach is greater by two orders of magnitude. (H. Peskin, personal communication, April 2005).

43 Seroa da Motta and Ferraz do Amaral (2000) further illustrate this point with data from Brazil's forest sector. Improved accounting to adjust for Brazil's rapid forest depletion has little effect because Brazil's forest is so large to begin with. The substantial quantity adjustment is multiplied by a very small price adjustment.

TABLE 6A.3 Modified Philippine income and product accounts for 1988 (in million pesos)

Income			Expenditure	
Compensation of employees		27,874	Personal consumption	558,765
Indirect taxes		56,763	Government consumption	77,183
Depreciation (produced asset)		67,162	Capital formation	147,515
Net operating surplus		399,747	Exports	226,431
			Imports (-)	(215,292)
			Statistical discrepancy	12,917
Charges Against Gross Domestic Product		802,519	*Gross Domestic Product*	802,519
Capital depreciation (-)		(67,162)	*Capital depreciation (-)*	(67,162)
Charges Against Net Domestic Product		735,357	*Net Domestic Product*	735,357
Natural resource inputs to unmktd household prod'n (+)		6,250	Unmarketed household prod'n (+)	6,250
(a) upland agriculture	1,950		(a) upland agriculture	1950
(b) fuelwood	4,300		(b) fuelwood	4300
Environmental waste disposal services (-)			Environmental damages (-)	(3108)
(a) air	(3,317)		(a) air	(381)
(b) water	(14,349)		(b) water	(2727)
Net environmental benefit (disbenefit)		16,138	Direct nature services (+)	1,580
			(a) diving (coral reefs)	1
			(b) visits to national forest parks	13
			(c) beach use	1566

Natural resource depreciation (-)		(2541)	Natural resource depletion (-)	(2,541)
(a) forests	(936)		(a) forests	(936)
(b) fisheries	(838)		(b) fisheries	(838)
(c) minerals	(387)		(c) minerals	(387)
(d) soils	(380)		(d) soils	(380)
Charges against Modified Net Domestic Product		737,538	Modified Net Domestic Product	737,538
Capital depreciation		67,162	Capital depreciation	67,162
Natural resource depreciation (+)		2,541	Natural resource depreciation	2,541
Charges against Modified Gross Domestic Product		807,241	Modified Gross Domestic Product	807,241

Charges against modified net domestic product without unmarketed household production – P731,288 million
Charges against modified gross domestic product without unmarketed house=old production = P800,991 million
Source: IRG with Edgevale Associates (1994)

in other countries. In particular, they encourage further inquiry into the net values of non-timber forest resource services and of the non-market consumption of forest-based goods that households produce for themselves.

Literature Cited

Amacher, G., W. Hyde, and B. Joshee. 1993. Joint production and consumption in traditional households: Fuelwood and agricultural residues in two districts in Nepal. *Journal of Development Economics* 17: 93–105.

Arnold, M., G. Kohlin, R. Persson, and G. Shepherd. 2003. *Fuelwood revisited: What has changed in the last decade?* Occasional paper 39. Bogor, Indonesia: Center for International Forestry Research.

Barlow, T., and G. Helfand. 1980. Timber giveaway — a dialogue. *The Living Wilderness* 44: 38–39.

Barlow, T, G. Helfand, T. Orr, and T. Stoel. 1980. *Giving away the National Forests: An analysis of U.S. Forest Service timber sales below cost.* Washington, DC: Natural Resources Defense Council.

Binkley, C. 1999. Accounting for forest assets. In W. Nordhaus and E. Kokkelenburg, eds., *Nature's numbers: Expanding the national economic accounts to include the environment.* Washington, DC: National Academy Press, pp. 202–206.

Boyd, R., and W. Hyde. 1989. *Forest sector intervention.* Ames: Iowa State University Press.

Contreras-Hermosilla, A. 2001. *Forest law compliance: An overview.* Rome: Food and Agricultural Organization of the UN.

Cooke, P. 1998. The effect of environmental good scarcity on own-farm labor allocation: The case of agricultural households in rural Nepal. *Environment and Development Economics* 3(4): 443–469.

El Sarafy, S. 1989. The proper calculation of income from depletable natural resources. In Y. Ahmad, S. El Sarafy, and E. Lutz, eds., *Environmental accounting for sustainable development.* Washington, DC: World Bank, pp. 10–18.

Hyde, W., and S. Daniels. 1988. Balancing market and nonmarket outputs on public forest lands. In V. Smith, ed., *Environmental resources and applied welfare economics: Essays in honor of John V. Krutilla.* Washington, DC: Resources for the Future. 135–163.

Hyde, W., D. Newman, and B. Seldon. 1992. *The economic benefits of forestry research.* Ames: Iowa State University Press.

IRG with Edgevale Associates. 1994. *The Philippine environment and natural resource accounting project (ENRAP Phase II).* Washington, DC: IRG.

Katila, M. 1995. *Accounting for market and non-market production of timber, fuelwood and fodder in national income framework: A case study.* IUFRO XX World Congress, Tampere, Finland, August.

Peskin, H., and E. Lutz. 1993. A survey of resource and environmental accounting practices industrialized countries. In E. Lutz, ed., *Toward improved accounting for the environment.* Washington, DC: World Bank, pp. 144–176.

Prince, R. 1996. *The challenge of incorporating environmental quality and natural resource availability into the national accounts.* Washington, DC: Global Planetary Change.

Repetto, R. 1992. Accounting for environmental assets. *Scientific American* 66: 94–100.

Repetto, R., and M. Gillis (Eds.) 1988. *Public policy and the misuse of forest resources.* Cambridge, UK: Cambridge University Press.

Seroa da Motta, R., and C. Ferraz do Amaral. 2000. Estimating timber depreciation in the Brazilian Amazon. *Environment and Development Economics* 5: 129–142.

Toppinen, A. 2002. *An environmental accountant's agony: Are stumpage prices reliable indicators of the scarcity of national forest resources?* Draft manuscript, Finnish Forest Research Institute, Helsinki.

UN Economic Commission for Europe. 2000. Forest resources of Europe, CIS; North America, Australia, Japan, and New Zealand. Geneva: UN.

United Nations, European Commission, International Monetary Fund, Organization for Economic Co-operation and Development, and World Bank. 2003. *Handbook of national accounting, integrated environmental and economic accounting.* New York: UN.

U.S. Department of Commerce. N.d Bureau of Economic Analysis website http://www.bea.gov/bea/dn/home/gdp.htm (accessed July 14, 2006).

Zhang, Y. 2007. *Green wealth: Assessing and accounting for social benefit of forest.* Beijing: China's Environmental Science Publisher. (in Chinese)

7
INDUSTRIAL FORESTRY

To this point, we have examined the general pattern of development in forestry and the characteristics of forests in three stages of their economic development. The remainder of the book reviews the characteristic structures and behaviors of the different classes of forest landowners: industrial landowners, institutional landowners, non-industrial landowners including forest farmers, and the public. These landowner distinctions are common around the world and they are classic to the analysis of forest policy and management.

The full forest industry includes forestry itself and the various industries that process wood and wood fiber in the manufacture of secondary forest products. This chapter examines these manufacturing industries, especially with regard to the factors that affect their decisions to own or control forest land. The chapter does not include the various industries that provide forest recreation, hunting, or fishing or other forest values that are often considered non-marketed. It also excludes the producers of fuelwood and other subsistence forest products. The former will receive some attention during the chapter 10 discussion of public land management. The latter will be a component of the discussion of community impacts in chapter 11.

Those vertically integrated industrial establishments that operate mills and also possess rights to some forestland are known as industrial forest landowners. They possess the means for supplying woody raw material for further production in the own manufacturing operations. A few firms own or possess rights to the timber on thousands of hectares of land, an area sufficient to supply all of the primary raw material consumed by their own mills plus some that they sell to other wood processing firms. Regardless of their ownership of forestland, the bulk of the capital investment of most firms is in their processing facilities. Their operational objective has to do with the profitability of this capital, or perhaps the profitability of their combined capital investment in both mills and forestland. Accordingly, the large industrial firms in the United States have divested themselves of their forests in recent years in order to focus on those greater capital investments in their manufacturing processes. The next chapter—on institutional forest landowners—discusses the reasons and procedures for these divestitures. Some industrial forestry firms in the United States and many others around the world own rights to smaller amounts of timber, not enough to satisfy the full requirements of their mills. Still others own neither land nor the rights to timber. They obtain their entire supply of raw material in market transactions. Indeed, corporate opinions differ on the optimal amount of land to hold, the best allocation of timber from this land, and the preferred performance of their timber managers.

In fact, the wood products industries and the firms within them are not homogeneous. Their use of capital, labor, and land, and their consumption of wood fiber varies widely across the three major categories of the full industry: lumber and wood products, furniture, and paper and allied products. It even varies between firms within each of these industry categories. It is not important for all firms to possess their own timber supplies and it is reasonable that the degree of control a firm exercises over its timber supply varies with the competition it faces for raw material as well as with its own capital intensity. The former tends to be lesser for firms that operate in regions in the first stage of forest development and greater for those that operate in regions in the third stage; and therein lies an important link to the discussion of the previous six chapters.

The chapter begins with an example of industrial development from Brazil's eastern Amazon. This example illustrates the differences in communities and the different forest industries and their access to forests and forest land as they progress through the three stages of development. The pattern of development in the eastern Amazon is similar to that described for the U.S. South in chapter 3. However, Brazil's example is more recent and its development has not progressed as far. As an example, it has the advantage of good data on the relative costs of land, labor, and capital for different categories of firms.

The evidence from both the U.S. South and the eastern Amazon encourages a closer examination of the specific characteristics of important categories of firms within the manufacturing industries, and subsequent sections of the chapter do just that. Those sections review the organization and technologies, the range in scale of operations, and the implications of these features of industrial structure for harvest operations and land management for firms in each of four industry categories. A final section summarizes, particularly with respect to the pattern of industrial land ownership and forest management, and an appendix adds comment on the proximity of the various wood product industries relative to the forest and to the markets for their products.

The Eastern Amazon

The gradual development of the wood products industry in Finland, in southern Ontario and Quebec in Canada, and in other developed economies as well, has followed a pattern similar to that of the U.S. South. The state of Para in Brazil's eastern Amazon provides a second example from a region that has developed differently, much more recently and much more rapidly than the U.S. South. Brazil is the world's sixth largest exporter of sawn hardwood and Para is home to half of the registered hardwood mills in Brazil. Stone (1997, 1998) traced the development of the forest industry in Para for a period in the 1990s with a comparison of three milling centers, the municipalities of Tailandia, Breves, and Paragominas, that describes a continuum of access to the frontier and a continuum of forest industrial development. Our discussion follows Stone's assessment.

Until the early 1980s, logging in the eastern Amazon was limited to fluvial areas and a few high-valued species. The history of the forest industry since then is one of continued extraction from the natural forest (comparable to the First Forest of the U.S. South), although some harvest of natural regrowth on previously cutover lands (comparable to the South's Second Forest) had occurred by the mid-1990s, and a few firms had begun making their first moderate investments in forest management by that time.

The region developed rapidly because it has good access to port facilities in the capital city of Belem and good overland connections with the population centers to the south via the Belem-Brasilia highway. The number of mills in the region expanded rapidly to over 1,500 officially registered mills employing almost 250,000 workers in 1995. In 1995, industrial operations in Para ranged from seasonal logging to permanent forest management as well as logging from the managed lands, and from uncounted rudimentary mobile sawmills in the informal sector to a few large integrated sawmills, plywood and veneer facilities, and multimillion dollar export houses.

Tailandia

Tailandia is the newest and least developed of the three municipalities. Tailandia was an isolated sawmill center in the 1990s approximately 200 km from the provincial capital of Belem. It had neither electricity, well-developed paved roads, nor inexpensive water transportation. It did have thirty sawmills and two veneer operations, and these supported numerous loggers.

The initial timber harvests around Tailandia were characterized by very selective logging. The common harvest technology was simple, involving only a chainsaw for felling and bucking and for opening a path for the logging truck, and a hand winch for loading. Loggers typically removed small volumes (14 to 19 m^3/ha) of the best trees among fewer than twenty high-valued species and timber removals accounted for only 8 percent of the forest canopy—although another 4 percent was damaged during the logging operation and what was damaged was left in the forest. About 88 percent of the original forest remained standing after the harvest operation was complete. The average logging team, composed of three men, loaded 1,500 m^3/yr at an operating cost of US$6.35/$m^3$. Unit-costs were even lower for the smallest operators.

The evidence of forest disturbance caused by logging later in the mid-1990s is incomplete, but loggers clearly removed a greater variety of species and greater volumes per unit of logged area than they did five years earlier. Logging operations became increasingly mechanized as most grew to include a US$70,000 bulldozer. Larger operators, on average, owned three bulldozers and two mechanical loaders by 1995. (The capital-labor ratio in logging increased from less than 0.9 in 1990 to greater than 1.1 in 1995.) The average logging operation, composed of nine men by 1995, loaded 10,000 m^3/yr at an operating cost of approximately $7/m^3. The increased mechanization permitted increases in the harvest volume/unit-area and per unit of labor, and decreased the unit cost of production. Stumpage prices remained low—at about $77/ha or $2/m^3.

Even over the short period from 1990 to 1995, logging proceeded farther into the natural forest. The average hauling distance to Tailandia increased to forty km by 1995. Capital investments in hauling increased accordingly. Most logging operations had invested US$200,000 in trucks by 1995. Larger operators invested more than $350,000 in the combination of loading equipment and trucks and they were much more capital intensive than their smaller counterparts (capital-labor ratios of 4.1 and 2.7, respectively). Unit transportation costs were low for all loggers but the cost per kilometer was high ($0.17/m^3/km), reflecting the low quality of roads and limited easy commercial access to the forest.

In the wood processing industry, small and mobile operators, each with a single circular saw, dominated in 1990. The number of these operations declined after 1990, but small mills still dominated in 1995. Smaller operations tended to be family-run enterprises with one or, occasionally, two production lines. Larger mills simply added a production line—without changing the production process. In 1990 only 23 percent of Tailandia's sawmills were vertically integrated to include their own logging and hauling operations. The remainder purchased from independent loggers. By 1995, 38 percent of small mills (one saw) and two-thirds of the larger mills (multiple saws) were vertically integrated to include their own logging operations. In 1990, the average mill produced 3,000 to 4,000 m^3 of rough cut lumber annually. By 1995, mill production increased to an average of 5,000 m^3 annually at an average cost of about $28/m^3 of sawnwood.

Breves

Breves is located in the Amazon River delta, accessible only by water and air. It had no independent source of electricity in 1990. Highly selective logging had been conducted in the general area for over 300 years and numerous mills were still operating in 1995.

Logging around Breves had been associated with commercially mature and accessible *virola* (*Virola surinamensis* Warb.) timber. Traditionally, loggers manually felled the mature *virola* that grew along the estuaries, hauled it to the river banks, and floated it to the mills. The harvest activity itself was very selective, harvest costs were low, and river-borne hauling costs were also low. Aside from the removal of a few mature trees, the remaining forest was largely undisturbed.

More recently, as the inventory of mature *virola* declined, loggers and mills began to exploit other tropical hardwoods growing farther inland. The capital equipment required for this operation is much greater. Hardwood logging in Breves became similar to mechanized logging in Tailandia, and the logging costs per cubic meter were roughly comparable for the two municipalities by the mid-1990s. (Capital-labor ratios for logging ranged a little higher in Breves). Stumpage prices and hauling costs, however, were much lower in Breves. Hardwood stumpage prices were about US$15/hectare, or well under $1/m^3. These lower prices reflect both an activity that only occurred near the geographic limit of profitable timber harvest, and also the large costs of the capital equipment used to transport the inland hardwoods. The first factor was decisive because the final unit processing costs (including all raw materials) for sawmills in Breves were only marginally less than those for Tailandia and Paragominas by 1995. See Table 7.1.

Hauling is generally the most capital intensive activity in the manufactured wood products industry, but river transportation diminished the hauling cost considerably for Breves—from capital-labor ratios of 2.7 or greater for Tailandia to ratios in Breves of 0.4 for raft transportation of *virola* and 1.3 for barge-loaded shipment of the inland hardwoods. (The latter are denser and do not float.) Barges represented a capital investment of more than US$1.7 million in 1995, twelve times the investment in the diesel-powered boats used to tow floating rafts of *virola*. Nevertheless, barge transportation did reduce shipment costs—to the very low level of $0.07/m^3/km. As a result, mills in Breves were able to ship logs from 74 km on average and some mills shipped from as far as 234 km.

Small family-run operations with a single band saw and 3–4 employees comprised about half of all processing establishments in Breves in 1990. They averaged less than 1,000 m^3 of rough cut sawnwood per year. Their share of all wood processing establishments remained approximately unchanged in 1995 but the scale of local operations increasingly diverged. By 1995 larger firms with multiple production lines produced 4,000–7,000 m^3 of sawnwood per production line per year. Each of the four largest sawmills in Breves had its own loading dock for ocean-going vessels and two were vertically integrated as well, possessing the capital equipment for inland hardwood logging. These two ran as many as four lines of band saws and they were horizontally integrated to include veneer and plywood operations as well as sawmills. Stone speculates that it was their logging capital that provided the cost advantage to build their larger processing capacities. The average production cost in 1995 for the larger sawmills was US$27.47/m^3 of sawnwood. Average production costs for the smaller sawmills were a little less than this and the production costs of both large and small sawmills in Breves were a little less than those for comparably sized mills in Tailandia.

The greater heterogeneity of the industry in Breves was due to the logging transition from *virola* to inland hardwoods and the returns to scale that a few firms with greater access to financial capital obtained when they invested both in mechanized logging and also in barges for the transportation of denser hardwood logs. The mobility of the smaller sawmills and, as family operations, their ability to avoid high labor taxes, continued to provide them an advantage in harvesting the small remaining pockets of *virola* that were of insufficient scale to be profitable for larger operators.

TABLE 7.1 Characteristics of the forest products industry in three municipalities in Brazil's eastern Amazon[1]

Characteristic (average measures)	Tailandia 1990	Tailandia 1995	Breves 1990	Breves 1995	Paragominas 1990	Paragominas 1995
Stumpage price ($/A)	—	77	—	15	—	183
($/m3)	—	2	—	≈ 0	—	< 4.50
Harvest volume (m³/ha)	14–19	—	—	—	38	—
Average Log haul (km)	—	40	—	73	—	94
Delivered log price ($/m³)	—	—	—	—	38	43
Unit costs of production ($/m³)[2]						
logging	—	5.34 – 8.92	—	6.12 – 10.26	—	8.72 – 8.15
hauling	—	7.55 – 5.63	—	8.05 – 11.53	—	10.04 – 8.19
sawmill processing	—	27.95 – 32.07	—	22.96 – 27.47	—	31.83 – 33.24
Value of capital investment[2]						
logging ($000)	< 0.84 –	69 – 288	< 0.84 –	22 – 518	—	168 – 409
hauling ($000)	—	97 – 186	—	99 – 715	—	227 – 376
wood processing facilities (no.)[3]						
small sawmills—circular saws	—	2	—	12	—	0
small sawmills—band saws	—	22	—	7	—	20
large sawmills	—	6	—	4	—	7
veneer and plywood operations	—	2	—	2	—	4

1 Vacant cells indicate no information
2 The first entree in each cell is for small operators, the second is for large operators
3 The sample for Tailandia is the full population. The sample for Breves includes all 24 mills that were readily accessible. The sample for Paragominas is a scientifically random sample of the total population of 238 mills. Additional mills operated in the informal sector in all three municipalities.

Source: Compiled from Stone (1997, 1998).

Paragominas

Paragominas is the most accessible and most developed of the three municipalities. It is also the farthest from the forest frontier. Its good infrastructure includes paved roads and commercial electricity. Over 230 registered wood processing firms operated in and around Paragominas in 1995.

Logging operations were much more mechanized in Paragominas than in either Tailandia or Breves. The investment in capital equipment by the smaller loggers in Paragominas was well over double that for smaller operators in Tailandia and eight times greater than in Breves. Larger operators invested half again as much as their counterparts in Tailandia and almost as much as those that harvested inland hardwoods in Breves.

Three-fifths of all wood processing firms in Paragominas were vertically integrated to include their own logging operations by 1990. Four-fifths were vertically integrated by 1995, and the value of capital equipment used in logging had increased further by the latter date.

Most of the lands around Paragominas had been logged selectively before 1990. Therefore, fewer high-valued trees and species remained. The greater mechanization of the Paragominas firms allowed them to overcome this disadvantage by profitably harvesting smaller trees, lower-valued species, and greater volumes per hectare—on average 38 m^3/hectare from over 90 different species even in 1990—at unit costs that were less than those for either of the other two municipalities, at least for the larger operators. Greater harvest volumes meant that less than half as much volume (less than 60 m^3/hectare) remained standing in the forest after the harvest. The technologically more advanced equipment in use in Paragominas in 1995 also meant that less damaged wood was left on the ground at each logging site.

The era of cheap raw material was over for Paragominas by 1995. Stumpage prices averaged US$183/hectare, or nearly $4.50/$m^3$, more than double the price in Tailandia and more than four times the price in Breves. The higher price was partly explained by increasing scarcity and partly by the better defined and more stable property rights for timber on previously cutover lands farther from the frontier. (Better defined property rights meant that loggers invested less in negotiating harvest rights and some of the savings transferred to higher stumpage prices for the landowners.) The greater harvest volumes per hectare obtained by Paragominas' more capital intensive logging operations helped compensate for the higher stumpage prices per hectare.

Capital costs for log shipment stand out in Paragominas as they did in Tailandia and Breves. The investment in logging trucks by the average firm in Paragominas almost doubled between 1990 and 1995 as very many loggers added larger capacity trucks to their operations. This investment in hauling equipment was greater in Paragominas than in Tailandia for both small and large firms in both 1990 and 1995. Delivered log prices were higher in Paragominas than in the other municipalities and they were higher in 1995 than in 1990 for all grades of logs. Both of these observations follow our expectations because hauling distances were greater in Paragominas than Tailandia (94 km on average in Paragominas vs. 40 km on average in Tailandia) and because smaller operators tended to close their local operations in order to relocate closer to the frontier where they are more competitive. Unit hauling costs were about 50 percent greater in Paragominas than in Tailandia, but unit costs per km were lower (US$0.10 vs. $0.17)—due to better roads and greater distances over which to average the fixed costs of hauling capital.[1]

Small sawmills comprised about half of all wood processing operations in Paragominas, as they did in Tailandia and Breves, and their annual production levels were similar. However, none of the sawmills in Paragominas employed the less efficient but more mobile circular saws. Some small establishments consolidated their operations between 1990 and 1995. Others departed Paragominas for regions closer to the frontier. Those small firms that remained tended to operate with highly depreciated equipment.

Larger mills found additional means of responding to the increasing competition and increasing costs that occurred farther from the forest frontier. By 1995 one-fourth of the larger mills in Paragominas had established satellite sawmills closer to the frontier and some firms with higher fixed costs were investigating the potential of plantation management closer to their mills. Annual output per production line in Paragominas remained comparable to that in the other two municipalities, but Paragominas had more large sawmills with multiple production lines, and these large mills were responsible for a larger share of total production. Sawmill processing costs in the neighborhood of US$32/m^3 were only marginally greater in Paragominas than in Tailandia, but about 25 percent greater than in Breves. Four plymills and four veneer factories also operated in Paragominas. These more technologically advanced operations represented capital investments of nearly $2 million apiece. They operated multiple shifts and they produced 40 percent of the municipality's final physical output. Finally, Paragominas also supported two export houses which did not process logs. They purchased export grade sawnwood for kiln drying, planing and packaging. These export houses represented investments of more than $0.5 million apiece. Gross revenues from their combined production exceeded $7 million annually in 1995.

Summary Observations

Table 7.1 captures many of Stone's observations for the three municipalities. The pattern of growth that Stone traces over the short period from 1990 to 1995 for each of the municipalities and spatially from less developed Tailandia to Breves to more developed Paragominas is general for the wood products industry almost anywhere in the world, with the caveat that river transportation at Breves adds a specialized element. Tailandia and Breves were in the second stage of forest development and Paragominas was already entering the third stage by 1995.

---

1 Unit hauling costs also declined in Paragominas, from US$0.20 in 1990 to $0.10 in 1995. Stone does not provide these costs for Tailandia and Breves in 1990.

It is clear from the table that delivered log prices, the sum of stumpage prices and logging and hauling costs, increased with distance and, what is essentially the same, they also increased over time as logging depleted the best timber at the frontier and pushed this frontier farther inland.[2] As delivered log prices increased, the wood processing industry responded by substituting lower-valued species and by increasing the harvested volumes of smaller trees of lower-valued species. The industry also consolidated into the larger firms that could harvest and process larger volumes. With larger volumes, opportunities to specialize arose at all levels of industry operation.

Larger, more specialized, firms made greater capital investments in the technologies designed for specialized tasks. As capital intensity increases, the return on capital becomes a more important indicator of successful operation. More permanent mills operating multiple shifts become the norm and a regular supply of raw material becomes more important to the entire operation. Therefore, by 1995 several mills in Paragominas had already established long-term commitments with landowners for their future log production and a few larger mills were considering investments in plantations. Greater capital intensity and more specialized equipment probably required more skilled labor as well—although Stone's observations are not as clear on this.

Meanwhile, the technological range of the industry remained broad. Small and mobile firms with less specialized capital continued to operate near the frontier. Their size and mobility provided an advantage where small pockets of less accessible timber remained. Smaller operations can be opportunistic, following the resource where they find it and taking advantage of good market conditions when they occur. The return on labor is relatively more important for these small operations, but they are often family enterprises and their family laborers often have the personal flexibility of other employment opportunities. The fluid labor opportunity of these small firms adds to the flexibility of their spatial and temporal mobility.

In sum, with the passage of time and with continued harvesting at Para's receding frontier, firms in the wood processing industries became larger and increasingly capital intensive. They also become increasingly vertically integrated. Both logging operations and wood processing facilities employed a range of technologies by 1995—from rudimentary chainsaws to expensive bulldozers and mecha-

---

2 Stumpage prices decreased with increases in average distance in the eastern Amazon but, as we discussed in the appendix to chapter 2, stumpage prices are not a good measure because they do not represent uniform market points. The stumpage prices quoted for a region like Para are an average of high prices for timber in more accessible managed locations nearer the mill and low prices in poorly roaded, less accessible, locations closer to the frontier. Therefore, there is an undetermined element of transportation implicit in each municipality's average stumpage price. Furthermore, each individual observation on stumpage price from which these regional average prices are calculated includes its own premium for property rights. Lands with more secure rights bring higher stumpage prices. Lands with less secure rights bring lower stumpage prices but often conceal an additional fee for negotiating the rights. Delivered log prices are a more useful measure for our comparisons because they do not conceal these differences.

nized skidders and loaders for logging, and from sawmills composed of a single circular saw operating at the forest frontier to those with multiple lines of band saws operating in more distant and developed market centers. Sawmills comprised the great majority of wood processing establishments in all three municipalities in the region, but a few plymills and veneer factories, some with multiple lathes, had begun operation by 1995. These had become an important part of the industry in the more developed municipality of Paragominas.

These patterns are generally similar to those we observed in chapter 2 for the U.S. South and we might anticipate that they are general for forestry and the industries manufacturing forest products almost anywhere. They show that the pattern of development may be gradual, as it has been for the U.S. South, or more rapid, as it has been for Brazil's eastern Amazon. It is also similar for both developed and developing regions and countries. However, our information on the pattern is not complete. Neither of the two specific examples to which we have referred provides detail on the labor used in wood processing. In fact, the forestry literature is not strong on this point. Nevertheless, there is a large amount of statistical detail on the various forest industries and the differences between them that lends conviction to our maturing insight.

Industrial Organization: The Manufacture of Forest Products

Government reports of industrial data generally follow the International Statistical Industrial Classification (ISIC, or just SIC in the United States). Canada, Mexico, and the United States converted to the North American Industrial Classification System (NAICS) in 1997, revised in 2002. The ISIC identifies 42 specialized forest product industries in three general categories: lumber and wood products, furniture, and paper and allied products. The NAICS separates logging from the lumber category (where it was previously classified) and reclassifies the 41 remaining SIC industries as 46 specialized industries in three similar categories: manufactured wood products, furniture, and paper manufacturing.[3]

Within these three general industry categories, sawmills, plywood, reconstituted wood products, furniture, pulp, and paper are, by far, the largest consumers of wood and wood fiber. The logging industry is a separate industry, whose characteristic structure and conduct are understood best in the context of the particular secondary industry served by each group of logging firms. Therefore, logging will be the seventh and final industry in a sequence of industrial reviews which follow.

The objective in this review is to build a fuller appreciation of the patterns observed in the previous discussions of the U.S. South and Brazil's Paragominas.

---

3   A U.S. Census Bureau website (http://www.census.gov/epcd/www/naicstab.htm) traces the correspondence between the SIC and NAICS systems. In the United States, the reassignment of the logging industry from SIC 2411 (a component of lumber and wood products) to its own NAICS category 113310 includes a transfer of the reporting and recording responsibility for this industry from the U.S. Census Bureau to the U.S. Department of Agriculture.

We will observe a range of operations with respect to the stage of forest development and the use of resource inputs. We will also observe that firms within each industry category apply a range of technologies in their production processes and the technologies each applies—and the resulting patterns of use of capital and labor—are closely related to each firm's and industry's demands on the forest and the stage of regional forest development in which each operates successfully. Each industry's range of technologies is also suggestive of the relative ability of establishments within that industry to adjust to changes in their timber supply.

Manufactured Wood Products (NAICS 321)

The manufactured wood products industries produce lumber, veneer and plywood, wood containers and wood flooring, trusses, and other wood-based prefabricated building material. These industries begin with logs, bolts (squared logs), and rough cut lumber which they saw, plane, shape, and laminate in the course of manufacturing their final products. Tables 7.2A and 7.2B summarize NAICS data for the United States for the six largest of the fourteen manufactured wood products industries in 2002.[4]

Sawmills and veneer and plywood are the largest, and also the most interesting of these industries for our purposes. In recent years, new technologies for using smaller logs, lower-valued species, and the woody residues from sawmill and plywood operations to produce particleboard, fiberboard, hardboard, and oriented strand board (OSB) have cut into the markets for lumber and especially plywood, particularly in some construction processes and even as both lumber and plywood industries continue to innovate.

The sawmill industry (NAICS 321113): Its unique rates of use of the three primary factors of production—manufactured capital, labor, and raw materials—are central to the description of any industry. In the sawmill industry, we observe a particularly wide range of technologies and, therefore, an unusually broad array of manufactured capital in combination with different numbers of laborers of various skill levels. This range of technologies makes the activity of sawing logs into lumber adaptable to almost any natural and economic conditions. At one end of the scale, in the least developed rural markets of Africa and Asia, two-person teams with little more than a spade for digging a hole and a crosscut saw operate pit saws wherever they find trees they can rip into rough cut boards for the least developed rural markets. Mobile circular saws attached to old trucks or tractors as their power sources operate in only slightly more developed economic conditions in similar locations and elsewhere too.

The smallest operations, even in developed economies, are mobile, both spatially and temporally. Their mobility allows them to operate at the frontier and,

---

4 Corresponding data for the same NAICS industries in Canada and Mexico are available from Statistics Canada and Instituto Nacional de Estadistica, Geographia e Informatica, respectively.

thereby, to avoid the cost of transporting that share of the initial log that becomes a residual to lumber production. They often operate in isolated plots of remaining mature timber that are too small to attract larger operators. Because these small operations are often built on wheeled platforms, the cost of closing operations as they deplete profitable harvest opportunities at one location and moving to new sources of logs is small.

These portable mills tend to restrict operation to periods when the market conditions for both material and product are good. They tend to lag the business cycle, entering into operation after the manager/owner recognizes and has confidence in the signs of strong demand for sawn lumber, and remaining in operation even as economic activity begins to slacken—as the manager is slow in perceiving up-to-date market signals or remains overly optimistic that the market will remain strong or recover rapidly.

The labor force for these operations is also flexible. It is composed of an owner/manager and as few as two employees, each of whom may have other income opportunities. The entire crew may use sawmill employment to supplement income from other activities.

More technologically advanced sawmills tend to operate at a distance from the forest itself, and they are not as mobile. They make use of heavier equipment and a modicum of skilled labor, but their initial investment in manufactured capital is still low relative to that for many other manufacturing industries and their labor force includes at least a few who are essentially unskilled workers. These establishments sort and store inventories of logs in ponds and yards awaiting transfer for processing. Their sawing operations range from a single production line with one band saw to multiple production lines each specializing in input diameter classes and final products. Each line in the more technologically advanced mills begins with a mechanical debarker and then a conveyer that transports raw logs past an optical laser that assists the head sawyer in determining the best log position for the initial cut. Additional saws, farther along the same production line, make subsequent cuts before the fresh boards are planed into smooth lumber and stacked for kiln drying and shipping. The entire operation is mechanized in the most technologically advanced mills and, for them, the manual handling of logs or lumber is absolutely minimal.

The economies to be gained from increasing size in the sawmill industry are small. Measures of the elasticity of return to increasing scale for sawmills in developed countries fall in the narrow and only slightly increasing range of zero to 0.11 (Nyrud and Baardsen 2003; Baardsen 2000). Furthermore, the industry's rate of technological change has not been rapid. While the technological advances associated with smaller logs, different species, changing energy sources, and the optical laser are notable, observed rates of technical change for this industry are low, ranging only to 0.6 percent per annum (Stier and Bengston, 1992; Nagubadi and Zhang 2006). The smaller magnitudes of these two factors, scale economies and technological change, anticipate the wide variety of manufactured capital that continues to operate successfully in this industry.

TABLE 7.2A Measures of industry and firm size for select manufactured wood products industries in the U.S., 2002

Industry[1]	Number of companies	Number of establishments	Employees per establishment			Concentration ratios (% of value of shipments)				Total number of employees[2]	Value of shipments[3]	Value added[3]
			1–19	20–99	≥100	C_4	C_8	C_{20}	C_{50}			
321113 sawmills	3,461	3,807	2,708	337	262	17.5	29.9	33.9	45.1	95.5	21,339	6,798
321211 hardwood veneer & plywood	303	335	152	127	56	33.1	44.8	61.5	77.4	19.6	3,195	1,323
321212 softwood veneer & plywood	85	147	34	36	77	57.3	72.2	88.6	99.2	24.5	4,704	1,336
321213 engineered wood members (except truss)	90	120	56	43	21	66.6	80.2	90.3	98.1	13.6	1,954	642
321214 truss mfg.	897	1,036	465	483	88	10.0	14.6	24.8	40.9	40.2	4,488	2,296
321219 reconstituted wood products mfg.	177	278	83	96	99	35.2	50.5	79.1	90.6	22.3	5,753	2,475

1 NAICS identifying number and name
2 in 000 of persons
3 in 000,000 of US$

Sources: U.S. Census Bureau 2004 and 2005b, c, i, j, k, l.

While the manufactured capital in use in the more technologically advanced sawmills is not mobile, the numerous employees who are less-skilled do tend to be mobile. They move easily to other jobs as the sawmills that employ them adjust their output levels or as new employment opportunities appear elsewhere.[5]

Delivered logs, the third primary factor of production, account for a large share of all sawmill costs. Their share of production costs is largest for the portable mills that operate at the source of their supply. The cost shares of labor and manufactured capital are relatively greater for larger operations, but the absolute cost share of delivered logs remains large, even for the largest and most capital-intensive sawmills. This means that transportation cost (generally the largest component of the delivered log cost) defines a geographic limit for any mill's supply of raw materials.

Therefore, it is not surprising that sawmill performance is generally more sensitive to delivered log prices than to the costs of either labor or manufactured capital. Nevertheless, some large mills operate in markets with few competing mills and a plentiful log supply—generally in the first two stages of forest development. These mills may exercise a degree of market power over their suppliers—with favorable price effects for themselves—and the production decisions of these mills are relatively more responsive to the prices of labor and manufactured capital (Nagubadi and Zhang 2006; Abt and Ahn 2003).[6,7] Nyrud and Bergseng (2002) speculate that it is the additional knowledge of their technology and their markets that these large mills possess that explains any (limited) observations of scale economies in the industry.

Others of the more technologically advanced mills that operate in more competitive log markets, or that have other reason for concern about the sup-

---

5 Stevens' (1978) survey of Oregon millworkers shows that the youngest and least experienced are the first to be laid off when mills decrease production, but these workers move easily to alternative employment, either in sawmills elsewhere in the region or in other rural activities. Some, in my own U.S. experience, even look forward to a seasonal period of unemployment and the government compensation that comes with it.

6 Mead (1966) provides statistical evidence of market power in the bidding history of sawmills in the Douglas fir region of Oregon and Washington. Stordahl and Baardsen (2002), with mill and market evidence from Norway, observe that the sawlog market of that country became less competitive during the 1970s and 1980s and they reject the hypothesis that Norwegian sawmills are price takers. Munn and Rucker (1994) approach the question from a different perspective. They observe that forest landowners in the U.S. South who employ consulting foresters to assist in the sale of their timber obtain higher stumpage prices than those landowners who do not use consultants. Their evidence suggests that the mills have market power built on better market information than that possessed by those less-informed landowners who do not employ consulting foresters.

7 Williamson, Hauer, and Luckert (2004) and Latta and Adams (2000) summarize production and demand elasticities for their own and previous assessments of various regions of Canada. The range of elasticities for other developed countries is comparable to those for Canada (e.g., Lewandrowski 1990 for the U.S. and Buongiorno et al. 2003 more generally). However, all published assessments of these elasticities are comprehensive for broad regional cases. None speculate on the difference between mills operating in varying regions of competitive or less competitive log markets—as we do in this chapter.

TABLE 7.2B Operating characteristics with respect to major input categories for select manufactured wood products industries in the U.S., 2002

Industry	% Capacity	Capital/labor[1] K/L	Capital/primary material[1,2]		Labor/primary material[1,2]		Output ratios[3]			
			Capital/ virgin wood K/R_1	Capital/all wood K/R_2	L/R_1	L/R_2	K/O	L/O	R_1/O	R_2/O
321113 sawmills	75	2.58	1.13	1.11	0.45	0.42	0.46	0.18	0.40	0.42
321211 hardwood veneer & plywood	67	1.10	1.76	0.64	1.74	0.58	0.25	0.22	0.13	0.38
321212 softwood veneer & plywood	86	2.19	1.24	1.04	0.58	0.46	0.52	0.24	0.41	0.50
321213 engineered wood members (except truss)	64	2.61	dz	1.31	dz	0.50	0.39	0.05	0	0.30
321214 truss mfg.	54	0.08	na	0.89	dz	1.17	0.23	0.31	0	0.26
321219 reconstituted wood products mfg.	72	5.47	11.06	4.06	2.02	0.79	1.02	0.19	0.09	0.23

1 Capital calculated as beginning of year gross assets minus beginning of year inventory. Labor is total labor compensation. Output is value of shipments.
2 Primary material: virgin wood is the sum of expenditures on stumpage, logs, bolts and pulpwood. All wood is the sum of expenditures on virgin wood plus expenditures on chips, slabs, shavings, mill residues, rough cut lumber, etc.
3 Capital, labor, and raw material measured as in previous columns. Output is total value of shipments. These ratios are not percentage shares and they may add to more or less than one, as capital is the total value of durable capital (not an annual value) and primary material includes annual expenditures only on the wood input and not that on other materials or energy.
dz: not available—the denominator is zero.
Sources: U.S. Census Bureau 2004 and 2005b, c, i, j, k, 40

ply of delivered logs, maintain vast log yards for storage, and they may establish arrangements with loggers and landowners to insure a limited regular flow of logs.[8] These arrangements may be as formal as employment of loggers and fee simple ownership of forestland in vertically integrated operations. Alternatively, they may be as informal as demonstrated preference for certain reliable providers. In either event, these arrangements seldom insure the total supply of logs necessary for full-scale long-term operation of the mill. Even the largest sawmills with the greatest investment in manufactured capital in the most competitive log markets have enough flexibility in their capital, labor, and raw material decisions that a fully guaranteed log supply is unnecessary. A dependable permanent supply sufficient to guarantee long-term operation of the mill would be more expensive than limited purchases of high-priced logs or modifications in the flow of production during periods of tighter log markets.

NAICS (or ISIC) data permit a more quantitative picture of the U.S. industry. It is a cumulative picture that conceals the many specialized differences that exist where the technological range of mill operations is so large. Nevertheless, these data are informative—particularly as we introduce comparisons with other forest product industries later in the chapter. The first rows of Tables 7.2A and 7.2B summarize these data for the U.S. sawmill industry in 2002.

Table 7.2A provides evidence of firm size and total industry size. 3,461 companies reported operating 3,807 sawmills in 2002. An unknown additional number of small mills either operated without reporting or were closed during the reporting period but remained ready to operate again whenever local market conditions improved. There is at least one sawmill in almost every county in the United States. These 3,807 mills employed a total of almost 96 thousand workers and shipped products worth about $24 billion, an average of $6.4 million per mill. The average mill's investment in manufactured capital was less than $250,000 in 2002 (not shown in the table), and the initial capital investment in many smaller mills was much less, as little as $11,000 for some portable mills. The purchase price of some ultra-portable mills is only $4,000.

Further employment data and the industry concentration ratios reinforce the previous observation of the generally small scale of sawmill operations. The average U.S. mill employed 25 workers in 2002—but more than 70 percent of all mills employed fewer than twenty full-time workers. While the average mill shipped $6.4 million of lumber, the added value of these shipments was less than $1.8 million per mill. The four largest sawmills accounted for only 17.5 percent of the total value shipped by the industry and even the fifty largest mills accounted for barely 45 percent of total value shipped in 2002.

---

8 Log yards with the capacity to store large inventories of logs may also occur in regions where logging is a seasonal activity or in regions beset by political and social instability or unpredictable government policy. With larger inventories, mills can continue operation during seasons when weather makes logging more difficult or despite other uncertainties which affect the flow of delivered logs.

Table 7.2B summarizes the operating characteristics of the industry with respect to its major input categories. The industry operated at only 75 percent capacity in year 2002. Of course, this number would be somewhat lower if the unknown number of non-reporting small mills were included. The capacity data suggest that starting up or shutting down operations is a relatively easy decision, as we anticipated.

The factor proportions show that manufactured capital and raw materials are approximately twice the expense of labor for an average mill in the United States. We will discover, as anticipated, that the capital-labor ratio (K/L) of almost 2.6 is very low in comparison with most other manufacturing industries, while the ratios of the primary resource, uncut or "virgin" logs, with either capital or final output ($R_1/K = 0.85$ and $R_1/O = 0.42$, respectively) are very high.[9] In fact, the cost of uncut logs is 25 percent greater than the value added in the production process of this industry (not shown, but calculated from the same tables and sources). These factor proportions are indicative of the greater importance of labor and an available inventory of logs relative to that of manufactured capital in the short-run operating decisions of sawmill managers.

Altogether, these data describe an industry comprised of many small firms, each too small to have an individual impact on the national market for their lumber product. It is an industry with low manufactured capital start-up costs and, therefore, an industry that displays ease of entry and exit, and an industry that adapts easily to a range of conditions regarding either the availability of uncut logs or favorable product prices. Favorable local markets for its primary raw material, uncut logs, are a crucial determinant of the location of successful operations. Favorable product markets as well are crucial to the uninterrupted operation of the smallest sawmills, although larger mills generally operate steadily throughout the year and throughout the business cycle.

Veneer and plywood (NAICS 321211 and 321212): The veneer and plywood industry possesses a narrower range of technological options than the sawmill industry and its fundamental technology is more expensive. Veneer and plywood operations depend on a lathe for slicing or peeling continuous thin sheets of wood (veneer) from a revolving log. Very similar machines, differing in little more than age and maintenance, appear in plymills throughout the world.

The plywood production operation begins, once more, with debarking. A conveyer transports debarked logs to the machine that peels sheets of veneer from the log, after which the sheets are patched as necessary, cut into 4 × 8 foot lengths,

---

9   The K/L ratio for U.S. sawmills has gradually increased over the last half century as the number of mills has steadily declined, capital investment and the value of shipments for the remaining mills has also increased, and the absolute level of labor input has declined. The skill level of sawmill labor has increased substantially in accordance with these trends. Nevertheless, and despite these trends, the K/L ratio even for the U.S. industry that is technologically advanced relative to the same industry in many other countries, remains low relative to that ratio for other manufacturing industries in the United States.

glued together in layers of selected thickness and orientation, pressed, dried, and stacked for shipment. As with sawmills, the veneer and plywood manufacturing process leaves a residual of woody material including log cores that are too small or of poor quality for peeling, and various scraps of wood. Some of this residue may be burnt to produce power for the mill, some is resold for other uses, and some is simply burnt as waste.

NAICS separates the veneer and plywood industry into softwood and hardwood components. We will focus on the softwood industry (NAICS 321212), the older and larger in terms of value shipped in the United States. The hardwood plywood industry is growing in the United States and in other developed countries, but it is more important in tropical countries where the forest is primarily deciduous. The plywood technology, while not highly skilled, does require more skill than the simplest sawmill operations and this restricts its operation in some developing countries.

In the United States, 85 companies operated 147 softwood plywood establishments in 2002. They employed 24.5 thousand workers and shipped $4.7 billion of goods, for averages of 167 employees and $32 million in shipment per establishment. These are still small operations in comparison with many manufacturing industries, but they are considerably larger than the average sawmill. The initial capital investment for a mill producing exterior-type sheathing-grade plywood was approximately $2 million in 2002, but most plywood mills represent greater levels of total capital investment. The capital investment of an average establishment in 2002 was $16 million, more than five times the capital investment in an average sawmill. The average plywood operation also employed its labor force more regularly throughout the year. The average total compensation, including benefits, for all plywood employees (including both managers and production workers) was more than $46,000 in 2002, approximately $6,000 per employee more than in the sawmill industry (not shown in Table 7.2A but calculated from the same sources). The average annual wage (excluding benefits) for plywood production workers only was $33,000, about $3,000 more than in the sawmill industry.[10]

The capital-labor ratio ($K/L = 2.19$) for softwood plymills shows that these mills, despite requiring a larger initial capital investment, are a little less capital intensive and more labor intensive than the average sawmill. The ratios involving the two measures of raw material input show that wood is still a major cost for plymills—although its cost share is slightly less than that for sawmills. As expected, plywood mills consume very little processed wood and, therefore, require good access to the forest. In fact, the cost of uncut logs (R_l) for this industry is 47 percent greater than the industry's value added, a number that is even greater for plywood than for sawmills. Clearly, a reliable supply of peeler quality logs is crucial to their operation.

---

10 Total compensation and benefits data are from Bureau of Census data identified in footnote 3 and recorded in the tables of this chapter. Annual wages for production workers are from the U.S. Bureau of Labor Statistics website www.bls.gov/oes/current/oes454022.htm (accessed June 28, 2008).

The softwood plywood industry is increasingly concentrated. The four largest companies produced 38 percent of industry shipments in the late 1980s but 67 percent by 2002. C_4 concentration ratios greater than 40–50 percent often indicate a non-competitive industry. However, the inability of plywood mills to differentiate their products and the increasing competition from reconstituted wood products like hardboard, particleboard, and oriented strand board prevent the larger mills in the softwood plywood industry from establishing significant control over prices in their product markets.

In summary, softwood plywood mills require more capital and labor that is slightly more skilled than most sawmills. The range of technologies available to plywood operations is narrower than for sawmills. The additional capital required for plywood mills discourages mobility and the skilled labor requirement encourages operation in towns where this labor is available, yet the expense of transporting uncut logs also requires proximity to the forest. Therefore, it is not surprising that plywood and veneer mills tend to operate farther from the forest than the most rudimentary sawmills. Indeed, this is exactly the experience we observed in both the U.S. South and Brazil's Para. However, it is also not surprising that mills in both the U.S. South and Brazil's Para still tend to operate in towns that are near large sources of timber. Raw materials remain a large component of plywood operating costs, and successful veneer and plywood operations rely on favorable local markets for uncut, peeler quality, logs.

Engineered wood members (NAICS 321213), trusses (NAICS 321214), and reconstituted wood products (NAICS 321219): Engineered wood products like laminated beams, truss frames constructed at the mill or lumber yard, and reconstituted wood products are rapidly attracting market share in the United States away from both the lumber and plywood industries. Each of these products incorporates a cost saving in raw material, which we have seen is a substantial share of all production costs for both the sawmill and the veneer and plywood industries.

Laminated beams combine wood strips and glue to replace the broad, long, solid beams of old. Truss frame manufacture is the lumberyard assembly of beams and joists into frames of uniform specifications. Mass assembly cuts down on what otherwise would be wasted odd-sized remainder pieces of lumber when the manufacture occurs in very small lots at each construction site. Reconstituted wood products combine wood residues from sawmills and plymills with glue to create 4 × 8 foot sheets that are often stronger than competing plywood products. Some reconstituted wood products also use fiber from plentiful but otherwise low-valued and underutilized hardwood logs. The net effect is a tremendous saving in the consumption of less plentiful and more expensive virgin softwood logs.

The distinctions between uncut logs (R_1) and partly processed wood products (R_2) in Table 7.2B reflect the savings in virgin wood. Uncut logs comprise more than 90 percent of the woody material consumed in sawmill and plywood operations. In contrast, none of the raw material used in the engineered wood members and truss manufacturing industries is virgin wood. The entire raw material of these latter two industries is either residual woody material from the sawmill and plywood industries or, to a lesser extent, rough cut lumber or low grade plywood. The raw material of the reconstituted wood products industry is, similarly, more than 90 percent sawmill or plywood residue. The remaining 9 percent is largely low-valued hardwood logs.

One result of the financial saving in raw material costs is a favorable effect on forest conservation. For example, the U.S. Forest Service calculated, as long ago as 1982, that the annual saving in wood consumption from the truss frame technology alone was greater than the annual volume of programmed timber harvests on 28.6 million hectares of forestland challenged for other non-timber uses.[11]

Furthermore, the trend is increasingly in the direction of substituting the products of these three industries for the products of the sawmill and plywood industries. Spelter's (1984) observations of declining price elasticities for softwood plywood (from 0.83 in 1970 to 0.10 in 1980) argue that the substitution continued and, even as early as 1980, other reconstituted products were replacing some demand for softwood plywood. Structural particleboard (SPB) was probably the major plywood substitute during the years of Spelter's examination, but Spelter observed that the price elasticity of SPB was also declining by 1980, a fact consistent with its replacement by still other reconstituted products, such as oriented strand board. Table 7.2C shows that this trend continues into the twenty-first century. The table shows production data for 1997 and 2002, the most recent two industrial reporting dates for the U.S. Census Bureau. The total value of shipments from both the sawmill and plywood industries declined over the five-year period, as did their consumption of woody raw material. Meanwhile, the total value of shipments from the three newer industries increased by more than enough to offset the decrease in production from the sawmill and softwood plywood industries. The consumption of woody raw material by these

TABLE 7.2C Growth in raw material consumption and shipment of products in select manufactured wood products industries in the U.S., 1997–2002

Industry	Increase in total raw material (R_2) consumption	Increase in total value of shipments	R_1/O (2002)
321113 sawmills	– 7 %	– 1 %	0.35
321211 hardwood veneer & plywood	+ 9 %	+ 12 %	0.13
321212 softwood veneer & plywood	– 15 %	– 20 %	0.41
321213 engineered wood members (except truss)	+ 50 %	+ 38 %	0
321214 truss mfg.	+ 5 %	+ 24 %	0
321219 reconstituted wood products mfg.	+ 9 %	+ 9 %	0.09

Sources: Same as tables 7.2A and 7.2B.

---

11 Calculation by H. Gus Wahlgren, February 15, 1989, provided by Robert Buckman.

newer industries also increased, and much of this increase was from low-valued sawmill and plywood residue. The hardwood plywood industry expanded as well and it too is a younger industry that relies to a substantial extent on lower-valued and previously-unused raw material.

In sum, for this book's perspective of industrial relationships with the forest, the most interesting characteristic of the three newer manufactured wood product industries is their increasing shares of the general manufactured wood products market and their substitution of wood residues from other manufacturing processes for uncut logs. The net effect is a decrease in the demand for uncut logs by the aggregated manufactured wood products industries and, therefore, a conserving effect on the natural forest.

Furniture and Related Product Manufacturing (NAICS 337)

The furniture industry cuts, bends, molds, laminates, and assembles wood, rattan, metal, glass, and plastic materials in the manufacture of furniture and articles such as mattresses, window blinds, cabinets, and fixtures. Design and fashion have an important role in the production of furniture and, therefore, specialized labor also has an important role in this industry.[12]

The furniture industry is composed of three four-digit industries (3371—household and institutional furniture and kitchen cabinets, 3372—office furniture, and 3379—other furniture and related products), which are further divided into thirteen six-digit industries. The full three-digit industry is an important consumer of wood but wood is an important input for only five of the thirteen six-digit industries (those described in Tables 7.3A and 7.3B) and even these five are consumers of lumber, plywood, and other manufactured wood products rather than the primary material.

The three-digit industry is composed of numerous small establishments. Eighty-four percent of establishments in the industry employed fewer than twenty workers in 2002 and many furniture manufacturers are individual entrepreneurs with fewer than three employees. The capitalization of the average establishment was only $5.5 million in 2002. The average firm added $1.9 million of value during its production and shipped products worth $3.4 million. These numbers are roughly comparable to those for an average establishment in the sawmill industry but much smaller than those for the average softwood plywood establishment. The wood kitchen cabinet and countertop industry (NAICS 37110) dominates the full furniture industry in numbers of establishments and it produces approximately 20 percent of the three-digit industry's final output value. The comparable measures for the size of its individual enterprises are even smaller—average capitalization of $0.2 million, average annual value added of $0.9 million, and average annual value shipped of $1.4 million in 2002.

---

12 The economics literature on this industry is particularly sparse. The Business and Institutional Furniture Manufacturer's Association (the North American trade association) is an alternative source of basic data.

TABLE 7.3A Measures of industry and firm Size for select furniture manufacturing industries in the U.S., 2002

Industry[1]	Number of companies	Number of establishments	Employees per establishment			Concentration ratios (% of value of shipments)				Total number of employees[2]	Value of shipments[3]	Value added[3]
			1–19	20–99	≥100	C_4	C_8	C_{20}	C_{50}			
337 furniture and related product mfg.	21,531	22,523	17,645	3,663	1,215	11.0	18.0	28.6	39.6	595.9	75,965	43,052
337110 wood kitchen cabinet and countertop mfg.	9,457	9,557	8,486	895	176	29.2	35.1	43.0	50.0	126.2	14,102	8,497
337122 nonupholstered wood household furniture mfg.	3,975	4,114	3,427	481	206	23.9	34.5	49.9	62.2	113.2	12,727	7,718
337129 wood TV, radio, & sewing machine cabinet mfg	201	203	178	18	7	64.4	75.4	84.6	93.1	3.3	486	295
337211 wood office furniture mfg	547	569	353	161	55	34.1	43.3	55.8	71.6	24.3	2,817	1,720
337212 custom architectural woodwork & millwork mfg	1,543	1,557	1,041	466	50	10.1	14.9	23.0	35.6	33.6	3,846	2,408

1 NAICS identifying number and name
2 in 000 of persons
3 in 000,000 of US$
na: not available for 4-digit industries
Sources: U.S. Census Bureau 2004 and 2005a, e, m, n, o

TABLE 7.3B Operating characteristics with respect to major Input categories for select furniture manufacturing industries in the U.S., 2002

Industry	% Capacity	Capital/ labor[1] K/L	Capital/primary material[1,2]		Labor/primary material[1,2]		Output ratios[3]			
			Capital/virgin wood K/R_1	Capital/all wood K/R_2	L/R_1	L/R_2	K/O	L/O	R_1/O	R_2/O
337 furniture and related product mfg	61	0.66	dz	na	dz	na	0.18	0.28	0	Na
337110 wood kitchen cabinets & countertops	73	0.50	dz	0.99	dz	1.98	0.15	0.31	0	0.002
337122 nonupholstered wood household furniture mfg	65	0.55	dz	1.08	dz	2.00	0.15	0.27	0	0.13
337129 wood TV, radio, & sewing machine cabinet mfg	67	0.68	dz	1.97	dz	2.92	0.15	0.23	0	0.08
337211 wood office furniture mfg	42	0.88	dz	2.52	dz	2.86	0.27	0.31	0	0.11
337212 custom architectural woodwork & millwork mfg	66	0.39	dz	1.66	dz	4.28	0.14	0.37	0	0.09

1 Capital calculated as beginning of year gross assets minus beginning of year inventory. Labor is total labor compensation. Output is value of shipments.
2 Primary material: virgin wood is the sum of expenditures on stumpage, logs, bolts and pulpwood. All wood is the sum of expenditures on virgin wood plus expenditures on chips, slabs, shavings, mill residues, rough cut lumber, etc.
3 Capital, labor, and raw material measured as in previous columns. Output is total value of shipments. These ratios are not percentage shares and they may add to more or less than one, as capital is the total value of durable capital (not an annual value) and primary material includes annual expenditures only on the wood input and not that on other materials or energy.
dz: denominator is zero
na: not available

Sources: U.S. Census Bureau 2004 and 2005a, e, m, n, o

The ratios of labor to capital, raw material, and output show that the role of labor is two to four times larger for the furniture industry than for either the sawmill or plywood industry. The average annual total compensation of $45,000 for all employees in the furniture industry is comparable to that for plywood and slightly more than that for the sawmill industry. The average wage for production workers alone was $25,000, or about 17 percent less than in the sawmill industry.

The ratios involving R_l in Table 7.3B confirm that the industry is not a consumer of uncut logs. Furthermore, its consumption of woody material from all sources is very small relative to that for either the sawmill or plywood industries. Wood-to-output ratios of 0.1 are common for those five furniture industries that are important consumers of wood. Of course, these ratios are even smaller for the remaining eight six-digit furniture industries whose production does not require significant amounts of wood or wood products. These ratios anticipate the observation that furniture manufacturers almost never include vertically integrated logging components, or even a regular association with either loggers or forest landowners.

In sum, the furniture industry is composed of numerous small establishments, many of which specialize in products of a particular style or design and others which specialize in one or a few products such as tables or chairs or kitchen cabinets. The industry seems to be distinguished by economies of agglomeration, or clustering of many small manufacturers in a location that is central for all of them, rather than by significant economies of scale or by proximity to its primary material (Hagenstein 1963; Scott 2006). Industry concentration in towns like High Point, North Carolina in the United States, Manzano in Italy, DaNang in Vietnam, or the central Java district of Japara in Indonesia are prime examples. Clustering in these locations permits specialized furniture manufacturers to maintain individual operations while participating in the marketing opportunities available to manufacturers of a diversity of complementary products that can be shown in a common geographic center.

Pulp and Paper Manufacturing (NAICS 322)

The paper industries use wood and other fibers to manufacture pulp, paper, and converted paper products. Their production involves either mechanical or chemical processes for separating cellulose fibers from impurities in wood, used paper, and other fibrous materials such as straw, bagasse, bamboo, and rags, and then matting these fibers into sheets of various papers and paperboard. The pulp and paper industries are growing world-wide and, as the income elasticity of demand for paper is generally considered to be in the neighborhood of 1.0, we can expect that they will continue growing along with global economic development. The rapidly growing Asian economies provide an example: The Asian demand for paper has exploded since 1985, and half of all new paper machines are now sold to China.

We will focus on the pulp (NAICS 322210) and paper and newsprint industries (NAICS 322121 and 322122). These three industries are consumers of the primary wood resource. Additional establishments in the converted paper products manufacturing industries (NAICS 3222) purchase paper for further processing into containers, bags, stationary, tissue, and laminated paper products (sixteen six-digit industries).[13]

Pulp (NAICS 322110): Many establishments in the three-digit pulp and paper manufacturing industry are vertically integrated, manufacturing pulp for further use in their own papermaking operations. A smaller number, 32 establishments in the United States in 2002, manufacturing about 15 percent of total U.S. production, produce pulp solely for market sale to other industrial manufacturers of paper. These 32 constitute the six-digit pulp manufacturing industry.

The structure of the pulp industry is in distinct contrast with the manufactured wood products and furniture industries. Pulp mills are described in terms of hundreds of thousands of tons of annual capacity and several U.S., Canadian, and Northern European mills exceed a million tons in capacity. These are, obviously, very capital intensive operations. The initial capital investment for a new market pulp mill was about $525 million and the average capitalization of existing U.S. mills was almost $190 million in 2002. The latter is 750 times the capitalization of the average sawmill and almost twelve times the capitalization of the average softwood plywood mill. The pulp industry is twice as capital intensive as any other U.S. manufacturing industry (Butner and Stapely 1997; Phillips 1997; Yin, Harris, and Izlar 2000).

Because of the industry's high fixed costs of operation, and also the relatively inelastic papermill demand for its product, the pulp industry tends to operate steadily at high capacity regardless of the cyclical condition of the general economy. It is less costly for pulp mills to operate for short periods when revenues might not cover the mill's relatively small variable costs than to close operations during an economic downturn and, later, absorb the greater costs of restarting again once market conditions improve.

Tables 7.4A and 7.4B support these contentions. The average establishment in the pulp industry added $46 million of value during its production and shipped products worth $110 million in 2002. The ratios of capital to labor, raw materials, and output affirm the industry's capital intensity. The capital-labor and capital-output ratios ($K/L = 9.64$ and $K/O = 1.88$) are approximately four times those in the sawmill and softwood plywood industries.

The industry employed only 7,700 workers or about 200 workers per mill in 2002. However, these were skilled workers, capable of operating the industry's

---

13 Although these industries attract increasing attention from forestry researchers in recent years, the economic literature on the industry remains relatively undeveloped. Gillis and Buongiorno (1987) may be the most thorough statement. The Center for Paper Business and Industry Studies at Georgia Institute of Technology and the American Forest and Paper Association (the trade organization for the industry) are good sources of information. The Food and Agriculture Organization of the United Nations publishes regular surveys of international trade and of industrial capacity for pulp and paper. IIED's (1996) Towards a Sustainable Paper Cycle is a fine global survey of the industry.

TABLE 7.4A Measures of industry and firm size for select paper manufacturing industries in the U.S., 2002

Industry[1]	Number of companies	Number of establish-ments	Employees per establishment			Concentration ratios (% of value of shipments)				Total number of employees[2]	Value of ship-ments[3]	*Value added*[3]
			1–19	20–99	≥100	C_4	C_8	C_{20}	C_{50}			
322110 pulp mills	21	32	1	9	22	61.1	87.7	na	100	7.7	3,531	1,678
322121 paper mills (except newsprint)	193	307	55	78	174	53.1	69.5	84.5	95.6	96.5	42,502	24,437
322122 newsprint mills	18	22	0	1	21	59.9	78.2	100	100	8.4	3,597	1,650
322130 paper-board mills	87	199	10	75	114	48.5	67.6	87.5	98.8	46.2	20,854	10,786
3222 converted paper product mfg	3,345	4,959	na	na	na	17.6	27.8	43.8	58.4	332.8	83,735	37,668

1 NAICS identifying number and product
2 in 000 of persons
3 in 000,000 of US$
na: not available

Sources: U.S. Census Bureau, 2005d, f, g, h

TABLE 7.4B Operating characteristics with respect to major input categories for select paper manufacturing industries in the U.S., 2002

industry	% Capacity	Capital/ labor[1] K/L	Capital/primary material[1,2]			Labor/ primary material[1] L/R_1	Output ratios[3]		
			Capital/virgin wood K/R_1	Capital/all wood K/R_2	Capital/all fiber K/R_3		K/O	L/O	R_1/O
322110 pulp mills	87	9.64	17.05	7.57	6.67	1.77	1.88	0.20	0.11
322121 paper mills (except newsprint)	90	7.06	28.55	18.81	10.38	4.04	3.17	0.50	0.11
322122 newsprint mills	88	8.16	36.78	15.57	9.23	4.51	2.17	0.27	0.06
322130 paperboard mills	93	8.56	21.62	11.32	6.88	2.53	2.36	0.28	0.11
3222 converted paper products mfg	73	na	na	na	na	na	na	na	na

na: not available
1 Capital calculated as beginning of year gross assets minus beginning of year inventory. Labor is total labor compensation. Output is value of shipments.
2 Primary material: virgin wood is the sum of expenditures on stumpage, logs, bolts and pulpwood. All wood is the sum of expenditures on virgin wood plus expenditures on chips, slabs, shavings, mill residues, rough cut lumber, etc. Does not include either market pulp or pulp produced in affiliated operations.
3 Capital, labor, and raw material measured as in previous columns. Output is total value of shipments. These ratios are not percentage shares and they may add to more or less than one, as capital is the total value of durable capital (not an annual value) and primary material includes annual expenditures only on the wood input and not that on other materials or energy.

Sources: U.S. Census Bureau 2004, 2005d,f,g,h

expensive computer-driven manufactured capital. The average mill employee was compensated $81,000 in 2002, three-fourths of which was salary and the other one-fourth fringe benefits. Total compensation per employee was 80–100 percent greater than in the sawmill and softwood plywood industries and, not surprisingly in view of the industry's steady employment, fringe benefits were 5–10 percent larger as a share of total compensation than for workers in the manufactured wood products industries. Wage compensation alone for production workers in the pulp industry was $57,000, about 85 percent greater than in the sawmill industry.

Of course, our greater interest in this book is in the industry's consumption of raw material. This industry's capital-all fiber (K/R_3), capital-wood fiber (K/R_2), and capital-virgin wood (K/R_1) ratios were 6.67, 7.57, and 17.05, respectively, in 2002. (The latter is fifteen times the comparable measure for the sawmill industry.) These ratios emphasize further the relative importance of manufactured capital and the comparison of these ratios reflects the industry's ability to substitute wood chips, wood residues from other manufacturing processes, and non-wood fibers for virgin wood. Indeed, woody material from all sources comprises only 44 percent of the total fiber consumption of American pulpmills, and virgin wood comprises only a little more than one-third of the total fiber consumption.

On the other hand, with high fixed costs imposing the practice of steady operation at close to full capacity and with the associated requirement for a stable labor force, the relatively much less costly fiber inputs remain the most important variable input to pulp production. Therefore, as the returns to this industry are largely returns to capital, individual mills are often willing to pay a premium for their fiber input in order to insure continuing operation of their manufactured capital and a steady return on their investment. This suggests that, where the industry operates in competitive markets for wood and other fiber, some individual firms may invest in forest land and management beyond the level indicated strictly by the market stumpage return to forestry. Others may offer preferential financial treatment to insure that a number of independent loggers or landowners remain reliable suppliers. Still others, in periods of tight markets for pulpwood, offer stumpage or pulpwood payments in excess of normal market prices and in excess of the mill's apparent marginal valuation for wood—simply to ensure a flow of wood.[14] In sum, raw material costs may be small compared to the other operating costs of pulpmills, but the raw material input is the crucial variable input for a mill's ability to operate without interruption, and pulp

---

14 This explains the superficially counter-intuitive observation that some southern U.S. mills pay more for more distant stumpage. They may have established long-term market relationships with near-by suppliers of pulpwood, and they may have a degree of market power as well with respect to these producers, market power that allows them to negotiate lower prices for local stumpage. In times of tighter markets and higher prices, these mills must extend their demand into more distant markets where the presence of additional competitors forces them to pay a higher price for the more distant raw material.

mills are willing to pay a premium where necessary to insure a minimum flow of wood fiber.[15]

These points about relative operating expenses and the industry's ability to substitute among sources of fiber are reinforced by the observation that, over longer periods of time, some mills remain operative despite diminishing local supplies of inexpensive wood fiber. The great cost of their fixed capital makes them immobile, but their much smaller variable costs allow them to increase payments and other incentives to local landowners to induce those landowners to grow wood fiber as the local supply of naturally grown fiber is depleted. Chapter 3 mentioned the example of the Roanoke Rapids pulp and paper operation in Virginia, built in 1919 and still operating (after numerous investments to upgrade its facility) despite a complete change in fiber sources. This mill once relied on a local supply of virgin wood fiber, most of it naturally grown. It now consumes a mix of wood from the managed forests of local small landowners, sawmill residues, and recycled material. We can conclude that an existing supply of naturally grown wood fiber is beneficial, but it is seldom a constraint on the long-term continuous operation of large, financially viable, pulpmills.

Paper (NAICS 322121) and newsprint (NAICS 322122): The paper industry includes integrated firms that manufacture pulp for use in their own manufacture of paper as well as non-integrated firms that rely entirely on the purchase of market pulp. Globally, the industry includes a vast range of operations—from the occasional very small operation using manual production to operations capitalized at hundreds of millions of dollars.

Smaller operations, such as the novelty hand-made paper operation in Chang Mai, Thailand, or the 1,100 Chinese mills of less than 30,000 tons of annual capacity, are integrated pulp and paper operations. Canadian and U.S. mills, which range up to three orders of magnitude larger, include both integrated and non-integrated operations. Manufacturers of specialty papers are often smaller and non-integrated. Larger operations, especially in the newsprint industry, integrate their pulp and paper operations in order to save the water and energy costs involved in drying pulp for shipment. They transfer pulp in slush form directly to an adjacent papermill. Approximately, one-sixth of U.S. papermills and two-thirds of U.S. newsprint manufacturers are vertically integrated.

The paper industry is a major consumer of wood fiber, and this is widely perceived as an environmental detriment. However, as with the pulp industry, its consumption of virgin wood is limited. Other environmental problems associated

---

15 As further evidence that the return on their large investments in manufactured capital is more important than timber supply for large pulp (and paper) operations, we observe that some firms, on occasion, have sold large segments of their timberland when they require short-term cash for other investments or when they find it important to show stockholders an apparent improvement in short-term returns on corporate equity. The sale of timberland exposes them to greater long-term uncertainty in the market for wood but it satisfies other objectives that are more closely tied to an immediate return on manufactured capital.

with its consumption of water and its discharge of suspended solids and chemical effluents have had more serious environmental consequences.[16]

In the United States, virgin wood provides less than 35 percent of the fiber for paper and less than 25 percent of the fiber for newsprint. Furthermore, by far the largest share of these percentages originates from the renewable fiber of managed woodlands, rather than from natural forests. The wood fiber share is no greater in developing countries. In China, for example, agricultural residues constitute three-fourths of the fiber input (Xu, Amacher, and Hyde 2003; Xu and Hyde 2007).

A range of environmental controls introduced since the 1970s has induced numerous technical changes resulting in much cleaner operations in Northern Europe and North America. However, the new technologies were not designed for the smaller operations and agricultural fibers typical of paper manufacture in many developing country operations. Therefore, the cleaner technology does not transfer easily and, in China for example, paper manufacture remains the source of 10 percent of all industrial wastewater emissions and one-fourth of all chemical oxygen demand. It is the largest source of rural environmental pollution.

Tables 7.4 describe the U.S. paper and newsprint industries in 2002. These industries are composed of a small number of large—and growing—firms. Mergers have increased the concentration ratios for the four largest firms from approximately 30 percent of shipments in 1983 to over 50 percent in 2002. In fact, increasing concentration has drawn the attention of the U.S. Department of Justice in the form of five major cases of antitrust litigation since the 1980s. During this period the number of individual non-integrated papermill establishments has decreased by one-third while total industry capacity has doubled. The average papermill and average newsprint mill were capitalized at over $160 million and over $270 million, respectively, by 2002. The initial investment required for a new pulp and paper facility in the United States is at least $500 million (Yin et al. 2000).

In 2002, the newsprint industry added $1.7 billion in value and shipped products worth $3.6 billion ($75 million and $123 million per establishment). These industry totals are intermediate to those for the sawmill and softwood plywood industries—but from far fewer establishments. The larger paper industry (except newsprint) added $42.5 billion in value and shipped products worth $24.4 billion ($80 million and $138 million per establishment).

Factor proportions in the paper and newsprint industries follow those observed for the market pulp industry, and these factor proportions are fairly constant regardless of mill production levels. (Constant returns to scale characterize these two industries for mills in excess of 80,000 tons of annual capacity [Yin 1998, 1999].) Even non-integrated firms in the paper and newsprint industries are highly

---

16 U.S. Environmental Protection Agency (2002) provides a thorough review of environmental issues for the industry, beginning with a discussion of papermaking production processes, historical trends of industry structure and technology, and continuing with discussion of environmental discharges and compliance.

capital intensive. Their fixed costs of capital are such that they operate at close to 90 percent capacity year round and regardless of the business cycle.

The industry is a large employer of skilled workers. The paper industry alone employed 96.5 thousand workers (approximately 300 per mill) in 2002, about the same number as the sawmill industry. These workers must be skilled to operate the industry's sophisticated manufactured capital and their employment is as steady as the regular operation of the mills. Average compensation for all employees of $73,000 in the paper industry and $85,000 in the newsprint industry brackets the $80,000 observed in the pulp industry. Fringe benefits are one-fourth of total compensation, as in pulp, considerably greater than the 18 percent shares in the sawmill and furniture industries. Wages for production workers averaged $52,000 and $60,000 in paper and newsprint, respectively, once more bracketing the average wage for production workers in the pulp industry.[17]

Once more, raw material is the least costly but most variable and, therefore, more manageable of primary inputs. The capital-resource and labor-resource ratios in paper and newsprint are even greater than those in pulp, largely because the fiber inputs are similar while the manufacturing process is more complete in paper and newsprint. The more complete process requires additional labor and manufactured capital, but no additional raw material. Very clearly, wood fiber is a crucial variable input for these operations but, just as clearly, they have a large degree of flexibility in both their sources of wood fiber and in their substitution of various other fibers for wood.

Independent Logging Operations (NAICS 113310)

For NAICS (and ISIC) reporting purposes, the logging industry is composed of independent firms that cut and transport timber. It excludes firms primarily engaged in transporting timber and it excludes the logging component of vertically integrated operations that combine logging and mill operations in the same firm, as well as the wood dealers that occupy a niche between loggers and mill operations in a few markets.

Logging begins with the harvest of the tree (felling), continues with the removal of unusable branches and other material (bucking), transporting the log to a landing (skidding), and loading it for transport to the mill. In the most fundamental logging systems, these are separate activities that use separate pieces of rudimentary equipment: chainsaws for felling and bucking and draft animals for

---

17 Indeed, the high costs of labor have been and continue to be an inducement for innovation in this industry. This industry is the largest of all forest industry participants in research and development, measured either by its absolute expenditure for R&D or by expenditure as a share of total revenues. A significant share of industry R&D is for new product innovation. A second significant share is for labor-saving innovation. As a result, the capital input in the pulp and paper industry has grown rapidly since the 1960s in the United States and Canada and, accordingly, productivity per unit of input has grown most rapidly for labor and material inputs (Oum and Tretheway 1992; Yin 1999). These trends will probably continue and some speculate on the operation of one-person paper machines by 2030.

skidding. More technologically advanced operations use one or more pieces of mechanized equipment ranging upward in total cost from $100,000. Some field operations also convert the bole of the tree and its branches into chips for pulping.

The logging operation accounts for the entire cost of delivering logs from the frontier to the mill for some regions in the first stage of forest development. In the U.S. South, a mature forest frontier with a competitive timber market, logging accounts for approximately 45 percent of the average delivered cost (Yin 1998). The remainder is the cost of stumpage itself.

Logging is an integrated operation for those rudimentary sawmills that operate at the frontier or in the forest itself. Some other, more technologically advanced, sawmills and pulp and paper mills maintain their own logging crews. However, many also outsource their logging operations, or a share of them, to independent loggers.[18] They outsource largely as a means for shifting the cost of the employee benefit package. Logging is a dangerous activity. For example, the risk of mortality for loggers is second only to that of security-related occupations (guards, watchmen, and doorkeepers) (Thaler and Rosen 1975). As a function of this risk, Occupational Safety and Health regulations and health insurance impose costs as great as $0.37 on every $1.00 of logging wages (LeBel 1996).[19] Smaller independent logging operations are not subject to the same level of regulatory scrutiny as the larger mills and the willingness of smaller operations to substitute other cost savings (e.g., in retirement programs) to offset some of the greater health insurance costs explains their advantage in this industry.

Nevertheless, independent loggers and mills often develop regular relationships. The assurance of a regular flow of raw material inputs is crucial for the steady operation of larger mills. A regular relationship with some number of loggers guarantees this flow. Independent loggers, who often have heavily mortgaged equipment, need the assurance of a steady income flow throughout seasonal swings of weather and market conditions. As a result, both larger mills and independent loggers benefit from regular logging operations and we observe a wide variety of long-term arrangements between loggers and mills. Some are formalized contractually. Some mills guarantee a level of regular demand. Others own the logging equipment and contract it to independent loggers. Still others assist independent loggers in the purchase of their own equipment.

In a survey of 2,217 full-time independent loggers in the U.S. South, de Hoop et al. (2002) observe that 53 percent have preferred supplier relationships with one mill—although most deliver to other mills as well. Some purchase timber directly from a landowner, others obtain their timber through a dealer, and still others contract with a mill that supplies them with either purchased or fee simple timber. Half the loggers in the de Hoop et al. sample owned their logging equipment.

---

18 In the United States, mills employ less than one-fourth of all fellers, graders, scalers, and logging equipment operators. Independent loggers employ the remainder (http://bls.gov/oes/current/oes454022.htm#, accessed June 30, 2008).

19 The comparable rate for carpenters, for example, is only $0.12 per $1.00 of wages.

The others contracted for some or all of their equipment. Meanwhile, the opportunity to take advantage of good market conditions is just as important to the mills as a secure resource supply. Therefore, de Hoop et al. also observe that preferred contractors provided only 20 percent of the mills' logyard inventories. In sum, both loggers and mills rely on preferred supplier relationships for some share of their log exchange, but both also rely on market conditions for a larger share.

Tables 7.5 provide an image of the structure of the U.S. logging industry. These tables are based on 1997 data, unlike the 2002 data of the previous tables for other forest product manufacturing industries. 1997 was the most recent year of U.S. Bureau of Census reporting—before the U.S. Department of Agriculture assumed responsibility for logging industry data and those data became less completely available.

More than 13,000 logging firms employed over 83,000 workers in 1997, for an average of 6.1 total employees per establishment.[20] 5.4 of these were production workers, hardly one logging crew per establishment. The remaining, non-production, employee was often a family member with bookkeeping and other office responsibilities. Clearly, these are small operations. The minimum capitalization of a full-time operator approached $200,000 in 2002 (de Hoop et al. 2002; LeBel and Stuart 1998; Stuart et al. 2007; Stuart, Grace, and Grala, 2010). LeBel and Stuart observe a wide range of profitable operations from 25 to 250,000 tons per year with constant returns to scale (and perhaps even increasing returns) in the upper range. The average firm added value of $456,000 and shipped product worth barely $1 million in 1997. By any measure, the average independent logging establishment is smaller than its counterparts in other forest products industries.

Estimates of capacity utilization in the logging industry are highly variable but they average in the neighborhood of 70 percent, even for full time independent loggers (de Hoop et al. 2002 for Maine and LeBel and Stuart 1998 for the South). This datum is yet another indication that loggers are highly dependent on local market conditions, as well as the weather, for successful operation. It is also suggestive of the benefit loggers obtain from establishing dependable long-term supplier relationships with mills, as those without long-term relationships may have to experience unprofitable periods of inactivity.

The factor proportions in Table 7.5B reflect the importance of labor to this industry. The capital-labor ratio ($K/L = 0.27$) is much lower and the labor-primary resource and labor-output ratios ($L/R_t = 9.29$ and $L/O = 1.48$) are much greater than for the other wood and paper product industries. For the most rudimentary operations in the U.S. South 70 years ago, access to available timber was

---

[20] The numbers of firms and establishments are almost identical in this industry. This, along with the small number of employees per firm or establishment, indicates that the vast majority of independent loggers operate as one economic unit with only one crew from one home location. The numbers of establishments and employees declined to approximately 11,000 and 68,000, respectively, by 2002.

TABLE 7.5A Measures of industry and firm size for the logging industry in the U.S., 1997

Industry[1]	Number of companies	Number of establish-ments	Employees			Concentration ratios (% of value of shipments)				Total number of employees[2]	Total cost of materials[3]	Value of ship-ments[3]	Value added[3]
			1–19	20–99	≥100	C_4	C_8	C_{20}	C_{50}				
113310 independent logging operators	13,461	13,533	12,899	607	27	na	~10	<20	<30	83	7,427	13,613	6,166

na: not available
1 NAICS identifying number and name
2 in 000 of persons
3 in 000,000 of US$

Source: http://factfinder.census.gov/home/en/datanotes/exp_econ97.html

TABLE 7.5B Operating characteristics with respect to major input categories for independent logging contractors in the U.S., 1997

Industry	% Capacity	Capital/labor[1] K/L	Capital/virgin wood K/R_1	Labor/primary material1 L/R_1	Output ratios[2]		
					K/O	L/O	R_1/O
113310 independent logging operators	70	0.27	2.52	9.29	0.40	1.48	0.15

1 Capital calculated as beginning of year gross assets minus beginning of year inventory. Labor is total labor compensation. Output is value of shipments.
2 Capital, labor, and raw material measured as in previous columns. Output is total value of shipments. These ratios are not percentage shares and they may add to more or less than one, as capital is the total value of durable capital (not an annual value) and primary material includes annual expenditures only on the wood input and not that on other materials or energy.

Sources: De Hoop et al. 2002 and LeBel and Stuart 1998 for capacity, http://factfinder.census.gov/home/en/datanotes/exp_econ97.html

crucial. These operations could rely on lower-wage seasonal labor from agriculture. This is no longer the case. The heavy equipment of the modern industry in the United States, Canada, and the Nordic countries requires skilled labor. Seasonal agricultural laborers are less available today and they generally do not have the specialized skills necessary to operate modern logging equipment. Nevertheless, weather continues to impose some seasonality on the industry and, as a result, employment varies as much as 7.5 percent from quarter to quarter (not shown in the table but computed from the same source). The seasonal variation would be even greater if we added those often inactive establishments that did not report to the Census Bureau in 1997.

Average annual wages in the industry were $31,000 in 1997, $22,000 for production workers. By 2002, the latter had increased to $30,000 for fellers and logging equipment operators. These are roughly comparable to wages for production workers in the sawmill and furniture industries, but much less than those for pulp and paper. Benefits are 22 percent of the total compensation for production workers, a larger share than the 18 percent for the sawmill industry—and an indication of the larger costs of workman's compensation insurance for logging. That the benefit share is smaller than the 25 percent share in pulp and paper is an indication of the savings that mills accrue by out-sourcing their logging.

In sum, the structure of the logging industry varies with the forward wood processing industry it serves. Logging operations associated with rudimentary sawmills at the frontier may be integrated components of those operations. More technologically advanced sawmills and, particularly, pulp and paper mills may keep a few logging crews on their payrolls to help insure a continued log flow during tight markets. However, these mills generally prefer to purchase delivered logs and concentrate on the specialized operation of their own capital equipment. The remaining three-quarters of loggers are independent entrepreneurs who generally operate no more than one crew. Some fell, buck, skid and load, many also do their own trucking, and some who deliver to pulp and paper mills also chip the product in the field before loading and delivering. Crew members are skilled operators of specialized heavy equipment. They are not highly paid, but most live and work in low wage regions of the country and they may be higher paid than many of their neighbors. Capital and labor do not substitute easily in this industry and the required specialized skills are a barrier to employment for some (Smith and Munn 1998). Available skilled labor can be a constraint on modern logging operations in some regions, particularly in periods of strong economic growth when better employment opportunities may exist elsewhere.

Conclusion: The Industry in the Context of the Three Stages of Forest Development

What can we say about a pattern of development for the forest products industry? Is it related to and consistent with the three stages of forest development set out in chapter 2 and, specifically, how does industrial ownership of forest land fit into the pattern of industrial development and the three stages of forest development?

The observations for the U.S. South in chapter 3, and from more recent and more rapid industrial development in Brazil's eastern Amazon, as well as the descriptions of U.S. industrial structure and conduct in this chapter, are instructive.

The initial industrial operations at the stage I forest frontier are rudimentary sawmills. Logging is an integrated activity for these sawmills. Indeed, for these operations, the same employees may be both loggers and sawyers at different times even within the same day, and their logging/millwork employment may be only a supplement to their primary activity as farmers. As loggers, they harvest predominantly from lands immediately adjacent to the sawing operation. The small sawmills that characterize these operations are mobile and, as they deplete the adjacent timber, their employees pick up their equipment and follow the receding frontier. However, as the frontier recedes and the mill is moved, its distance to the local market must increase and the cost for shipping its rough sawn lumber product to the market must also increase.

Eventually, somewhat larger sawmills that employ manufactured capital more intensively become competitive by locating closer to their labor supply and their lumber market but a distance from the mature forest frontier. The operation of these somewhat more technologically advanced sawmills coincides with the beginning of the second stage of forest development and the expanding open access region of depleted forest that appears between the local market and its adjacent lands with more secure property rights and the more distant natural forest. These mills offset the higher cost of their delivered log inputs by employing more specialized technologies and obtaining greater productivity from their labor and their manufactured capital. Some of these mills may be vertically integrated to include their own logging crews.

As timber harvesting continues and the forest frontier recedes farther, delivered log prices must continue to rise and lower-valued substitute log species eventually become competitive. Rudimentary mobile sawmills continue to operate where they find small pockets of high quality timber near the frontier or within the region of open access. The more technologically advanced sawmills closer to the market center continue operating as well but they may increase their capitalization as they continue to search for ways to utilize smaller logs and lower quality species. Plywood mills—producing a substitute for some uses of lumber—eventually become profitable. They require greater capitalization yet. Their millworkers are more skilled and they demand somewhat higher wages than those paid to sawmill workers.

Some sawmills and plywood operations in some markets still may not experience much competition for their input of delivered logs. For these, the accessible natural forest promises a reliable supply for the mill. The rate of harvest to supply these mills and, therefore, the rate of mill production, depends on two key factors, the confidence that managers have in their timber supply situation and the expected lifetime of the mill. If the accessible natural forest is plentiful, yet the manager has concerns that policy or other exogenous factors may alter the mill's control of its timber resource, then managers harvest more rapidly—and closer to the short-term economic frontier of the forest where the marginal log harvest and delivery costs just equal their immediate delivered value at the mill. On the other hand, where the general market and policy environment is more stable, the mill managers are not as myopic. Rather, in this case, managers tend to harvest at a rate that insures a continued flow of delivered logs for a period comparable to the expected lifetimes of their mills.

Therefore, we observe mills in some regions and countries with a plentiful inventory of standing timber that produce as rapidly as possible in order to insure recovery of the initial capital invested in the mill before any potential change in the local policy environment alters the availability of delivered logs. This is often the case for new industrial ventures in countries with less predictable long-term forest, industrial, or macroeconomic policy. In contrast, we also observe mills in other regions of plentiful timber, and with or without formal timber rights that operate at a slower rate designed to increase the longer-term return on their manufactured capital—as is the case for the capital intensive pulp and paper industry in the more certain policy environment of eastern Canada, for example.[21]

We can make the same two observations over time and even within some countries as the market and policy environment changes. For example, a plywood mill manager in Indonesia in the mid-1990s explained the decision to limit the harvest rate on lands controlled by his firm as a decision to protect the long-run availability of timber for his modern mill. However, times changed. The 1997 financial crisis generated financial uncertainty and the collapse of the Suharto family's political control generated general political uncertainty and numerous changes in the claims to Indonesia's forests. The same mill manager responded to the new, less predictable, environment by increasing his harvest rate and mill production. This meant that his loggers were harvesting closer to the short-run forest financial forest frontier in the late 1990s and in the earliest years of the twenty first century. He was attempting to capture what revenue he could before further change occurred and the operation of his mill became less certain. That is, he made a decision to forego some uncertain long-term return in exchange for obtaining a more certain financial return now.

The industries' structure with respect to control of forest land is altogether different where the market for delivered logs is competitive. The more technologically advanced sawmills and plywood mills may need to find ways to protect their capital investment. Managers may begin looking for ways to establish some control over their inflow of logs as a means of insuring that the mills can operate regularly and obtain a regular return on their costly manufactured capital. They may purchase some timberland outright or they may establish relationships with loggers and landowners on whom they can depend to deliver a limited volume of logs when the local markets are tight. This signals the beginning of the third stage of forest development when managed forests and forest plantations become economically feasible.

---

21 In economic terms, mills in this second situation perceive a user cost to timber harvests that are too rapid. This user cost is related to the expected long term financial loss should the operation have to close while the mill itself is still technologically and economically viable.

As harvesting continues, the natural forest frontier recedes still farther. The industry continues to compensate by substituting ever lower-valued species, smaller logs and residual wood fiber from other industrial processes. It also continues introducing more advanced logging and processing technologies which recover more product per unit of log input. These advanced technologies effectively substitute manufactured capital for wood fiber.

At some point, the pulp and paper industry enters the region. In some ways, raw material is less important than the other inputs for this industry. Virgin wood may be less than one-third of the industry's raw material. The ratio of manufactured capital to virgin logs in a modern pulp and paper facility is more than twenty times that in the average U.S. sawmill. The ratio of manufactured capital to final output in pulp and paper is more than eight times that of an average sawmill, while the ratio of virgin logs to output is only one-fourth that of the average sawmill. Of course, these comparisons are even more extreme for the differences between pulp and paper mills and the small portable sawmills that still exist on the few remaining pockets of accessible mature natural timber in the United States.

For the most capital intensive industrial operations, the pulp and paper facilities, raw material is a small share of total operating costs, but it may be the most important variable cost. These operations are often willing to pay a premium to guarantee some minimum flow of raw material during times when the stumpage market is tightest. This explains why we generally observe that pulp and paper operations are more active than sawmill and plywood operations in negotiating preferred supplier arrangements with loggers and with owners of timberland. It is the reason that pulp and paper operations in competitive log markets have been more likely than sawmill and plywood operations to own some timberland themselves and it is the reason that pulp and paper operations may invest more per acre on forest management than either their neighboring non-industrial land owners seem to invest or than the simplest timber market analysis would suggest is profitable. It is also why these industrial operations may delay harvesting their own lands, consuming available market timber and saving their own for those periods of tighter markets. Finally, it is one reason why pulp and paper operations have spent more on forest management research than any other segment of the wood products industry.

The importance of these small raw material variable costs to the capital intensive pulp and paper industry is reinforced by what has been called the asset specificity of the industry (Yin et al. 2000). The industry's assets are overwhelmingly concentrated in pulp and paper production. Firms in the industry do not tend to be horizontally diversified into other productive activities that could help insure cash flow in times of weak paper markets. This has reinforced the motivation for many pulp and paper firms to control a share of their timber flow and, together with their high fixed costs, it explains their oligopsonistic market behavior (Murray 1995; Brannlund, Johansson, and Lofgren 1985; Kallio 2001; Bernstein 1992). The U.S. pulp and paper industry was approximately 30–40 percent self-sufficient in timber in the late 1990s, although the self-sufficiency of individual companies ranged from 17 to 60 percent (Yin 1998). The product market inflexibility associated with asset specificity causes some mills to increase production in periods of weak markets. They increase production in order to maintain revenues, corporate share prices and dividends, and also to avoid the high fixed start-up costs following a period of mill closure, but the additional production during these weaker economic times exacerbates the cyclic downturn in product and stumpage markets.

Nevertheless, firms in the pulp and paper industry are clearly not uniform in their behavior and not all of them operate in regions of competitive pulpwood markets. Those that operate in regions with less competitive pulpwood markets exhibit fewer of these characteristic behaviors with respect to their raw material supply. The contrast between the pulp and paper industry in Georgia in the United States and Quebec in Canada is illustrative. The industry in both regions is modern and technologically advanced, but the market for pulpwood is much more competitive in Georgia. Therefore, it is not surprising that we observe a much broader array of contractual arrangements to insure the flow of raw material to the mill in Georgia, and it is not surprising that industrial investments in forest management and forestry research have been somewhat greater in Georgia than in Quebec.

It is also not surprising that the structures of the forest product industries become increasingly complex as their product markets continue developing and the local markets for their wood resource become ever-more competitive. The greatest specialization in each of the wood product industries occurs in the most developed regions. The same highly developed regions experience the greatest variety in the logging operations that support these diversified industries and also the greatest variety in the contractual arrangements between wood processing companies and forest landowners.[22]

In sum, manufactured capital, not timber, is the fundamental investment for most large firms in the modern forest products industry, and the return on this manufactured capital is a guiding criterion for their planning and continued operation. Therefore, those firms that do own their own timberlands rely on these lands for only a small share of their total flow of wood fiber. Public and non-industrial landowners supply 80 percent of the full wood products industry's annual consumption of industrial roundwood even in the U.S. South, and the remaining smaller 20 percent share from industry-owned land is declining rapidly with recent growth in the new class of institutional forest managers.

By the time the industry has matured to include a broad range of technologically advanced industrial wood and paper product operations, the mature forest frontier no longer has an easily observed physical boundary. Some pockets of less accessible, lower quality mature timber will remain standing even within the smaller open access region that tends to occur in developed economies. In contrast, individual pockets of high-valued and accessible timber deep within the mature forest will already have been removed. In some places illegal loggers will

---

22 We have emphasized these distinctions in the U.S. South. Nurminen and Heinonen (2007) discuss the many distinctions in Finland's very developed and modern logging industry.

have managed to remove timber all the way to the short-term financial margin where its harvest and delivery costs equal its delivered value. Meanwhile, financially motivated managers of the timberlands divisions of large integrated wood processing operations will have designed roads into their own forest properties that enable them to monitor access, effectively control entry, and delay harvests of financially mature timber until some future date when the wood markets are tighter or the demands for corporate cash flow are greater and, therefore, harvests from these lands become more important to the full corporate operation.

Literature Cited

Abt, R., and S. Ahn. 2003. Timber demand: Aggregation and substitution. In E. Sills and K. Abt, eds., *Forests in a market economy*. Dordrecht, Netherlands: Kluwer Academic,,pp. 133–152.

Baardsen, S. 2000. An econometric analysis of Norwegian sawmilling, 1974–1991, based on mill-level data. *Forest Science* 46: 537–574.

Bernstein, J. 1992. Price margins and capital adjustment: Canadian mill products and pulp and paper industries. *International Journal of Industrial Organization* 10(3): 491–510.

Brannlund, R., P.-O. Johansson, and K.-G. Lofgren. 1985. An econometric analysis of aggregate sawtimber and pulpwood supply in Sweden. Forest *Science* 31: 395–406.

Buongiorno, J., S. Zhu, D. Zhang, J. Turner, and D. Tomberlin. 2003. *The global forest products model*. Waltham, MA: Academic Press.

Butner, R., and C. Stapley. 1997. Capital effectiveness of the paper industry. *Tappi Journal* 80(10): 155–165.

de Hoop, C., A. Egan, W. Greene, and J. Mayo. 2002. *Surveys of the logging contractor population—8 southern states and Maine*. Working paper 55. Baton Rouge: Louisiana Forest Products Development Center.

Gilless, J., and J. Buongiorno. 1987. PAPYRUS: A model of the North American pulp and paper industry. *Forest Science Monograph* 28.

Hagenstein, P. 1963. *The location decision for primary wood-using industries in the northern Appalachians*. Unpublished doctoral dissertation, University of Michigan, Ann Arbor.

Hyberg, B., and D. Holthausen. 1989. The behavior of non-industrial landowners. *Canadian Journal of Forest Research* 19(8): 1014–1023.

Kallio, M. 2001. Analyzing the Finnish pulpwood market under alternative hypothesis of competition. *Canadian Journal of Forestry Research* 31: 236–245.

International Institute for Environment and Development. 1996. *Towards a sustainable paper cycle*. London: IIED.

Latta, G., and D. Adams. 2000. An econometric analysis of output supply and input demand in the Canadian softwood lumber industry. *Canadian Journal of Forest Research* 30: 1419–1428.

LeBel, L., and W. Stuart. 1998. Technical efficiency evaluation of logging contractors using a non-parametric model. *Journal of Forest Engineering* 9(2): 15–24.

Lewandrowski, J. 1990. *A regional model of the US softwood plywood industry*. Unpublished doctoral dissertation. North Carolina State University, Raleigh.

Munn, I., and R. Rucker. 1994. The value of information services for factors of production with multiple attributes: The role of consultants in private timber sales. *Forest Science* 40(3): 474–496.

Mead, W. 1966. *Competition and oligopsony in the Douglas fir lumber region*. Berkeley: University of California Press.

Nurminen, T., and J Heinonen. 2007. Characteristics and time consumption of timber trucking in Finland. *Silva Fennica* 41(3): 471–487.

Nyrud, A., and S. Baardsen. 2003. Production efficiency and productivity growth in Norwegian sawmilling. *Forest Science* 49: 89–97.

Nyrud, A., and E. Bergseng. 2002. Production efficiency and size in Norwegian sawmilling. *Scandinavian Journal of Forest Research* 17: 566–575.

Oum, T., and M. Tretheway. 1992. A comparison of productivity performance of the US and Canadian pulp and paper industries. In P. Nemitz, ed., *Emerging issues in forest policy.* Vancouver, Canada: University of British Columbia Press, pp. 212–236.

Phillips, R. 1997. Impact of capital spending on paper industry profitability. *Tappi Journal* 80(10): 145–152.

Scott, A. 2006. The changing global geography of low-technology, labor-intensive industry: Clothing, footwear, and furniture. *World Development* 34(9): 1517–1536.

Smith, P., and I. Munn. 1998. Regional cost function analysis of the logging industry in the Pacific Northwest and Southeast. *Forest Science* 44(4): 517–525.

Spelter, H. 1984. Price elasticities for softwood plywood and structural particleboard in the United States. *Canadian Journal of Forest Research* 14: 528–535.

Stevens, J. 1978. *The Oregon wood products labor force.* Unpublished manuscript, Department of Agricultural Economics, Oregon State University, Corvallis.

Stier, J., and D. Bengston. 1992. Technical change in the North American forestry sector: a review. *Forest Science* 38(1): 134–159.

Stone, S. 1997. Economic trends in the timber industry of Amazonia: Survey results from Para state, 1990–95. *Journal of Developing Areas* 32: 97–122.

Stone, S. 1998. Evolution of the timber industry along an aging frontier: the case of Paragominas (1990–95). *World Development* 26(3): 433–448.

Stordahl, S., and S. Baardsen. 2002. Estimating price-taking behavior with mill level data: The Norwegian sawlog market, 1970–1991. *Canadian Journal of Forest Research* 32: 401–411.

Stuart, W., L. Grace, C. Altizer, and J. Sith. 2007. *2005 Logging cost indices.* Mississippi State: Mississippi State University, Wood Supply Research Institute.

Stuart, W., L. Grace, and R. Grala. 2010. Returns to scale in the eastern United States logging industry. *Forest Policy and Economics* 12(6): 451–456.

Thaler, R., and S. Rosen. 1975. The value of saving a life: evidence from the labor market. In N Terleckyj, ed., *Household production and consumption.* New York: Columbia University Press for the National Bureau of Economic Research.

U.S. Census Bureau. 2004. Survey of plant capacity: 2002. MQ-C1(02). Washington: U.S. Government Printing Office.

U.S. Census Bureau. 2005a. *Custom architectural woodwork and millwork manufacturing: 2002. EC02-311-337212 (RV).* Washington, DC: U.S. Government Printing Office.

U.S. Census Bureau. 2005b. *Engineered wood members (except truss): 2002. EC02-311-321213 (RV).* Washington, DC: U.S. Government Printing Office.

U.S. Census Bureau. 2005c. *Hardwood veneer and plywood: 2002. EC02-311-321211 (RV).* Washington, DC: U.S. Government Printing Office.

U.S. Census Bureau. 2005d. *Newsprint mills: 2002. EC02-311-322122 (RV).* Washington, DC: U.S. Government Printing Office.

U.S. Census Bureau. 2005e. Nonupholstered wood household furniture manufacturing: 2002. EC02-311-337122 (RV). Washington, DC: U.S. Government Printing Office.

U.S. Census Bureau. 2005f. *Paper (except newsprint) mills: 2002. EC02-311-322121 (RV).* Washington, DC: U.S. Government Printing Office.

U.S. Census Bureau. 2005g. *Paperboard Mills: 2002. EC02-311-322130 (RV).* Washington, DC: U.S. Government Printing Office.

U.S. Census Bureau. 2005h. *Pulp mills: 2002. EC02-311-322110 (RV).* Washington, DC: U.S. Government Printing Office.

U.S. Census Bureau. 2005i. *Reconstituted wood product manufacturing: 2002. EC02-311-321219 (RV).* Washington, DC: U.S. Government Printing Office.

U.S. Census Bureau. 2005j. *Sawmills: 2002. EC02-311-321113 (RV).* Washington, DC: U.S. Government Printing Office.

U.S. Census Bureau. 2005k. *Softwood veneer and plywood: 2002. EC02-311-321212 (RV).* Washington, DC: U.S. Government Printing Office.

U.S. Census Bureau. 2005l. *Truss manufacturing: 2002. EC02-311-321214 (RV).* Washington, DC: U.S. Government Printing Office.

U.S. Census Bureau. 2005m. *Wood kitchen cabinet and countertop manufacturing: 2002. EC02-311-337110 (RV).* Washington, DC: U.S. Government Printing Office.

U.S. Census Bureau. 2005n. *Wood office furniture manufacturing: 2002. EC02-311-337211 (RV).* Washington, DC: U.S. Government Printing Office.

U.S. Census Bureau. 2005o. *Wood television, radio, and sewing machine cabinet manufacturing: 2002. EC02-311-337129 (RV).* Washington, DC: U.S. Government Printing Office.

U.S. Environmental Protection Agency. 2002. *Profile of the pulp and paper industry, 2nd edition.* EPA/310-R-95-015. Washington, DC: U.S. Government Printing Office.

Williamson, T., G. Hauer, and M. Luckert. 2004. A restricted Leontief profit function model of the Canadian lumber and chip industry. *Canadian Journal of Forestry Research* 34(9): 1833–1844.

Xu, J., G. Amacher, and W. Hyde. 2003. China's paper industry: growth and environmental policy during economic reform. *Journal of Economic Development* 28(1): 49–79.

Xu, J., and W. Hyde. 2007. *Shadow pricing pollutants for China's paper industry.* Unpublished working paper. Peking University.

Yin, R. 1998. DEA: a new methodology for evaluating the performance of forest products producers. *Forest Products Journal* 48(1): 29–34.

Yin, R. 1999. Production efficiency and cost competitiveness of pulp producers in the Pacific Rim. *Forest Products Journal* 49(7/8): 43–49.

Yin, R., T. Harris, and B. Izlar. 2000. Why forest products companies may need to hold timberland. *Forest Products Journal* 50(9): 39–44.

Appendix 7A:
A Note on Factor Costs and Industry Location

Why are the markets for primary wood products, e.g., logs and fuelwood, so geographically limited? Even where they have the advantage of free backhaul, these products are seldom shipped as far as 300 km. While the markets for lumber, paper, and other finished forest products may be international, the markets for primary forest products tend to be many and local. But why?

Bulk, defined as the ratio of weight to value, is a crucial factor in determining the financial advantage of transporting any product. The bulk of the primary wood product, cut timber, and also of some secondary wood products (e.g., charcoal, lumber) is substantial. Therefore, bulk, and its effect on relative factor costs, is a determinant of both the location of industrial production facilities and of the geographic limits of the markets for many wood products.

These relative factor costs explain why charcoal producers and rudimentary sawmills operate near or even within the original natural forest and they anticipate how the more technologically advanced sawmills that operate a distance from the forest frontier must restructure in order to successfully compete with those more rudimentary mills operating at the frontier. Similar comparisons of relative factor costs explain the differences in production location and shipment between products such as fuelwood and charcoal or between lumber and plywood or pulp and paper.

Input Proportions

Consider the production or supply functions for processed forest products, then consider the relationships between their factors of production and the costs of delivering their products to the market.

In most general terms, production and supply are functions of:

p: the exogenous delivered price of the processed wood product;
c_x: the cost of transporting this product from the mill to its final market;
c_r: the delivered cost of the woody resource input at the site of primary processing, the charcoal kiln, sawmill, plywood operation or pulp and paper mill; and
c_o: the sum of the other input costs (for labor, manufactured capital, non-virgin wood, non-wood materials) of the processing facility.

The relative input proportions suggest two conditions that indicate production near the forest.

First, when the ratio of primary resource costs to other input costs (c_r/c_o) is large, primary production occurs near the source of that primary resource; that is, near the forest. Compare fuelwood production with furniture production, or sawmills with pulp and paper mills. For fuelwood, the cost of the delivered wood resource is significant, but other processing costs are virtually nonexistent. For furniture, we can refer to the discussion of the U.S. industry in the body of this chapter. The U.S. furniture industry uses very little unprocessed wood. Partially processed wood and other inputs account for virtually all of its production costs. Therefore, $(c_r/c_o)_{fuelwood} > (c_r/c_o)_{furniture}$ and, of these two products, fuelwood production occurs closer to the forest and the primary wood resource. In fact, we observed that economies of agglomeration for other inputs and for marketing opportunity seem to be more important than forest access as determinants of location for firms in the furniture industry.

In a different example, Tables 7.2B and 7.4B show that $(c_r/c_o)_{sawmills} > (c_r/c_o)_{pulp\ \&\ paper}$ for the United States. We can anticipate that this comparison remains true for lumber and paper production anywhere in the world. Therefore, it is not surprising that pulp and paper facilities locate farther from the forest than most sawmills in the U.S. and most anywhere else in the world as well.

Second, when the cost of shipping the product c_x is relatively large compared with other production costs c_o, then shipment over large distances will not be financially rewarding and the product market will be geographically contained. Once more, production will occur near the forest. Compare fuelwood and charcoal. Fuelwood production costs are minimal but the cost of transporting a unit of fuelwood certainly limits its market. In fact, fuelwood markets are often contained within local rural communities with easy access to local woody plants. Charcoal production costs are also small as a kiln handmade at the site may be the only implement. However, the cost of transporting an energy unit of charcoal is much less than that of a similar unit of fuelwood. Therefore, both may be produced near the forest but $(c_x/c_o)_{fuelwood} > (c_x/c_o)_{charcoal}$ and charcoal is often shipped five, ten, or even 100 times farther than fuelwood produced in the same local forest area.

Alternatively, compare charcoal with paper which, for these factor proportions, is the industrial extreme. Shipment costs per unit of product value are clearly greater for charcoal. Therefore, $(c_x/c_o)_{charcoal} > (c_x/c_o)_{paper}$ and charcoal must be produced closer to a mutual source of the primary wood input for both products and its markets are more geographically limited as well.

Of course, for products for which both ratios are relatively large, as they are for fuelwood, then production must occur near the forest and each fuelwood market must be very limited in its geographic extent. In contrast, for products for which both ratios are relatively small, as they are for the manufacture of paper, then mills have much greater flexibility with regard to the geographic source of their primary woody resource and, for them, the woody resource is not as important a determinant of their location. The Roanoke Rapids paper mill in Virginia in the United States, once more, serves as an example. As we observed in chapter 3 and again in the body of this chapter, this mill has adjusted to a declining local forest base and increasing resource costs while continuing in operation since 1919. It seems undeterred by higher wood prices and greater shipping distances for a share of its primary raw material.

Factor-Output Ratios

Factor-output ratios provide a similar means for anticipating manufacturing location relative to the forest—although they are less informative with respect to the geographic extent of any product's market.

When the resource cost per unit of output is large relative to its delivered market price (c_r/p), production occurs near the forest. For example, the delivered resource cost and price are almost identical for fuelwood, the value of this ratio is almost one, and fuelwood is always produced locally in this almost trivial case. The ratio of resource costs to market price is greater for lumber than for plywood and greater yet for paper. Therefore, it is not surprising that many sawmills are located closer to the forest than most plywood operations and that plywood operations tend to be located closer to the forest than most pulp and paper facilities.

The same comparison is predictive of the structural adjustments that managers of more technologically advanced sawmills make in order to be competitive with the rudimentary sawmills that operate at the forest frontier. Delivered resource costs, the numerator in the ratio, must be greater for the more technologically advanced mills as these mills tend to be located farther from the forest. Therefore, the delivery costs for their resource inputs are greater even if the stumpage price is identical. So—how do these technologically advanced mills compete? By finding ways to decrease their resource costs per unit of output—effectively decreasing the numerator in the ratio—or by increasing the market price, the denominator in the ratio. They decrease unit resource costs by employing improved manufactured capital and skilled labor that enable them to increase output per unit of resource input. They obtain higher market prices in some cases by adding value through additional processing and, in others, by producing more specialized, higher value, lumber products.

8
INSTITUTIONAL INVESTORS

Institutional investors are establishments that hold forestland, along with numerous other assets, in large financial portfolios. Unlike industrial landowners, institutional landowners are not vertically integrated with their own wood processing facilities—although they may hold contracts to provide timber to mills under other ownership. Financial institutions comprise the majority of these landowners, and a broad range of financial institutions, including pension funds, insurance companies, banks, endowments, and foundations, invest in forestland. A few of these manage their own forestland investments but most contract with firms that specialize in managing land for financial institutions.

These institutions have become an important class of forest landowners in the United States only since the mid-1980s. In fact, first the emergence and then the growth of this new class of land owners has been the most dramatic development in U.S. forestry over the last thirty years. In 1975, thirty vertically integrated industrial wood products establishments controlled over 22 million hectares of forestland. Since then, most of these large industrial landowners have divested themselves of much of their forestland, and only one owned as many as 400,000 hectares in 2007. They often sold their lands to the emerging institutional landowners, and those institutions or the firms that manage land for them now include ten of the thirteen largest forest landholders in the United States. By 2007, they controlled over twelve million hectares of timberland worth roughly $50 billion (Clutter 2007; Mortimer 2009). Their U.S. landholdings are still expanding but, as they reach the limit of large blocks of land available for purchase, some are beginning to look at international markets for forestland. About 9 percent of the total global institutional investment in timberland was in South American, Australia, and New Zealand by 2006 (Hagler 2006).

Financial conditions in the wood processing industry had to change and a new understanding of the financial characteristics of forestland had to arise before either the forest industry divested or the financial institutions chose to invest in forestland. The next two sections of this chapter review the investment environment that justified these changes. The first discusses the changes in U.S. federal tax law and the perceived weak return on industrial investments that served to diminish the traditional incentives for industrial land ownership. The second describes the financial characteristics of markets for forestland that appeal to institutional investors. A concluding section discusses the position of institutional investors within the three-stage pattern of forest development, the broader impacts of the transfer of timberlands from industrial to institutional landowners and, finally, the prospects for further growth in this new class of landowners. A brief mathematical appendix introduces the capital asset pricing model (CAPM) that has been instrumental in quantifying the previously unrecognized financial characteristics of forestland.

Revenues and Taxes

In summary, the justifications for industrial ownership, or at least limited control, of forestland discussed in chapter 7 remain, particularly for firms in the capital intensive pulp and paper industry. However, two new features of the U.S. tax code and of general investment policy became factors in the market for large holdings of forestland in the mid-1970s. These competed with and weakened the industrial justifications for land ownership and created an emerging interest in forestland on the part of real estate limited partnerships (REITs), pension funds, and insurance companies. Subsequent financial assessments revealed previously unidentified characteristics of the level and distribution of risk associated with forestland, and these reaffirmed the interest of large financial institutions in holding forestland among numerous other assets in their investment portfolios.

First, the Employee Retirement Income Security Act of 1974 (ERISA) permitted pension funds and institutional investors holding pension funds to diversify from their traditional holdings of fixed income assets such as government and corporate bonds. They began to include real estate in their portfolios.

Shortly thereafter, the Tax Reform Act of 1976 removed the corporate capital gains advantage for holding forestland. Capital gains are the appreciation in asset value that comes from the simple act of holding an asset during a period of rising value. They are unrelated to any investment in capital improvement—such as silvicultural investments in forestland. Capital gains were taxed at a lower rate than other corporate income and, as such, they may have been responsible for as much as 20 percent of corporate earnings in the forest industry in the 1960s and early 1970s (Russakoff 1985). The removal of the capital gains tax advantage clearly lessened the integrated industry's justification for holding forestland.

Meanwhile, the perception was growing that the stock market undervalued the timberlands of integrated industrial firms. Many companies in the forest industry had purchased their timberlands in the distant past and held these lands at a book value comparable to the original purchase price. Furthermore, the corporate objective of holding land and its standing timber to insure an uninterrupted flow of future timber to its mills in periods of tight stumpage markets meant that the revenue flows from corporate timberlands were less than the average return on comparable market-valued investments. Both were reasons that the perception of undervalued timberland may have been accurate. In fact, subsequent evidence showed that the return on corporate timberland for the 1970s and 1980s, and even beyond, was less than either the return on other corporate forest industry investments or the more general average return on common stocks (Binkley, Raper, and Washburn 1996; Sun and Zhang 2001; Zinkhan 2007). Furthermore, subsequent evidence also showed that there was no correlation between corporate timberland ownership and the share prices of forest industrial corporate common stocks (Yin, Harris, and Izlar 2000).

The recognition that corporate timberlands were undervalued led to hostile takeovers from outside the forest industry and to capital restructuring within the industry. Hostile takeovers were successful where the cost of purchasing sufficient corporate stock to takeover a company was less than the true corporate value. After the takeover, the new management typically divested itself of some corporate assets, often including the timberland, for a price sufficient to compensate its purchase costs. After this divestiture, the new owner still held possession of the remaining corporate assets but without the debt associated with the costs of the takeover.

Meanwhile, technological change in logging and transportation systems enabled many mills to extend their wood procurement regions, thereby increasing the amount of wood available to them and diminishing (but certainly not eliminating) the old incentive to insure against periods of tight stumpage markets.

Thus, two factors were redefining the best level of timberland ownership for any industrial forestry operation, and timberland ownership became a debatable issue subject to a wide range of opinions among managers of vertically integrated firms. Many managers felt that their corporate needs could be satisfied with lower than historical levels of corporate ownership of timberland. As a result, some companies placed a share of their timberlands in limited partnerships. This strategy allowed them to escape the double taxation of corporate earnings (first on corporate profits and then again on dividend income received by individual corporate shareholders) from these lands.

Others attempted to sell some lands outright. These sales enabled some firms to consolidate corporate debt after a period of weak financial performance and enabled others to raise capital for additional investment in their mills. However, the sellers were cautious in their selection of prospective buyers. Their sales were too large for individual private purchasers. Sales to other firms in the wood and paper products industries would not have been appealing because those firms' purchases might place the selling firm at a competitive disadvantage in times of tight stumpage markets. Pension funds looking to diversify their assets were the answer. These and other institutional landowners generally hire a large forestry firm to manage on their behalf. By the mid-1990s, five large timberland investment management organizations (TIMOs) controlled timberland worth almost $2.8 billion for these emerging institutional landowners (Binkley et al. 1996).

Risk and Return

Financial institutions only became willing participants in the market for timberland as they developed an understanding of that land's specialized contribution to their diversified investment portfolios. Portfolio managers seek investments that promise high returns, but no manager can afford to be caught with a strong portfolio that has a temporarily lower cyclic value at a time when the manager's clients demand substantial payouts. Since predicting the timing of such demands is very difficult, portfolio managers seek low inter-temporal variability, along with a high return, for their full investment portfolios. They can satisfy both objectives by selecting investments with high potential yields but countervailing risk factors—such that when the risk factors of one investment cause a temporary decline in its return, similar risk factors induce an offsetting temporary increase in the return to a second investment. The result is a steady high yield for a portfolio containing both investments.

The average returns to investments in forestland were known, but one could only speculate on the relationship between risk and return in timberland and the correlation between the occurrence of risk in timberland and in other investments. Institutional managers and analysts began to assess this relationship for timberland in the mid-1980s.

For purposes of portfolio analysis, risk is the statistical variation from some benchmark risk-free return to capital. If risk free investments generate a return r_f, then riskier investments must generate ever greater returns to compensate for the greater risks. For example, investments that demonstrate variation in the pattern of their returns equal to σ on average must earn an additional return equal to $r_\sigma - r_f$ (for a total return of $r_f + r_\sigma$) to compensate for their risk. Investments that demonstrate variation greater than σ must earn an additional return even greater than $r_\sigma - r_f$.

A comprehensive market in financial investments includes assets with different risk factors and a wide range of individual riskiness. This implies the market risk-return profile described by Figure 8.1. Investments whose risk-return coordinates fall above and to the left of the curve out-perform the market, while those that fall below the curve and to its right are underachievers.

The return on U.S. government Treasury bills is often used as an empirical estimate of the risk free return because the U.S. government guarantees this return, and most accept that the long-run viability of the U.S. government is more certain than that of any other institution issuing financial securities. Common stocks, for example, are riskier than T-bills. Therefore, the returns on common stocks must be greater than the return on T-bills. If a portfolio of common stocks is an approximate measure of the market average, then the risk-return coordinates for that portfolio must also fall on the curve described by Figure 8.1, but its position on that curve must be above and to the right of the coordinates for T-bills.

If timberland provides an efficient opportunity for diversification in broad financial portfolios, then its coordinates must fall to the left of the coordinates for

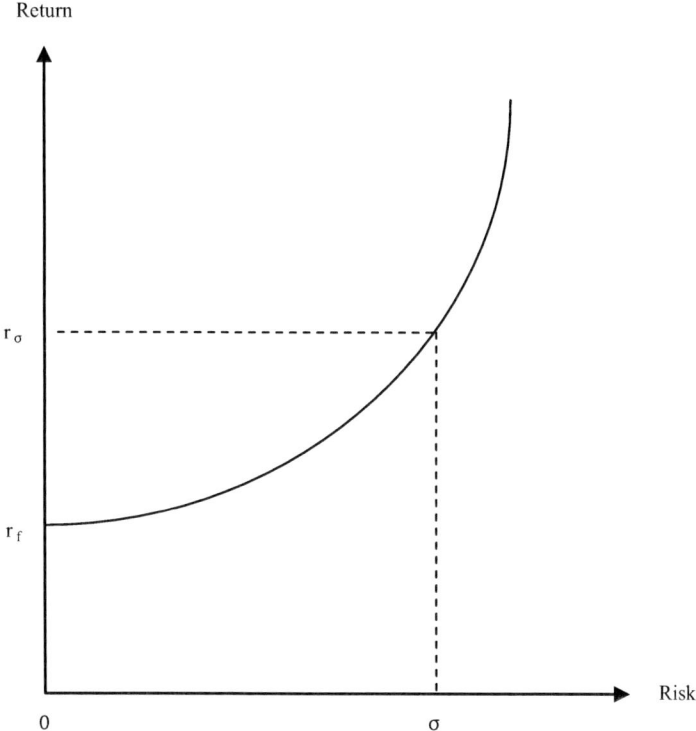

FIGURE 8.1 Asset performance: Risk vs. average return

common stocks. In fact, Zinkhan (2007) confirms that this accurately describes the market since the beginning of institutional interest in timberland in the mid-1980s. He observes that between 1987 and 2006 the average returns on timberland (11.2 percent) were similar to those on common stocks (11.0 percent), while the standard deviation for returns on timberland (6.4 percent) was much less than for common stocks (16.0 percent). Therefore, the risk-return coordinates for timberland did, indeed, fall to the left of those for common stocks, and timberland has provided a beneficial opportunity for portfolio diversification.

Zinkhan's (2007) measures are for timberland in the U.S. South, the region that experienced the greatest number of large market transactions in timberland over the twenty-year period of his analysis. First, Binkley et al. (1996) and Washburn and Binkley (1993), then Sun and Zhang (2001) made inquiries of other timbered regions of the United States and of comparisons with other measures of the general market for financial securities. Their assessments provide a more complete understanding of the various factors explaining the profile of timberland risk.

Binkley et al. (1996) compared historic returns on portfolios of timberland from three regions of the United States with returns on portfolios composed of four different broad categories of financial assets. They followed the statistical regression approach of the CAPM model (described fully in the appendix as eq.

TABLE 8.1 CAPM comparisons of regional timberland portfolios with portfolios of various financial assets, 1960–1994

Asset	α	β	R^2	DW
Timberland				
Pacific Northwest	10.22★	−0.88★	0.14	1.40
Southeast	5.89★	−0.54★	0.15	0.81
Northeast	2.80★	−0.21★	0.09	0.82
All regions	7.31★	−0.65★	0.18	0.99
Financial assets				
Common stocks	0.00	1.67★	0.85	2.17
Small company stocks	3.04	1.98★	0.48	1.21
Long term corporate bonds	−0.01	0.87★	0.53	1.85
U.S. government bonds	−0.01	0.60	0.42	1.96

★ Statistical significance at the 0.05 level of confidence
Source: Binkley et al. (1996)

8a.4) to measure each portfolio's non-diversifiable risk (the CAPM β), and also to differentiate the return each portfolio generated from the return justified by its unique level of risk (the CAPM α); that is, its distance to the left (−) or right (+) of the curve in Figure 8.1 that characterizes the market average. Table 8.1 summarizes their results. The βs (column 3) indicate that investments in timberland in each region reflected less non-diversifiable risk than any of the four broad market averages. Indeed, the negative β coefficients mean that timberland in each region performed even better in this respect than the zero coefficient that indicates a risk free portfolio. The αs (column 2) indicate that timberland in all three regions earned a return greater than that necessary to justify its diversifiable risk—and this implies that, indeed, timberland was undervalued in the market. Furthermore, timberland in two of the three regions outperformed all four more general portfolios in this respect. Even in the third region, the U.S. Northeast, only one of the four general portfolios, a portfolio of common stocks, outperformed timberland.[1]

Washburn's and Binkley's (1993) findings are roughly comparable and they, too, observe that regional distinctions are important. However, their greater interest was in the relationship between timberland value and inflation. They observe that timberland in the U.S. West and South—but not in the Northeast—has been overvalued during periods of greater than expected inflation. Therefore, timberland from the first two regions has provided an effective hedge against greater than expected inflation. This means that timberland from these two regions has contributed balance to portfolios with financial assets such as common stocks, corporate bonds, and T-bills that hedge lower than expected inflation. Washburn and Binkley speculate that the regional differences they observe are due to two

---

[1] Redmond and Cubbage (1988) obtained similar results.

features of market structure, the greater share of forestland in industrial ownership and the larger pulpwood market share in the Northeast. Both could be sources of oligopsonistic pricing in local sawtimber markets. Of course, the more general lesson to be taken from these observations is that regional market structure and conduct can affect the prices of timber and timberland and, therefore, one should consider the regional effects of these variables before automatically assuming that all timberland assets have beneficial risk sharing effects in general investment portfolios.

Sun and Zhang (2001) provide more recent and more detailed confirmation. They compared CAPM results for eight different forest-related investments with ten measures of broader market performance. Their results show that timberland alone among forest-related investments has provided a low risk opportunity for portfolio diversification. Other forest-related investments (lumber futures, stumpage prices, forest industry common stocks) have not shared the same risk-return advantage and, in particular, industry-owned forestland has not earned a risk adjusted return. Of course, many anticipated this last observation. It was one reason for both hostile takeovers and industrial divestiture of timberland of the 1980s.

However, Sun and Zhang also observe that the CAPM approach is limited to one explanatory factor. Surely the market is more complex than this. By extending the CAPM model to include additional arrangements of risk factors (an arbitrage pricing model, also described in the appendix), Sun and Zhang were able to explain a larger share of the variation in return for all eighteen of the portfolios they examined. Their observations are that:

- Institutional timberland returns for the period between 1987 and 1997 were (a statistically significant) 0.5 percentage point greater than average market returns for the same level of risk.
- Returns to investments in timberland have a low correlation with returns to other financial investments.
- Furthermore (confirming Washburn and Binkley), while there are differences in regional markets, timberland has been a hedge against inflation.

Finally, Sun and Zhang caution that their observations are based on historical data. The future performance of investments in timberland may be different. Indeed, we should expect that industrial sales and institutional purchases of timberland will continue in the United States, but at a declining rate, as fewer and fewer advantageous transfers remain. Eventually, an equilibrium will obtain, an equilibrium described by risk-return coordinates for further transfers of timberland that are comparable to those for the broader market.

Summary, Implications, and Future Prospects

Institutional investors assemble portfolios of financial assets with the objective of obtaining high and regular returns for their clients. They have aggressively invested in U.S. forestlands over the last twenty years as they recognized not only the satisfying returns from these assets, but the countercyclical nature of the returns and their characteristic lower-than-market risk. Over fifteen million hectares changed ownership in the ten years between 1995 and 2005 as a result of institutional interest (Clutter 2007).

Typically, institutional landowners either contract with timberland investment management organizations (TIMOs) to manage their lands or they invest in real estate investment trusts (REITs) that manage the land. The former focus on returns to the financial asset. The latter focus on the total return, including both capital appreciation of the asset and returns from the use of the land (stumpage fees, the sale of hunting rights, etc.). Neither the institutional landowners nor their managers are integrated operations with their own wood processing facilities. Rather they contract with firms in the forest industry for the sale of their mature timber.

Since the lands of these new institutional investors were very largely accumulated as transfers from industrial forest landowners and since the objectives of these new landowners are commercial, their forestlands fall within the managed component of the third stage of forest development. However, the impacts of the new institutional ownerships on land management and on the stumpage market are somewhat different from the impacts when those same lands were within their previous possession by the forest industry. Institutional landowners seem to invest less in silvicultural management and they do not participate in forestry research. In particular, their expenditures for fire management are, perhaps, 50 percent less than previous forest industry expenditures for the same lands. Individual firms in the forest industry invested in both intensive forest management and forestry research, in part to insure a ready private timber supply to their mills in periods of tight markets. Institutional landowners, since they possess no mills of their own, have no such need.

Some industrial landowners also tended to increase their timber harvests during slack markets. Revenues from additional harvests during these periods helped offset declining revenues from other parts of the industrial operation and, thereby, helped maintain a flow of total corporate revenues and corporate dividends. Institutional landowners, however, are diversified into other sectors of the economy. Downturns in the markets for either stumpage or forest products have less effect on them. Therefore, they are more likely to decrease their timber harvests in periods of weak stumpage prices and, as a result, timber prices and the overall timber market have probably been less volatile since the involvement of institutional landowners.[2]

The rapid entry of financial institutions into the market for U.S. timberland, together with the decreasing availability of large blocks of timberland, beg ques-

---

2 Individual industrial landowners also protected against fire and insect and disease epidemics by owning a diversified collection of forestlands. Will the new institutional landowners need to diversify their landholdings for the same reason—or will their broader portfolios of various non-forest investments provide sufficient protection against natural hazards and the resulting interruptions in the returns to their forestlands?

tions about the future development of the institutional landowner class. Will the involvement of financial institutions, and of TIMOs and REITs, stabilize—or can we anticipate their further growth? More financial capital is becoming available as private investors and families recognize the advantages of investments in TIMOs and REITs. However, timberland prices and transactions costs are increasing, even as stumpage prices are stagnant. The increasing monetization of the forest environment adds some value to timberland and some institutional landowners have been able to sell hunting rights, for example, while maintaining the timber rights on their lands. Nevertheless, it is clear that the opportunities for the exchange of large tracts of U.S. timberland are becoming scarcer (Zinkhan 2007).

Global investments may be the new option for this landowner class. In fact, the world's forest industry, including its U.S. segment, is becoming more globally integrated, and the land area in the managed forest stands that are attractive to financial investments is growing as more regions around the world move into the third stage of forest development. Therefore, we might anticipate greater international opportunity for those financial institutions that invest in timberland as well.

Washburn and Binkley (1993), and also Sun and Zhang (2001), warned that not all regions of the U.S. provide equally good opportunities for institutional investment in forestland. Surely their caution is good advice for international investors, and just as surely, the different risk profiles of individual countries will be the crucial determinants for investment in one country rather than another. Nevertheless, Zhang (2003) observes risk-return portfolio diversification opportunities for timberland in China that are somewhat comparable to those in much of the United States. It is likely that opportunities exist elsewhere around the globe as well. We can anticipate that financial institutions from the United States and elsewhere will take advantage of them. However, these are sophisticated investors. We can also be certain that they will consider the risk-return characteristics of the available timberland assets in each region, and they will recognize that these are affected by the stability of the general economic environment and by local market structure and conduct, as well as by the particular characteristics of the timberlands in question.

Meanwhile, in regions like the U.S. South where institutional ownership has very largely replaced the former industrial ownership, the differences between forest management under the two ownership classes raise questions about future timber supply, particularly from the lands of the REITs, some of which already have been converted from timber production to alternative higher-valued uses. Surely, if the land available for timber production declines, then the market prices for both timber and forestland will adjust upward and, just as surely, the market adjustment will be an incentive to shift new forest into the third stage region of managed and even plantation forestry. This means that the relevant questions become how much of a shift in forest production will occur, how soon, where, and what other effects on the industry will result from such market adjustment? The history is brief, the evidence is sparse and, therefore, the answers must be conjecture. However, chapter 7 has shown that in the sawmill industry where the cost of wood is crucial, adjustment is often rapid. In the pulp and paper industry, the cost of wood fiber is a relatively smaller share of all production costs, and adjustment may be slow and perhaps only negligible.

Notes

1 Redmond and Cubbage (1988) obtained similar results.
2 Individual industrial landowners also protected against fire and insect and disease epidemics by owning a diversified collection of forestlands. Will the new institutional landowners need to diversify their landholdings for the same reason—or will their broader portfolios of various non-forest investments provide sufficient protection against natural hazards and the resulting interruptions in the returns to their forestlands?

Literature Cited

Binkley, C., C. Raper, and C. Washburn. 1996. Institutional ownership of US timberland: Historical rationale and implications for forest management. *Journal of Forestry* 94(9): 21–28.

Clutter, M. 2007. Current and future trends in U.S. forestland investment. In *Investing globally in forestland*. Portland, OR: World Forestry Center.

Hagler, R. 2006. Why do pension funds invest in forests? Summary of presentation to New Zealand Institute of Forestry, April. Boston: Hancock Timber Resources Group.

Mortimer, J. 2009. *Investing in timberland: another means of diversification*. New York: J.P. Morgan Investment Analytics and Consulting.

Redmond, C., and F. Cubbage. 1988. Risk and returns from timber investments. *Land Economics* 64: 325–337.

Russakoff, D. 1985. Timber industry is rooted in tax breaks. *Washington Post* (24 March) pp. A2 ff.

Sun, C., and D. Zhang. 2001. Assessing the financial performance of forestry-related investment vehicles: capital asset pricing vs. arbitrage pricing theory. *American Journal of Agricultural Economics* 83(3): 617–618.

Washburn, C., and C. Binkley. 1993. Do forest assets hedge inflation? *Land Economics* 69(3): 215–224.

Yin, R., T. Harris, and B. Izlar. 2000. Why forest products companies may need to hold timberland. *Forest Products Journal* 50(9): 39–44.

Zhang, D. 2003. Policy reform and investment in forestry. In W. Hyde, B. Belcher, and J. Xu, eds., *China's forests: Global lessons from market reforms*. Washington, DC: Resources for the Future, pp. 85–108.

Zinkhan, C. 2007. *Timber: An asset class for all seasons. Investing globally in forestland*. Portland, OR: World Forestry Center.

Appendix 8A: CAPM

The capital asset pricing model, or CAPM, was developed by Sharpe (1964) and Lintner (1965) to separate the riskless rate of return on an investment from the return to its market risk. CAPM states that the expected return on the *i*th individual investment or class of investments r_i is equal to the expected and general market risk-free rate of return $r_{m,f}$ plus a premium on the risk $r_{i,p}$ that is unique to the asset in question.

$$r_i = r_{m,f} + r_{i,p} \tag{8a.1}$$

The risk premium itself is some function of the difference between the expected market return r_m and the expected return on the investment in question.

$$r_{i,p} = \beta_i(r_m - r_{m,f}) \tag{8a.2}$$

Combining the two equations,

$$r_i = r_{m,f} + \beta_i(r_m - r_{m,f}) \tag{8a.3}$$

where $r_{m,f}$ is generally taken as the return on some risk-free investment such as government T-bills in the United States and r_m is taken from a broad measure of the entire financial market, a measure such as the average return on S&P 500 common stocks.

The final step converts eq. (8a.3) into a form that is amenable for empirical estimation:

$$r_i - r_{m,f} = \alpha_i + \beta_i(r_m - r_{m,f}) + \mu_i \tag{8a.4}$$

where the μ_i are randomly distributed error terms. The intercept α_i, known as the CAPM alpha, estimates the value an asset that is due to factors other than conditions that are general throughout the market. A positive (negative) alpha indicates an expected return that is greater (less) than the market return. The coefficient β_i, known as the CAPM beta, indicates the asset's level of risk relative to overall market risk. A beta greater (less) than one indicates an asset that is more (less) responsive than the market and, therefore, more (less) risky than the market in general.

CAPM identifies only a single summary risk factor. A straightforward variation in eq. (8a.4), known as the arbitrage pricing model, allows for the identification of multiple factors that may influence risk and it discriminates among these (Ross 1976; Roll and Ross 1980).

$$r_i = \gamma_0 + \beta_{i1}\rho_1 + \beta_{i2}\rho_2 + \ldots + \beta_{in}\rho_n + \varepsilon_i \tag{8a.5}$$

where γ_0 is the risk free rate and the ε_i are randomly distributed error terms. The ρ_i are independent risk factors. It is not surprising that, with its more complete specification, the estimated results of the arbitrage pricing model are generally more robust than those obtained with the simpler CAPM model.

Literature Cited

Lintner, J. 1965. The valuation of risk assets and the selection of risky assets in stock portfolios and capital budgets. *Review of Economics and Statistics* 47: 13–37.

Roll, R., and S. Ross. An empirical investigation of arbitrage pricing theory. *Journal of Finance* 35: 1073–1103.

Ross, S. 1976. The arbitrage pricing theory of capital assets pricing. *Journal of Economic Theory* 12: 341–360.

Sharpe, W. 1964. Capital asset pricing: A theory of market equilibrium under conditions of risk. *Journal of Finance* 19: 425–442.

9
NON-INDUSTRIAL PRIVATE LANDOWNERS

Non-industrial private forest (NIPF) landowners are owners of generally smaller parcels of forestland and trees who have no permanent association with a wood processing facility. Their individually small forests may support occasional, even periodic, but seldom regular, timber harvests. Their lands include farm woodlots in both developed and developing parts of the world where those may be the sources of both commercial reward and support for household subsistence. The NIPF lands also include conservation woodlands, such as shelterbelts around agricultural properties, and recreational forestlands, such as those on properties used primarily for second homes or tourist lodges and those belonging to hunting or fishing clubs.

Individually, these parcels may be small but, in the aggregate, they are very large. In the United States, Finland, China, Chile, and South Africa, for examples from each continent, the NIPF lands account for 56, 61, 51, 59, and 52 percent, respectively, of all forests (Butler 2008; Smith et al. 2004; Hänninen and Sevola 2009; *China Forestry Yearbook* 2000; INFOR 1997; GovUSAf 1953). In Europe, twelve million families own forests which average only eleven hectares each (Jeanrenaud 2001). In the United States, ten million families own forests averaging an even smaller ten hectares each (Butler 2008). In China, more than 50 million farm households manage forests averaging hardly more than one hectare apiece (Hyde, Xu, and Belcher 2003). Globally, NIPF landowners probably account for more than 25 percent of all forestlands, and this large global share is one reason the NIPF lands are among the most studied of forest resources.

These non-industrial forests tend to occur in smaller individual parcels than those of the other three characteristic landowner classes, but other distinctions are apparent as well, and we can identify some of these with reference to the three-stage model. The first section of this chapter examines those distinctions.

A second section of the chapter examines the various motives of NIPF landowners. The multitude of individual motives and the variance between landowner objectives and policymaker expectations is another reason that the NIPF lands have been widely studied. Landowners' motives range from financial or subsistence gain, either periodic or as an occasional income supplement; to land conservation and the provision of recreation and amenity values and ecosystem services that may not be as directly measured in the market; to an even less easily assessed desire to bequeath something of tangible personal value to one's heirs. Assessments of the latter two, the conservation/recreation/amenity and bequest motives, are additionally complex as landowners often hold these motives in combination with a desire for some amount of financial gain from extractive forest products (usually timber) from the same land.

The concluding section of the chapter summarizes the evidence regarding the general market responsiveness of the non-industrial category of forest landowners. An appendix will address two of the traditional justifications for public policy designed to assist non-industrial private forestry: the arguments that small forest landowners do not make rational decisions regarding long-term silvicultural investments, and that public programs that assist NIPF landowners satisfy the broader income distributive objectives of national or regional policy. Both are long standing parts of forestry lore, but they are lore that is not supported by the empirical evidence.

Distinctions Among NIPF Lands and Landowners

NIPF landowners are a more diverse group than either the industrial or institutional landowners of the last two chapters or the public landowners who will be the topic of the next chapter. Non-industrial landowners are more varied and complex in their motives for owning forestland and also in their experience within the three-stage pattern of forest development.

Industrial and institutional landowners are profit seekers. As such, the lands they own outright tend to be concentrated in the commercially productive forest that occurs only in stage III—although the former also compete for harvest rights to concessions at the frontier in any stage of forest development. The public lands, in contrast, are heavily concentrated in regions that were financially unattractive to private claimants before some date in local history when the public established its permanent responsibility for these lands. The public may have invested in other forest lands since that time (e.g., Acadia, Grand Teton, and Redwood National Parks in the United States) but the lands that remain beyond the modern commercial forest frontier are overwhelmingly public.

Both the historic and the current locations of the non-industrial private lands are an altogether different story. Non-industrial private landowners possess some productive timberland as well as some marginal and even sub-marginal forest, often in small parcels intermixed with agricultural land, and both of which they may manage for whatever financial gain they can eventually obtain. They also hold smaller parcels of forestland for other reasons. The latter occur anywhere that offers good conservation or aesthetic opportunity. Some may be along watercourses within regions of commercially productive agricultural land or timberland where the ownership objective may be to protect the owner's agricultural lands from soil or wind erosion. Or the objective may be to minimize the agricultural contribution to damaging runoff. Others of these parcels may be in regions of previously cutover and abandoned timberland that remain attractive for recreational use, and certainly some are highly-valued non-timber in-holdings surrounded by public lands well beyond the commercial timber frontier.

Within the context of the three-stage model, all of the new settlers of regions in the first stage might be categorized as non-industrial private landowners—although few of them may possess clear rights to forestland. As with our original discussion in chapter 2, these settlers claim land out to point B in Figure 9.1a, and their homegardens on some of this land may include a few fruit trees. The land between points B and D may support some agricultural activities intermixed with the forest, and some local settlers generally extract some use from the forest (e.g., grazing for their domestic livestock, naturally growing fruits and nuts, medicinal plants, etc.) all the way out to point C.

Non-industrial landowners in regions in stage II possess rights to small parcels of lower-valued forest surrounded by more productive agricultural land out to point B in Figure 9.1b.[1] Some also obtain agricultural use from the previously cutover lands out to point C and some forest use from cutover lands to point D. Still others build recreational homes and tourist lodges and establish ski resorts and private hunting and fishing clubs wherever they find lands that are exceptional for these purposes.

The forested possessions of non-industrial owners in regions in stage III are similar to those in stage II—except that the lands between points B' and B'' in stage III are commercially viable and permanently managed timberlands. Non-industrial owners possess some of these lands, usually in parcels that are small in comparison with those owned by industrial or institutional landowners. Many of these are owned for their commercial timber value, but a few are valued for their unusual flora or fauna or their exceptional aesthetic value or recreational opportunity.

The small individual parcels and the fragmented and dispersed geographic pattern of even the better parcels of non-industrial forestlands should raise questions about both the minimal operating scale for non-industrial landowners and their scheduling of timber harvests and financial returns. We will speculate on both. In addition, the possession of small parcels of less productive forestland

---

1 They hold small parcels of sub-marginal forestland only because they would find few willing buyers for these small parcels, because the costs of holding these lands are slight, and because divestiture would create new management costs for the maintenance of the longer boundaries that would be necessary in order to exclude these small and isolated parcels from the surrounding private agricultural lands.

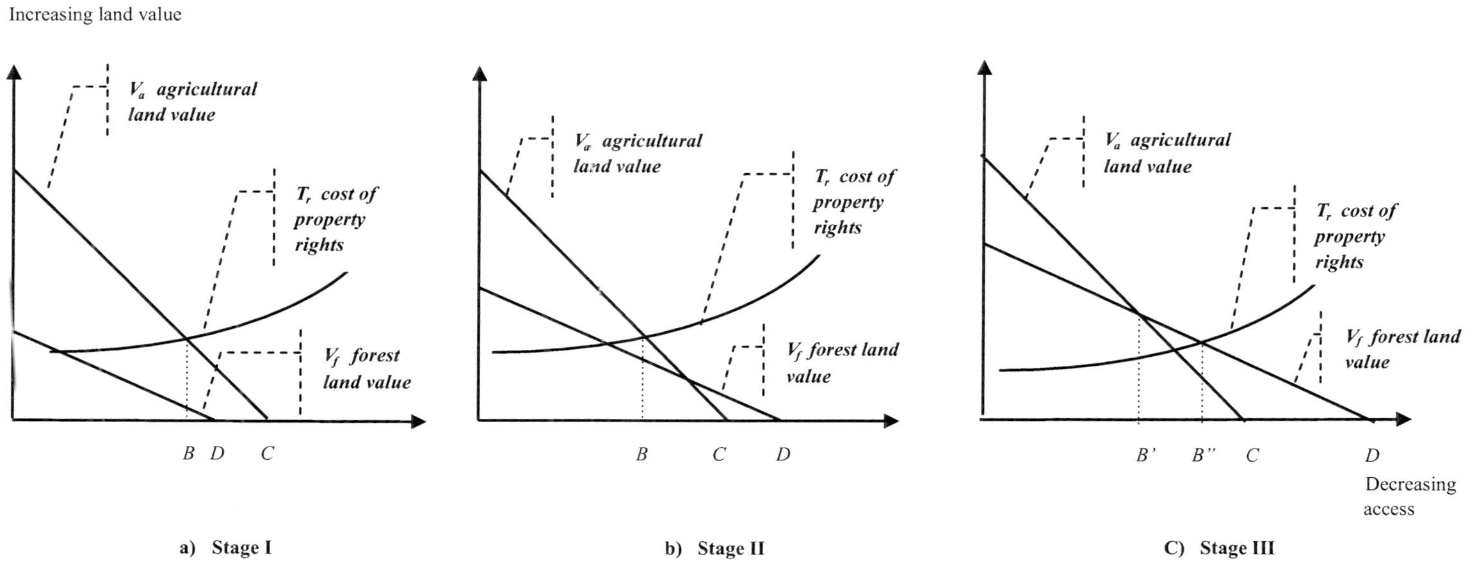

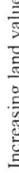

FIGURE 9.1 The three stage model, repeated

contained within the boundaries of larger possessions of financially productive land—but figuratively beyond points B or B" in stage II or stage III, respectively, as these small parcels are less productive—create occasional financial opportunity for their NIPF landowners, and we will comment on this opportunity as well.

Landowner Objectives

When considering NIPF landowners as a group, the rationality of their decisions can be difficult to understand. Different landowners hold their properties with different motivations and the aggregate of all NIPF landowner behavior in any region is a complex mix of several motivations. This mix of motivations causes differences in forest management decisions regarding planting, stand treatment, and harvesting, as well as different responses to market incentives and public forestry programs.

The usual list of NIPF landowner motivations includes:

- commercial gain from the sale of extracted resources, or subsistence gain from extraction and household consumption in some cases in developing countries;
- household gain from conservation and amenity values—including recreation and leisure opportunity;
- intergenerational bequests; and
- various combinations of the first three motives.

The following pages review each of these motivations in turn.

Commercial or Subsistence Gain

Some landowners possess their forests as a source of income or income supplements, generally from the sale of mature timber. Others, particularly in regions of subsistence agriculture, hold very small forested properties as a source of fuelwood and other subsistence goods.[2] In both cases, landowners intend to maximize the financial or subsistence return from the production of extracted resources. This motivation may be constrained, however, by the landowner's knowledge of forest management and markets for forest products. The primary occupation of very many NIPF landowners, especially in Europe and North America, is increasingly in some activity other than forestry. These landowners may be farmers who possess a small forested plot in addition to their agricultural lands, or they may be absentee landowners who live and work in a nearby town but who still enjoy use

---

2 Even in the United States, firewood production is one objective of 12 percent and the collection of non-timber forest products is an objective of 7 percent of all family forest owners. Family owned forests comprise 71 percent of all NIPF forestlands in the country (calculated from Butler 2008).

of their forests as a second occupation. In economic terms, their "comparative advantage" is in activities different from forest management.[3] Therefore, their knowledge of forest management may not be as complete or up-to-date as that of industrial or institutional landowners who specialize in forestry, and they may be slower to respond to market signals. Nevertheless, their intention is to do as well as they can with what they do know of the local market opportunities.

The generally smaller scale of NIPF properties means that, regardless of the market, harvests from any one landowner's property must be irregular. It is unlikely that a small property can support a forest with a full range of evenly spaced age classes from new seedlings through mature standing timber—what foresters call a "regulated forest." However, it is common for NIPF lands to contain at least some timber within the broad range of harvestable ages. Without a regulated forest, harvesting must be irregular in both timing and volume but, within any region, there are numerous NIPF landowners who could harvest some amount of timber in almost any given year. For them, the question is when to harvest.[4] Some harvest in response to their perception of good market prices. Others are more responsive to an unusual personal or family need for income.

Market opportunity, harvest scheduling, and financial returns: NIPF landowners are price takers in markets with information asymmetries. That is, individual landowners act as perfectly competitive sellers in markets in which their own information is limited. Each individual landowner's knowledge is concentrated in the different market of his or her own comparative advantage and his or her participation in the market for forest products may be too limited to justify the effort to obtain additional information.[5] However, the buyers of their products, loggers and millowners, are specialists in these markets. Because of their full-time specialized involvement in these markets, it is natural that they would assemble more complete market information. Therefore, in any one region, the much smaller number of loggers and millowners are likely to exercise

---

3 In the United States, for example, 20 percent of the lands in family forests are part of a family farm. Forty percent of the lands in family forests are owned by absentee landowners (Butler 2008). The forest economics literature of the 1950s and 1960s repeated often and well the fact that both farmers and absentee landowners have comparative advantage in activities different from forestry. See James (1950), Stoltenberg (1954), Yoho and James (1958), Stoltenburg and Webster (1959), James and Schallau (1961), Stoddard (1961), Lord (1963), Conklin (1966), Muench (1966), and Gregersen, Houghtaling, and Rubenstein (1979).
4 Twenty percent of family forests in the United States harvested some timber within a ten-year period (Butler 2008).
5 Lonnstedt (1997) argues that landowners in the Nordic countries rely on timber market information from neighbors and their own prior experience. (Their neighbors are only a limited sample of the market and the neighbors' experience may be dated because their previous harvests might have been a few years back.) They are not especially well-informed observers of timber markets.

oligopsonistic market power over individuals among the much greater number of NIPF landowners.

The small size and irregular periodic offering of any landowner's timber operation encourages landowners to delay harvests until higher than usual local timber prices present an unusual financial incentive to harvest. Indeed, Gould and O'Reagan (1963), in a simulation of randomly adjusting market prices, show that, rather than harvesting either in regular periods or as their forests reach some biologically mature age, NIPF landowners with smaller parcels of forestland improve their long-run net financial gain by holding mature timber until the local prices are favorable and, in particular, by delaying their harvests until they observe an unusual spike in local timber prices. By doing this, they lose current value at one price, but gain some additional physical growth for their timber as well as the greater discounted financial gain from the higher future price.[6]

However, recognizing good market opportunity may be a problem. Landowners may delay harvests while waiting for a period of strong prices but the brevity of their market information means that they are likely to be slow in recognizing the best prices and they may lag in their responses, eventually selling their timber only as the market passes its peak.[7] The effect may not be trivial because market peaks and troughs for stumpage prices have been known to vary by a factor greater than four in the short period of a couple years.

One solution is for landowners to establish personal reservation stumpage prices as a guideline minimum that triggers their willingness to sell timber. A second alternative is for landowners to form collectives to do their marketing. Collectives, similar to those for some dairy farmers and citrus growers in U.S. agriculture, employ well-informed specialists to do the marketing and product distribution for all members of the collective—leaving the private land management responsibilities to each individual landowning member.

Forestry has but few examples. Jeanrenaud (2001) and Hultkranz (n.d.) describe a successful marketing collective for timber producing small landowners in Sweden, Hammett (1994) describes a successful marketing collective for producers of non-timber products in Nepal, and RECOFTC (2008) describes the remarkable success that the formation of marketing groups had in raising the price of cardamom seventy-fold for one local group of Indian tree farmers. Snelder, Klein, and Schuren (2007) provide the counter-example of Philippine farmers whose experience with tree crops would have been much more financially rewarding if they had better networks for market information.

Alternatively, price fluctuations and the information asymmetry create a market niche that has been filled at one time or place or another by at least three dif-

---

6 Kangas, Leskinen, and Pukkala (2000) provide a more recent illustration of the advantages of adaptive timber-selling behavior.

7 The study of market responses to business cycles has receded as a topic of economic inquiry over the last 40 years. This is no less true in forest economics than for other economic specialties. Zivnuska's (1949) early work on this topic still stands alone in forestry.

ferent services: forestry consultants, a price reporting service, and preferred buyer contracts. One or more of these three tend to occur in many regions with vibrant timber markets and numerous NIPF landowners.

Forestry consultants are specialists in forest management and timber markets. They sell their services in regions with active markets for forest products and numerous NIPF landowners and, as one result, obtain better prices for their landowner clients. Munn and Rucker (1994), with evidence from the U.S. South, show that the value of these consultant services is a close approximation of the size of this market niche. That is, while landowners who hire consulting foresters do receive more for their timber, the value of contracts between NIPF landowners and forestry consultants absorbs most of the difference between the gross timber prices received by these landowners and the prices received by those who do not use the services of consulting foresters.

Price reporting services can also fill this niche by providing timely market information that is otherwise not readily available to many landowners. Agriculture has an extensive history with price reporting services. Anyone who has listened to the radio while driving through the U.S. Midwest knows about daily reports of corn and hog prices. The experience with price reporting in forestry is more limited. Timber Mart South, a price reporting service in the U.S. South, is the prime example. Boyd and Hyde (1989) demonstrate the value of this service to forest landowners and speculate that the opposition of millowners to public financing for this service was due to the information asymmetry and the millowners' loss of the information advantage once Timber Mart South was established.

Finally, many landowners in regions with active stumpage markets negotiate long-term contracts with local mills. Typically, these contracts offer the mill the right of first refusal at the price existing whenever the landowner does decide to sell. These contracts do not insure a high price, but they do guarantee a willing buyer under most market conditions.[8]

Each of these three services might help the NIPF landowner improve financial returns from his or her forest, but they do not alter the likelihood that harvests occur on an irregular schedule—and this is from productive forestland. Many NIPF landowners also possess pockets of less productive land. These marginal properties reinforce the irregularity of harvest responses to market prices. Landowners possess these properties, described by our figures as between points B and D in stages II and III, as small units within larger forest or agricultural properties. They are financially sub-marginal in normal times, but the maintenance cost of their continued possession is small. In fact, divestiture of these wholly contained sub-marginal properties would often raise the costs of maintaining boundaries to protect the adjacent larger agricultural properties. Furthermore, ownership of these marginal forest properties does have its occasional benefit. In times of unusually high prices, their harvestable resource becomes infra-marginal.

---

8 Desmond and Race (2000) provide a global survey of such arrangements. Also see Baumann (1998), Curtis and Race (1998), and Roberts and Dubois (1996).

Therefore, it can add another occasional increment to the landowner's irregular pattern of harvests and the irregular flow of financial returns.[9]

The aggregate flow of marketable products from all NIPF lands in any region has the potential to be very irregular, and, great unevenness of regional flow of timber could have a negative effect on mills and local communities. However, it is more likely that the rapid entry of NIPF landowners into regional markets in times of greater demand and their easy exit from the same markets in times of slackened demand or greater supply has the important beneficial effect of moderating the widest swings in the prices of primary forest products.

Landowner need, harvest scheduling, and financial returns: Rather than attempting to maximize the return from their forests, some NIPF landowners schedule their harvests to coincide with the cycle of personal need over the landowner's lifetime. Accordingly, they might increase harvests and decrease their standing forest stock when, as young adults, income from other activities might be lesser and the costs of establishing themselves and their young families are greater. In subsequent years, income from other sources may grow and the proportionally larger initial demands for family finances may lessen. Landowners can allow their forest stocks to recover without large harvests during this second period. Harvests might occur occasionally—to coincide with unusual personal need—rather than with peaks in market prices. Otherwise, their forest stands continue to grow until they may be at their maxima for the oldest NIPF landowners. Kuuluvainen (1989) first showed this "life cycle" experience to be a reasonable depiction of landowner behavior in Finland. (In Kuuluvainen's observations, the ability to borrow funds is also a crucial determinant of landowner harvest behavior.) Perz and Walker (2002) more recently described similar life cycle behavior among colonists in Brazil's Amazon.

Farmers in southern China may behave similarly. Observations of behavior over individual farm foresters' life cycles are unavailable, but their own reference to their trees as "green banks" suggests that they manage their trees to store and grow wealth until it is available in time of unusual need. Similarly, Lofgren, referring to Swedish farmers who sell timber to provide the financial means to purchase a new automobile, calls this the "Volvo effect." Of course, the Volvo effect could occur just as easily when the farmer's home or barn needs renovation, or when the farmer's family faces any large and unexpected, or just irregular, financial outlay. In fact, financial need is the primary reason for harvests on 15 percent of family-owned forests in the United States (calculated from Butler 2008).

---

9 Empirical estimates of the economic productivity of farm foresters often describe low rates of return. One reason could be due to combining the forest landowners' commercial and marginal forestlands in the same estimate. If separated, the productive commercial lands would obtain a larger return to forest management. Of course, the sub-marginal lands, held at very little cost to the farmer and held only because they are contained with the physical confines of more valued agricultural properties, would obtain a lower return—and diminish calculations of the combined return from both classes of land.

In a related way, the development economics literature, with specific focus on poorer landowners in developing countries, refers to the "social safety net" provided by trees or their products. These landowners recognize the availability of a few trees or their products as a source of emergency income or other sustenance that can be called upon during occasions of unusual family hardship—crop failure, unemployment, etc. (e.g., Scheer 1995; McSweeney 2005; Andersson, Mekonnen, and Stage 2008; Salam, Noguchi, and Koike n.d.; Neumann and Hirsch 2000, review this literature).

Of course, the green bank, the Volvo effect, and the social safety net are similar in their implications for an irregular harvest and irregular return reserved for a time of unusual household usefulness. The improvement brought to our understanding by the life cycle hypothesis lies in its potential to predict when some unusual needs occur. In all cases, the harvest and sale of trees or other forest products provides a welcome supplement to the regular flow of wealth the landowner obtains from his or her primary non-forest occupation.

This idea should be familiar. The flow of cash or household-consumed forest products in time of need for the NIPF landowner is similar to the benefit from those forestlands that industrial firms hold to provide occasional flows of timber to their mill in times of short supply and high timber prices. It is also comparable, in a similar way, to the counter-cyclical return that forests provide for portfolios of large financial investors. Indeed, Dewees (1995) observes that the most favorable forest investments for farm foresters in Malawi are low-risk investments—just as large institutional investors in developed countries find forest properties most attractive when their risks are low and counter-cyclical. In all three cases—NIPF, the industry, and institutional portfolios—forest ownership provides a store of wealth and a form of insurance against unusual need.

Discretion in forest management: The previous two sections of this chapter focused on landowner discretion in harvest scheduling. Discretion in management as well is an attractive feature of forestry for NIPF landowners but the smallest size of some forest properties can be a constraint on efficient NIPF activity.

For farm landowners, there is a strict seasonality to agricultural inputs and outputs. The primary occupations of many absentee NIPF landowners have their own scheduling obligations. Teachers are busy during the school year. Lawyers are busy whenever their clients demand. Businessmen and women and shopkeepers have their own unique schedules and certainly have little discretionary time during tax and inventory seasons.

Many forest operations, however, can be implemented a year early or a year later with little impact on production, and some can be introduced in virtually any convenient season of the year. This discretion creates a tremendous advantage for forestry as a supplemental occupational activity. Scheer (1995) observes in Kenya, for example, that when subsistence farmers find off-farm labor, they have to forego tending crops but they can convert some cropland to forest, tend that forest only when their off farm employment permits temporary visits back to the land, and still obtain a measure of return from their land.

For another example, women in Malawi stated that if men will perform the heavier planting activity when they have time available from other activities, then the women will tend the young trees, watering them and protecting them from grazing livestock, while the men are occupied elsewhere (Hyde and Seve 1993). Amacher, Hyde, and Joshee (1993) observe a similar gender difference in the household allocation of tasks in Nepal, with men tending the trees and collecting fuelwood from them when these tasks can be conducted jointly with their agricultural activities. Women are more involved in forest management and collection activities when they have discretionary time and, in this case, fuelwood collection is often a joint activity with child care. Cooke (1998) observes, furthermore, that the fuelwood collection activity of these Nepali women is a seasonal occupation, conducted when their assistance is not required for peak agricultural activities.

These observations have been from developing countries and usually from subsistence agricultural households. However, either convenient seasonality or discretionary timing is also a feature of forest activities in developed countries. Spring and late summer/early fall are the crucial planting and harvest seasons for most farmers. Winter and mid-summer, however, are slack seasons for many agricultural activities. These seasons, winter when the ground is frozen or summer after porous soils have dried out from spring rains, are convenient times for allocating underutilized farm labor to timber management and harvest activities. In fact, the hard ground of either winter or summer is a requirement for timber harvesting in many parts of the world.

In an interesting case of one specialized forest product, the sap of sugar maple trees in eastern Canada and northeastern United States flows in the early spring before farmers begin planting. Farmers collect this forest product and reduce it to a sweet marketable syrup at a time when there are few other opportunities for the employment of their labor.

Scale as a constraint on forest management: Flexible timing is an advantage for the owners of the generally smaller parcels of NIPF land. The exceedingly small scale of some operations, however, can be a disadvantage. Some non-industrial forest ownerships are as small as a fraction of a hectare and the costs of fencing their boundaries and protecting these properties from trespass, vandalism, and theft can overwhelm any benefit from their management.[10]

There are two interesting cases. For the first, the forested area is a lower productivity fragment of land entirely contained within a somewhat larger agricultural property. In this case, the boundary controls for the surrounding agricultural property can be sufficient protection against trespass on the encircled forest property.

---

10 Trespass is not a trivial concern. In Butler's (2008) extensive survey, 44 percent of all family forest landowners in the United States expressed concern with trespass. A similar 44 percent expressed concern with vandalism or dumping on their properties and 24 percent expressed concern with timber theft.

Free-standing forest properties in individually scattered parcels are the more interesting second case. As these are scattered, their oversight and protection from trespass and theft can be a problem. Fencing, for example, increases with perimeter, and the unit cost of fencing increases with the ratio of perimeter to area. Therefore, this unit cost is greatest for the smallest properties. Furthermore, establishing an efficient level of management can also be a problem for smaller parcels as the unit costs of forest management plans and some silvicultural activities also increase with declining size of the operation. For the many NIPF landowners for whom forestry is not a primary occupation, the cost of acquiring the technical skills of a knowledgeable forester may not be worth the additional revenue recovered on a smaller property.

China's experience in the 1980s illustrates the problem—and the solution. China's collectives reallocated responsibility for 31 million hectares of forest to 57 million agricultural households in the early 1980s. The distribution of forested land was egalitarian. Forestland was first separated into plots according to species, age, stand density, site and soil quality, and distance from the village, among other criteria. These plots were reallocated such that households of similar family size received similar mixes of high and low quality forest, accessible and less accessible forest, and so forth (Sun 1992; Lu et al. 2002). As a consequence, land on one slope was distributed among many households in one village and each household in the village received several very small forest plots on different hills and different slopes of any hill. The average household in southern and southwestern China obtained four or five plots totaling just one or two hectares. The average household in Zhaizao township in Guizhou Province, for example, received fourteen plots with a total area of just 4.2 hectares. The largest plot in this township was 2.3 hectares, while the smallest was just 0.06 hectares. In the village of Yixiang in Jinggu County of Yunnan Province, 45 hectares were distributed in 169 plots that averaged only 0.27 hectares each. Each household had to assign a family member to look after its numerous forest plots, where only one or two guards had been necessary to protect the entire forest under the previous arrangement of collective management. Providing daily protection from theft and fire was a burden, especially for households with smaller families (Liu and Edmunds 2003).

Nevertheless, the gains from the conversion from collective to household forest management were spectacular. The households themselves began their own afforestation and reforestation activities, increasing the total forested area of the collectives by 33 percent by the time of China's third forest survey conducted between 1984 and 1988.

Despite the obvious gains, the management of numerous small plots with long boundaries was clearly a problem for many households and some eventually responded by combining their plots with those of their neighbors to form larger management units. Some households transferred rights to other households and the latter became specialists in forestry as they acquired more forestland. Yet other households combined their forested properties in what amounted to local shareholding companies, each with its own responsible manager. And in still other cases, the collective that had been responsible before the redistribution, reasserted its claim to the land, reasoning that individual households were unable to manage successfully (Xu and Hyde 2005).[11]

Clearly, very small parcels of land are a disadvantage for effective forest management. There are a number of examples in addition to these from China where local community groups manage private forest properties for the common good.[12] Nevertheless, the numerous individual farm households from countries around the world that do manage only slightly larger parcels of forest, regardless of greater occupation with their primary agricultural activities, suggest that scale ceases to be a deterrent to independent private forest ownership, even for its timber value, for forest lands of even very moderate size.[13]

In conclusion, for the very smallest properties, there is a tradeoff between the cost reductions obtained by combining parcels of land with lengthy perimeters under some form of collective management, and the foregone opportunity for individual decision making by a heterogeneous group of independent landowners, each with his or her own lands and preferences. The diseconomy of individual management of the very smallest parcels of forest land can be one determinant of collective action. However, above some minimal scale, the preference of the landowners themselves seems to be for independent private land management, although with collective arrangements for selective activities such as fire protection that affect large areas or for activities like marketing and distribution of forest products that may be outside the specialized experience of most individual landowners.[14]

The Recreation/Leisure and Conservation Motives

The previous pages focused on landowners who use their forests to generate extractable products such as timber or fuelwood that have commercial or house-

---

11 The problem persists in China even in 2009. Many households hold management rights to multiple dispersed small forested plots. The government is reluctant to permit the free transfer of the rights despite the well-known success of a few farmer initiatives.
12 Jeanrenaud (2001) surveys the western European experience.
13 In the United States, 92 percent of family forest owners prefer to make their own land management decisions. Other family members make half of the remaining 8 percent of all land management decisions (Butler 2008).
14 The forestry literature on cooperative landowner ventures shows that Finnish landowners who are dependent on regular forest income and who are faced with capital market imperfections may enter into cooperative ventures for planning, or even for joint strategizing on harvests, as a way of hedging risk. Nevertheless, these landowners still manage their own lands (Uusivouri and Kuuluvainen 2001). Others in the United States enter into cooperative ventures as a means of preserving wildlife corridors that extend beyond any one landowner's property boundaries (Jacobson 2002). We previously identified cooperative landowner ventures in Nepal and Sweden for the marketing of forest products (Hammett 1994; Hultkranz n.d.). However, the diversity of landowner preferences apparently makes it difficult to assemble a management package that creates sufficient financial reward to attract many NIPF landowners into coordinated management of their lands (Sample 1996; Koslowski et al. 2001; Kurtilla et al. 2001; and Eid, Hoen, and Okseter 2001).

hold subsistence value. Other landowners possess their NIPF lands for conservation or amenity values such as the erosion protection offered by windbreaks or the recreation and leisure activities the owners enjoy while vacationing or living on these properties. Second homes nestled within forested environments and generally used for holiday or vacation relaxation are an example of the latter. Lodges for hikers, skiers, and eco-tourists built in forest environments recognized for their scenic beauty or unusual flora or fauna are another. Hunting and fishing clubs, often with lodges as well as possession of, or access to the use of, larger adjacent forest properties, are a third.

Generally, the individual forest properties in this category are small because the opportunity costs of holding larger properties are great and the landowners, nevertheless, generally obtain considerable aesthetic and recreational value from the adjacent forest lands of other owners. Even for commercial recreation activities, the main lodge at the core of the recreational and leisure use is contained within a small part of the full forested property. Therefore, scale may not be as crucial an issue as it is for small properties owned for their timber value. These recreational properties are higher-valued and their higher values justify the precaution of locks, gates, fences and other means to enforce the property rights where this is necessary.

The benefits that landowners obtain from conservation improvements such as windbreaks are best measured by the improved productivity these improvements bring to adjacent agricultural lands, and even poor farmers in less developed countries seem willing to devote land and effort to these and similar conservation improvements. Mekonnen (1998), for example, uses Ethiopian data and an agricultural production framework to demonstrate the gains obtained from converting a small amount of cropland to bunds and trees to protect the remaining cropland from wind and water erosion. Yin and Hyde (2000) use a similar approach to calculate the 5.4 percent gain in agricultural productivity that farmers in China's northern plains obtained from intercropping trees with agricultural crops and from planting shelterbelts against wind erosion.

The benefits that landowners obtain from recreation and leisure properties are different, and they are better measured by the owners' willingness-to-pay for the properties themselves than by the improved flow of market-valued goods from the properties. The willingness-to-pay for these properties seems to grow with higher incomes. Certainly the private properties associated with recreation and leisure use increase in number as well as in individual market value as regional personal incomes grow, and certainly private hunting and fishing clubs and lodges for skiers and eco-tourists attract higher income participants no matter where the clubs and lodges may be located.[15]

---

15 Jeanrenaud (2001) describes these trends for Europe in general. Hugosson and Ingemarson (2004) identify the same shift specifically for Sweden. Jacobson (2002) identifies similar income-associated shifts for Florida and Zhang, Zhang, and Schelhas (2005) identify them for Alabama. Along with these shifts is a shift from forests owned by farm households, often with financial objectives, to absentee ownership of forests to satisfy recreational and leisure objectives.

Two examples illustrate the importance of recreational and leisure forest properties. Butler's (2008) extensive survey finds that more than seven million family forest landowners in the United States obtain some value from the scenic beauty or hunting, fishing, or other recreational opportunity on their forestlands. Of these, 15 percent or over one million family foresters own almost twenty million forested hectares containing vacation homes or cabins.

China's recent experience is especially illustrative of the growth in demand for forest recreational opportunity—and property—that coincides with higher incomes. The opportunity for individual private landownership is not as great in China. However, no visitor to China's forests can fail to be amazed at the rapid expansion in forest tourism and the increase in the number of hotels, restaurants, developed trails, shelters, chairlifts, and curio shops designed to support it. As real per capita income rose more than six-fold since 1978, tourism rose from 777,710 person-visits for all forms of tourism in 1982 to over 83 million visits to forest parks alone in 2002. Forest visits contributed 37 billion yuan (US$4 billion) to the economy in 2002, an increase of 25 percent in only two years (Sayer and Sun 2003).

The Bequest Motive

Some NIPF landowners hold their lands with the intention of bequeathing something of value to future generations. In the United States, for example, 20 percent of family-owned forests were inherited and 48 percent of current owners desire to bequeath their forests to their heirs (Butler 2008). Other landowners are happy to maximize the financial proceeds from the land as they manage according to motives we have discussed previously, and then bequeath the revenues to their heirs. A recent literature examines the first alternative and concludes that the decision to bequeath forestland is greatly dependent on the assignment of inheritance taxes.[16]

If bequests are taxed, then the sale of part or all of the forestland may be necessary in order to recover the funds necessary to pay the tax. The size of the final transfer received by the next generation will be smaller than the original bequest, smaller by the size of the tax. Therefore, higher inheritance taxes result in smaller net transfers.

Inheritance taxes, however, capture a share of market value alone. Some NIPF landowners and their heirs attach personal value to the land in excess of the market value of the land or its timber. This is the case when the land includes, for example, a second home that is the source of many happy family memories. These memories have personal value that cannot be transferred in market exchange—

---

16 Amacher et al. (2002) review this literature. Pan et al. (2003, p. 168), with a sample of Alabama landowners, suggest that higher death rates, therefore greater exposure to inheritance taxes, are a "driving force behind fragmentation and parcellization" of forestlands. DeCoster (1998) and Mehmood and Zhang (2001) concur with this observation.

and inheritance taxes do not capture a share of this personal value. Therefore, the likelihood of a bequest increases when landowners and their heirs attach such very personal values to their forest property.

Amacher et al. (2002) examine these explanations for a sample of NIPF landowners from Virginia. Twenty-five percent of landowners in the sample inherited their forestlands. The important observations from their sample are that timber price has a significant negative effect on bequests, while days spent on the property in non-consumptive activities and absentee ownership both have significant positive effects. All three observations are consistent with expectations. Greater timber prices mean greater taxable value and, therefore, reduced likelihood of a bequest of the land. Non-consumptive activities and absentee ownership both suggest a personal attachment to the land, a personal value in addition to that reflected in the market and, therefore, a value that can be bequeathed without risk of capture by inheritance taxes, and a greater likelihood of a bequest.

Multiple Motives

Some NIPF owners of commercial timberland also enjoy the hunting, fishing, or other recreation and amenity opportunities of their properties, and others, while they intend to harvest some timber or other extractable products now, also intend to bequeath to their heirs a property with a remaining abundance of both standing timber and outstanding non-market personal value. Similarly, many owners of recreation and leisure properties also harvest timber or other forest products, and some of them also desire to bequeath to their heirs a property with characteristics in addition to those that would maximize the continuing recreation and amenity value for the current owners. In some developing countries, we might anticipate that tradeoffs between timber/fuelwood and agricultural land use or timber/fuelwood production and watershed protection are more important than the tradeoff between timber and recreation/leisure that must be more common for NIPF landowners in developed countries. Nevertheless, it is clear that many NIPF landowners in both developed and developing countries desire multiple products from the use of their lands.

This means that, NIPF landowners are less likely than industrial and institutional landowners to be single-minded in their motivations for land management and use. As a result of their multiple motivations, the group of all NIPF landowners is not as responsive as industrial or institutional landowners to the price of any one forest product or to policies designed to modify the production or use of any one product. This section of the chapter examines the problem of landowners with multiple motives in more detail and then reviews two assessments that consider the effect of multiple landowner motives on the single output of timber.

The combination of uses for any landowner's property depends on the labor and capital available to the landowner and the natural characteristics of the land, as well as the landowner's own personal preferences. Each unit of land and each landowner will be different—as we can show with Figure 9.2.

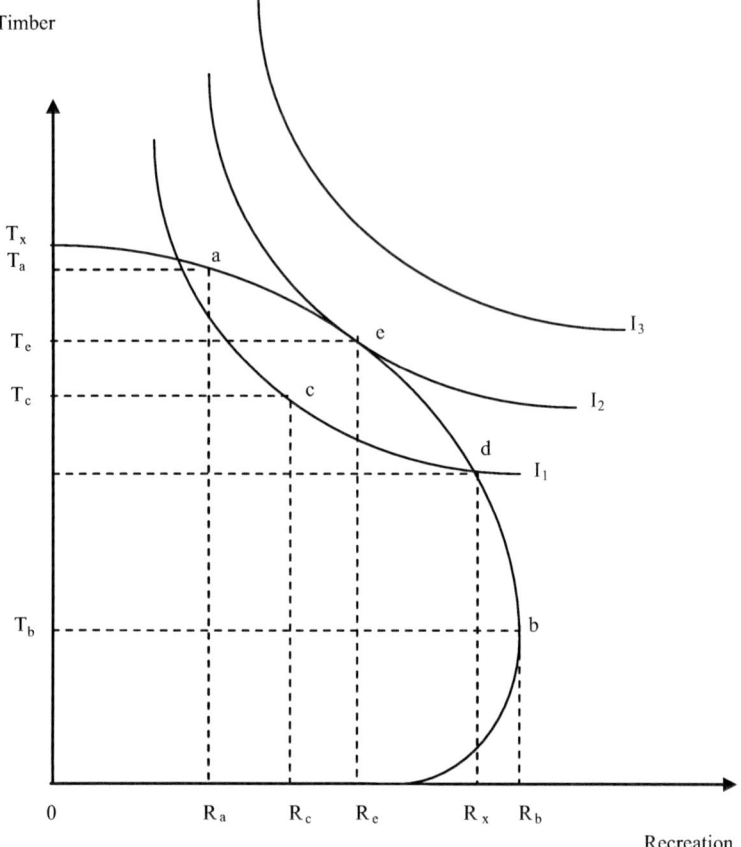

FIGURE 9.2 Multiple motives

The landowner can allocate all of the household's available labor and capital and all of its forestland to produce one product; for example, timber—in which case the land produces commercial or subsistence products equal to T_x. The same landowner could allocate all available household and land resources to produce a different output, say recreation opportunity equal to R_x. In fact, with the same household and land resources, the landowner can produce any combination of timber and recreation outputs described along the production possibility curve between T_x and R_x, for example timber value equal to T_a in combination with recreation equal to R_a. Stated a different way, the landowner can produce combinations of timber and recreation opportunity in excess of those described by this production possibility function only with additional household labor or capital or with additional forestland.

Over the greatest range of production possibility, between points b and T_x, the landowner can obtain an increment of one output, say timber, only by forgoing more of the other, recreation opportunity. However, on some properties, and for some selective uses, there may be a range of output over which additional timber production can also improve the recreation opportunity. For example, harvesting a modicum of timber can create additional habitat for certain flora and fauna. The range of production possibility between R_x and b in Figure 9.2 describes this case. When timber production increases from zero to T_b, then the recreation opportunity associated with the additional flora and fauna also increases—from R_x to R_b.

To this point, Figure 9.2 only describes the landowner's production possibilities. Landowners also have preferences in consumption which are different than their production opportunities. For any one landowner, these preferences in consumption might be described by the family of indifference curves, I_1, I_2, and I_3. Each indifference curve describes a single level of consumer satisfaction and all the different combinations of timber and recreation benefits for which the landowner is indifferent at that level of satisfaction. For example, the landowner obtains one level of satisfaction from the output combination T_a with R_a. The same landowner obtains the same level of overall satisfaction by giving up some timber income in return for partaking in more or improved recreation activity—as described by point c and the consumption pair, T_c and R_c, also on the same indifference curve. As the landowner forgoes more timber income, he or she must obtain ever greater increases in recreation in order to maintain the same level of overall satisfaction I_1 (as in movement from points a to c and then from points c to d). This explains the concave nature of the landowner's indifference curves.

The landowner obtains greater satisfaction from the combination of timber and recreation only by moving to a point on a higher indifference curve. The highest point possible with the given amount of household and forest resources is e, the point at which indifference curve I_2 is just tangent to the production possibility frontier and the landowner produces and consumes T_e units of timber and R_e units of recreation.

Of course, each landowner has a unique household endowment of available labor and capital for use on his or her forest property, and each forest property has its own productive characteristics. This means that each landowner's production possibility frontier will be different. Furthermore, each landowner's own preference for the consumption of timber and recreation will be unique. Therefore, each landowner's indifference map will be different from that of all other landowners. Together, the many different individual production possibility and indifference mappings mean that we should observe a vast array of different situations, each unique to a different NIPF landowner, with some landowners only producing timber, some using their lands for recreation and leisure exclusively, and yet others producing and consuming a broad range of individual combinations of timber and recreation. Furthermore, as production opportunities and consumer preferences change over time (and as individual NIPF lands change ownership), we can anticipate that the aggregate combination of timber and recreation production and consumption for all NIPF landowners must also change.

Binkley (1981) and Boyd (1984) began assessments of NIPF landowners in New Hampshire and North Carolina in the United States, respectively, with the general equilibrium approach outlined above. Both were strictly interested in NIPF timber production. Therefore, rather than comparing the timber motive with another single landowner objective such as recreation, they contrasted timber with the collection of all other motives/products that might cause landowners to forgo some timber opportunity. They did not assess the explicit NIPF tradeoff between timber and another single motive, and few others since Binkley and Boyd have done so either.[17] Nor have there been comparable assessments for NIPF landowners in developing countries.

Despite the very specific nature of their focus, the Binkley and Boyd assessments are broadly interesting for the NIPF landowner responses they do observe. Binkley examined harvest behavior for both a single year and also over a twenty-six-year period. Boyd examined behavior only for the short run of one year, but he also examined investments in timber stand improvements as well as timber harvest behavior. A crucial underlying observation for both Binkley's New Hampshire and Boyd's North Carolina landowners is that they do respond to timber prices but neither is as responsive to timber market incentives as they would be if timber were the only forest product.[18] Of course, this is exactly our expectation for landowners with multiple objectives, objectives in addition to maximizing timber revenues.

Binkley observes that greater stumpage prices in particular (but also pulpwood prices in some cases) are positively related to the likelihood that NIPF landowners will harvest timber. Indeed, those New Hampshire landowners who do harvest are very responsive to price, with an elasticity between two and five in the short run.[19] This strongly positive price effect is an indication that, for most landowners, additional income from timber harvests outweighs a broad range of loss in other forest values that might occur in association with timber harvesting.

Furthermore, Binkley observes that nominal prices are better predictors of landowner behavior than constant-dollar prices. This suggests that, indeed, many NIPF landowners do operate with a reservation price in mind. The better performance of nominal prices suggests a lag in landowner perceptions of inflation and this means that the "landowner establishes in his mind what he considers to be a 'good' price ... then sells if that price is reached or exceeded, irrespective of the real price at the time of sale" (p. 60). The very elastic short-run response is further evidence, suggesting that when landowners do respond to the nominal price, they do so aggressively.

---

17 See Newman and Wear (1993), Kline, Abt, and Johnson (2000), and Pattanayak, Murray, and Abt (2002).

18 Romm, Tuazon, and Washburn (1987), Max and Lehman (1988), and Hyberg and Holthausen (1989) each follow a similar approach to that of Binkley and Boyd. None observe an unambiguously positive effect of price on either NIPF timber management or timber harvests.

19 Kuuluvainen and Tahvonen (1999), with evidence from Finland, confirm Binkley's observation that price has a significant short-term effect on non-industrial private timber harvests. They too observe a short-term supply price elasticity greater than two for NIPF landowners.

Landowners with larger holdings in Binkley's sample are more likely to harvest timber, but those with larger incomes are less likely to harvest. Binkley reconciles the possible contradiction here by suggesting that the positive relationship between harvests and forest area may be a greater indicator of loggers' concerns with scale of their operations than of landowners' willingness to harvest. A different explanation could be that wealthier landowners are more likely to be motivated by recreational than timber values and are, therefore, less likely to harvest. Furthermore, we know that recreational properties, generally owned by those who are better off, tend to be smaller than those of landowners whose timber profits are a greater motivation.

The landowner's age has a positive effect on the likelihood of harvest in Binkley's sample—and this observation reflects on the bequest motive. Is the increased likelihood of harvest with advancing landowner age due to the prospect of imminent inheritance taxes? Or, as Binkley suggests, does the life-time income hypothesis provide a better explanation? The latter is very similar to the life cycle hypothesis. It suggests that landowners invest early, then hold their investments, including land and timber, as a hedge against financial need in mid-life years. With even more age, the landowner's financial horizon shortens and the need for a hedge against uncertainty diminishes. Therefore, with advanced age, the likelihood of harvesting increases and this, rather than the bequest motive may be the reason that the likelihood of harvest increases for some older landowners.

Finally, Binkley examined the differences between farm and non-farm landowners. Of course, non-farm landowners tend to be absentee landowners with greater occupational commitment to other activities away from their forestland. Their uses of their forests are more likely to be associated with a variety of leisure activities. Binkley's observations are consistent with this expectation. He finds that the oldest and wealthiest non-farm landowners are the most likely members of this category to harvest. The harvest decisions of farm landowners, in contrast, are unaffected by age or individual wealth. Their harvest decisions are significantly responsive only to stumpage price among Binkley's several independent variables and their price response is twice as elastic as that of non-farm landowners.[20]

Boyd's (1984) similar analysis with a sample of landowners from an altogether different region of the United States leads to similar conclusions with regard to price, income, and the size of landholding as determinants of harvest decisions. Boyd adds that landowners do respond to government programs designed to assist with silvicultural management (programs such as forest incentive payments, or FIP, discussed in chapter 4)—even when they have no intention of eventually

---

20 Loikkanen, Kuuluvainen, and Salo (1986) draw similar conclusions about the price responsiveness of NIPF landowners in Finland where farm forest landowners are price responsive while absentee forest landowners are not. However, Loikkanen et al. also note that fluctuations in timber prices are the most important short-term determinant of the decision to sell timber for either farm or absentee landowners.

harvesting their timber. And he adds that distance is an important harvest disincentive for non-farm landowners—an observation that is consistent with the three-stage development of commercial forestry described in this book.

The Binkley/Boyd observations of non-farm landowners may become increasingly important as those landowners are an increasingly large share of all NIPF landowners in many countries. If the likelihood of harvesting timber decreases as landowner incomes increase, as both Binkley and Boyd observe, and also as absentee ownership undoubtedly increases with regional development and personal income, then we can anticipate that the tradeoff between motivations for timber revenues and recreation/leisure opportunity will shift toward the latter with time and with regional growth and development. Will this mean that less timber is available for harvest and, if so, will the reservation prices of many landowners rise and will market clearing stumpage prices rise? Will the resulting increasing scarcity improve the market position of NIPF landowners relative to the monopsonistic position generally enjoyed by the millowners who purchase their timber today?

Synthesis

NIPF landowners are defined as owners of small parcels of forestland who do not also possess their own wood processing facilities. They are a diverse lot, and they are also a class in transition. Some manage their lands for their commercial or subsistence wood value, while others manage for a variety of environmental, recreational, amenity, or other non-timber values. Perhaps a majority manage for a combination of both timber and environmental, recreational, or amenity values.

Some are farmers whose forests are mixed with their agricultural properties. Others are absentee landowners whose occupational specialties differ from either agriculture or forestry. The latter are growing in number, as the former are declining, in both developed and developing countries. In the developed countries, as personal and regional incomes grow, the demand for leisure activities also grows—and a substantial part of the growth in this demand finds expression in forest-based recreational activities such as hunting, fishing, camping and skiing, and in increased demand for recreational second homes on forested properties. Many of these properties have been purchased from farmers and the likelihood of timber harvests from these properties declines with the transfer of property rights to the recreational landowner.

In developing countries, regional growth attracts workers away from agriculture and, as it does, some subsistence farmers obtain employment some distance from their homes and, thereby, become absentee managers of their agricultural lands. Some convert a portion of these lands to forest. Forest management is not as demanding of labor inputs as agriculture and the timing of the labor demand is largely discretionary—unlike the strict seasonality of the planting and harvesting activities for agriculture. Therefore, conversion to forestry is a way to obtain some continuing return from the land without placing an undue management burden on the landowner whose employment has taken him away from the home property.

With economic growth in most regions of the world, both developed and developing, we can anticipate that these transitions will become more common.

Small landowners exist within all three stages in the pattern of forest development. The small, generally subsistence, farmers in the first stage do not exercise property rights to forests but some may manage trees in their homegardens and all of the pioneer settlers in the first stage have access to the resources of the open access forest. Farm landowners in the second and third stages of forest development may also manage trees in homegardens or in shelterbelts along the periphery of their agricultural properties and some possess small parcels of commercially sub-marginal forestland contained within the boundaries of their agricultural lands. Some farm landowners in the third stage possess full rights to additional small commercial forest properties. And, of course, others possess rights to recreational forest properties wherever the natural landscape attracts recreational use within any of the three stages of development.

By the third stage of forest development, the distinction between farm and non-farm landowners is crucial.[21] In the developed countries of North America and Western Europe, in the Union of South Africa, and perhaps in other places as well, commercial gain from their timber seems to be the primary motivation for farm owners of forestland. Similarly, for many developing country farmers a combination of commercial and household subsistence gain seems to provide the primary motivation for holding forestland. The limited evidence is that these farm foresters are very responsive both to timber prices, particularly in the short run, and also to special household need. In the case of the former, their own greater familiarity with agriculture than with forestry causes many farm foresters to set an implicit threshold "reservation" price and to delay timber harvests from their productive commercial forestlands until they recognize a market price at this level. Since their holdings are small, their harvests must be irregular, but they harvest aggressively once the market attains the threshold price. Their additional holdings of isolated parcels of marginal timberland are a second reason for aggressive harvesting at the threshold price. Throughout most of the business cycle, harvests from these marginal forests are not commercially attractive. They

---

21 The available empirical literature often combines farm and non-farm landowners in the single broader category of all NIPF landowners, although econometric assessments may include a variable to distinguish between the two groups. This literature tends to focus on the timber supply responses of these landowners. Assessments of non-farm forest landowners alone or of strictly non-timber motivations for possessing forestland are virtually nonexistent. See Amacher, Conway, and Sullivan (2003) and Kuuluvainen, Karppinen, V. Ovaskainen (1996) for surveys of the literature, or Stordahl, Lien, and Baardsen (2008) and Zhang and Owiredu (2006) for more recent assessments for countries as different as Norway and Ghana, respectively. While these various assessments focus on different aspects of NIPF decision making, and the analytical approaches and the countries and regions that supplied the data may vary, the results of all of them are largely consistent with those of Binkley and Boyd summarized in the body of the chapter.

become commercially attractive at higher than normal stumpage prices and alert landowners do respond to the unusual opportunity for financial gain by harvesting from these lands as well when the prices are great enough.

Household need, whether for an unusual purchase by developed country farm foresters or in times economic hardship for those in developing countries, is well-documented as a second basic determinant of the timing of harvests of extractive resources by farm foresters. In this case, the forest provides an insurance against unusual circumstances for the NIPF landowner much like the insurance that the forests of industrial and institutional landowners provide in times of higher than usual input prices or adverse cyclical portfolio returns, respectively.

Non-farm landowners behave differently. Most non-farm forest landowners are absentee landowners and they are more often motivated by the recreation/amenity values associated with their forestlands. These landowners tend to possess smaller landholdings on average than farm foresters, and they tend to be above average in personal wealth as well as from regions of greater economic development. Therefore, non-farm forest ownership tends to be a larger share of all NIPF ownership in the more developed countries of the world and the recreation/amenity motivation for NIPF ownership also tends to be more important in developed countries.

As a group, non-farm landowners do harvest some timber from their lands but they are unlikely to be very responsive to timber prices. Better estimates of the values of their lands are captured in prices observed in the real estate market where unusually attractive natural landscapes, with their individually characteristic flora, fauna, and visual amenities are worth a premium to buyers searching for personal or family opportunity for forest recreation. However, for the sellers within this landowner group, a history of personal experiences on these lands, experiences such as special family occasions or successful hunting and fishing trips, adds unique personal value. This personal value is additional to the market value for the lands' recreational opportunity. As the real estate market cannot incorporate the additional personal value, it becomes a disincentive to sell, a justification for holding and bequeathing the property to an heir who shares the same experiences and the same additional non-marketed value for the particular land unit in question.

In sum, non-industrial private landowners are a complex group, including farm and non-farm landowners, some of whom live on or adjacent to their forest properties and some who do not, and some who intend to profit from commercial timber opportunities and some with additional or even very different motivations. Because of these complexities, the most revealing assessments of NIPF owner motivations or of their responses to market or policy incentives are likely to come from two-step analytical processes, the first step distinguishing the sub-group or motive in question (e.g., farm forest landowners or stumpage suppliers, absentee landowners or those with recreation/amenity motives) and the second step assessing the determining characteristics for this select sub-group or for those sharing their primary motive (e.g., stumpage supply or recreation/amenity values). Assessments of the full undifferentiated class of NIPF landowners, in contrast, as they lump a variety of landowners with disparate motives, are not likely to be as revealing of any particular pattern of behavior or as instructive for market analysis or policy.

Literature Cited

Amacher, G., W. Hyde, and B. Joshee. 1993. Joint production and consumption in traditional households: Fuelwood and agricultural residues in two districts of Nepal. *Journal of Development Studies* 30(1): 206–225.

Amacher, G., W. Hyde, and K. Kanel. 1999. Nepali fuelwood consumption and production: Regional and household distinctions, substitution, and successful intervention. *Journal of Development Studies* 35(4): 138–163.

Amacher, G., E. Koskela, M. Ollikainen, and M. Conway. 2002. Bequests and forest landowners: theory and empirical evidence. *American Journal of Agricultural Economics* 84(4): 1103–1114.

Amacher, G., M. Conway, and J. Sullivan. 2003. Economatic analyses of nonindustrial forest landowners: is there anything left to study? *Journal of Forest Economics* 9: 137–164.

Andersson, C., A. Mekonnen, and J. Stage. 2008. Impact of the productive safety net program on the livestock and tree holdings of rural households. Unpublished research paper, Environmental Economics Unit, Gothenburg University, Sweden.

Baumann, P. 1998. *Equity and efficiency in contract farming schemes. The experience of agricultural tree crops.* London: Overseas Development Institute.

Binkley, C. 1981. *Timber supply from private nonindustrial forests. Bulletin no. 92.* New Haven, CT: Yale University School of Forestry and Environmental Studies.

Boyd R. 1984. Government support of nonindustrial production: The case of private forests. *Southern Economic Journal* 59: 89–107.

Boyd, R., and W. Hyde. 1989. Forestry sector intervention: The impacts of public regulation on social welfare. Ames: Iowa State University Press, pp. 90–119.

Butler, B. 2008. *Family forest owners of the United States, 2006.* USDA Forest Service general technical report NRS-27. Newtown Square, PA: Department of Agriculture, Forest Service, Northern Research Station.

China Forestry Yearbook. 2000. Beijing: China Forestry Press.

Cooke, P. 1998. Intrahousehold labor allocation responses to environmental good scarcity: A case study from the hills of Nepal. *Economic Development and Cultural Change* 46: 807–830.

Conklin, M. 1966. The new forests of New York. *Land Economics* 42: 203–204.

Curtis, A., and D. Race. 1998. *Links between farm forestry growers and the wood processing industry: Lessons from the Green Triangle, Tasmania, and Western Australia.* RIRDC publication no 98/41. Canberra, Australia: Rural Industries Research and Development Corporation.

DeCoster, L. 1998. The boom in forest owners—a bust for forestry? *Journal of Forestry* 96(5): 25–28.

Dewees, P. 1995. Trees on farms in Malawi: Private investment, public policy, and farmer choice. *World Development* 23(6): 1085–1102.

Desmond, H., and D. Race. 2000. *Global survey and analytical framework for forestry out-grower arrangements.* Canberra, Australia: Department of Forestry, Australian National University.

Eid, T., H. Hoen, and P. Okseter. 2001. Economic consequences of sustainable forest management regimes at non-industrial forest owner level in Norway. *Forest Policy and Economics* 2(3-4): 213–228.

Gould, E., and W. O'Reagan. 1963. *Simulation: A step toward better forest planning.* Harvard Forest Papers no. 13. Petersham, MA: Harvard Forest.

Government of Union of South Africa. 1953. Forest policy in South Africa. *Unasylva* 7(4).

Gregersen, H., T. Houghtaling, and A. Rubenstein. 1979. *Economics of public forestry incentive programs: A case study of cost sharing in Minnesota.* Agricultural Experiment Station Technical Bulletin no. 315. St. Paul: University of Minnesota.

Hammett, A. 1994. Developing community-based market information systems. In J. Raintree and H. Francisco, eds., Marketing of multipurpose tree products in Asia. Proceedings of an international workshop held at Bagio City, the Philippines. Bangkok, Thailand: Winrock International, pp. 289–300.

Hänninen, R., and Y. Sevola (Eds.). 2009. *Finnish forest sector economic outlook 2008–2009*. Helsinki: METLA, the Finnish Forest Research Institute.

Hugosson, M., and F. Ingemarson. 2004. Objectives and motivations of small-scale forest owners: theoretical modeling and qualitative assessment. *Silva Fennica* 38(2): 217–228.

Hultkranz, L. (n.d.). *Commitment, irreversible investment & the utilization of forest resources: The role of forest owners associations in the development of paper pulp production in Sweden 1959–1985*. Arbetsrapport 103, Sveriges Lantbruksuniversitet Institutionen for Skogsekonomi.

Hyberg, B., and D. Holthausen. 1989. The behavior of non-industrial private forest landowners. *Canadian Journal of Forest Research* 19: 1014–1023.

Hyde, W., and J. Seve. 1993. The economic role of wood products in tropical deforestation: the severe experience of Malawi. *Forest Ecology and Management* 57(2): 283–300.

Hyde, W., J. Xu, and B. Belcher. 2003. Introduction. In W. Hyde, B. Belcher, and J. Xu, eds.. *China's forests: Global lessons from market reforms*. Washington, DC: Resources for the Future and Center for International Forestry Research, pp. 1–21.

INFOR (Instituto Forestal). 1997. National de Estudisticas, vicenso National Agropecuario 1997 [National Statistics]. Santiago, Chile: INFOR.

Jacobson, M. 2002. Ecosystem management in the United States: Interest of forest landowners in joint management across ownerships. *Small Scale Forest Economics, Management, and Policy* 1(1): 71–92.

James, L. 1950. Determining forest landownership and its relation to timber management. *Journal of Forestry* 48(4): 261–264.

James, L., and C. Schallau. 1961. Forestry applications under the Agricultural Conservation Program. *Land Economics* 37: 142–149.

Jeanenraud, S. 2001. *Communities and forest management in Western Europe*. Gland, Switzerland: International Union for the Conservation of Nature.

Kangas, J. P. Leskinen, and T. Pukkala. 2000. *Integrating timber price scenario modeling with tactical management planning of private forestry at forest holding level*. *Silva Fennica* 34(4): 399–409.

Kline, J., R. Abt, and R. Johnson. 2000. Fostering the production of nontimber services among forest owners with heterogeneous objectives. *Forest Science* 46: 302–311.

Koslowski, R., T. Stevens, D. Kittredge, and D. Dennis. 2001. Economic incentives for coordinated management of forest land: A case study of southern New England. *Forest Policy and Economics* 2(1): 29–38.

Kurtilla, M., K. Hamalainen, M. Kajanus, and M. Pesonen. 2001. Non-industrial private forest owners' attitudes toward the operational environment of forestry: A multinomial logit model analysis. *Forest Policy and Economics* 2(1): 13–28.

Kuuluvainen, J. 1989. *Nonindustrial private timber supply and credit rationing*. Umea, Sweden: Swedish University of Agricultural Sciences, Department of Forest Economics. Report No. 85.

Kuuluvainen, J., H. Karppinen, and V. Ovaskainen. 1996. Landowner objectives and nonindustrial private timber supply. *Forest Science* 42: 300–309.

Kuuluvainen, J., and O. Tahvonen. 1999. Testing the forest rotation model: evidence from panel data. *Forest Science* 42(4): 539–549.

Liu, D., and D. Edmunds. 2003. Devolution as a means of expanding local forest management in South China: Lessons from the last 20 years. In W. Hyde, B. Belcher, and J. Xu, eds., *China's forests: Global lessons from market reforms*. Washington, DC: Resources for the Future, pp. 27–44.

Loikkanen, H., J. Kuuluvainen, and J. Salo. 1986. *Timber supply of private nonindustrial forest owners: Evidence from Finland*. Tutkimutsia research report no. 30. Institute of Economics, University of Helsinki.

Lonnstedt, L. 1997. Nonindustrial private forest owners' goals, timber perspective, opportunities and alternatives: A qualitative study. *Scandinavian Forest Economics* 36: 89–98.

Lord, W. 1963. A reconsideration of the farm forestry problem. *Journal of Forestry* 61(4): 262–264.

Lu, Wenming, N. Landell-Mills, L. Jinlong, J. Xu, and L. Can. 2002. *Getting the private sector to work for the public good: Instruments for sustainable private sector forestry in China*. London: IIED.

Max, W., and D. Lehman. 1988. A behavioral model of timber supply. *Journal of Environmental Economics and Management* 15(1): 71–86.

McSweeney, K. 2005. Natural insurance, forest access, and compounded misfortune: Forest resources in smallholder coping strategies before and after Hurricane Mitch, northeastern Honduras. *World Development* 33(9): 1453–1471.

Mekonnen, A. 1998. *Rural energy and afforestation: case studies from Ethiopia*. Unpublished doctoral dissertation. Environmental Economics Unit, Gothenburg University, Sweden.

Mehmood, S., and D. Zhang. Forest parcellization in the United States: a study of contributing factors. *Journal of Forestry* 99(4): 30–34.

Muench, J. 1966. The impact of public versus private ownership on timberland in a rural economy. *Journal of Forestry* 64(11): 721–724.

Munn, I., and R. Rucker. 1994. The value of information services for factors of production with multiple attributes: The role of consultants in private timber sales. *Forest Science* 40(3): 474–496.

Neumann, R., and E. Hirsch. 2000. *Commercialisaton of non-timber forest products: Review and analysis of research*. Bogor, Indonesia: Center for International Forestry Research.

Newman, D., and D. Wear. 1993. Production economics of private forestry: a comparison of industrial and nonindustrial forest owners. *American Journal of Agricultural Economics* 75(3): 674–684.

Pattanayak, S., B. Murray, and R. Abt. 2002. How joint is joint forest production: An econometric analysis of timber supply conditional on endogenous amenity values. *Forest Science* 48(3): 479–491.

Pan, Y., Y. Zhang, and I. Majumdar. 2003. Population, economic welfare and holding size distribution of private forestland in Alabama, USA. *Silva Fennica* 43(1): 161–171.

Perz, S., and R. Walker. 2002. Household life cycles and secondary forest cover among small farm colonists in the Amazon. *World Development* 30(6):1009–1027.

RECOFTC (Regional Community Forestry Training Center). 2008. *Is there a future role for forests and forestry in reducing poverty?* Unpublished research paper. Bangkok, Thailand: RECOFTC.

Roberts, S., and O. Dubois. 1996. *The role of social/farm forestry schemes in supplying fibre to the pulp and paper industry*. London: International Institute for Environment and Development.

Romm, J., J. Tuazon, and C. Washburn. 1987. Relating investment to the characteristics of nonindustrial private forestland owners in northern California. *Forest Science* 33(1): 197–209.

Salam, M., T. Noguchi, and M. Koike. n.d. *Understanding why farmers plant trees in the homestead: Agroforestry in Bangladesh*. Unpublished manuscript. Faculty of Agriculture, Shinshu University, Japan.

Sample, A. 1996. Sustainability in forest management: An evolving concept. *International Advances in Economic Research* 2(2): 165–173.

Sayer, J., and C. Sun. 2003. Impacts of policy reforms on forest environments and biodiversity. In W. Hyde, B. Belcher, and J. Xu, eds., China's forests: *Global lessons from market reforms*. Washington, DC: Resources for the Future, pp. 177–194.

Scherr, S. 1995. Economic factors in farmer adoption of agroforestry: Patterns observed in western Kenya. *World Development* 23(5): 787–804.

Smith, W., P. Miles, J. Vissage, and S. Pugh, 2002. Forest Resources of the United States, *General technical report NC-241*. St. Paul, MN: U.S. Dept. of Agriculture, Forest Service, North Central Research Station.

Snelder, D., M. Klein, and S. Schuren. 2007. Farmers preferences, uncertainties and opportunities in fruit-tree cultivation in Northeast Luzon. *Agroforestry Systems* 71(1): 1–17.

Stoddard, C. 1961. *The small private forest in the United States*. Washington, DC: Resources for the Future.

Stoltenberg, C. 1954. Rural zoning in Minnesota. *Land Economics* 30: 153.

Stoltenburg, C., and H. Webster. 1959. What ownership characteristics are useful in predicting response to forestry programs? *Land Economics* 35: 292–295.

Stordahl, S., G. Lien, and S. Baardsen. 2008. Analyzing determinants of forest owners' decision-making using a sample selection framework. *Journal of Forest Economics* 14: 159–176.

Sun, C. 1992. Community forestry in south China. *Journal of Forestry* 90(6): 35–40.

Tomich, T., 1991. Smallholder rubber development in Indonesia. In D. Perkins and M Roema, eds.,

Reforming economics systems in developing countries. Cambridge, MA: Harvard University Press, pp. 250–270.

Uusivoury, J., and J. Kuuluvainen. 2001. Benefits of forest-owner collaboration and imperfect capital markets. *Forest Science* 47(3): 428–436.

Xu, J., and W. Hyde. 2005. From centrally planned economy to vigorous rural enterprise: China. In M. Goforth and J. Mayers, eds., *Plantations, privatization, poverty and power: Changing ownership and management of state forests*. London: Earthscan, pp. 154–174.

Yoho, J., and L. James. 1958. Influence of some public assistance programs on forest landowners in northern Michigan. *Land Economics* 34(4): 357–364.

Zhang, D., and E. Owiredu. 2006. Land tenure, market, and the establishment of forest plantations in Ghana. *Forest Policy and Economics* 9: 602–610.

Zhang, Y., D. Zhang, and J. Schelhas. 2005. Small-scale non-industrial private forest ownership in the United States: rationale and implications for forest management. *Silva Fennica* 39(3): 443–454.

Zivnuska, J. 1949. Commercial forestry in an unstable economy. *Journal of Forestry* 47(1): 4–13.

Appendix 9A:
Myths and Fallacies in the Tradition of Non-Industrial Private Forestry

The discussion in this chapter establishes the rationality of both the motivation and the behavior of non-industrial private forest landowners. In spite of this, a long list of forest policymakers have argued that these landowners fail to recognize their own advantage from either long-run private market timber investments or from environmental or conservation investments in their forests and forestlands. Their time horizons are too short for forest investments and their private gains from conservation investments are insufficient—or so the argument goes. As a result, NIPF landowners underinvest in their forests and this underinvestment is a justification for public programs designed to encourage silvicultural activity.

Furthermore, some of these same policymakers contend that public programs designed to assist NIPF landowners also have distributive merit because these landowners, as small scale farmers, are deserving of wealth transfers. Both the underinvestment and the distributive components of the argument appear in policy discussions in both developed and developing countries.

This appendix reviews the merit of both parts of the argument and, first reflects on the best arrangements of NIPF landowner data to address these beliefs and then asks a new question. That is, if the two components of this argument are not well-founded, then might economic reasoning support any other justifications for public programs designed to assist NIPF landowners?

NIPF Landowners Underinvest in Silviculture

The first contention is that non-industrial private forest landowners underinvest in silvicultural management. The usual reason given is that NIPF landowners are smaller operators who do not share the long-term investment horizons that go with the long production period common to forestry.

There may have been two sources for this point of view, the trained foresters' concern with the longer time horizons and the particular focus of U.S. forest policy in the 1930s. Foresters have always been concerned with long-run timber supply. They are trained to be concerned that reforestation should follow all timber harvests. Since, as we have seen, local markets, together with the natural environment create a three-stage development sequence in forestry, rather than narrow sustained yield on all lands, professional foresters and politically active environmentalists have encouraged public intervention to insure sustained forestry on logged over lands, like those in stages I and II, that are not immediately reforested.

The depleted public forests were their first focus but, with the establishment of the U.S. Forest Service in the 1890s, professional foresters eventually turned their focus to sustained timber production from the remaining (private) lands. Consideration was given to public regulation to enforce reforestation of cutover private lands but, at about the same time, some industrial firms began reforesting on their own—as we discussed in chapter 3. This left the non-industrial private lands as a focus for public action. Various U.S. states began regulating their private forestlands in the early part of the twentieth century to insure timber supply and, after 1968, for environmental protection. Some western and northern European countries and Brazil, Chile, and Ghana have introduced their own regulations in similar sequence—with mixed impacts as discussed in chapter 4.

The focus of enforcement always seems to be on the non-industrial forestlands. We can speculate that, because these lands are individually so small and numerous, enforcing some form of regulation would have been exceedingly difficult, especially in the early twentieth century when these regulations were first introduced. Furthermore, over time, a number of hypothetical farm forest budget studies seemed to support the contention that the earning potential of these lands and forests was greater than many landowners seemed to recognize and, therefore, that NIPF landowners underinvest. The expected difficulty of regulation and these farm budget observations became the bases for support for public programs like forest incentive payments (FIP) and extension forestry in the United States. Similar focus on sustained management, along with farm budget studies and the professional foresters' acceptance of them, has lent support to programs like FIP in other countries as well.[22, 23]

These may have been the historical reasons for the argument—but what is the evidence that NIPF landowners, and particularly farm forest landowners, do not have the long time horizons necessary for investment in forestry?

---

22 A senior forest development officer with the World Bank once made the case to me that Chile's FIP program was the source of the success of its commercial forest industry. He then took his argument a step farther and suggested that, with a similar FIP-like program, Peru could also develop a commercially viable forest industry. His contentions overlook the importance of Chile's favorable climate, its access to easy transportation to world markets, and its macroeconomic financial stability, each of which are probably more important factors for successful forestry and some of which were absent in Peru then and now too.

23 Tomich (1991) relates a history similar to the U.S. experience for the view that that small forest landowners in developing countries in general and Indonesia in particular underinvest in forest management.

Of course, the superficial evidence that some agroforestry tree species begin to yield either products or conservation benefits within periods as short as five years suggests that investment time periods are not necessarily long. The further evidence that farmers do make even longer term investments in facilities, such as barns and irrigation systems, and in equipment, such as tractors, contradicts the argument that their time horizons are insufficient.

The analytical evidence is at least as strong. Karpinnen's (1998) assessment is particularly careful, distinguishing between types of NIPF landowners in Finland. He notes that landowners who express multiple objectives for their forest lands are the most active silvicultural managers. Those landowners who focus on financial returns do not tend to actively introduce silvicultural improvements—perhaps because those are not financially rewarding. Those with recreational or amenity objectives are selective in their silvicultural decisions—as they must be in order to satisfy specific non-timber objectives such as improved wildlife habitat or improved scenic vistas. And regarding the financial return, Nyrud (2002) observes that the return on timber for Norwegian forest farmers is a very low 1.2 percent per annum. Carter, Newman, and Moss (1996) observe that the returns to industrial and NIPF lands in the U.S. Southeast are similar and that management efficiency of NIPF lands may even be slightly greater than for industrial lands. (Of course, we can anticipate that returns on industrial lands are diminished to the extent that some are held past narrow financially optimal dates in order to protect against future industry supply uncertainty. And narrow timber returns on NIPF lands are diminished to the extent that landowners have both other, non-timber objectives and also some financially marginal lands.)

The evidence from developing countries is all the more compelling because the farm forest landowners in these countries are almost all poor and, therefore, very much concerned with the importance of immediate returns for the daily livelihoods of their families. Yet an extensive literature shows that even these poor landowners do make longer-term investments, some of them in forestry. Schultz (1964, 1968) won a Nobel prize, largely for demonstrating the fundamental economic rationality of farmer motivation and behavior in developing countries. More selectively, Alemu (1999) shows the willingness of subsistence farmers in Ethiopia to forego current production and consumption in order to make watershed conservation investments that will yield longer-term agricultural returns. Yin and Hyde (2000) show that northern Chinese farmers earning, on average, only US$60 per annum, invested in trees for windbreaks that had an improving effect on agricultural production. In another region of China, farmers invested in trees with the expectation of eventual timber returns just as soon as they were confident of the longer term investment climate (Yin and Newman 1997). Amacher and his colleagues (Amacher et al., 1993; Amacher, Hyde, and Kanel 1999) show that Nepali farmers began to manage trees on their agricultural lands wherever the fuelwood prices were sufficient. Shively (1998) reports similarly for the Philippines.

We can only speculate that those who overlook this literature and still expect greater than observed returns from investments in forestry on the NIPF lands fail to recognize underlying distinctions in the data and the objectives of the landowners. They may overlook that some NIPF lands are economically sub-marginal throughout most of the normal economic cycle. Including these lands along with a farmer's more productive forestlands must decrease average expected yields on any investment, forestry or otherwise. Furthermore, they may overlook the comparative advantage of most NIPF landowners in some activity other than forestry and, therefore, the additional opportunity cost that landowners must incur to obtain full knowledge of their forest management opportunities. Finally, they may fail to separate farm forest landowners from those other NIPF landowners whose objectives tend toward preferences for amenity and recreation opportunity rather than timber production. After incorporating these adjustments in data and landowners classes, there should be no doubt that NIPF landowners do make longer term investments in forestry where those investments are justified by expectations of either conservation or commercial and subsistence returns.

Public Assistance to NIPF Landowners Has Distributive Merit

Are NIPF landowners among a region's or a country's poorer citizens and, therefore, justifiable targets for public programs with income redistributive merit? The affirmative contention probably originated as an additional justification for forestry assistance programs already dependent on the previous contention of underinvestment. It may have been valid in the United States as recently as the early part of the twentieth century. Be that as it may, is this second contention accurate today? Indeed, some NIPF landowners are not wealthy but most are not among the poorer citizens of the region either.

Landowners in general are not among the poorer populations that are usually associated with distributive programs in developed countries. In fact, both Binkley (1981) and Boyd (1984) and, more recently, Butler (2008) observe that the household wealth of NIPF landowners in the United States is above the national average. Butler reports that the median household income of family forest owners in the United States was between $50,000 and $100,000 in 2006. Boyd observes that wealthier landowners are more likely than those who are less well off to accept financial assistance from FIP and those who accept the assistance of extension foresters are likely to be even more financially secure than those who accept public financial assistance. Binkley adds that wealthier non-farm forest landowners are less likely than farm foresters to harvest the timber on their lands. Therefore, the best we can say is that FIP and technical forestry assistance programs do not successfully target a generally poorer segment of the U.S. population. Furthermore, the most likely recipients of these programs are not likely to be significant contributors to future timber supply—which was the original concern of the earliest, as well as many of the current, supporters of these government programs.

Of course, many farm forest landowners in developing countries are very much poorer than those in countries like the United States. As such, they may be reasonable beneficiaries of the bilateral and multilateral donor programs originating in developed countries. However, the desperately poor rural population in these countries is landless. The landless are absent from most of our assessments of the role of forestry or forest products in the rural development of these countries.[24] If their neighbors who own forestland are deserving of public assistance, then many of the landless are even more deserving.[25]

Conclusion

If neither the underinvestment nor the redistributive argument has much merit, then are there other public programs that might have justifiable merit for NIPF landowners? The brief sections of the next chapter that address forestry research and technical assistance add comment on this question.

Literature Cited

Alemu. T. 1999. *Land tenure and soil conservation: Evidence from Ethiopia.* Unpublished doctoral thesis, Environmental Economics Unit, Gothenborg University, Sweden.

Amacher, G., W. Hyde, and B. Joshee. 1993. Joint production and consumption in traditional households: Fuelwood and agricultural residues in two districts of Nepal. *Journal of Development Studies* 30(1) 206–225.

Amacher, G., W. Hyde, and K. Kanel. 1999. Nepali fuelwood consumption and production: Regional and household distinctions, substitution, and successful intervention. *Journal of Development Studies* 35(4): 138–163.

Binkley, C. 1981. *Timber supply from private nonindustrial forests. Bulletin no. 92.* New Haven, CT: Yale University School of Forestry and Environmental Studies.

Boyd, R. 1984. Government support of nonindustrial production: the case of private forests. *Southern Economic Journal* 59: 89–107.

Butler, B. 2008. *Family forest owners of the United States, 2006.* USDA Forest Service General Technical Report NRS-27.

Carter, D. D. Newman, and C. Moss. 1996. The relative efficiency of NIPF and industry timberland ownerships in the southern U.S. In J. Greene, ed., *Redefining roles in forest economics research: Proceedings of the 26th Annual Southern Forest Economics Workshop.* Starkville, MS: Mississippi State University, pp. 359–368.

Hyde, W., and G. Kohlin. 2000. Social forestry reconsidered. *Silva Fennica* 34(3): 285–315.

Karppinen, H. 1998. Values and objectives of non-industrial private forest owners in Finland. *Silva Fennica* 32(1): 43–59.

Nyrud, A. 2002. *Analyzing Norwegina forest management using a stochastic Euler equation approach.* Unpublished research paper. Department of Forest Sciences, Agricultural University of Norway, Aas.

Schultz, T. 1964. *Transforming traditional agriculture.* New Haven, CT: Yale University Press.

Schultz, T. 1968. *Economic growth and agriculture.* New York: McGraw-Hill.

Shively, G. 1998. Economic policies and the environment: tree planting on low income farms in the Philippines. *Environment and Development Economics* 3(1): 83–104.

---

24 Hyde and Kohlin (2000) survey a segment of this literature, the economic literature on fuelwood, and observe that the landless are absent from virtually all raw data used in this literature, despite their roles in supplying wood fuels from the public lands, in transporting it from all sources, and in consuming it. Surely, these landless are more deserving of distributive public programs. Yet our forestry literature seems to overlook them.

25 As Tomich (1991) argues, it is difficult to claim that a program that benefits a very small share of the population is equitable. Of course, his point is even more valid if that small share is not poor by local or national standards.

Tomich, T., 1991. Smallholder rubber development in Indonesia. In D. Perkins and M Roema, eds., *Reforming economics systems in developing countries*. Cambridge, MA: Harvard University Press, pp. 250–270.

Yin, R., and W. Hyde. 2000. The impact of agroforestry on agricultural productivity: The case of northern China. *Agroforestry Systems* 50: 179–194.

Yin, R., and D. Newman. 1997. The impact of rural reform on China's forestry development. *Environment and Development Economics* 2(3): 289–303.

10
PUBLIC LANDOWNERS

Public ownership, among the four major landowner classes, accounts for the largest share of all forests, perhaps 70 percent, or nine billion from a global total of more than 13 billion forested hectares.[1] Central government forestry agencies are the most widely recognized managers of these public lands, but national park and wildlife agencies, as well as provincial and local agencies in many countries, also manage extensive areas of forest.

Many public forests are the residual of all forests that existed without significant market value and prior formal private claim at some date when a central authority established its own permanent responsibility for their management. Of course, the date that central authority was established varies from country to country, but many developed countries established their central forestry agencies in the second half of the nineteenth century. Even as time has passed and values have changed since then, some of the lands managed by these agencies remain today beyond the margin of significant private commercial value. Meanwhile, other lands from which all commercially valuable resources have been extracted have reverted to public responsibility. For all of these lands beyond current commercial viability, the government forestry agencies have a caretaker or stewardship responsibility.

Nevertheless, also as time has passed and local values have changed, others of the original public forestlands have taken on new and significant value, some for their extractive resources like timber, minerals, and grassland; some for their outstanding natural beauty and recreational opportunity or for their historical or other cultural value; and some for their *in situ* ability to protect important environmental values. Still others have been transferred from private to public management, usually because their outstanding natural characteristics qualify

---

1 See footnote 1 in chapter 1. Protected areas alone, many of them forested, cover 12 percent of the earth's surface (UNEP 2008).

them as national treasures. Redwood, Acadia, and Grand Teton National Parks in the United States are examples. More productive public lands similar to these generated timber sales of 4.8 billion ccf (ccf = hundred cubic feet) worth more than US$104 million in 2009 (USDA Forest Service 2010). They produce crucial watershed services near cities ranging geographically from New York in the East to Portland in the West, and most of the 37 million hectares in the National Park System that hosts more than 270 million visitors each year are forested (National Park Service 2009) Government forest, park, and game agencies around the world, serving developed as well as developing countries, have responsibility for a similar diversity of values originating from the forestlands they manage. For all of them, the assignment of the lands they manage among their different potential uses, and the allocation of their budgets and personnel, remains a complex, and sometimes contentious, issue.

Public land managers face a third challenging responsibility, controlling the effects of natural hazards like wildfires, floods, typhoons, and insect and disease epidemics. The first of these, fire, may be the most difficult of all land management problems. The effects of these natural hazards can spread rapidly and the boundaries of private property do not deter them. Therefore, their control is truly a problem for entire communities of landowners spread over large regions. The public agencies are at the center of any solution, however, because the public lands are a source of fuels for wildfires, include bottom lands and coastal areas that are the first to be flooded, and host the stagnant older forest stands that often serve as vectors for disease. Furthermore, the public forestry agencies are among the few with the accumulated resources to combat many of these hazards.

The first part of this chapter sorts through these three public agency responsibilities: stewardship, particularly of the non-commercial lands; allocation among numerous financially more valuable land and resource uses; and protection and control of natural hazards. Many public forestry agencies have additional responsibilities as well, but they are most widely recognized for these three responsibilities.

The three standard justifications for public intervention in a market economy are efficiency, stabilization, and distribution. As for all other government activity, these are the economic bases for government ownership of forestland and for other government forestry services. Market failure and its opposite, improved efficiency, is the economic justification for providing non-market valued services on many public lands. Stabilization is more relevant for macroeconomic and broad regional policy that is generally outside the capability of government forestry agencies—although we will briefly discuss two less common government activities that fall under this justification, the price reporting services mentioned in the previous chapter and the community stability objective that will be discussed in the next. The distribution argument is a justification for two additional responsibilities common to most government forestry agencies, technology transfer and forestry research. Clearly, these two activities, unlike those previously mentioned, extend benefits to private lands and landowners. They reach beyond the agencies' responsibilities for the public lands

The technology transfer responsibility is like the extension responsibility of many government agriculture agencies. In fact, the branch of the government forestry agencies with this responsibility is generally known as forestry extension. Its objective is to share with private landowners the latest and best information regarding good forest management. Small private landowners may be the greatest beneficiaries of the new information, but getting appropriate information to them and encouraging its timely implementation is the challenge. We observed, in the discussion of technical assistance in chapter 4, that a clear understanding of the three-stage model is a crucial component in the determination of what information is "appropriate" and for which landowners.

For most central government forest agencies, the research responsibility includes research about new technologies to support the full forest sector, including all four landowner classes discussed in this book and often extending to the forest products industry's processing activities as well. Once more, there are lessons to be learned from the experience of public agricultural research. As in agriculture, the forest industry participates with its own research and the most effective allocation of uniquely public research money and personnel is an important question. That is, which categories of forestry research have the potential to yield benefits that are more completely captured by individual private investors, and which have the potential to yield a broader array of non-proprietary benefits for producers as well as consumers of forest products and services? Surely private forestry research can accommodate the former and public forestry research should concentrate on the latter. Public forestry research has an additional responsibility, however, that is unlike that of agriculture. Forestry is unlike agriculture in its extensive public lands, many of unique environmental and aesthetic value. Therefore, public forestry research has an additional role in producing answers to some of the specialized questions of managing these dispersed environmental and aesthetic resources.

Finally, as the public forestry agencies struggle with satisfying the public with its diverse opinions as to the best allocations of public forestland, and also with the natural hazards that impact the lands they manage and those of other landowners as well, the agencies themselves periodically reexamine their own most effective bureaucratic organization. A final section of the chapter will briefly examine this question as well. It is a question that has become particularly important in recent years for the forest agencies in those countries undergoing transition from centrally planned economies to more open markets, but surely the public agencies of all countries could benefit from periodic inquiry into their own most effective operation.

Public Land Management

The public agencies have an underlying stewardship responsibility for all of the lands within their jurisdiction, as well as responsibility for allocation among the different human uses on their economically valuable lands. The stewardship responsibility has to do with protecting these lands and their resources, whether they are of current economic value or not, from undesirable activity, whether caused by mankind or by nature. The two crucial components of this responsibility are the protection of formal boundaries against human trespass and the unwanted activities that often follow, and restricting the damage caused by natural hazards. The next section of the chapter begins by reviewing the difficulty of protecting boundaries and the public lands within them and continues with a review of the question of management for the multiple of acceptable and desirable economic activities that occur on a portion of the public lands, before returning to the question of protecton against natural hazards.

Stewardship

The public lands include some commercially valuable forestland—but they also include virtually all of the lands anywhere in the world that are beyond the boundaries of net commercial value. In terms of the three-stage model, they include much of the land beyond point B and all of the land and resources beyond points C and D in each of the figures. These are the lands beyond the point where the costs of protecting their property rights, described by the function T_r in the figures, are easily justified. While these lands are of negative net value, they do include some resources of positive value less than the costs of protecting their property rights (land between points B and C in stage I and between points B or B'' and D in stages II and III, respectively), including resources at the frontier of current exploitation of the mature natural forest at D. This makes these lands attractive for short-term exploitation and susceptible to trespass for activities that are often detrimental to the landscape.

Government agencies and private contractors both recognize the short-term advantage from exploiting some timber, grassland, mineral, or other resources out to point D where their net *in situ* value diminishes to zero, and governments may contract for the exploitation of some of these lands within a design that protects the long-term condition of the land. Some individuals seek to exploit these resources illegally, without the government agencies' permission, while others simply seek adventure by exploring even beyond point D. The government forestry agencies have the responsibility to protect all of these lands from long-term damage. In fact, this is the agencies' full responsibility for those lands beyond point D. They have no responsibility other than stewardship for these lands as the lands themselves are of no immediate commercial value.

In many cases, one component of stewardship is the responsibility for rescuing those who venture beyond point D but lose the means to make their way back, and for correcting the damages that some leave behind. These are the lost or disabled hikers and mountain climbers we read about in the news and the camp trash, and even old vehicles and vandalism that some leave as evidence of their presence. Resource values do not justify the agencies' expenses for these stewardship responsibilities. Rather, the public agencies justify these expenses as part of their duty to protect public property and its resources for the citizens of the country and for the unknown future potential of these lands and their resources.

Asserting this stewardship responsibility can be a challenging task for public agencies in both developed and developing countries around the world—as very many countries possess some lands of current value less than the costs of their protection. Furthermore, we have seen that activities like illegal logging occur to some degree in almost all forested regions and countries. Nevertheless, the stewardship responsibility can be more difficult for developing countries for at least three reasons: public agency budgets tend to be more limited in developing countries, and both lower wages and also less-developed institutions characterize these countries. We discussed the latter differences in the appendix to chapter 4 and described their effects within the three-stage model in Figure 4A.2, and again in chapter 5.

To repeat in brief: The physical boundaries of public responsibility may be formal and some boundaries are easily identified by roads and geographic barriers such as rivers and mountain ridges. Nevertheless, these boundaries, even when of a physical nature, are easily surmounted. The low-valued and dispersed nature of many of the resources at risk means that effective monitoring and enforcement of the boundaries and of the properties within them is expensive relative to the reward a few perpetrators can obtain from exploiting them—even if their exploitation also exposes the perpetrators to the risk of detection and payment of a penalty. Monitoring and enforcement are a drain on public agency budgets and, even if the agencies collect fines from those trespassers they detect, these fines are generally insufficient to offset the agencies' expenses.

Regarding wages and institutions, we recall that wages in less-developed regions tend to be lower relative to the dispersed forest resource values and, therefore, these resource values are a relatively greater attractant than they would be in more developed, higher wage, regions. We also recall that public institutions in general, and the institutions that assure property rights in particular, tend to improve with regional and national economic development. They tend to be less developed and, therefore, the institutions themselves, provide less assurance for local property rights in many developing countries.

This said, there are examples of improved, if still partial, success with the stewardship responsibility when the public agencies use local employees to assist with monitoring and enforcement. The involvement of volunteers in maintaining park facilities and in search and rescue operations in the national parks and forests of the United States is an example. But even willing and able volunteers cannot prevent all vandalism and other illegal activity, and they cannot cover all the costs of rescuing lost or disabled hikers.

In conclusion, determining the best allocation of public agency finances and personnel to the stewardship responsibility remains a challenging assignment. In fact, Robinson, Mahaputra, and Albers (2009) in their survey of the topic conclude that while the general academic literature on enforcement, fairness, and conflict is well-developed, its application to the special case of natural resource protection is not—and much remains, both for academic inquiry and for our understanding of improved on-the-ground policy and application.

Lands and Resources of Greater Commercial Value

Of course, the public lands also include numerous natural resources of positive net value, those on lands between points B' and B'' in the three-stage model or on lands in the neighborhood of point D that, because of their timber, grassland, minerals, watershed value, or attraction for recreational use, contribute to the current well-being of the human community. Some of these are best managed for one or another single resource use. For example, it is difficult to imagine combining either a grazing operation for domestic livestock or an active hardrock mineral operation on the same site and at the same time as an active high density tourist attraction such as an alpine ski resort or the Inn at Old Faithful in Yellowstone National Park. It is difficult to imagine the easy combination of many other resource uses with a watershed that provides drinking water for human consumption. However, there are altogether different occasions where two of these resources may be managed jointly to the benefit of both. For example, as timber harvests open the forest canopy, they create additional forest edge and grassland, thereby creating more opportunity for both browsing and grazing wildlife and domestic livestock.

Markets do exist in one place or another for many of the economically valuable services produced on the public lands. However, the public agencies seldom charge market prices for these services, and public land is not generally bought or sold. Therefore, additional information is often necessary and decisions regarding allocation between resource uses are more difficult for public than for private forestlands. The next two sections of the chapter review this problem, first examining the allocation of public lands to each of several single resource uses, then to combinations of uses on the same land unit.

Management for a single dominant resource value: The agencies contract with private agents for many of the uses of their resources—rather than conducting the timber harvesting, livestock, or mining operations themselves. Therefore, the concept for the allocation of land and resources between different uses of potential economic value is straightforward: follow market offerings for the land use in question—within the restrictions for the specific characteristics of each site and within management priorities defined by the public agency. These two restrictions are environmental in focus, designed to protect the surrounding land and resources and to assure future land use options, including safe and sustainable use after the contracted resource use is complete. This means insuring that the land is in good condition after the timber harvest is complete, the livestock have finished grazing, or the mine is depleted.[2] Protecting future land use options is consistent with

---

2 The agencies and their publics generally agree that depleting the timber or grassland and then leaving the land is unacceptable on most public lands—even if short-term economics justifies it. Similarly, depleting a mine and leaving an unsafe shaft or an unrecovered open pit, or leaving mine tailings to spill where they might is no longer acceptable on the public lands of most countries.

the stewardship responsibility that public agencies have for all of the lands they manage—including both those of no net economic value discussed in the previous section and those of commercial value discussed in this section.

Following the market means obtaining the highest bid price among alternative uses and among various bidders for the same use. For example, where contracts for mineral extraction would return more than contracts for timber extraction, then mineral extraction is the higher-valued and, therefore, the preferred use. Among competing bids for any particular resource use, the highest bid wins the contract.

The selection of winning bids, however, is more complex in practice, and there are important distinctions between assessments of the bids and, therefore, between contracts for different uses of forestland. The markets for some activities are not especially active and, where they are not, managers must rely on indirect methods to estimate the values associated with resource use.

The assignments of the various management activities associated with any particular resource use also tend to vary, and this adds further complexity—highlighting the distinction between the perspective of private bidders for use of the public resource on the one hand, and the perspective of the responsible public land management agency on the other.

For example, the construction and maintenance of roads necessary to access the resource is sometimes assigned to the winning contractor, sometimes assumed by the public agency, and sometimes excluded from any part of the agency's calculation of net resource value. This is crucial. If roads are the responsibility of the contractor, then we can be certain that contractors will decrease their bids by the amount of these and any other costs they expect to incur. If roads, or any other activity, are the responsibility of the public agency, then private contractors will ignore them in their bids, but the agency should be unwilling to accept any bid that is insufficient to cover the agency's assumption of these costs. If no competing bid covers these agency costs, then the agency should be unwilling to allow to the resource use that benefits from the road at this location at this time. If the agency disregards the cost by, for example, building and maintaining a road into a mineral lease or timber sale, but neglecting to account for the road costs as it reviews the various bids from potential logging or mining contractors, then how can it know whether or not the offer of the public resource is financially viable—and what justification does the agency have for offering public resources to private contractors without regard for these costs, costs that the agency must absorb in its budget and the public must absorb in its taxes?

Finally, the environmental restrictions written into the contracts for resource use vary as mineral operations, for example, impose different potential environmental risks than timber harvests and timber harvests impose different risks than recreational use. For each of these reasons; differences in the assessments and allocations of price, cost, and environmental risk; the actual evaluations of economic merit tend to be different for the different uses of the public land. The remainder of this section of the chapter considers these differences for timber, grazing, minerals, and recreation in turn.

Timber: Chapter 5 reviewed the fundamental characteristics of timber contracts. Obtaining fair timber prices is not generally a problem.[3] However, we might profitably add to the discussion of chapter 5 with further perspective on the cost accounting practices of some public agencies and on the changing assessment of environmental impacts. It remains important to separate the perspective of the public agency from the perspective of the contractor.

The public forestry agencies of many governments sell timber without regard for their own costs of managing the timber sale itself or of monitoring the performance of the successful concessionaire. For example, in some cases more common to developed country agencies with less limited budgets, the public forestry agency has built the roads necessary for timber removal—but not charged the contracting loggers for their use. In one estimate for a timber sale in southwestern Colorado in the United States, the administrative costs for the sale itself, plus the road costs, together amounted to over $30 per thousand board feet of harvestable timber that would have returned only one dollar of revenue per thousand board feet (Hyde 1981).

Furthermore, public agencies seldom include the costs of silvicultural activities over the years of the timber rotation in their calculations of financially viable timber operations. Some public agencies in the United States, France, Germany, and China, for example, perform many of these activities themselves. Others, in Canada and Indonesia, for example, include this responsibility in the successful concessionaire's contract. For those agencies that contract these activities, potential contractors adjust their bids accordingly, but these public agencies should include the administrative costs of monitoring to insure successful satisfaction of the agencies' silvicultural requirements in their calculations of the viability of long-term public timber operations. These costs can be substantial. After accounting for all of its own direct costs of long-term timber management and short-term timber harvest activities, many believe that the U.S. National Forest System as a whole has, not one year in its history, obtained a net financial gain from its timber sales (Barlow and Helfand 1980; Barlow et al. 1980; Zimmermann and Collier 2004).

The environmental risks associated with timber harvesting and the related management constraints written into timber contracts impose additional difficulty because they are so uncertain. At one time, uncontrolled fire was the greatest environmental risk associated with timber harvesting. The Peshtigo and Hinkley fires in Wisconsin in 1871 and Minnesota in 1894 were the results of indiscriminate disposal of the slash left after harvesting large trees and inattention to the small fires consuming some of this slash (Gess and Lutz 2002; Brown 2006). Both fires burned hundreds of thousands of hectares, destroyed entire towns, and killed unknown hundreds of people within only a few hours. Fortunately, by the twentieth century, this particular source of wildfire is well understood and, with few exceptions, timber harvesting is no longer a source of such great fires.[4]

---

3 Chapter 5 discusses analytical approaches for those cases where small markets and limited numbers of bidders may distort efficient prices.

4 Nevertheless, logging was one of several possible sources for the million hectare fire in Indonesia's Kalimantan in 1997.

The greater environmental risks from modern timber harvesting tend to be related to the impacts on water courses and aesthetics. Hillside logging and logging roads are sources of erosion and sedimentation in the local streambeds. Hillside logging, and clearcutting in general, can also leave unattractive scars in the forest visible to those who use the forest for other purposes. These vary with the local soils and topography. Therefore, predicting their potential impacts and controlling for them varies with each timber harvest site. This means that the environmental bond (discussed in chapter 5) or the form of the any contractual restrictions on harvesting must also vary from harvest site to harvest site.

It is clear that the science of timber harvesting has advanced and the probabilistic environmental costs associated with timber harvests no longer need to be as great as they once were. It is also clear, however, that some risks still do exist and that someone has to absorb the costs they impose. The public forestry agency is responsible for identifying the local environmental risk on forest lands it manages, protecting against these risks by writing management restrictions into timber contracts, monitoring to insure that successful private contractors follow the restrictions, and including the administrative costs of these activities in its own public agency timber accounts.

In sum, some public forestland is sufficiently productive that revenue from its timber sales is sufficient to cover the direct costs of both short-term timber harvests and longer-term sustained timber management. In these cases, timber harvesting is a benefit to the local economy and the national treasury. In other cases, as in some of our examples, the revenues from timber harvests are insufficient to compensate the agency's direct management costs. In either event, there can be no justification for accounting systems that fail to assign management activities and their costs to the outputs they help produce. The agencies are spending public money and managing public lands. They have a responsibility to their public to make their accounting complete and transparent.

Grassland and grazing for domestic livestock: Grassland is another marketable resource, one with limited, if important, environmental impacts and recognizable direct costs. The differences between valuing and administering contracts for timber concessions and those for grazing rights largely arise from the integrated pattern of grazing on multiple land units and the traditional access of some users of the public grazing rights.

Grass is a consumable, but renewable, resource. Therefore, the usual pattern of its use involves moving livestock onto one land unit where they consume its grass down to some acceptable level, and then moving the same livestock to a second land unit where they have access to fresh grass, while that on the first unit recovers. For example, cattle ranchers in the U.S. West often rely on their private lands and a store of feed in the winter, but herd their livestock to a sequence of higher elevation public lands for summer grazing.

The grazing rights on public land were generally established in the past, with either formal or informal permits for individual private landowner use. In the U.S. West, the permit is formal and presumed to be attached to the private property with which its use is integrated. The private landowner pays an annual fee for the land use, but the permit generally transfers along with the private property when ownership of the latter changes. Therefore, the value of the grazing permit has become capitalized in the private value of the ranch belonging to each permit holder. Establishing a competitive market value for the grazing permit would require an estimate of the difference in the value of the private ranch with and without the associated public grazing rights.

While grazing rights on these public lands tend to remain attached to an original private land unit, their continued use is conditional on the quality of the grassland. Public agency managers may limit these rights if the quality deteriorates. They may restrict the permitted amount of grazing or even withdraw the rights altogether where the land can no longer sustain it. This can be a crucial. Overgrazing is the greatest environmental risk associated with this land use. It is responsible for long-term changes in the pattern of vegetation as well as serious erosion in various places around the world. Protecting against overgrazing and its negative impacts is, once more, part of the stewardship responsibility of the public land management agency.

The public agency could impose land management requirements on the private permit holder, requirements such as water tanks for livestock and fences to control their movement, both designed to maintain the quality of the land and the grazing opportunity. In this case, the private permit holder would include the costs of the tanks and fences in the permit holder's own calculation of the financial merit of the full opportunity. Often, the public agency absorbs some of these costs as well. For example, in the U.S. West, the agency reseeds some grassland and constructs the fences that define grazing land units and restrict livestock use to approved areas. In these cases, the costs of seeding and fencing should be included in the agency's calculation of the financial merit of its offer of the grazing opportunity. For the agency, financial merit would require that the annual fee for the permit to equal or exceed the public agency's annual costs of administering and monitoring each land segment within the permit system, plus its annualized costs for capital improvements like seeding and fences.

The nineteenth century encouragement for western settlement, together with the continued political strength of agricultural interests, has created an imbalance in the United States in this regard. By some accounts, ranchers have 26,000 permits to graze livestock on the U.S. public lands. The federal government's annual collection of approximately $21 million in grazing fees contrasts with its expenditure of approximately $135 million to manage these lands (Borrell 2011). Of course, the alternative, wildlife use of the land, suffers and the public treasury and, therefore, the U.S. taxpayer make up the financial difference.

Minerals: The broad category of minerals; including hardrock minerals, oil and gas, and even gravel; is different than timber and grazing. Locating mineral deposits is generally a probabilistic activity because, unlike timber and grassland, the recoverable economic material lies hidden beneath the earth's surface. However, potential contractors for mineral rights are able to improve the probability of discovery and, therefore, the expected value of any specific sub-surface mineral deposit with modern knowledge of the region's geologic strata and modern technologies for detecting the depth, direction, quantity and quality of the sub-surface resource.

For the most part, where minerals exist in recoverable quantities, they are relatively high-valued and concentrated—in contrast with timber and grassland which are lower-valued and dispersed across the landscape. The greater value of mineral deposits justifies, first exploration, then removal, even at locations well beyond the economic frontier for timber and grassland. Because minerals are higher-valued, potential bidders for mineral concessions on the public lands tend to come from farther afield to bid for each deposit and their bids are generally competitive.[5] Finally, because the deposits are both high-valued and concentrated, the incentive to protect the rights to them is greater and the protection activity itself is more likely to be successful than for timber or grassland. That is, monitoring access to a mine and controlling what is taken from it is much simpler and less expensive than building a fence around an expansive mountain pasture and it is more likely to be successful than any effort to control illegal logging, grazing, or cattle rustling.

The assignment of costs is more consistent across mineral leases than for timber contracts or grazing permits. Mineral concessions simply assign all direct operating costs to the holder of the concession. Concessionaires build their own roads and erect their own fences.

These characteristics of mineral assessments make their evaluations straightforward—for both the prospective private concessionaires and the public agency. The public land manager has greater difficulties, however, assessing the environmental risks, both present and future, associated with mineral development. The mines, quarries, and wells of mineral, oil, and gas development interfere with both the aesthetic value and wildlife use of the landscape. Furthermore, the long-run effects of the tailings left after mining exhausts the economically recoverable resource are uncertain and they can be very large. For example, the hazardous effects of mine tailings from the apex of silver and gold mining in the U.S. West 150 years ago were unknown at that time. They were still not understood during the uranium boom in Colorado and Utah in the 1950s and 1960s. Yet, we know today that the U.S. Environmental Protection Agency and local communities spend large amounts of money controlling and cleaning up the erosion of hazardous materials from old mine tailings into local watercourses, and we hope that the public land management agencies will not overlook the risk to human health that comes with the potential re-emergence of uranium mining and exposure to radioactive material that could arise as we search for alternative sources of energy today. In general, the effects on wildlife are contentious and the long-run effects

---

5   In many historical cases, there was no contract. Miners searched the landscape (prospected) and established claims wherever they found the desired mineral. In other cases, the government land management agency auctions the right to search and mine. In the U.S. West, the former is still the case for most hardrock minerals. The latter is the case for oil and gas.

on watercourses and human health can be great, but their magnitudes are subject to serious debate. These uncertain effects seriously complicate the task of any public agency as it establishes environmental restraints within the contracts for mineral development on the public lands.

Watersheds: The economic evaluation of public watersheds is a still different problem. The benefits of the hydrological services of forests are highly variable and, while no one doubts their importance in any number of cases, direct evidence of the value of watershed services is sparse—although of increasing interest.[6] Furthermore, those local examples of watershed services that do trade in the market usually trade in bulk. That is, their evidence is not of a nature that can be converted into a delivered price per unit of water of specified quality from a unit like a hectare of the watershed.

The alternative is to focus on the desired product in any specific local case; water for irrigation, for industrial use, or for domestic consumption; and then assess the cost of producing water for this purpose by some different means. The different means could be a deep well, a treatment plant for water from a different source, seawater conversion, or whatever alternative means and sources are available locally. This alternative cost approach provides a first measure of the minimum value necessary to make the watershed a financially viable substitute resource.

The net value of managing a particular resource for its watershed services is, then, equal to this alternative cost minus the cost of other opportunities for that resource that must be forgone. These are the timber harvests, agriculture products, grazing for domestic livestock, or whatever would occur on the same land if it were not allocated to producing watershed services. The environmental restrictions necessary in order to deliver the expected quantity and quality of water determine which opportunities must be forgone.

Two substantial examples are illustrative. New York City began aggressive management of its traditional watershed in the Catskill Mountains in 1997. Had it not chosen to do so, its alternative would have been a $4–6 billion initial outlay for new water treatment facilities, plus $250 million in annual operating costs for these facilities. These alternative costs establish the minimum gross value of Catskill Mountain watershed management for the city.

The opportunity cost to landowners in the Catskills for satisfying the environmental restrictions necessary to meet the city's water management standards is equal to the $250 million real estate cost of those lands purchased by the city and removed from other production, plus another $100 million the city spends in annual payments to convince farmers of other lands in the same watershed

---

6 Chomitz and Kumari (1998) discuss the issue for forest watersheds in the humid tropics. Landell-Mills and Portas (2002) identify over 180 cases of markets for watershed services from countries around the globe and in a multitude of local institutional arrangements. The website of the Katoomba Groups (www.katoombagroup.org) and Johnson, White, and D. Perrot-Maitre (n.d.) identify still more examples.

to refrain from erosive and polluting activities (Kenny 2009). The net value for the city of its Catskill watershed is the difference between these alternative and opportunity costs. A positive net value indicates the feasibility of allocating this portion of the Catskills to provide watershed services for the city.

The two watersheds on either side of the Panama Canal are a second example. As these are deforested, they create a surge of water and silt into the canal, and the subsequent long-term regular flow of water to the canal declines. This puts the long-term operation of the canal at risk. This is a particular concern to large firms like Wal-Mart and those Japanese automobile manufacturers that would have to absorb the much greater costs of shipping through the Straits of Magellan at the tip of South America if the canal were no longer available. These firms currently purchase insurance against this event. The cost of the insurance is their alternative cost. The opportunity cost of the watershed is the expense that must be incurred to reforest and protect the two watersheds from subsequent human activities that cause deforestation. Is the alternative cost greater than these opportunity costs? ForestRe, a global insurance firm, believes that it is. ForestRe has put together a bond to cover the costs of reforestation. It is asking the companies that would benefit from continued operation of the canal to buy the bond (ForestRe 2009)

For some smaller and less spectacular private cases, a more direct calculation of the benefits and costs of water conservation investments like improvements in the irrigation system or shelterbelts to protect against desiccating winds is possible. However, for most publicly-owned watersheds or watersheds that provide benefits to entire communities, the alternative and opportunity cost approaches are generally necessary to establish an appropriate measure for decisions about land allocation to watershed services.

Forest recreation: Forest recreation includes an exceedingly wide range of activities, from those that make use of highly developed facilities like plush lodges for alpine skiers and international environmental tourists through less developed but still expensive equipment such as that used for floating wild and scenic rivers all the way to the least developed campsites and picnic grounds. And even among campsites, there is a range from those with modern shelters, toilets and showers, and electrical hookups for massive recreational vehicles to primitive backcountry sites for wilderness backpackers. There is yet another category of activities for hunters and fishermen, and this category too extends across a similar vast range from generously turned-out member-only clubhouses through expensive guide services to activities as inexpensive and common as a young boy hunting squirrels or a parent and child fishing in a neighborhood stream. At one end, forest recreation is a big industry. At the other, it is a parent's and child's simple afternoon relaxation.

Markets exist for many of these activities and in many countries—where they are offered on private land. But even where markets are known to work well, the public forestry, park and game agencies generally prefer to rely on a range of both market and non-market instruments for allocating these forest-based resource services. Table 10.1 summarizes this range of activities in column one and matches

TABLE 10.1 Forest recreation on public lands: Activities and economic instruments for allocation

Activity	Instruments for allocating access for the public resource
Alpine ski facilities Concessions (hotels, restaurants, shops) at natural marvels like Yellowstone or the Serengeti	Auctions for each major concession. The concessionaire charges the consumer a market price.
Guided mountain treks Float trips on wild and scenic rivers Wild animal safaris	Free to the guide service until congestion sets in. Thereafter, often ration to the companies that provide guide services according to their prior use. Guide companies charge the customer a market price. In some cases, the company transfers a small fee per customer to the public agency.
Developed picnic and campsites	The public agency sometimes charges a nominal fee for entry, sometimes no fee—then distributes permits till all sites are full.
Undeveloped and backcountry picnic and camp sites	Similar to developed picnic and campsites, except the demand is generally less for these numerous sites. Fewer nominal fee sites. Agencies distribute permits until the sites are full.
Hunting and fishing	Periodic fee for a license for adults, plus an additional fee for some highly-valued species. Ration by lottery when demand exceeds the number of permits available for these species.

them in column two with the instruments that the agencies commonly use for the allocation of lands and resources to each activity. The instruments range from competitive bids for the right to operate high-end concessions at alpine ski resorts, through nominal admission fees for some activities and various systems for rationing permits where demand still exceeds the resource capacity, to free admission to those resources with excess capacity. Fees for licenses are common for hunting and fishing, with a combination of additional fees and rationing for the most popular varieties of big game.

It should be clear that where fees are charged, then both the level of the fee and the location of its collection are important. If the fee is site specific, as when it is collected at the entrance to the site, then a market-based fee is instructive both for the allocation of a site among its potential alternative uses and for determining the level of management for the selected use. If, instead, the fee is charged without reference to a selected site, then it cannot be instructive for the management of any particular site. Kenya, for example, anticipates the many visitors to its large game parks by charging a fee for international tourist visas that is greater than for that for business visas or visiting family members. Tourists purchase their visas either in the tourists' home countries or upon arrival at an airport in Kenya. This system is easy to administer, but it does not reflect the tourists' preferences between different game parks and, therefore, is of little assistance to managers who must allocate budgets and personnel among parks. The U.S. National Park

System, in a similar case, sells Golden Eagle lifetime passes to senior citizens that permit entrance to all of the 392 national parks, monuments, seashores, battlefields, and other recreation areas it administers.

From Table 10.1, it should also be clear that many of the consumers of forest-based recreation are not poor. Some of these activities cost well over US$100 for a single days' recreational activity as well as an additional amount for restaurant and hotel services for tourists who stay more than a day. Others are less expensive by the day but the reusable equipment required for the activity may cost hundreds or even thousands of dollars.

Table 10.1 also shows that the public agencies very often charge less than the market price for the resource services of their lands. The agencies would not have to ration some activities if they did charge market prices. The market would adjust till the capacity of each facility just equals the consumers' desired level of consumption. This raises two questions. First, if the agencies do charge market prices to lease land and buildings to concessions like hotels, restaurants, and shops at natural marvels like Old Faithful geyser at Yellowstone National Park, then why not charge market prices for other forest-based recreation services? And, second, how can the agencies make efficient allocation decisions in the absence of reliable market information?

The allocation instruments in Table 10.1 follow agency tradition and the public land management agencies, like many large organizations, may be hesitant to change. However, the objections to market-based systems seem to originate largely from recreational users themselves and the organizations that represent them. Some object that it is their right as citizens to use the public forest resources for free. But is it their right to consume agency budgets and taxes paid by others? Some argue that charging market fees would exclude those who are less well off and cannot afford the fee. There are two responses to this argument. First, those who are less well off do not participate in many of these activities and for those in which they do participate, activities like picnicking and camping at less developed sites, there are many alternative sites. Market-based fees for these sites would be small. Second, many hunters and fishermen are not especially well off. Yet they expect to pay for their licenses. Why should other less-well-off users of these resources be any different? Finally, some argue that it is difficult to charge for widely distributed and low density activities like backcountry hiking and camping. This may be true in some cases, but the public agencies have no difficulty administering a permit system for those backcountry sites that are in greater demand and, in the United States, the various public agencies have begun to charge nominal fees for overnight camping at those backcountry sites that are in greatest demand.

Most would agree that the public forestry agencies are funded at levels that are insufficient to provide an acceptable level of maintenance for these recreational activities, and that the quality of the basic resource may be declining as a result. Most also know that the hunters and fishermen who pay for their licenses take pride in the fact that these fees constitute a large component of the budgets of public game agencies and, therefore, that their license fees contribute to the quality of their resource experience. Can we not imagine that other recreational users of forest lands would be similarly pleased to see an increase in the financial allocations to forest-based recreation on the public lands and take satisfaction in their own contributions to the increase if the fee were used to improve the quality of the resource and of their own experience using that resource? In fact, in the United States we do see gradual increase in the willingness of consumers of forest-based recreation to accept the idea of user fees, and the public land management agencies are gradually increasing the fees for the use of even their more dispersed campgrounds. Nevertheless, the accumulated fees remain less than the management costs for most sites.

All this aside, what about the second question regarding allocation decisions in the absence of market-based fees? These decisions—which sites and how much development—are largely the responsibility of experienced public land managers. Economists have a variety of techniques for assessing the non-market demand for recreation. The literature is extensive, but the techniques are expensive, time consuming, and beyond the experience of most public resource managers.[7]

The economics literature focuses on intermediate-value higher-volume activities at well-known sites. It also suffers from a weakness. It focuses on demand while showing almost no recognition that supply is also important for management decisions.[8] Yet all managers know that recreation management is costly. It is costly to build and maintain campgrounds and picnic sites, and the personnel who staff them demand significant wages and benefits. Even the smallest cost decreases the optimal scale of even the smallest campground—as Figure 10.1 shows. Ignoring costs, the optimal size of the campground in Figure 10.1 is at q_1. After consideration of even a very low level of variable management costs, the optimal size is the smaller q_2. Of course, for the more substantial variable costs that must characterize large campgrounds that employ many seasonal workers, the economically optimal scale of operation may be much lower than the zero cost scale comparable to q_1. (The decrease in optimal scale is smaller where demand is inelastic or supply is elastic, and greater for the opposite.)

Daniels interesting assessment (1986; Hyde and Daniels 1988) does examine the interaction of both demand and supply. Daniels examined three U.S. Forest Service campgrounds, each with multiple campsites in the Swan Valley of Montana. He used the travel cost methodology to assess demand for these campgrounds, and compared his results with historical annual and seasonal management costs plus salaries and benefits for the campground personnel. He draws the satisfying conclusion that, even without the benefit of a technical economic assessment such as his own, U.S. Forest Service managers made approximately optimal decisions regarding which campgrounds to open for seasonal use, when to open them, and

---

7   Mitchell and Carson (1989) and Freeman (1979, 1993) are classic references.
8   The textbook by Loomis and Walsh (1997) is instructive for recreation cost and supply but even this comprehensive volume fails to find an illustrative example that considers both demand and supply together for even one single site.

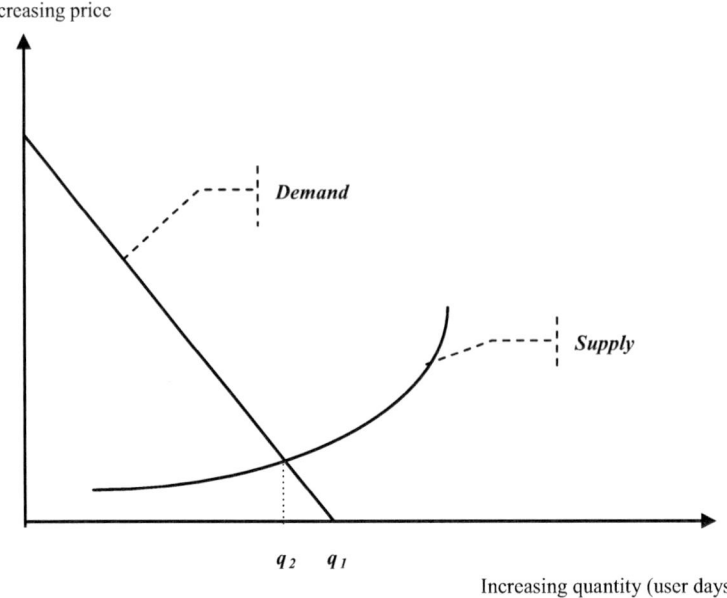

FIGURE 10.1 Recreation demand and supply

which campsites within the campground to open. More assessments like that of Daniels would help us understand the analytical problems of integrating recreation supply and demand, and teach us more about the bottlenecks that confront on-the-ground managers. Additional similar assessments would either confirm that the good intuitive economic judgment managers showed in the Swan Valley is the standard for U.S. Forest Service recreation managers—or show just how remarkable the allocation judgments of those Swan Valley managers really were.

In conclusion, recreation management is a difficult task that is becoming more difficult with increasing demand and smaller public agency budgets. At high-end facilities like the Inn at Old Faithful and most alpine ski resorts, auctions for concession rights probably allocate efficiently. At other sites and for other forest-based recreation activities some system of rationing permits is a common practice for controlling demand. Changing this practice and charging a market-based fee would simplify the process and add revenues for public agency management. Examinations of the elasticity of demand for a few activities and sites like these would provide an indication of just how much revenue might be collected and, therefore, provide an indication of the impact on agency budgets and recreation management. Of course, getting the public agency managers and those who use their services to accept market-based pricing is the challenge.

Multiple Use: Multiple use has been an important phrase in the lexicon of public agency forestry since the late 1950s when the U.S. National Forest System acknowledged increasing public interest in forest-based activities other than timber and began its effort to be perceived as something more that an agent for timber production. Times have changed. There can be no doubt today that the U.S. National Forest System, other federal and state agencies that manage forestland, and public forestry agencies around the world as well, are multiple use managers. Timber is still important for many, but the growth in demand for forest recreation has been phenomenal. The raw numbers are hard to imagine: more than 460 million visitors annually to U.S. national forests and parks, plus an untold additional number at state parks, fish and game reserves, etc. (National Park Service 2009; U.S. Forest Service 2010). The numbers for some other countries are sketchy but still impressive: 83 million visitors to China's forested national reserves in 1983, and rapid increases as China's population has grown wealthier and found more leisure time since then. Recreational visits to China's forests grew 25 percent in only two years between 2000 and 2002 (Sayer and Sun 2003). Even a smaller and less wealthy country like Laos recorded 226,000 ecotourism visitor days in 2006 (Whiteman, Noulak, and Broadhead 2006).

Meanwhile, minerals, oil, and gas remain important products, and leases of public lands allowing the removal of these resources are an important source of public revenues for some governments. Grazing on the public lands remains important in a few cases, and the value of the watershed services of the public lands continues to grow.[9]

At some times and on some forestlands the various uses come in conflict with one another. In these cases, the efficient economic recommendation is to allocate to the higher-valued use. In fact, concentrating higher-valued uses on the lands best suited for them increases their production on these lands and diminishes the demand for their additional production on other lands—leaving more of those other lands and forests for other resource uses. This is an important point, and one that is frequently overlooked. There is a tradeoff between managing all forests according to certain environmental and aesthetic standards or, alternatively, concentrating timber production on a few more intensively managed stands. Producing more timber and fiber on a few stands leaves less total land in timber and fiber production and more land available elsewhere for special environmental and aesthetic consideration.

Nevertheless, there will always be some public lands that are highly valued for two competing uses, much as Redwood National Park in California was high-quality timberland before it was set aside as a national park in 1968, or as China's panda reserve contains plots that continue to be important to the local human community for their agricultural production.

There will also be forests that can support two uses on the same land. Some of these will yield products and services of greater value when produced jointly

---

9 In the United States, the Bureau of Land Management and the U.S. Forest Service combined to collect $6.1 million in grazing fees in 2000—while spending $465 million for grassland management (Moscowitz and Romaniello, 2003).

than when produced separately. An extensive literature beginning with Gregory's (1955) article in the very first issue of Forest Science, still the major journal for international forestry research, reviews the conceptual economics.[10]

The practical accounting necessary to support a decision for multiple use ("joint production" is the economics terminology) is straightforward and simple. It requires separate and independent accounting of the revenues and other benefits obtained from each independent output, separate accounting of those costs that contribute to each output independently, and an additional collection of those common costs that contribute jointly to the multiple outputs of the land unit in question.[11]

The net benefits from the comparison of the first two accounts for each independent output must be positive to justify that productive activity:

$$B_{si} > C_{si} \quad \text{for } i = 1, 2, \ldots, n \quad (10.1)$$

where the B_{si} are the separable benefits associated with one output i, and the C_{si} are the separable costs that contribute only to output i and none other. Eq. (10.1) must yield positive net benefits for each potentially acceptable land use. If eq. (10.1) is positive for two land uses but not a third, then the third is not an economically acceptable activity on the land in question.

For the second step in the decision to jointly allocate land to more than one use, the sum of net benefits from all uses that pass the test of eq. (10.1) must exceed any production costs C_c that these outputs share in common.

$$\Sigma_i (B_{si} - C_{si}) > C_c \quad (10.2)$$

For example, if a unit of forestland produces valuable timber but the road into it also opens the area for good hunting, then the road construction and maintenance costs are common costs. If the net timber revenues are positive and the net benefits from hunting are also positive, then the only question is whether the combined net effects of the two exceed the common road costs. If they do, then multiple use timber and hunting is justified. If the net benefits from only one of the two activities, say hunting, are positive, then timber is not an economically viable use of the land unit. Hunting may or may not be a viable single use depending on whether its net benefits, calculated from the first equation, exceed the road costs which can now be assigned to only the one output.

Fortunately, the measures of separable and common costs are easily recognized in most public agency accounting systems and the only challenging question in an example like ours is whether the gross hunting benefits are sufficient to carry the decision. Unfortunately, errors are often made because managers overlook even a cursory approximation of this two-step calculation

---

10 Zhang (2005) includes a recent list of citations on multiple use and related topics.
11 Krutilla (1958) first laid out this calculation for multiple purpose river projects.

Natural Hazards

We are all aware of natural hazards—hurricanes, typhoons, and tornados, floods and avalanches, etc.—and of their effects on forests and human communities. In the case of at least one of these, tropical storms, trees in wetlands along the coastline, and in the riparian zone along large rivers can have a beneficial effect, absorbing some of the power of the storm and the sea and mitigating some of the impact farther inland.

More generally, natural hazards destroy mature forests. Of the various hazards, wildfire is probably the most destructive of forests. Of the hazards affecting forests, wildfire certainly has the greatest effect on adjacent human communities.[12] Furthermore, its probabilistic nature, the severity of the worst fires, the ingrained legacy of historical fire management practice, and the public attention that wildfires receive combine to make it the most difficult of all problems for forest management. It is a problem for private as well as public managers but, traditionally, it has been seen as a greater responsibility for public land managers because so much of the total land area in drier, more fire prone areas, is in public ownership and because public agencies have taken the lead in fighting forest fires.

Some fire is caused by man, intentionally or otherwise, and mainly in more populated regions. In the less populated Rocky Mountains of Canada and the United States, for example, natural (lightning) and man-caused fires occur approximately in a 3:1 ratio. As an indication of natural sources, in the days of fire towers in the northern Rockies (through the late 1960s), observations of 300 lightning strikes per night within the vision of one tower were common for the months of July and August. Some of these succeeded in igniting fires and, in the absence of fire control, some fires would burn along the same area of forest floor almost annually, thereby preventing any great built-up of dead fuels on the forest floor and any great risk to the surrounding resource. John Muir (1912) tells of actually walking through such fires in the Yosemite Valley in California in the early twentieth century. However, with fire control, the fuels do build up and the potential for large fires, already an occasional result of extreme weather conditions, becomes even greater.

Natural biotic conditions also have an effect. Insect and disease are endemic in the forest and, as individual trees age, they become weaker and more susceptible to these hazards. Species such as the spruces, firs, and pines tend to grow in even-aged stands and, therefore, entire stands tend to grow old together. As insects such as the spruce budworm or the western pine beetle successfully overcome the defenses of a few older trees, their populations grow and they more successfully

---

12 Wildfire is known on six continents—all except Antarctica. The international news often includes stories of vast forest fires, e.g., those in Canada, the United States, Spain, Greece, Siberia, and Australia in 2009. In one year, 2006, wildfire burned almost 15,000 square miles across the United States. In 2007, the twenty-seven largest fires in the western part of the country burned 1.2 million hectares. The cost of fires suppression alone was greater than $10 million for each of these 27 fires—for a combined total of $547 million (Brookings Institution 2008).

attack other older and weaker trees until vast areas of timber within northern New England and eastern Canada (spruce budworm) or the Rocky Mountains (western pine beetle) may be dead or dying at the same time.[13] This means that the forest fuels build up in cycles and when a heavy fuel load combines with unusual weather, the conditions for fire become severe and extreme fires occur. Unusual fire seasons tend to occur naturally every 60 years in spruce-fir forests, in perhaps similar periods in intermediate elevation ponderosa pine forests, but perhaps only every 500 years in some higher elevation pine and spruce forests such as those that burned in Yellowstone National Park in 1998.

The great problem, of course, is that great losses can occur when unusual weather and biotic conditions combine to create great fires. The losses to the natural environment are obvious and most recognize that extreme fires can be fatal to fire fighters and even to some local inhabitants. But even these are not always the full costs. In the very worst conditions, as in Indonesia's fires in 1997–98, the resulting haze and particulates covered Singapore and Malaysia as well as Indonesia and created approximately $1 million of respiratory damages to human health (Glover and Jessup 1999).[14]

The redeeming natural factor is that forests (if not human structures and communities) recover rapidly. Many pine species, for example, regenerate best when their cones are exposed to both direct sunlight and the scarified soil that occurs after fire removes the fuels on the forest floor and creates openings in the forest itself. The question is how to manage with the expectation of periodic unusual weather and biotic conditions and the resulting extreme fire conditions.

The U.S. Forest Service has managed with the policy objective to control all fires by 10 a.m. of the first morning after their detection and, failing that, to suppress any remaining fire by 10 a.m. of the second morning, etc. Warmer and drier late morning air and winds that tend to pick up after 10 a.m. each morning provide the rationale for this policy. Late morning and afternoon winds tend to fan the flames, increase fire intensity, and extend the damage.

Budget decisions have supported this policy. The U.S. public land and wildland fire management agencies submit budgets for fire suppression to Congress and Congress always approves, regardless of the budget level, regardless of its rapid increase in recent years, and regardless of the general opinion that greater funding for fire prevention (rather than fire suppression) would be more effective. For example, expenditures for fire suppression were 13 percent of the U.S. Forest Service budget in 1991. By 2008, suppression expenditures had increased to $1.9

---

13 A western pine beetle epidemic that began in the mid-1990s, has ravaged the inland regions of western Canada and the United States ever since. For example, by 2010 the infestation covers one million hectares of lodgepole pine in Colorado and over half of the lodgepole pine in British Columbia.

14 Also see Rittmaster et al. (2005) for an assessment of air quality health effects from wildfires in western Canada and Siminov and Dahmer (2008) for comment on fires and the spread of human diseases in northeast China.

billion or 48 percent of the total Forest Service expenditures that year for timber, forest recreation, roads, and all other activities (U.S. Forest Service 2010).

Clearly, this is not a satisfying system. As many fires are controlled, fuels continue to build up on lands which have not experienced recent fire, and when fires finally do extend to these areas, extreme fires do occur, losses increase, and expenditures for fire suppression continue to increase. Furthermore, as long as Congress approves all expenditures, fire managers have no little incentive to manage their costs.

The land management agencies now attempt to modify the build-up of fuels by allowing some fires to burn under controlled conditions. Nevertheless, some fires escape even in the presence of the most aggressive efforts to prevent, and then to suppress, them. Once they escape, many continue to burn until they reach geographic barriers or, more likely, until the weather changes. Meanwhile, over time, human structures have been erected in increasingly risky locations and the public's expectations for success with fire control have increased. Eight million new homes were built in the wildland-urban interface in the 1990s in the United States. The number of homes in the wildland-urban interface of the Rocky Mountain West increased by almost 70 percent during this period (Steelman 2008). Wildfire will eventually effect many of these. In fact, wildfire destroyed at least 900 homes in southern California in November 2008 alone (Portland Fire and Rescue 2008).

One response could be to prohibit expansion into this wildland-urban interface. This restriction, however, would not reduce the risk to the many dwellings already in the threatened area. A few communities have begun to take some responsibility themselves. Sixty-seven communities in Colorado, for example, have created their own wildfire protection plans and a few even impose restrictive covenants on all homeowners, requiring fuels management on their properties and even fireproof shingles on their houses (Bunch 2007).

Nevertheless, when the inevitable fire is not easily controlled, the public, its news media, and its political leaders use their hindsight to object—strenuously—to whatever errors they perceive in the practice of fire management.

What can be done about all of this? All prior fire policy and planning has been based on knowledge of the physical characteristics of fire behavior. Can the addition of economic insight help? The fundamental economics are simple. The value of the resource at risk is whatever it was before the fire and any effort to control it. The relevant variables are the losses due to fire and the costs of fire prevention, detection, and suppression activity. Losses and costs are inversely related—as increased expenditures on fire management control fires at smaller size and diminish the resource losses due to fire. Figure 10.2 shows this relationship.[15] The optimal level of fire management E^* is the level at which total losses and total costs are at a combined minimum, or where marginal losses equal marginal management costs for the combination of prevention, detection, and suppression.

---

15 Appendix 10a outlines the mathematics for the simple deterministic economic model.

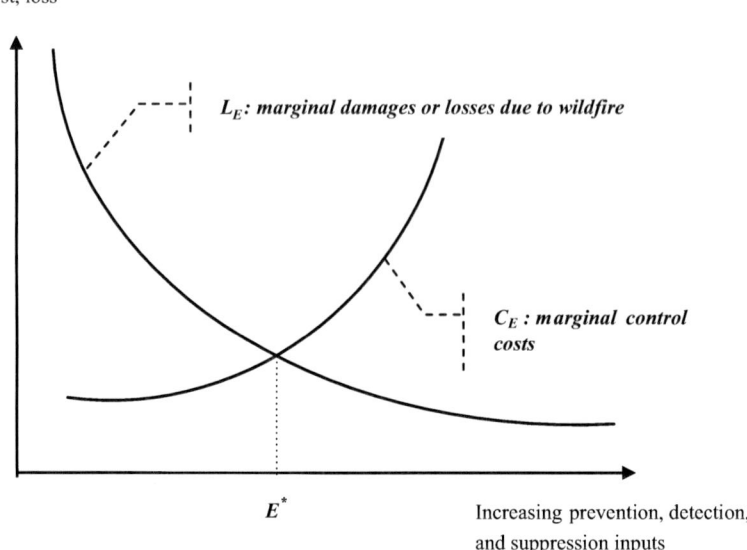

FIGURE 10.2 The wildfire optimization problem

This description is, indeed, simple and straightforward. However, it overlooks two most serious complexities. First, the several activities involved in fire control are themselves inter-related—as, for example, more effort spent on prevention should decrease the required level of expenditure and effort on suppression. Similarly, improved methods of detection reduce the necessary level of suppression associated with any level of loss. Second, the full impacts of each of these activities are uncertain—as is the weather and the behavior of the fire itself.

These complexities could be introduced in the specification of a more detailed model—if the correct specifications were known. In fact, the U.S. government land management agencies have developed sophisticated mathematical models to help them anticipate the physical characteristics of the latter stages of this experience. These models anticipate the rate and direction of spread under specific conditions of fuel, terrain, and weather. One example, the Wildland Fire Decision Support System of U.S. government agencies, converts its model into a program for hand-held computers that maps local conditions and anticipates fire behavior. On-the-ground fire managers use the program for tactical decisions, even while actively fighting fires, to assist in their suppression activities (WFDSS 2010) However, even these physical models of fire behavior cannot predict perfectly.

As a result, while managers know more about wildfire than they did even 20 years ago, and while their management options have increased, a most difficult problem remains. When the inevitable unexpected great fire does occur, we can anticipate that public opinion will challenge the wisdom of fire managers and force increases in suppression activities regardless of whether or not more suppression would successfully control the extreme event.

Finally, there is some evidence that the problem will only increase as global warming increases the risk of extreme fire conditions—and as wildfires release increasing amounts of carbon into the atmosphere (Westerling et al. 2006; Hennessy et al. 2008).

In sum, addressing the problem of managing for natural hazards in general, and wildfire in particular, remains a daunting task for the public forestry agencies.

Technology Transfer and Public Forestry Research

The public forestry agencies in many countries have two responsibilities, technology transfer and forestry research, in addition to their responsibility for managing the public lands. These added responsibilities serve the public by creating and sharing information that otherwise would be unavailable to consumers or to smaller forest landowners and smaller producers of forest products in competitive markets.

Larger landowners and larger producers can afford the costs of gathering specialized information about their products and they are more likely to be able to use this information to obtain advantage in their markets. Therefore, the justification for public assistance for these larger landowners is not as valid. The next two sections of the chapter examine these considerations more thoroughly.

Furthermore, as the long-run quality of the environment has taken on increasing importance, the environmental quality of forests and forestlands has become an increasing component of the public agencies' technology transfer and research responsibilities. Information on environmental quality is important to private producers, but it is also crucial for the managers and users of public forestlands as many of the forest lands and resources at greatest risk are in public ownership. This creates one difference between the similar technology transfer and research responsibilities of public forestry agencies and their counterparts in the agricultural ministries. The public does not generally own agricultural lands in most countries but it does own forestland. Therefore, it is a unique responsibility of public forestry agencies to develop and transfer improved information for public land management, and especially improved information about the specialized ecosystem services that these lands provide and the specialized environmental risks that occur on them.

Technology Transfer

The objective of the technology transfer responsibility is to share knowledge of the best current practices with consumers and with on-the-ground managers. Most landowners, for example, know the specialized characteristics of their own lands but many are not fully informed of the latest management technologies. Extension foresters, like agricultural extension agents, fill this gap as the bearers of information about new technologies. The task for these extension agents is to allocate their own effort efficiently, focusing on regions where the adoption of new technologies is most likely, on the few among many new technologies for which adoption is most likely, and on the producers within any region who are most likely to adopt them and demonstrate their benefits to neighbors who may follow with their own adoption of the newer technologies.

Two examples can help to illustrate these points. The first is generalized market price information and the second is information about improved silvicultural practices.

Market prices obtained in recent timber sales are known to the professional buyers who represent themselves or the mills that employ many of them. Market prices are also known to managers of large forestlands who are fully and permanently involved in producing timber and selling stumpage. Market log and stumpage prices are not as well known to the many small forest landowners who, in combination, produce the largest share of timber in many regions but who often find primary occupation outside of forestry and who individually sell only small and occasional volumes of timber. The previous chapter explained that these small producers are not as responsive to market trends and, when they do sell, that they often negotiate at a disadvantage relative to the better informed professional buyers of their timber.

Government extension services that broadcast the latest market prices to all buyers and sellers correct this information asymmetry in agriculture. Boyd and Hyde (1989), in an early assessment of Timber Mart South, a public sawtimber and pulpwood price reporting service, show that similar gains are likely in forestry. Specifically, they show that the gains come in the form of decreased volatility in the sawtimber market—and, in the case of Timber Mart South, these public gains exceeded the government agency costs of providing the service. Boyd and Hyde observe that, while both producers and consumers gained from more stable market prices, sawtimber producers gained over 3½ times as much as sawtimber consumers. Small landowners dominate production in the southern U.S. sawtimber market. It is likely that they received the largest share of the benefit from the price information made available by Timber Mart South.

For the second example, improved silvicultural practices, the questions of efficient allocation and likely adoption are complex but, once more, reference to the three-stage model of forest development provides important insight.

Numerous technologies; including many variations on site preparation, weeding, thinning, fertilizing, and pruning; while known to researchers and also to practicing foresters, are applied sparingly, if at all. Is the absence of more widespread adoption of these technologies simply a matter of sharing better information with on-the-ground land managers? Not at all. We have seen that landowners in regions in the first two stages of forest development have no market incentive to adopt these silvicultural practices. Sufficient inventories of natural timber are readily available for inexpensive harvest and these silvicultural technologies would bring additional cost with no additional financial gain for local landowners in the first two stages of forest production.

Landowners in regions in the third stage of forest development do have the market incentive to apply certain more intensive management practices. As product prices and the relative costs of more intensive technologies shift, the owners of forestland in regions in the third stage, even as they have adopted some more intensive silvicultural practices, may find that even more intensive practices are becoming commercially rewarding. Forest extension agents who monitor the trends in product prices and input costs are able to anticipate and recommend appropriate new technologies in response to these market shifts.

In addition, regions that are about to enter the third stage provide likely targets for the introduction of silvicultural technologies that are known elsewhere but, as yet, have not been adopted in these regions. Managers of forest lands in these regions have not had previous reason to apply these silvicultural practices, but they are likely to gain from knowledge of limited new silvicultural applications. Forest extension agents can identify the regions about to enter the third stage and advise these land managers as well.

Many who might benefit from adopting rewarding new technologies are cautious about doing so. Therefore, successful extension agents target their initial efforts in any region toward land managers and land owners who can afford an uncertain chance with a new technology. These are often larger, better educated, and wealthier landowners. However, forestry extension is an activity of public forestry agencies. Their responsibility is to assist wherever the market fails or wherever there is some inequitable distribution of information about forest technologies. Therefore, other landowners who are less able to afford the initial risk are the more appropriate target of public forestry extension. If they follow rapidly upon first observing successful adoption by the local leaders who can afford the investment in the new technology, then forestry extension has supported an "appropriate technology" transfer and the transfer has been an effective use of public agency finances and personnel.

In sum, this means that effective forestry extension monitors market trends in both the prices of primary forest products and the costs of applying current and alternative technologies. These trends are indicative of those products and regions that either are about to enter stage III of their development or are already within stage III and are likely candidates for more advanced technologies. Forest extension agents should expect their greatest impact from introducing new technologies that save output or delivery costs per unit production for those who satisfy two characteristics: they are the most likely early adopters of new technologies and they are community leaders who will share their successful experiences. Nevertheless, the real target audience for this public activity is those smaller and less-informed land managers and who are less able to afford the initial risk of investment in an unfamiliar new technology but who will follow along in adopting the new technology as they observe the success of their neighbors.

Furthermore, alert forest extension agents who are aware of local market trends are also in good position to identify both bottlenecks in the marketing of forest products and significant cost barriers to improved production. These bottlenecks and cost barriers suggest useful areas of research for the development of improved technology. By identifying these and sharing the insight with public forest researchers, forest extension can provide crucial feedback to the third (research) arm of the public forestry agencies.

Public Forestry Research

The objective of research is to create a means to produce more or better output from any given amount of input; for example, to produce more lumber or paper from a fixed area of forest or a fixed volume of harvested logs, or to produce the same amount in a less costly way.

Forestry has a long and graphic history of progress of this sort. Two examples (from many) show the immense differences in production that can occur. In the early twentieth century, minimum useable log size in the Pacific Northwest was 32 feet in length and 11½ inches in diameter. By 1980, sawmills in the region were manufacturing lumber from logs as small as eight feet in length and four inches in diameter. Research in the late 1930s showed that young southern pines are free of resin. The impact was rapid and large as mills began producing pulp from southern pines. The South became the largest paper producing region of North America by the 1950s.

Some technological changes are due to insightful modifications made by manufacturing personnel working within the mill. Others are the result of successful well-designed research conducted by engineers and other scientists. Lazers on sawmill headrigs that show the best orientation for the initial log cut, the powered back-up roller in plywood mills that distributes pressure evenly all along the log as a long parallel knife peels veneer from it, and various modifications now applied in the manufacture of pulp and paper to control pollution are all results of carefully designed research conducted at either government or private forestry research facilities around the world. All reduce either the private or the social/environmental costs (or both) of production and, as they save on inputs, many also decrease the demand for logs, and thereby decrease the pressure on the forest environment.

The savings to the forest itself from these technological modifications can be phenomenal. Consider just two U.S. examples. The powered back-up roller increased production per unit of log input in plywood mills by 17 percent. If the quantity demanded remained unchanged, then this translates to a smaller but still significant decrease in overall demand for raw logs from the forest in the 1980s when the backup roller was developed and introduced. For the second example, Wahlgren (G. Wahlgren, personal communication, February 22, 1989) calculated that the truss frames technology, developed at about the same time as the powered back-up roller, sharply reduced the demand for sawlogs. Truss frames assembled in quantity in manufacturing yards use 30 percent less lumber than structural framing cut and assembled at the construction site. The annual saving was equivalent to the entire potential annual timber harvest on all controversial (RARE II) wilderness areas.

The general criteria for economically successful research investments are familiar. The net gains from new products must exceed their research and development costs. And for existing products for which research contributes an improvement in the manufacturing technology, the discounted time stream of future cost savings in production due to the research must offset the discounted time stream of all expenditures made in accomplishing the research.

An extensive literature measures the economic gains from research in agriculture. It finds a range of returns on specific *public* research investments varying upward to 110 percent per annum on cotton research in Brazil and rapeseed research in Canada.[16] This literature encouraged more recent inquiry into the economic returns to forestry research where Hyde, Newman, and Seldon. (1992) observe marginal annual returns on publicly financed research ranging from the neighborhood of 300 percent on softwood plywood research in the 1980s downward to negligible for investments on research designed to discover improved practices for commercial timber management. Of course, the smaller observed returns to timber management research are consistent with our knowledge of the three-stage model, and our observations of the plentiful inventory of accessible natural forest that still exists in most forested regions of the world. That is, research in forest management for timber production has created new insight. The plentiful inventory of accessible natural timber has just meant that these insights remain "stored" and ready for implementation where and when the accessible natural inventory dwindles, more regions move into the third stage of forest development, and the demand for a substitute source of wood increases.

Both the agriculture and the forestry literature have tended to focus on public research and this encourages a question about the distinction between appropriate roles for private and public research activity. The distinction is dependent on two characteristics of the distribution of gains from research. Do consumers or producers gain more from the research and product improvement and, if producers gain more, can the producers who fund successful research claim sufficient returns to cover their research and development costs?

Consumers are the greater benefactors from research on products with very inelastic demands. Thus, even producers of those products who do obtain sufficient returns to cover their R&D costs on such products are likely to invest at less than a socially optimal level. Many agricultural products fit this description. In forestry, lumber has a moderately inelastic demand price. Therefore, research investments in the sawmill industry may fit this description.

Public investment is also justified in markets with many competitive sellers. The largest share of benefits from R&D breakthroughs in these markets may accumulate to producers, but the producers may be so numerous and the product so undifferentiable that no single producer can recover sufficient revenue to justify its own research program. Once more, the sawmill industry with its many small firms is an example in forestry.

Producers with larger market shares—who are also often producers of specialized and differentiable products—can control the largest share of either cost

---

16 Ruttan (1980) summarizes this extensive literature.

savings from technological improvements or revenue gains from new products that result from their R&D efforts. The pulp and paper industry is a clear example in forestry. In these cases private operators possess the market incentive to conduct their own research, and it is more difficult to find an economic justification for public research expenditures.

Finally, the economic justifications for public research are convincing for two increasingly important cases that are not widely captured in existing markets: first, research having to do with unique characteristics of the vast land areas managed by the public forestry agencies and, second, research having to do with the very broadest social services of forests. Research on biodiversity and research on carbon sequestration and global change are examples of the latter. Private investors find little financial reward from research investment in these areas of inquiry. The public in general, however, has much to gain from improved knowledge about diverse habitats and their protection and from knowledge about climate change and the means of limiting it. In the United States, for example, U.S. Forest Service researchers are involved in both of these activities. Research having to do with forest fire and the widespread effect of smoke in much of Southeast Asia is another more specific example.

A Note on Public Organization and Administration

Everyone has an opinion about public organization and administration. Either the government is too big, or it should grow to do more about some preferred specialized activity. In forestry in particular, perhaps the government forestry agencies are too large or too small. Perhaps they should spend more on timber and less on recreation, or vice versa. Perhaps, within their timber activities, they should spend more on the equipment used in fire control less on personnel—or more on prevention and less on suppression.

Economic analysis has a long history of examining similar opinions for the producers of market-based goods and services. Only recently, with the development of two new analytical techniques, has objective assessments about resource allocation for non-profit institutions such as hospitals and for the government agencies that largely produce non-marketed outputs become more feasible. These techniques, known as data envelopment analysis (or DEA, a linear programming approach) and stochastic frontier analysis (an econometric approach) review the inputs of multiple similar establishments to determine

- which establishments operate on the production frontier and which operate less efficiently below the frontier, and
- which inputs to production are used efficiently and which tend to be used in inefficient excess.[17]

---

17 Fare et al. (1989), Greene (1993), and Lovell (1993) are standard references. *The Journal of Productivity Analysis* features applications of these two techniques for assessing productivity.

For natural resources, Kao and Yong (1991), Vittala and Hanninen (1998), and Rhodes (1986) used DEA to assess resource allocation across Taiwan's public forests, Finnish Forestry Boards and U.S. National Parks, respectively. They observe potential savings in the use of inputs in the range of 20 percent from more widespread adoption of the practices of the more efficient forestry or park units in each of the three countries; that is, if the less efficient administrative units of these agencies adopted the practices and procedures of their most efficient colleagues in the same agency and country. Similar assessments of government forestry agencies could provide information about any number of questions having to do with the alternatives of more or less centralized administrative structure or the allocation of personnel or other resources to various tasks.

One assessment to date uses the stochastic frontier approach to examine comparative timber production efficiency in 40 compartments of Poland's state forestry agency (Siry and Newman 2001). We might expect to find some range of differences between these 40 compartments as they, like Poland's entire economy, undergo the transition from Soviet-style centralized planning to, perhaps, greater market orientation. Siry and Newman observe that twenty-six of the forty operate more than 40 percent below the production frontier established by the most efficient compartments. In particular, many compartments are inefficiently small and they suffer from an excess of administrative personnel (but not an excess of forest workers). Their operation gains efficiency when they out-source some timber management and timber harvest activities.

Imagine a 40 percent difference in efficiency between different offices in the same agency. Imagine a 40 percent increase in efficiency simply by observing and adopting the best administrative structures and management practices of one's own colleagues. Forty percent is a phenomenal improvement. Of course, the Siry-Newman observations are valid only for Poland and, since Poland is in a period of rapid economic and policy transition, perhaps more recent data would cause these authors to revise their observations. However, this does not change the fact that the Siry-Newman results for Poland truly stand out. They almost beg us to ask similar questions of operational efficiency, to apply similar analytical techniques, elsewhere for other forestry agencies around the world. Surely the discussions of public forestry agency organization and administration in many countries could benefit from similar analyses—for their timber activities and for other agency activities as well.

Conclusions

Two crucial characteristics define the public forestlands and their management: first, their vast size and diversity, and then the difficulties their managers have ascertaining the public will and the additional difficulty they have reconciling this with public law and institutional forest agency preferences.

The public forestlands are the largest in total area of any of the four ownership categories (industrial, institutional, non-industrial private, and public), and the central government forest management agencies tend to be larger individual land holders than any other single private or corporate forest landholder in most countries. The public forestlands are more diverse because, while they often include some commercial forest, their greatest total land area tends to be in the open access lands beyond points B or B'' in the descriptive figures and in the even less accessible lands beyond points C and D. As many public forestlands are less accessible and less commercially desirable, they generally include a greater variety of terrain and, therefore, greater biotic variety as well.

Their diversity would be a disadvantage for industrial landowners with their more singly focused and specialized objectives, and probably a disadvantage for institutional landowners for the same reason. The greater numbers and variety of non-industrial private landowners suggests that some of this latter group would individually find advantage in some, but not all, of the diverse features of the public lands.

This diversity can also create difficulties for public land managers. For example, diversity means that the public lands are most certainly a source of greater environmental risk than that experienced by managers of private forestlands. However, the diversity of the public lands also provides the means for satisfying the very widest range of public agency responsibilities: including endangered species protection, a great variety in outdoor recreational activities, the widest possible range of habitat for native fish and game, etc. In this respect, diversity is also an advantage for the public forestlands and the social responsibilities of their managers.

The second fundamental characteristic of the public forestlands has to do with their management, with (1) ascertaining the public will for their management and (2) reconciling this with both the statutory and regulatory obligations of the land management agency and the institutionalized preferences of its managers. Certainly the private decisions of the managers of the other three categories of forestlands are simpler, whether within the strictures of corporate organization (for industrial and institutional landowners) or smaller family operations (for non-industrial private landowners).

The public demonstrates its will through market-like economic information and also by democratic or representative democratic vote. Commercial markets, as in auctions for public timber or payments for some livestock use and mineral extraction, show some economic preferences. Others must rely on indirect assessments using more complex techniques for identifying the social valuation of many non-market activities and services of the public forestlands. Displays of democratic preference also come in multiple and often conflicting forms: the voices of elected public officials, the shared perspectives of those participating in organized fora on management alternatives, and the perspectives of selected members of community oversight boards.

The public agencies are generally very good about seeking a wide range of input from the various perspectives of the broadest collection of users of the public lands—but conflicts in these perspectives often leave managers with little more than their own best judgment, something they had in good measure before the time consuming and expensive effort to obtain the variety of market-based and political inputs. In fact, this conclusion should not be surprising. It would be almost incomprehensibly unique for a system relying solely on economic criteria to arrive at the same conclusion as a system that relies solely on democratic vote. That is, assessments based on one dollar one vote or, alternatively, those based on one person one vote are most unlikely to yield similar decisions. Decision making becomes more difficult yet when land managers try to satisfy multiple voting schemes; for example, local users of the forest polled at one meeting and more distant, less frequent, users with their different preferences held with different intensity polled at a second meeting, or loggers polled at one meeting and environmentalists at another.

What are the criteria for combining economic information with political insight from even one user community, let alone several? A system for combining these different inputs should identify the criteria for their combination; that is, the weights assigned to each category of economic and political input.[18] However, it is probably safe to say that no public agency has ever established such criteria and, furthermore, most public land managers do not understand the concept—even while they are most clear about the great difficulties they experience while trying to satisfy their many constituents.

Clarity at the outset regarding the weights given the various public inputs to the decision-making process would ease some of the complexity. However, even with this, statutory obligations and bureaucratic preferences would add yet another layer of difficulty. All land managers, public and private, must abide by the statutory requirements of public law. The public lands must abide by all the laws that affect private lands, plus an additional set of statutory requirements created specifically for them alone. On all of these, the managers of public lands experience a greater burden of oversight than that experienced by private land managers. This is as it should be. The public has reason to guard its own lands closely. Nevertheless, the additional oversight is yet another challenge that makes public land management more difficult than the management of private lands.

The public agencies add their own layer of institutional regulations. Some of these are useful directives that help maintain internal organizational consistency. Others only succeed in protecting the agencies' status quo, sometimes in the face of overwhelming economic evidence and public opinion to the contrary.

An example from U.S. Forest Service experience is illustrative. The U.S. Forest Service, as many other government forestry agencies in the United States and elsewhere, persists in its reliance on the "allowable cut" model for assessing timber harvest levels and persists in acting as if this model represents a fair economic appraisal despite the model's association of revenues from timber harvests with management costs on different lands and regardless of the resulting extreme estimates for rates of financial returns on silvicultural investments that should be a warning that something is amiss (see appendix 3b.) In fact, the agency's planning

---

18 See Maass et al. (1962) for a conceptual discussion of this problem.

does not include any comparative accounting of timber revenues with full direct costs of timber management, whether at the level of the individual timber operation, the ranger district, the national forest, or agency-wide. It is fair to say that the agency has revised its fundamental management perspective over the years, using ever-newer terms like multiple use, roadless area review, community stability, and ecosystem management to describe the revisions. It has also developed proficiency with a succession of large computer models designed to provide guidance for its timber managers. Nevertheless, it has also insisted on consistency with an institutionalized timber accounting practice that its famous founding chief, Gifford Pinchot, argued against from his retirement in the 1930s, and which both the forest industry and environmentalists showed to be detrimental to both of their interests as long ago as the 1970s (Walker 1974; Kutay 1977; Hyde 1980). The Reagan administration required the Forest Service (as all other agencies) to review all internal economic regulations in 1990, but the agency rejected this opportunity to examine anything other than a few mineral and gravel leases. As one result, the Wilderness Society continues to use the net revenue losses from Natonal Forest timber operations as an argument for recruiting new members and as justification for its critical oversight of the Forest Service even in the early years of the twenty-first century.

This is one example. Timber operations on the public lands of other countries that use the allowable cut model may be another. Grazing operations on the U.S. public lands are still another, and surely there are others.

Market signals and market performance eventually force correction of the most misguided activities of private corporations and private individuals. Sometimes, as in the U.S. public timber example, public agencies can create well-intended internal rules that isolate themselves from market forces and even from political action for extended periods of time. However, even in cases such as this, external political forces may eventually force change—as we might argue that a succession of court cases in the 1990s having to do with the northern spotted owl forced on U.S. Forest Service timber management.[19]

In sum, there can be no doubt that the task of a public agency manager is difficult. We can only hope for modification and improvement of an agency's guiding criteria over time and for honest internal appraisals of the agency's professional and institutional biases. The threat, however, is that, without these, management options may eventually be imposed on the agency externally, as they were in the spotted owl case and, while some public land users may be satisfied, many on all sides of an issue may end up feeling that good professional judgment would have been better than that of an incompletely informed judicial process—if only the agency and its managers had been more alert to both economic principles and democratic preferences at a somewhat earlier date.

---

19 Marcot and Thomas (1997) provide a comprehensive history of the spotted owl controversy. Wear and Murray (2004) trace its impact on timber and final product consumers across North America.

Literature Cited

Barlow, T., and G. Helfand. 1980. Timber giveaway — a dialogue. *The Living Wilderness* 44: 38–39.
Barlow, T, G. Helfand, T. Orr, and T. Stoel. 1980. *Giving away the national forests: An analysis of U.S. Forest Service timber sales below cost.* Washington, DC: Natural Resources Defense Council.
Borrell, B. 2011. Free-ranging market could save to wolves. *Denver Post* (August 28) 1D, 6D.
Boyd, R., and W. Hyde. 1989. *Forestry sector intervention: The impacts of public regulation on social welfare.* Ames: Iowa State University Press.
Brookings Institution. 2008. *2007 U. S. Forest Service & Department of Interior large wildfire cost review. A report on 2007 wildland fires by the independent large wildfire cost panel.* Washington, DC: Brookings Institution.
Brown, D. 2006. *Under flaming sky: The great Hinkley firestorm of 1894.* Guilford, CT: Lyons Press.
Bunch, J. 2007. Castle Rock aims to prevent flames in its valley. *Denver Post* (December 3) pp. B1, B4.
Chomitz, K., and K. Kumari. 1998. The domestic benefits of tropical forests: A critical review. *World Bank Research Observer* 13(1): 13–35.
Daniels, S. 1986. *Marginal cost pricing and efficient resource allocation: the case of public campgrounds.* Unpublished doctoral thesis, Duke University, Durham, NC.
Fare, R., S. Grosshopf, C. Lovell, and C. Pasurka. 1989. Multilateral productivity comparisons when some outputs are undesirable: A nonparametric approach. *Review of Economics and Statistics* 71: 90–98.
ForestRe. 2009. http://www.forestre.com/main.php (accessed June 14, 2009).
Freeman, A. 1979. *The benefits of environmental improvement: Theory and practice.* Baltimore, MD: Johns Hopkins University Press.
Freeman, A. 1993. *The measurement of environmental and resource values: Theory and methods.* Washington, DC: Resources for the Future.
Glover, D., and T. Jessup (Eds.). 1999. *Indonesia's fires and haze: The cost of catastrophe.* Singapore: Institute of Southeast Asian Studies.
Greene, W., 1993. The econometric approach to efficiency analysis. In H. Fried, ed., *The measurement of productive efficiency.* London: Oxford University Press, pp. 68–119.
Gess, D., and W. Lutz. 2002. *Firestorm at Peshtigo.* New York: Henry Holt.
Gregory. G. 1955. An economic approach to multiple use. *Forest Science* 1(1): 6–13.
Hennessy K., R. Fawcett, D. Kirono, F. Mpelasoka, D. Jones, J. Bathols, P. Whetton, M., et al. 2008. *An assessment of the impact of climate change on the nature and frequency of exceptional climatic events.* Aspendale, Victoria, Canada: CSIRO Marine and Atmospheric Research.
Hyde, W. 1980. *Timber supply, land allocation, and economic efficiency.* Baltimore, MD: Johns Hopkins University Press for Resources for the Future.
Hyde, W. 1981. Timber economics in the Rockies: Efficiency and management options. *Land Economics* 57(4): 630–639.
Hyde, W., and S. Daniels. 1988. Balancing market and nonmarket outputs on public forestlands. In V. Smith, ed., *Environmental resources and applied welfare economics: Essays in honor of John V. Krutilla.* Washington, DC: Resources for the Future, pp. 135–161.
Hyde, W., D. Newman, and B. Seldon. 1992. *The economic benefits of forestry research.* Ames: Iowa State University Press.
Johnson, N., A. White, and D. Perrot-Maitre. n.d. *Developing markets for water services from forests: issues and lessons for innovators.* Washington, DC: Forest Trends.
Kao, C., and C. Yong. 1991. Measuring the efficiency of forest management. *Forest Science* 37(5): 1239–1252.
Kenny, A. 2009. Ecosystem services in the New York City watershed. Available at http://www.ecosystemmarketplace.com (accessed June 27, 2009).
Krutilla. J. 1958. *Multiple purpose river development.* Baltimore, MD: Johns Hopkins University Press for Resources for the Future.
Kutay, K. 1977. Oregon economic assessment of proposed wilderness legislation. In Oregon Omnibus Wilderness Act. Publ no 95-42, part 2, pp. 29–63. Hearings before the Subcommittee on Parks and Recreation of the Committee on Energy and Natural Resources, United States Senate, 95th Cong, 1st session, April 21, 1977. Washington, DC: Government Printing Office.

Landell-Mills, N., and I. Portas. 2002. *Silver bullet or fool's gold: A global view of markets for forest environmental services and their impacts on the poor.* London: International Institute for Environment and Development.

Loomis, J., and R. Walsh. 1997. *Recreation economic decisions: Comparing benefits and costs* (2nd ed.). State College, PA: Venture Publishing.

Lovell, C. 1993. Production frontiers and productive efficiency. In H. Fried, ed., *The measurement of productive efficiency.* London: Oxford University Press, pp. 3–67.

Maass, A., M. Hufschmidt, R. Dorfman, H. Thomas, Jr., S. Marglin, and G. Fair. 1962. *Design of water-resource systems: New techniques for relating economic objectives, engineering analysis, and government planning.* Cambridge, MD: Harvard University Press.

Marcot, B., and J. Thomas. 1997. *Of Spotted Owls, old growth, and new policies: A history since the interagency scientific committee report* (general technical report PNW-GTR-408). Portland, OR: U.S. Department of Agriculture Forest Service. general technical report PNW-GTR-408.

Mitchell, R., and R. Carson. 1989. *Using surveys to value public goods.* Baltimore, MD: Johns Hopkins University Press/Resources for the Future.

Moscowitz, K., and C. Romaniello. 2002. *Assessing the full cost of the federal grazing program.* Tucson, AZ: Center for Biological Diversity..

Muir, J. 1912. *The Yosemite.* New York: The Century Company.

National Park Service. 2009. *Frequently asked questions.* http://www.nps.gov/faqs.htm. (accessed June 11, 2009).

Portland Fire and Rescue. 2010. *Southern California November wildfire of 2008: One of the 25 largest fire losses in U.S. history.* http://www.portlandonline.com/fire/index.cfm?a=326554&c=53961 (accessed December 30, 2011).

Rhodes, E. 1986. An exploratory analysis of variations in performance among U.S. national parks. In P. Silkman, ed., Measuring efficiency: *An assessment of data envelopment analysis.* San Francisco: Jossey-Bass, pp. 47–71.

Rittmaster, R., W. Adamowicz, B. Amiro, and R. Pelletier. 2005. *Economic analysis of health effects from forest fires.* Unpublished research paper. Department of Rural Economy, University of Alberta, Edmonton, Canada.

Robinson, E., A. Mahaputra, and H. Albers. 2009. Optimal enforcement and practical issues of resource protection in developing countries. Discussion paper 9-08. Washington, DC: Resources for the Future Environment for Development.

Ruttan, V. 1980. Bureaucratic productivity: The case of agricultural research. *Public Choice* 35(3): 529–547.

Sayer, J., and C. Sun. 2003. Impacts of policy reforms on forest environments and biodiversity. In W. Hyde, B. Belcher, and J. Xu, eds., *China's forests: Global lessons from market reforms.* Washington, DC: Resources for the Future, pp. 177–194.

Siminov., E., and T. Dahmer (Eds.). 2008. *Amur-Heilong River Basin reader.* Hong Kong: Ecosystems Limited.

Siry, J., and D. Newman. 2001. A stochastic production frontier analysis of Polish state forests. *Forest Science* 47(4): 526–533.

Steelman, T. 2008. Communities and wildfire policy. In E. Donoghue and V. Sturtevant, eds., *Forest community connections: Implications for research, management, and governance.* Washington, DC: Resources for the Future, pp. 109–126.

United Nations Environment Programme (UNEP). 2008. The world's protected areas: Status, values and prospects in the 21st century. New York: UN.

U.S.D.A. Forest Service. 2010. *Cut and sold (new)-cut S203S.* Washington, DC: USDA Forest Service.

U.S. Forest Service. 2010. *100 years of caring for the land and serving people.* http://www.fs.fed.us (accessed May 13, 2010).

Vittala, E., and H. Hanninen. 1998. Measuring the efficiency of nonprofit forestry organizations. *Forest Science* 44 (2): 298–307.

Walker, J. 1974. *Timber management planning.* San Francisco: Western Timber Association. August, 1974.

Wear, D., and B. Murray. 2004. Federal timber restrictions, interregional spillovers, and the impact on US softwood markets. *Journal of Environmental Economics and Management* 47: 307–330.

WFDSS (Wildland Fire Decision Support System). 2010. *Welcome!* http://www.wfdss.usgs.gov (accessed May 7, 2010).

Westerling, A., H. Hidalgo, D. Cayan, and T. Swetnam. 2006. Warming and earlier spring increase western U.S. wildfire activity. *Science* 313: 940–943.

Whiteman, A., V. Noulak, and J. Broadhead. 2006. *The current and potential contribution of forest-based ecotourism to poverty alleviation in Laos.* (A report submitted to Asian Development Bank RETA-6115). Manila, Philippines: Asian Development Bank.

Zhang, Y. 2005. Multiple-use forestry vs. forestland-use specialization revisited. *Forest Policy and Economics* 2: 143–156.

Zimmermann, E., and S. Collier. 2004. Road wrecked: why the $10 billion Forest Service road maintenance backlog is bad for taxpayers. Washington, DC: Taxpayers for Common Sense.

Appendix 10A:
Least Cost Plus Loss

Least cost plus loss models address a problem that is common in the management of all natural hazards. A resource of given value $R°$ exists in the absence of the hazard. Both the hazard and efforts to control it impose costs and the levels of both of these costs are affected by management inputs E. That is, as human efforts to control the hazard increase, their total costs also increase. Increases in these costs have the opposite effect on the damages or losses due to the hazard. That is, increasing the level of management decreases the amount of damage due to the hazard.

More formally, the management objective is to minimize the combination of management input costs and losses imposed by the hazard.

$$min_E \pi(E) = R° - L(E) - C(E) \qquad (10a.1)$$

where the $L(E)$ are losses due to the natural hazard and the $C(E)$ are management costs. In the case of wildfire, these are the combined costs of prevention, detection, and suppression activities.

The first order condition is

$$\partial \pi(E) / \partial E = -L_E - C_E = 0 \qquad (10a.2)$$

or

$$-L_E = +C_E \qquad (10a.3)$$

where subscripts indicate derivatives with respect to the subscript and $L_E \leq 0$ and $C_E \geq 0$. The optimal level of effort E^* occurs where the combined total costs are minimized or where declining marginal losses due to the hazard equal increasing marginal control costs of effort to control it. Refer again to Figure 10.2.

This is a simple problem that is much, much more difficult in its application to wildfire because the various series of costs: protection, detection, and suppression are interdependent and because both costs and losses are stochastic.

11
FORESTS AND LOCAL HUMAN COMMUNITIES

Two important and related questions remain for our consideration. First, to what extent can forestry be a source of local economic development and, second, what do we know about local communities and their reliance on the forest?

More than 350 million people live within or adjacent to the world's forests and as many as 1.6 billion people may be dependent on forests for a portion of their livelihood (World Bank 2001, 2004). An understanding of the interactions between these peoples and their forests is essential to a fair appreciation of the potential role of the forest in local economic development.

Many of those living within or adjacent to the forest are poor, and their condition is one important reason for our interest. Improving the welfare of the poor—including these rural poor—is an issue of international concern and also a focus of domestic policy in many countries. Successful design of such policy requires an understanding of the demands that the poor and their communities place on their environment and its resources and also an understanding of what these resources can offer for them.[1]

The poor are where you find them, and forest-based communities are only one place where they are found. Certainly, we are well aware of the existence of both an urban poor and also a rural poor population that, while rural, does not live within easy access of the forest.[2] Furthermore some of the poor who do live in forested rural regions live where they do out of personal choice and regardless of other

---

[1] An important part of the discussion, beginning with Ekbom and Bojo (1999), points to a "poverty-environment nexus," a set of mutually reinforcing links between poverty and environmental damage. Dasgupta et al. (2005) summarize the discussion and, drawing on evidence from Southeast Asia, urge caution in generalizations. They conclude that "the nexus is quite different in each country". It should not be seen as a "general formula for policy design."

[2] Chomitz (2007) is particularly clear about this point.

opportunities available to them. Both Power (1980) and Whitelaw, Niemi, and Batten (1990) make the point that this latter group receives an unmeasured "second dividend," an increment to their personal welfare from living where, and as, they do. The second dividend is an addition they would have to forgo if they lived elsewhere. Therefore, their decision to continue living where they do, regardless of better financial opportunities elsewhere is an indication that some rural inhabitants are not as poor as their incomes alone would indicate. As this is the case, policies that target rural and forested regions are not always the best choice for assisting any particular country's poorest population and arguments for policies favoring rural and forested regions do not automatically have distributive merit.

Whatever the well-substantiated merits of these arguments, it is true that forests and rural poverty often do go hand in hand. Where we find one, we often find the other. For example:

- In China, more than 90 percent of the poor live in rural areas and 496 of 592 officially designated poverty stricken counties are in mountainous, forested regions (Zhang, Saint-Pierre, and Liu 2004; MOF 1995).
- Rural poverty exceeds 35 percent in both Papua New Guinea and Mongolia, the countries with the largest forest areas per capita in East Asia and the Pacific (Magrath 2004).
- In four of five continents the poorest countries possess more forest cover than the continental averages.[3]

Even in developed countries such as the United States and Canada, Sweden and Finland, typically the most remote parts of the country are forested and the local populations are often poor. In many countries, these regions of forest and poverty are also the home of native or ethnic minorities who are not only poor but also disadvantaged in modern society on basis of their ethnic differences.

So—what is the role of a standing forest in the welfare of adjacent human communities and in regional economic development? Reference to the three-stage model and experience from the U.S. South and Brazil's eastern Amazon, as discussed in chapters 3 and 7, respectively, provides some insight. The further example of bamboo production in China in this chapter adds the experience of yet another region and another forest product amidst the background of another political and economic experience. Bamboo is a forest and farm product that shares many characteristics with wood in its production and consumption. Its example is especially useful for the evidence of impacts on household welfare in general and on those who work in the bamboo sector in particular, topics that have not been as fully examined as some others in the forestry literature.

Finally, if the illustrations of these first three cases (the U.S. South, Brazil's eastern Amazon, and bamboo in China) are instructive, then we might inquire how their experience can be useful for predicting forest and community development

---

3 Countries with less than US$150 per capita annual income in Africa, less than US$400 in Asia, and less than US$1,000 in the Americas (compiled from FAO 2001).

in yet other regions of the world where the forestry sector is largely in the early stages of development. Three countries of the Mekong River Basin, Vietnam, Laos, and Cambodia, provide this opportunity. All three are well-forested, and their forest sectors are important to each country's future. The differences in national economic growth, government performance, and current levels of forest sector development across the three countries illustrate a range of opportunity and impose breadth on our assessment. In anticipating forest-based development in the Mekong Basin, it will be important to understand the effects imposed on the local communities and their forest sectors by the broader economic and political situation—as well as the characteristics of the local communities' relationships with their forests. In fact, these broader national economic and political effects are most useful in assessing the forest's potential contribution to economic development almost anywhere.

A second part of this chapter takes a step back from inquiring into the relationship between forestry and general regional development and refocuses attention on the condition of local communities, and particularly communities in the early stages of forest development. Their experiences range from communities reliant on the forest for household subsistence goods and services, to other communities with almost total dependence on the forest for commercial logging and sawmill operations, to still other communities where logging and forest-based tourism are a smaller share of total community economic activity but where these or other forest-based activities remain crucial for the livelihood of some members of the community. Those most affected may be members of particular demographic groups—the poor, the landless, women, or ethnic minorities—or they may be consumers of specialized forest products like woodfuels or non-timber forest products. Global policy interest in these demographic groups and these products begs our review of their characteristic experiences, and of the impact that a change in the availability of forest resources has upon them.

Modern empirical assessments of the production and consumption of specialized products with less vibrant markets or of specialized demographic effects often rely on one of two technical economic approaches, models of household production and consumption or general equilibrium models of the regional economy. Mathematical appendices to the chapter briefly introduce the merits of these two approaches and summarize their formulation.

In general, we will observe that a forest resource can be a primary means to sustainable development, but only under specialized conditions of

- access to markets and external demand for forest products and
- appropriate incentives for local market participants, including
  - secure rights to the product of their investment and to the means of producing and processing their forest products, and
  - regional and sectoral economic and political stability.

We will observe that the poor and other disadvantaged groups often benefit under these conditions—with greater personal wealth and, eventually, better employment opportunity as one result of a strong forest sector. However, an important caution is in order. That is, while the most disadvantaged do benefit—and that is crucial for our interest—they are not generally the greatest benefactors. Other income and demographic groups tend to benefit more. This has been the case in numerous examples, including our example of bamboo in China, and it will probably hold true for our second broad example from the Mekong River Basin of Southeast Asia as it too develops.

The General Experience

Can the forest resource be a source of economic development and a means for reducing rural poverty? In fact, various examples demonstrate that forest extraction can contribute to successful local economic development. However, it is clear that the simple existence of extensive forest cover does not assure opportunity for financial gain and development.[4] Furthermore, it is also clear that financial gain from forest extraction does not automatically guarantee regional development. The interior regions of northern Canada, Siberia, and the Amazon have almost endless expanses of forest. Yet those resources have minimal market value and minimal impact on regional development today. In other regions and times, large quantities of valuable timber were removed from the Lake States of the United States in the 1880s, from Cameroon in the 1990s, and from the region of Siberia bordering Heilongjiang Province of China more recently, all with great financial gain to some participants but little immediate impact on local development.

Therefore, the first step in addressing our question is to identify the characteristics of successful forest-led development and, in contrast, the underlying characteristics that were absent from those examples of forest extraction that have not been a part of successful regional development. Questions about selective disadvantaged demographic groups follow. Did they benefit? How much? What conditions describe those cases of forest-led development that have had the most favorable impacts on disadvantaged populations?

We will begin with the discussion of bamboo production and processing in six counties in China. Bamboo is a forest product that substitutes in many of

---

4 In fact, some, beginning with Prebisch (1950), have suggested a "resource curse," that is, countries with rich resource endowments grow slower than those with lesser endowments of undeveloped natural resources. Bulte, Damania, and Deacon (2005) reviewed this literature and tested the hypothesis empirically for 97 countries. They find that the relationship between resources and development is complex. Nevertheless, they conclude that resource-intensive countries do tend to experience lower levels of overall development. More recently, van der Ploeg (2011) reviewed the empirical experience and the literature, and tested various hypotheses regarding the resource curse. He concludes that "the best available empirical evidence suggests that countries with a large share of primary exports in GNP have bad growth records and high inequality, especially if quality of institutions, rule of law, and corruption are bad." "Resource rich countries are also vulnerable to the notorious volatility of commodity prices, especially if their financial institutions are not well developed."

the same uses as timber. Furthermore, in China its markets are less constrained by government administration than the markets for wood and wood products. Therefore, China's bamboo sector should be roughly illustrative of what its timber markets would be if they were less administered.

Following this example from China, we will call on the previous examples from the U.S. South, from the Eastern Amazon, and from other examples as well, to draw summary conclusions about the pattern of forest development and its effects on local household welfare, before demonstrating how these experiences can be used to anticipate the potential for forest development in yet other regions of the world.

Bamboo Development and Local Communities in China

The experience of China's bamboo sector is relevant because timber and bamboo share the same factors of production and their markets also share many similarities. On the supply side, they compete for similar land, labor, and capital resources, either for extraction from the natural forest or for farm forest or bamboo management. On the demand side, bamboo and timber compete in many of the same markets—panels, flooring, plywood, fiber for paper, and even some construction framing materials. China's logging ban, imposed in 1998, provides evidence of the product similarities between timber and bamboo. The demand for timber substitutes increased simultaneously with the imposition of the logging ban. The bamboo market response was immediate and emphatic. Bamboo prices increased almost 10 percent in 1999 and the land area in new bamboo plantations increased by 17 percent. China's State Forestry Administration recognized the substitution between the wood and bamboo and, as the ban was imposed, anticipated that that bamboo would substitute for 29 million m^3 of wood by 2010 (CFIC 1998).

Ruiz-Perez and his colleagues (Ruiz-Perez et al., 1996; Ruiz-Perez et al. 1999, 2001; Ruiz-Perez and Belcher 2001) examined the growth and development of China's bamboo industry through the 1990s—before the logging ban—with particular emphasis on the effect on household incomes. They relied on surveys of individual farm households and industrial firm managers, on interviews with key informants, on farm records, and on county price and production records. They focused, first, on Anji County in Zhejiang Province and, subsequently, reviewed experience in five additional counties.

Zhejiang is an economically dynamic coastal province near Shanghai. It is one of four provinces that account for two-thirds of China's bamboo production. This province, like all of China, has developed rapidly in response to the reforms of the last 30 years. Coastal provinces such as Zhejiang began to develop earlier and they developed more rapidly than some others. Zhejiang's bamboo sector developed particularly rapidly.

The other five counties follow an east-west grid inland to about 1,700 km west of Anji County. Production increased in all six counties, and key informants in all six identified a stable policy environment as a precondition for the increase.[5] Productivity per land unit was greater in the eastern counties and the richest county, Anji, experienced the greatest average productivity. The poorest county, Pingjiang, experienced the least average productivity, less than one-third of that in Anji. In general, farm households in all income classes and in all counties benefited, but middle and higher income households benefited most and improved off-farm labor opportunity was the source of their greatest improvement.

Anji's experience: In 1975, 51,000 hectares in Anji were managed for bamboo, 99 percent of them in farm collectives. Bamboo was sold through a state marketing collective at fixed procurement prices. Household Responsibility contracts, introduced in Anji in 1983–1984, brought an immediate change. By 1984, individual households managed 40,000 hectares of bamboo. The government first revised, and then eliminated, its procurement price system in 1985 and, by 1988, total output in the county had increased by 63 percent with little adjustment in the land area devoted to bamboo. By this time, 117,600 individual farmers had become responsible for 91 percent of the county's total production. Prices and production continued to increase through the 1990s. Altogether, prices increased almost 300 percent (in constant terms) from the late 1970s while production increased more than 90 percent.

In 1978, 96 percent of Anji's bamboo production was sold in unprocessed form and shipped out of the county. Local processing was scant. Nineteen local establishments employed 460 workers in producing 960,000 yuan (US$105,000) of finished goods. This too changed rapidly. Various marketing reforms created new opportunity and, by 1998, 1,182 processing establishments employed 18,914 workers. Annual production exceeded 875 million yuan (US$96 million) in value in 1998.

The growth in the processing sector was responsible for some of the upward pressure on the price of the primary bamboo resource and the marketing system grew more specialized in response. By the mid-1990s more than 200 bamboo traders had become part of the system. Meanwhile, some processing establishments, concerned for the supply of their own raw material, began negotiating production agreements with farmers at prices agreed upon before the harvest period. Some of these agreements included cash advances. Eventually, processors

---

5   Policies designed to identify property rights had changed four times over a period of thirty years and, therefore, farmers in China were particularly aware of the effect of policy change on investment and management. Any longer term investment, including investment in either forestry or bamboo, is difficult to justify under this kind of uncertainty. The new Household Responsibility System (HRS), beginning in 1978, was the first step in China's modern reforms. It established improved property rights for individual farm households. Nevertheless, farmers remained alert to the possibility of additional modifications in their property rights until they had experienced a period of stability regarding these rights and until most of these HRS contracts were renewed in the 1990s (Hyde, Xu, and Belcher 2003).

found the local supply insufficient for their needs and they began importing raw material from other counties.

Despite this impressive growth and diversification, little change occurred in the number of participating farm households or in the land area they devoted to primary production. Approximately 120,000 farm households remained the foundation of the bamboo industry in Anji into the late 1990s.

To examine the household impact of the increase in bamboo production, Ruiz-Perez et al. (1999) examined incomes for a random sample of 200 Anji farm households in 1994–1995 and compared their evidence with that from 1989–1990. They divided households into five income classes with the objective of comparing the sources of income and income growth for each class. Average household incomes rose almost 7 percent in real terms over the five-year period but income from bamboo remained at a fairly constant 24–25 percent share of total income. Agriculture also contributed a relatively constant, but much larger, share to household income—across all income classes in all 200 households. Therefore, Ruiz-Perez et al. conclude that agriculture provided the greater economic foundation for all households throughout the five-year period.

The absolute level of income from bamboo did increase between 1989–90 and 1994–95 for all household income classes, and bamboo was an increasingly important source of on-farm income for the higher income classes. However, off-farm employment (some of it in bamboo processing establishments) was the largest source of income differences between classes of households and it was especially important for households in the highest income class.[6] Figures 11.1A and 11.1B show the *relative* importance of income from bamboo to each income class in the two periods. Both figures display a convex shape, indicating that bamboo was a *relatively* more important source of income for middle income groups. It was the source of 30 percent of income for households in the second and third quintiles, but barely 20 percent of income for households in the extreme first and fifth quintiles.

Table 11.1 shows the results of a regression assessing the sources of differences in per capita income levels in the two years. The number of male laborers in the household, arable land, land in bamboo, and off-farm employment all had significant and increasing effects on household income. The income elasticities of bamboo land and off-farm labor are 0.32 and 0.48, respectively. It is clear that improvements in bamboo production opportunities benefited household income, but additional off-farm labor opportunity was the crucial ingredient for the overall income growth that occurred between 1989–90 and 1994–95.

---

6 Chapter 9 discussed the observation that forestry is a convenient land use for those with off-farm employment. Forestry is less labor-intensive than most agricultural crops or livestock and the timing of forest management activities is more flexible than the strict seasonal requirements of many agricultural activities. Bamboo shares these characteristics with forestry. Both timber and bamboo production permit those with off-farm employment to continue their employment without sacrificing the productivity of their land.

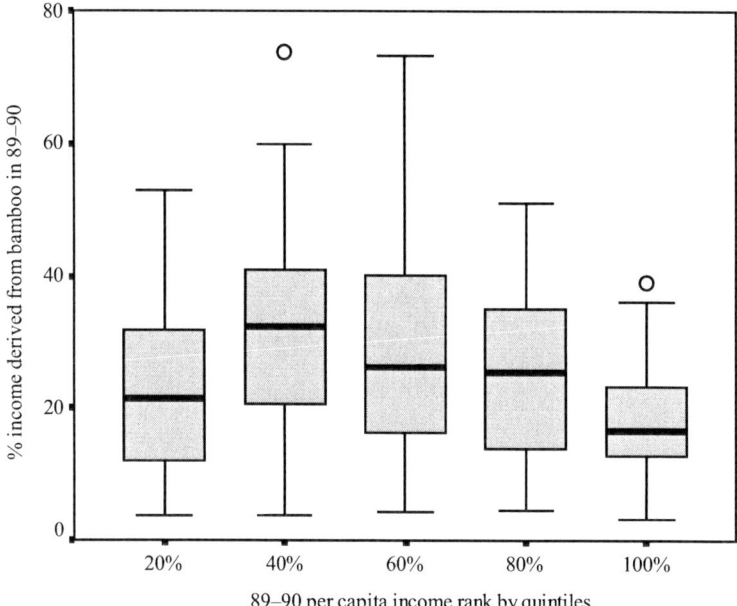

FIGURE 11.1A Relative Importance of Bamboo Income in Anji County—by Household Income Class, 1989-1990. *Source:* This table was published in *World Development* 27(1), M. Ruiz-Pérez, M. Zhong, B. Belcher, C. Xie, and M. Fu, The role of bamboo plantations in rural development: the case of Anji County, p. 105, copyright Elsevier (1999).

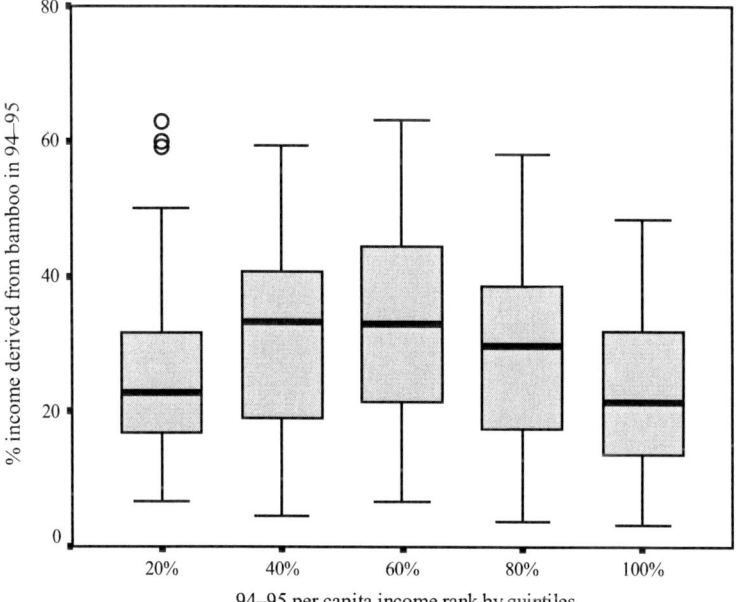

FIGURE 11.1B Relative Importance of Bamboo Income in Anji County—by Household Income Class, 1994-1995. *Source:* This table was published in *World Development* 27(1), M. Ruiz-Pérez, M. Zhong, B. Belcher, C. Xie, and M. Fu, The role of bamboo plantations in rural development: the case of Anji County, p. 107, copyright Elsevier (1999).

TABLE 11.1 Sources of difference in per capita farm income in Anji County, 1989–1990 to 1994–1995

Variable	B-coefficient	β-coefficient	T-value
Age, head of household (yrs)	−19.8	−0.127	−1.996
Family size	−297.8	−0.225	−3.198
Male/total labor ratio	1,361.6	0.152	2.244
Arable land	114.6	0.151	2.373
Bamboo land	53.9	0.390	5.997
Off-farm/total work ratio	855.2	0.157	2.420
Township	107.5	0.178	2.717
Constant	1,632.8		2.863

$R^2 = 0.341$

Adjusted $R^2 = 0.317$

$F = 14.178$

Probability $F < 0.0001$

Source: This table was published in *World Development* 27(1), M. Ruiz-Pérez, M. Zhong, B. Belcher, C. Xie, and M. Fu, The role of bamboo plantations in rural development: the case of Anji County, p. 101, copyright Elsevier (1999).

The six-county comparison: Between 1980 and 1998, the number of farmers growing bamboo in the broader six-county sample increased by 91 percent and the share of all farmers growing bamboo increased from 19 percent to 26 percent. The bamboo processing sector expanded in five of the six counties, doubling the share of primary processing conducted locally and offering better markets for farmers who increased their bamboo production, as well as more opportunities for both household-based pre-processing and off-farm employment in the processing industry itself.

The profile of the processing sector differed, however, across the counties. Three of the counties are illustrative:

- Muchuan County in the west is not as economically diverse as Anji County, either generally or in its bamboo industry. Papermaking accounted for 99 percent of all bamboo processing and 36 percent of county output from all industrial sources. The high profit rate (24 percent) in this industry indicates a growing industry that continued to expand after 1995. Bamboo production and its processing into paper clearly was a development leader for this county.
- Taojing County is geographically more central and industrially more diverse. Bamboo processing accounted for 12 percent of Taojing's industrial output in 1998. The industry's 19 percent profit rate indicates an industry that was still growing. Taojing had the lowest wage and lowest per capita income of the six counties, and its primary producing sector made a greater contribution to county GDP than for any of the other five counties. The smaller bamboo processing sector in Taojing was taking advantage of low wages to expand more rapidly than in the other counties.

- Longyou is the most developed of the six counties. Per capita income was twice that in the second most developed county and Longyou's industry was the most diversified of the six counties. Bamboo processing (for paper, plywood, and various other products) was only 7 percent of total industry output. The bamboo industry grew over the twenty-year period but, by 1995, its 8 percent profit rate approximated that of the other industries in the county. Therefore, bamboo processing no longer had a competitive seven advantage in Longyou and it no longer operated as a leading source of economic growth.

The profiles of farm incomes, wages, and employment also differed across the counties. Farm incomes were growing in 1995 but the disparity in farm incomes was also growing—both within counties and across counties. Bamboo production and pre-processing conducted in the household added to household income in all counties. Longyou's farm households recorded the largest income shares from bamboo. They also reported the widest spread of income—in general as well as from bamboo—and they attained the highest income levels from all sources for households in any of the six counties. Their farm production was more specialized and more of Longyou's households were able to obtain off-farm employment—as might be anticipated in the most developed and diversified county. Muchuan's farmers achieved the most rapid gains in income in the later years—probably as a result of rapid growth and increasing demand in Muchuan's paper industry. In general, and across all counties, poor farmers benefited with increasing incomes from bamboo, but they benefited proportionally less than farmers in higher income groups.[7]

General Observations and Further Insights

The pattern of forest development is similar—whether traced over longer time periods as in the U.S. South or across space and shorter time periods as in eastern Brazil and China. Furthermore, the pattern is similar whether we follow the forestry and wood products industry to its full development—as in the U.S. South—or a more restricted portion of the industry as it progresses along the path toward full development—as in the more limited development observed in eastern Brazil or in the more specialized case of bamboo in China.

Who benefits from this pattern of development? Of course, the benefactors include those who have access to the large sources of capital necessary to establish the modern capital intensive wood processing facilities that appear later in this pattern of development. But who else benefits?

---

7 As Ruiz-Perez and his colleagues have continued to examine the economic development of these counties, they observe that, with further development, the role of bamboo continues to change and the number of income options continues to expand. Tourism in particular becomes more important (Gutierrez et al. 2009). This finding is consistent with our more general observation regarding increasing leisure use of the forest and our hypothesis 5a about an environmental Kuznets' curve in chapter 6.

The original workers in an infant forest industry already live near the forest frontier. Most are local farmers and other agricultural workers who seek to supplement their income with part-time or seasonal employment. Some begin to operate their own rudimentary sawmills, but the wages and profits from these sawmills also tend to supplement household income as they, too, operate either seasonally or when market conditions are most favorable.

Eventually, as a local wood processing industry develops, opportunities for employment increase as well. As logging itself extends farther into the forest frontier and away from the local agricultural communities, it becomes a full-time occupation for some and employment in the mills in town becomes a full-time occupation for others. Evidence from Nepal (Bluffstone 1995; Amacher, Hyde, and Kanel 1999), India (Foster, Rosenzweig, and Behrman 1997), Sri Lanka (Gunatilake 1998; Illukpitiya and Yanagida 2008), Peru (Escobal and Aldana 2003) and Vietnam (Tachibana, Nguyen, and Otsuka 2001) shows that those who seek employment in the industry at this stage of its development often had been subsistence farmers who now place fewer demands of their own on the natural forest, and the forest environment may even benefit as a result.

Greater capital intensity means more specialized demands for labor and increasing demand for skilled labor. Therefore, the initial employment benefits from forest development probably accrue to a few entrepreneurial individuals and households who establish the first rudimentary sawmills. Later, the local benefactors are those who seek more permanent employment in the growing logging and processing industry and who develop the specialized skills required to operate increasingly technical logging and mill processing equipment. These entrepreneurial individuals and skilled laborers may not be wealthy—but they also are not usually among those whom we would identify as the rural poor.

Of course, landowners benefit as well. As the industry continues its development, some small landowners begin to benefit from the sale of timber from their own land. For them, the security of their property rights is a crucial characteristic of rural development. Any investment requires confidence that the expected future return from the investment will be the property of the investor. Therefore, longer term investments, such as those in trees, require confidence that their property rights and the overall investment climate will remain reliable for a longer period—the period it takes for trees to grow to a size that can yield marketable benefit to the investor. Where the local institutions can insure long-term secure property rights, some small landowners will begin to find personal advantage in growing timber as a source of income.

Examples of the importance of secure tenure are numerous. China's experience is illustrative. China's modern market reforms began in 1978 with 25-year household contracts for agricultural land. Agricultural production grew 8½ times over in six years. Lin (1992) shows that improved land tenure was responsible for the majority of the increase in production—although other reforms had their beneficial effects as well.

Chinese farmers were cautious. The rules for land tenure had changed four times in the previous 20 years. Nevertheless, even with the uncertainty generated by this experience, some farmers planted trees and the resulting improvement in the agricultural environment explained 10 percent of the growth in agricultural production over those first six years of reforms (Yin and Newman 1997). Of course, the beneficial impact of these new trees could only continue to increase as they matured beyond the juvenile growth of their first six years and as forest cover on China's former collective agricultural lands continued to increase. By 1998, China's forest cover was 60 percent greater than it had been in 1978 (Hyde, Xu, and Belcher 2003).

A secure investment climate, however, requires more than reliable contracts for forestland. Argentina provides an example for this point. Where the mountain passes between Argentina and Chile are low, Argentina experiences many of the same growing conditions as those that have made Chile's forest sector internationally well-known for its productivity. Yet Argentina has not shared the experience of high forest productivity. The difference is one of general economic conditions. Chile's economy has been stable for 30 years. Argentina has experienced wide fluctuations in inflation that make it difficult for landowners to predict the profitability of their timber investments. In addition, an unpredictable labor market makes it difficult for Argentine forest landowners to predict the cost of log shipment and, thereby, adds to the uncertainty surrounding the value of the landowners' timber investments.

In sum, there are opportunities at every stage forest development, opportunities for local households and even for those with the least specialized skills. Local workers and local households do benefit. However, the simple presence of a forest stock is an insufficient basis for development and local opportunity. The appropriate market conditions and arrangements for land tenure must also be in place.

Furthermore, where the forest sector is a basis for development, the industry's development does not necessarily mean that those who do benefit most will be the poorest. There is not a lot of empirical evidence on the income distributive effects of forest development but the evidence from China's bamboo development suggests that, while households from all income classes do benefit from new forest-based opportunities, those who benefit most seem to be those who are more entrepreneurial, those who own sufficient amounts of land to risk holding a share of it in a longer term investment like forestry, and those who can develop the specialized skills required in the emerging and increasingly more technical logging and processing industries.

Two examples from the United States, Boyd and Hyde's (1989) evidence from North Carolina and Stevens' (1978) evidence from Oregon, are consistent with these observations from China. Boyd and Hyde observe that larger landowners receive proportionately larger benefits from timber growing opportunities. Stevens observes that the most marginal woodsworkers and millworkers are the least likely to benefit from any opportunity for regular forest sector employment. On the other hand, these marginal workers are more mobile. They are able to move wherever the labor opportunities are most satisfying, whether in the wood industry or in other sectors, and whether locally or further "down the road." In sum, where forestry is a basis for regional development, the poor apparently do benefit from development but their gains are not as great as those of some of their better off neighbors.

Anticipating Development

If, as it seems, this pattern of development is general for forest sectors everywhere, then we should be able to look at forestry in most any region of the world, establish its position within the common three-stage pattern, and reliably anticipate the next steps for forest sector development in that region as well as for the local human population affected by that development. Vietnam, Cambodia, and Laos, three countries in the Mekong River Basin, provide an opportunity to examine this proposition. Improvements in the economic and political conditions of all three countries within the last 15-20 years have created interesting prospects for forestry, and each makes an interesting case—for itself and as a further illustration of expectations for similar situations elsewhere.[8]

Vietnam

The Vietnamese government began easing its central control and allowing greater reliance on the market about twenty years ago. Subsequently, the entire Vietnamese economy embarked upon a period of remarkable vitality. Its per capita annual economic growth rate of 7.2 percent for this period is among the very most rapid in the world. The Vietnamese forest sector, particularly its wood products industries, and especially the furniture industry within that sector, has shared in this vitality. Table 11.2 summarizes Vietnam's growth, both for its full economy and for the forest sector and its component parts.

Vietnam's forest sector production has grown at an annual rate of 5.4 percent, or almost as rapidly as the full economy. However, growth in the forest sector has been concentrated in the wood products industries as forest production itself has only grown at a lesser rate of 3.9 percent, or approximately half as fast as the aggregate economy. These comparisons, plus the recognition that Vietnam's domestic market consumes much of the forest sector's production, suggest that the forest sector has been a participant in Vietnam's growth, but not a driving force. Apparently, growth in the forest sector does support overall economic growth, but forestry has not been a leading sector for economy-wide growth.

Growth in the wood products industries has progressed to the point that some industries (pulp and paper, wood-based panels, furniture) have begun to suffer a shortage of woody raw material. Some firms (mostly state-owned firms) have begun to invest in forest plantations and those investments are growing rapidly—at a rate of 10 percent or 200,000 hectares per year. This is the strongest of several indicators (capital intensity, labor productivity, industry diversification,

---

8   This section of the chapter summarizes the more detailed assessment contained in Hyde (2006).

TABLE 11.2 Vietnam

Land area		32.928 million ha	deforestation rate: -0.5%
Forest area		12.094 million ha	forest plantations: 2.089 million ha
Paved highways		23,418 km	
Railroads		2,600 km	
Population	78.705 million	9.02/forested ha	growth rate: 1.6%
Gross domestic product:		$31.344 billion	growth rate: 8.8%
GDP/cap		$398	growth rate: 7.2%
Poverty rate		37%	
Agricultural share of GDP[1]		21.8%	
Trade			
Exports-total		$14.49 billion	grew 3x over decade
Forest products		$47.3 million	80% decrease over decade
Imports-total		$22.50 billion	
Forest products		$133 million	grew 8x over decade
Forest sector, excl. furniture	Gross value added		
Full forest sector		$555.9 million	grew 60% over decade
Forestry	Incl. logging & rel. activities	$421 million	grew 50% over decade
Woodworking		$58.2 million	grew 6x over decade
Pulp & paper		$76.6 million	grew 5x over decade
Contrib. to GDP		2.0%	remained relatively constant
Employment	Full time equivs.		
Full forestry. sector		106,155	grew 3.5x over decade
Ind'l forestry	Incl. logging	11,531	essentially constant over decade
Woodworking		59,600	grew 6x over decade
Pulp &paper		35,024	variable, but grew 3-4x over decade
% nat'l employment		0.3%	grew 3.5x over decade
Production			
All roundwood		30.9 million m^3	
Ind'l roundwood		4.2 million m^3	35% increase in decade productivity/worker: 363 m^3

(*continued*)

TABLE 11.2 Continued

Swnwd & panels	0.3 million m³	increase from zero within decade productivity/worker: 50 m³
Paper	377 MT	4½ x increase in decade productivity/worker: 20 MT

Source: Hyde (2006)
1 This and all subsequent values in 2000 US$.

and specialization of both labor and capital) that Vietnam has entered the final, mature, stage of forest development.

The forest sector has not been a major employer (0.3 percent of national employment in comparison with a 2.0 percent share of GDP), and those workers it does employ tend to reside in moderate to well-populated areas where the wood products industries have established themselves. Vietnam's forest plantations, whether large block plantations established by the industry or industry associations, or many small farm woodlots, do provide some economic support for rural populations and we can anticipate that this support will grow as the land area in plantation management grows. Forest plantations may even support a meaningful share of the population in regions such as the Central Highlands where many of the new plantations are concentrated. However, even the Central Highlands is primarily agricultural (largely tree crops such as cashews, tea, coffee, and rubber) and agriculture will probably continue to be the single largest source of rural household employment and income in that region.

Meanwhile, Vietnam's forest imports, mostly wood and pulp, have increased rapidly, more than eight-fold over the last decade. We can probably anticipate that the wood products industries that have begun to import raw material will continue to do so and the households that gain from producing those imports will be in the countries that export to Vietnam—although the fact of increasing Vietnamese imports does suggest upward pressure on domestic raw material prices and, therefore, additional incentive for local forest plantations and additional opportunity for small woodlots and the Vietnamese farm households that manage them.

In sum, the Vietnamese forest sector has grown rapidly as the entire economy has grown even more rapidly. Household welfare has also increased rapidly for those involved in the forest sector, but this is a very small share of the Vietnamese population. The number of benefiting rural households will probably increase as the land area and the management tasks involved in plantation forestry increase, but we can anticipate that these numbers will remain small as the wood products industries will continue to import large shares of their woody raw material. Vietnam's neighbors, Cambodia and Laos, and their rural populations, as they export wood to Vietnam, may be the greater benefactors of Vietnam's increasing demand for unprocessed wood as a raw material for its rapidly expanding wood processing industries.

Cambodia

The general conditions in Cambodia are very different from those in Vietnam, but still very interesting. Many factors are favorable for forest development—and for the benefit of a population that is poorer than Vietnam's. However, at least one absolutely crucial factor intervenes.

Cambodia, like Vietnam, is a country and an economy in transition. Its economy is growing but it only recently, in 2005, began to demonstrate the vitality of Vietnam's economy. Uncertainty has probably been the defining term for Cambodia, and especially for its undeveloped forest sector. Cambodia's twenty-year period of civil war is now history and everything about the country is better for that change. The general political environment does exhibit more stability. Nevertheless, the recent period of stability has not been sufficient to entice significant investment in long-term economic enterprise. Investment in either forest plantations or new wood processing facilities or even new logging equipment for the forest sector is nil.

Cambodia is extensively forested (53 percent of land cover is forest), and it has a history of exporting logs. It has road access to expanding markets in neighboring Vietnam and Thailand and rail access to world markets through its own port at Sihannoukville. Despite these advantages, Cambodia's wood products industry is almost nonexistent (many small sawmills and furniture factories, but only a declining US$0.4 million in gross value added in 2000 in comparison with US$77 million in logging) (see Table 11.3.) Even logging, which was the source of 40 percent of the country's export earnings in 1994, has declined until it was the source of only 2 percent of official export value in 2000. The decline in logging is due to a number of factors—the end of civil war and a decline in logging as a means of financial support for the rival militaries, declining demand in neighboring Thailand after the East Asian financial crisis in 1997, a domestic logging ban imposed in 2002, and inconsistent enforcement of logging policy today. Nevertheless, it is clear that some logging still does occur, some of it (perhaps a very large share) is illegal, and some that is not illegal occurs because some loggers manage to obtain official approval to circumvent government logging policy and regulations.

Under these conditions, the regular use of any particular logger's equipment must remain uncertain and the delivery of logs to any domestic processor or international consumer must be irregular. The local demand does exist for some private farm production of timber but even that is irregular at best because farmers cannot be certain of obtaining the required log shipment permits when their timber is mature. Short-term economic activity is the only option. No one invests under these conditions. Loggers and wood processors only use existing

TABLE 11.3 Cambodia

Land area		17.642 million ha	deforestation rate: 0.6%
Forest area		9.335 million ha.	forest plantations: 0.082 million ha.
Paved highways		1,996 km	
Railroads		602 km	
Population	10.945 million	1.17/forested ha.	growth rate: 2.3%
Gross domestic product		$3.183 billion	growth rate: 1.0%
GDP/cap		$303	growth rate: -1.3%
Poverty rate		36%	
Agricultural % of GDP[1]		35%	
Trade			
Exports-total		$1.327 billion	grew 2x since 1995
Forest products		$28 million	80% decrease over decade
Imports-total		$2.124 billion	
Forest products		$6 million	rapid, several-fold increase
Forest sector, excl, furn.	Gross value added		
Full forest sector		$77.9 million	declined 60% over decade
Forestry	Incl. logging & rel activities	$77.4 million	60% decline over decade
Woodworking		$0.4 million	70% decline over decade
Pulp & paper		0	
Contrib. to GDP		2.7 %	declined by 2/3 over decade
Employment	Full time equivs		
Full forest sector		6,642	declined 2/3 over decade
Ind'l forestry	Incl logging	775	declined 80% over decade
Woodworking		5,147	declined 70% over decade
Pulp & paper		720	
% nat'l emp'mt		0.1%	declined 2/3 over decade
Production			
All roundwood	10.3 million m^3		
Ind'l roundwood	0.2 million m^3	productivity/worker: 231 m^3	
Swnwd & panl	4 0.06 million m^3	productivity/worker: 12 m^3	
Paper		4 0	

Source: Hyde (2006)
1 This and all subsequent values in 2000 US$.

and continually depreciating equipment when the irregular opportunity arises. The industry declines, and the use of deteriorating equipment and the absence of a predictable return on forest management are a source of decline in the natural environment as well.

In sum, the prospects for a viable, if rudimentary, forest industry do exist in Cambodia. The industry's potential contribution to the overall economy and to employment in the logging and sawmill industries, whether regular employment or as an income supplement for small farmers, is probably significant. However, the policy environment has been debilitating. Furthermore, one could suggest that the intervention of the very many international forestry advisors representing an array of different interests and perspectives has been more hindrance than help. Many of these advisors focus their attention on narrow physical measures of forest protection, while disregarding economic incentives and overall Cambodian welfare. Furthermore, the variety in their advice, along with the strong platform from which they deliver it, reinforces economic uncertainty in the forest sector. The sector cannot achieve its sustainable potential under these conditions and Cambodia's rural poor cannot benefit as they might.

Laos

The prospects for forest sector development in Laos are somewhat better. The Lao economy is not as vibrant as that of Vietnam and the transportation infrastructure and access to international markets is not as developed as that in Cambodia. However, the overall Lao economy is growing and the domestic policy environment has been stable for several years longer than that of Cambodia. The government has accepted many market reforms in the last decade—although it could still further deemphasize its own market administration to the benefit of overall economic growth and, along with it, growth in the forest sector. Furthermore, the standing forest resource is larger, at 54 percent of Laos' total land area, than in either Vietnam or Cambodia and Laos' small population means that inexpensive land is available for any good opportunity—such as forestry.

The forest sector currently accounts for a 2 percent share of Lao GDP and a smaller share of employment. Nevertheless, the sector is important, as royalties from logging account for approximately 20 percent of the government budget. Forest production, 80 percent of which is exported, 90 percent in the form of relatively unprocessed logs, is the second largest (30 percent) source of the country's export earnings. The sector has experienced sufficient growth in demand that a few small wood processing establishments have begun investing in their own forest plantations. This indicates that the timber production component of the sector is approaching the mature stage of development—even if the wood processing component of the sector remains largely undeveloped (see Table 11.4).

It is Laos' potential for a much larger international market share that is most interesting. The overland market in China to the north is small but growing rapidly. Demand in Thailand could also grow—although monopsonistic buying practices in northeastern Thailand will limit that growth. A new bridge across the Mekong

TABLE 11.4 Laos

Land area		23.080 million ha.	deforestation rate: 0.4 %
Forest area		12.561 million ha.	forest plantations: 0.006 million ha
Paved highways		9,664 km	
Railroads		0	
Population	5.297 million	0.45/forested ha	pop. growth rate: 2.6%
Gross domestic product[1]		$1.709 billion	growth rate: 5.9%
GDP/cap		$289	growth rate: 3.0%
Poverty rate		33%	
Agricultural % of GDP		50.4%	20% decline over decade
Trade			
Exports-total		$330 million	50% increase since 1995
Industrial rdwd		$10 million	
Forest products	Excl ind'l rdwd & furn.	$3.5 million	
Imports-total		$507 million	sharp decline since 1996, has been as much as 36% of all exports
Forest products		0	$1-2 million in recent years
Forest sector, excl, furn.	Gross value added		
Full forest sector		$31.9 million	variable, $32-56 over million over decade
Forestry	Incl. logging & rel activities	$30.4 million	highly variable over decade
Woodworking		$1.4 million	variable, $1-5 million
Contrib. to GDP		1.9%	1.2-1.9% over decade, often more volume, less value/unit
Employment	Full time equivs.		
Full forest sector		6,861	as much as 2x this over decade
Ind'l forestry	Incl logging	3,262	
Woodworking		3,599	as much as 2x this over decade
Pulp & paper		0	
% nat'l emp'mt		0.3%	as much as 2x this over decade

Production			
All roundwood		6.4 million m³	relatively constant over decade
Ind'l rdwd	prod/worker: 174 m³	0.6 million m³	0.5-1.0 million over decade
Swnwd & panls	prod/worker: 59 m³	0.2 million m³	110-685 thousand over decade

Source: Hyde (2006)
1 This and all subsequent values in 2000 US$.

River may cut shipping costs to Bangkok and, thereby, to world markets by as much as 25–50 percent of fob log prices and 7–9 percent of fob sawnwood prices.

Probably the greatest potential, however, is over the mountains to the east to Vietnam. The elasticities of Vietnamese, Chinese and, for that matter, global demand for Lao logs and sawnwood beg assessment—and reliable predictions are dependent on them. However, it is clear that current log and sawnwood imports in both Vietnam and China dwarf the sum of all current Lao exports. A modest reduction in log shipment costs to either of these markets could have an immense impact on the Lao economy. Both China and Vietnam could easily absorb imports equal to Laos' total current production, thereby tripling current Lao production and employment in logging and inducing additional demand for new forest plantations and employment in forest management. The effect might be largest in the south central region of the country, the region on either side of Highway number 9 leading to Vietnam. This is a region of low population, some forest, and good land for additional forest plantations.

Thus, with a stable domestic policy environment—although one which would still benefit from a lesser government role—with continued strong markets in Vietnam and China, and with improved overland access, the forest sector's share of GDP could increase from 2 percent to 5 percent or more and export logs and minimally processed sawnwood would become the country's largest source of export earnings. Moreover, these may be cautious estimates that might be greater if we knew more about the demand elasticities for Lao exports.

Summary

In sum, it is not possible to develop a forest sector and, with it, add welfare for the local population without an available forest resource—but a forest resource by itself is insufficient for development. The Mekong Basin examples are further evidence that development requires an appropriate market and policy environment, including stability or, at least predictability, in the local markets or in markets within the range of inexpensive transportation. With a standing forest resource and with the appropriate market and policy environment, the forest sector can be an important source of development in some regions. It can even be a leading sector, as it might be in Laos. However, forest sector development is more likely as a source of income supplement for local households and as a long-term contributor to rather than leader in economic growth for most regions of the world.

Local Communities and Forest Dependence

The second part of this chapter turns to the question of smaller local and regional communities and either the effect on them of an active forest sector or their dependence on this forest. The global experience ranges from company towns where forest operations are the entire source of employment to somewhat more diverse communities whose primary activity is generally agricultural but for which forest opportunity is one, perhaps one of several, sources of employment and welfare for many local households.

The forest opportunity in some of these communities is commercial, focusing on logging or logging and primary wood processing, or on forest-based recreation and tourism. In others, non-timber forest products (NTFPs) are more important. In exceptional circumstances, NTFPs may provide as much as 95 percent of household cash income (Schreckenberg et al. 2005). More often, the collection and exchange of NTFPs provide either an income supplement or an important component of domestic consumption for some households in the community. In all cases, the forest-based opportunity is crucial for those households dependent on it, and often these include the poorer or minority households that motivate many of our public policy decisions.

There are two parts to this question. The first reviews the experience of local communities that are dependent on the forest for a variety of resources. These are often subsistence agricultural communities, and the role of women as collectors of fuelwood or NTFPs has been an important question within the literature examining these communities. Measures of the distance to the resource, the time spent collecting it, and the relative availability of either natural or managed forests and trees have been central to the analysis—and this is consistent with the importance of distance or access and the distinction between natural and managed forests within our three-stage model.

The second part of this inquiry examines the income and employment impacts of commercial forest operations, such as timber production or commercial recreational opportunity, on local communities. This has been a greater question in developed countries like the United States and Canada where a substantial natural forest is often viewed as a source for regional economic development or where management of a flow of timber from public lands is sometimes seen as a means of moderating the boom and bust swings in economic vitality that many forested communities experience.

Subsistence Community Impacts

Peters, Gentry, and Mendelsohn (1989), with data from the Peruvian Amazon, suggest that local communities might earn more money from the collection of wild fruits than from timber operations in their forests. Numerous observations since Peters et al. provide undeniable evidence that rural households do benefit from non-timber opportunities in adjacent forests—the opportunity to collect fuelwood or NTFPs, to graze their livestock, and to harvest wildlife as a source of protein. For some communities, we have estimates of the numbers of individuals or households involved or estimates of the volume of material they collect from the forest, and the numbers can be very large.[9] For example, the sale of charcoal alone is responsible for 38 percent of cash income for households in two regions of Tanzania. Adding sales of honey, fuelwood, and wild fruit increases the contribution of NTFPs to 58 percent of these households' cash income (Monela et al. 1998). Elsewhere in central Africa, bushmeat is the source of as much as 80 percent of the protein and fat for rural households. Poorer households depend on bushmeat as a source of cash income, as well as for personal consumption (Nasi et al. 2007). Artisanal activities, such as furniture making and basket making from ruttan, contribute still more, employing more than 140 million people worldwide, approximately 10 times as many as are employed in the forest industry in developing countries (*ITTO Tropical Forest Update* 2007; Byron and Arnold 1999). Prebble (1999) reports that 150 different NTFPs are "of significance" for international trade. Thailand alone exported US$18 million of bamboo, ruttan, gums and resins, medicinal plants, industrial insects, agarwood, and other NTFPs in 1996.

These measures encourage the general expectation that the collection of non-timber products offers an important alternative to timber harvesting and the decline of natural forests while, at the same time, benefiting the poorer households who are the greater collectors of these forest products. These favorable effects are undeniable for some select communities. However, the evidence supporting them tends to be drawn from anecdotal case studies, and other, less optimistic, estimates of the values obtained from fuelwood and NTFPs in other communities challenge their generality (Godoy and Bawa 1993; Godoy et al. 2000; Sheil and Wunder 2002).

More general observations: Three recent assessments attempt to sort out the more general role of NTFPs in local welfare across a wide range of natural and economic situations. Two of them also inquire into the effect of these communities' dependence on their forests for the conservation of natural forests.

Vedeld et al. (2004) reviewed 54 household studies from 17 countries in East and South Africa and South Asia. Their summary observations are that approximately one-fifth of annual household welfare originates with NTFPs, including about 8 percent from wild foods and another 6.5 percent from fuelwood. The one-fifth share of household welfare is split between domestic consumption and cash income from products collected for market sale. Wealthier households collect larger volumes, but NTFPs provide a greater share of total welfare for

---

9 Neumann and Hirsch (2000) review this literature.

poorer households. Households in villages that are farther from markets and are, therefore, less reliant on them, obtain greater shares of their own welfare from their own collection and consumption of forest products and smaller shares from market sales.

Dewi, Belcher, and Puntodewo (2005) draw similar conclusions from a more geographically select sample of 73 villages in Indonesia's East Kalimantan,. They add that the diversity of a village's economic base is also a factor in its reliance on the forest. Of course, increasing economic diversity is often correlated with market access and greater distance from the forest, and households in more economically diverse villages tend to be better off than households in less diverse villages.

A third assessment by Belcher, Ruiz-Perez, and Achdiawan (2005) provides more comparative insight by separating villages into categories according to their forest use. Belcher et al. examined 61 communities from a range of natural and economic situations spread across Africa, Asia and Latin America. They began by measuring the contribution from collection and market sale of NTFPs to household income as a function of local integration into the cash economy. Specifically, for each of the communities, they regressed income from the sale of NTFPs as a share of total household income on the household's cash income as a share of its total income. They consider the latter, their independent variable, to be an indication of a community's integration into the cash economy. Figure 11.2 shows their most general summary regression, a logarithmically increasing function of the NTFP contribution related to the community's relative integration into the cash economy.

Closer examination of Figure 11.2 reveals that the 61 communities separate neatly into five categories according to their use of the forest. Sixteen subsistence communities (denoted by circles) all appear within the lower left quadrant of the figure. NTFPs contribute less than 50 percent of their household income and, as subsistence communities, they are dependent on the cash economy for less than 50 percent of their total income. These communities are not well-integrated into a cash economy.

Two groups of communities that rely on NTFPs either for supplementary income (triangles, 22 communities) or as part of an integrated household economic strategy (diamonds, nine communities) both fall within the lower right quadrant of the figure. NTFPs still contribute less than 50 percent to their cash income, but these communities are better integrated into a local cash economy.

Finally, the communities that rely on either specialized natural (stars, eight communities) or cultivated (squares, six communities) techniques for growing and obtaining products from their trees and forests fall within upper right quadrant. These final 14 communities obtain 50 percent or more of their cash income from the sale of NTFPs and they too are better-integrated into a local cash economy.

Closer examination yet shows the underlying relationships between the five categories of communities and (a) their infrastructure development, and particularly market access, and also (b) their separate dependence on either natural or managed forests. These relationships are very similar to those predicted by our

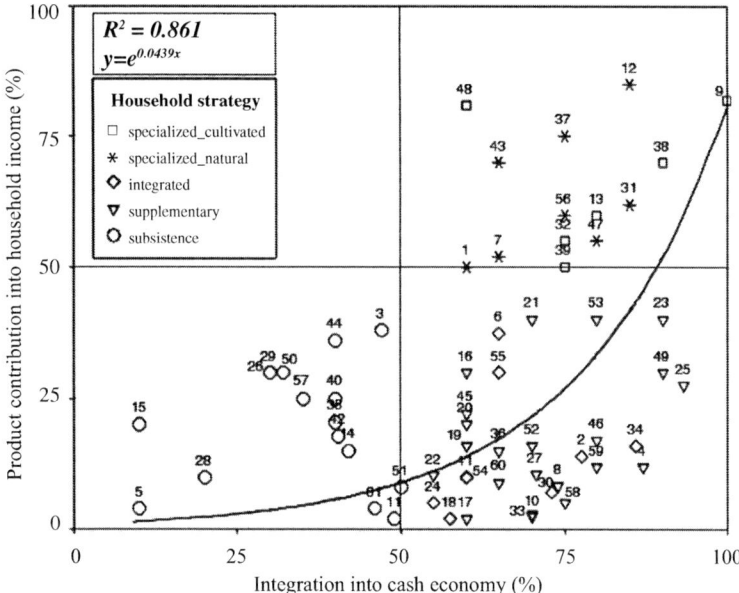

FIGURE 11.2 Relationship between integration into the cash economy (%) and NWFP contribution to household income (%) for 61 communities. *Source:* This table was published in *World Development* 33(9), B. Belcher, M. Ruiz-Perez, and R. Achdiawan, Global patterns and trends in the use and management of commercial NTFPs: implications for livelihoods and conservation, p. 1439, copyright Elsevier (2005).

three-stage description of economic activity in the forest, and they are consistent with the summary observations of Vedeld et al.

Infrastructure and market access improve progressively from the first to the fifth category. Communities within the first three categories rely entirely on unmanaged natural forests. They tend to be less dependent on the cash economy as the nearest markets are, presumably, more distant from many of them and household collection from the natural forest probably accounts for a larger share of household welfare. Communities within categories four and five may depend on some use of managed natural forests and a few households in the fifth category also manage their own stands of planted trees. Therefore, NTFPs from managed forests have begun to contribute to the cash income of these latter categories of communities with better market access.

Belcher et al. also considered the differences in the household labor input and forest productivity across the five categories of communities. Their own words say it best. For subsistence communities (the first category) "Wild gathering tends to occur in areas with less infrastructure. Labor inputs and productivity yields per unit area are low and production is typically a male dominated activity" (p. 1440). "Forestry is often the default option" (p. 1447). "The open access (natural forest) resource offers opportunities for people with limited resources, (but) the inability to exclude competitors often prevents (households) from making a good living from those resources.... There is a tendency for (the) overexploitation of marketed NTFPs" (p. 1447).[10]

"Managed systems (communities in categories two and three) tend to occupy an intermediate position between wild gathering and cultivation" (p. 1440). Women are greater participants in these intermediate systems because even a minimal level of "forest domestication" requires greater labor input.

The final two, more specialized, categories "tend to be associated with the highest levels of infrastructure development, population density, and relative development, offering the highest (household) incomes ... and the highest productivity per hectare" (p. 1443). Secure private land tenure tends to be the norm in these two categories and "it leads to more stable production systems for a given (forest) product" (p. 1443). Households in communities in these categories receive greater prices for their marketed NTFPs and, therefore, the rewards for management are greater. Perhaps because of their greater financial attraction, the managed forests of these two categories, once more, require greater inputs of male labor. Cultivated managed forests appear only in communities in the fifth category.

Belcher et al. were motivated to provide guidance for policy designed to help the rural poor. They conclude that "it is simplistic, and often wrong to assume that because an NTFP is important to the poor, efforts to develop it will help the poor" (p. 1446). And "ultimately, if NTFPs are to be useful in efforts to reduce poverty ... it will have to be through increased and more efficient commercial production and trade" (p. 1447). However, in all 61 communities of their sample, NTFPs contribute only a portion to household income, generally only a small portion. Therefore, they suggest that helping people move into alternative activities may be a more beneficial strategy, for both the poor and the forest environment, than policies that encourage increased commercial collection of NTPFs.

More detail on fuelwood, other NTFPs, gender and poverty: Belcher et al. were restricted to the effects on income from market sales of NTFPs. For many poorer rural communities, a more complete assessment of the welfare effects of an active forest sector would incorporate measures of both market sales and also direct domestic household consumption of self-collected wood and non-wood forest products. Direct household consumption of these products is larger than the market sales of the same products for many households in some communities.

A technique known as the "new household economics" has been developed to assess household activities in just such cases where production and consumption occur jointly and cannot be separated in the measures of household activity.[11]

---

10 Kusters et al. (2006), with observations of NTFP sales in 55 communities from Africa, Asia, and Latin America, also conclude that market opportunities for these products lead to their depletion in the natural forest. Scherr, White, and Kaimowitz (2004) comment further on forest markets and poverty.
11 Singh, Squire, and Strauss (1986) is the standard technical reference.

The technique is complex, as an appendix to this chapter shows, and it has been applied in only a few forestry cases, generally for fuelwood and its substitutes in South Asia. The results, however, are revealing, particularly for the participation of poor households and especially women, and for the dependence of households and communities on either natural or managed forests and trees.[12]

The household economic assessments of fuelwood and its substitutes were all taken from subsistence communities where fuelwood and other products collected from the forest are a small share of total household production and consumption. Agriculture, whether for household subsistence or market sales, occupies a much larger share of all activity for the sampled households and communities. Poorer households in these communities collect fuelwood and some other products (e.g., water, dung for fuel, broadleaves as eating utensils—depending on local practice and availability) from the natural forest. Better-off households—still poor, just not as poor—may collect some fuelwood but they more consistently rely on small formal markets for the same goods or else prefer to use fuelwood substitutes.[13]

Some who collect fuelwood and other forest products walk considerable distances to collect—but market conditions are important. They are more likely to participate in fuelwood collection when the price in the small local market is higher and, when the price is higher, these households tend to collect more for both their own domestic consumption and also for market sale. They are also more likely to collect when their own off-farm employment opportunities are not great or when their other labor opportunities are lower-valued. Therefore, it is not surprising that collection increases during the off-season for more rewarding activities. Since women are often the primary collectors, their participation may be greater when the time spent collecting can also be an input to some other household responsibility. Child rearing may be an example, as women often take small children with them to collect fuelwood, but the fuelwood that small children collect is negligible, and the combined collection by women and their accompanying children is sometimes less than that collected by other women who are not accompanied by children during the collection activity.

Some assessments of household production and consumption show that higher fuelwood prices and greater collection time, both evidence of greater scarcity, induce substitution. Evidence on the substitution of agricultural residues like dung and straw for fuel is weak. The evidence for, first, substituting improved stoves (that burn less fuel) and, then, alternative sources of woodfuel is better. Where

---

12 Hyde and Kohlin (2000) survey this literature. Mekonnen (1998) and Hegan et al. (2003) add examples from eastern and southern Africa

13 The existence of small formal markets is widespread, but the assessment of these markets is almost non-existent. In the existing assessments, poor households seem to supply less to these markets than better-off households purchase. Therefore, someone else must also be supplying the rudimentary local markets. The very poorest households, usually landless households, must make up the difference. We can hypothesize that landless households are significant suppliers of market fuelwood—yet these households apparently have been largely absent from the data of this literature.

the local scarcity is greatest, households are more likely switch from dependence on the natural forest to growing and managing trees for fuelwood on their own agricultural lands. In these cases, men, as they provide the greater share of field labor, seem to be greater participants in fuelwood collection. In fact, Linde-Rahr (2005), with evidence from Vietnam, finds that the returns to labor in agriculture and in the collection of general NTFPs are not significantly different.

These observations are all consistent with general economic expectations. Households allocate between their productive resources and make consumption choices that are equal at the margins. These observations, mostly for fuelwood with a general geographic focus on South Asia, should cause us to anticipate similar experience for household dependence both on fuelwood and on other NTFPs in other communities and other parts of the world.[14] In fact, the more limited analytical evidence for other NTFPs suggests that their price elasticities, like that of fuelwood, are less than one and, for a few, the income elasticities may occasionally be negative—indicating that these few are inferior goods that households aggressively replace as their incomes increase sufficiently to allow them to switch to preferred substitutes (Hyde and Kohlin 2000; Linde-Rahr 2005; Gopalakrishnan et al. 2005).

Robinson and Kajembe (2009) caution us, however, that the full household responses to increasing scarcity of the various NTFPs are not all similar. While households respond to increasing prices or other indicators of scarcity by generally decreasing consumption, the full household response differs with the different NTFPs. For example, evidence from Tanzania suggests that, in the presence of increasing scarcity, households substitute other sources of fuel for fuelwood and collect fewer wild fruits but increase their collection time in order to continue their access to materials for weaving and construction.

Household wealth is, clearly, a crucial determinant in the collection and use of NTFPs—as the evidence on income elasticity suggests. Poorer households are more dependent on NTFPs as a component of their regular consumption and, in times of greatest need, the natural forest provides an accessible and low-capital source of available NTFPs. It becomes a "safety net" for these households.[15,16]

---

14 Arnold and his colleagues (Arnold et al., 2003; Arnold, Kohlin, and Persson 2006) do generalize for fuelwood.
15 Sunderlin, Angelsen, and Wunder (2004) and Fisher and Shively (2005) refer to many of the numerous discussions of the social safety net. There is a small distinction from the safety net discussion in the previous chapter. Chapter 10 referred to the reliance on small private forestlands for their owners' own financial gain in unusual times, but not necessarily times of economic hardship. The reference in this chapter is more often to the use of open access or village forests, rather than private forests, and more commonly to use by poorer households during times of their economic hardship.
16 Most of the discussion of the forest safety net refers to rural households. Stoian (2005), with an example from the Bolivian Amazon, shows that poorer peri-urban households also gain from being able to increase their activity in NTFP processing, specifically palm hearts and Brazil nuts, during their times of limited financial opportunity.

The effects of consumption by subsistence communities: Forest degradation and improvement: What is the effect of local community consumption on the forest, and is there a means for improving the lot of those poor communities and households that are more dependent on the products of the natural forest? These questions seem to generate three different policy responses: conservation strategies to correct forest degradation; devolution of property rights as a means to improve forest protection, the availability of scarce forest products, and the lot of poor households dependent on them; and improved opportunities for marketing locally available forest products. Each of these were discussed in general terms, in one form or another, in this and previous chapters. However, this time the questions and these policy responses specifically relate to subsistence communities and poor households. It may be useful to review with regard to these communities and households in particular.

Both the wood and non-wood forest products that these communities and households use follow a three-stage pattern of consumption and development similar to that of commercial timber and this pattern, once more, is central to understanding each of the three policy responses. To review, households initially collect their wood and non-wood forest products from the natural forest. As collection begins to deplete accessible resources, further collection moves on to, removes, and degrades ever less accessible naturally growing products. If consumption is to continue, then prices must rise, and they continue to rise until they eventually overcome the costs of establishing property rights and managing the domestic production of the product or products for household consumption as substitutes for similar products from the natural forest. Chapter 2, referred to the examples of Brazil nuts in the Amazon and fuelwood in Nepal's hills (Viana et al. 1996; Amacher Hyde, and Joshee 1993).

The discussion of conservation strategies arises because many observers recognize the initial depletion within the natural forest and, particularly, the additional difficulty it creates for poor households who are disproportionately dependent on its wood and non-wood products. Any conservation strategy requires limiting access, at least for a period, while the resource recovers, and limiting access does not help poor households with immediate needs.[17]

The devolution of property rights from central to local institutions and households can improve the condition of the resource if local knowledge makes protection and management of the forest and those of its resources used locally less costly and more effective than more central administration of these properties. Devolution will not have an effect, however, on less accessible forest resources still within the range of open access exploitation—between points B and D in the figures in chapter 2—because the costs of establishing rights by any agent, local or otherwise, are too great. Furthermore, where devolution does successfully improve forest management, it will only help the poor if they obtain a share of the

---

17 See discussions by Reddy and Chakravarty (1999), Jodha (2000), and Dangi and Hyde (1991) for examples from Utter Pradesh and Rajasthan in India and Nepal's hills, respectively.

rights or access under the newly devolved management system. Often, better-off households and community leaders dominate in the new property arrangement and, while the forest and its various products may be better managed, the poor are not the real beneficiaries.[18]

Improved market opportunities can be a better approach to improving both the condition of the forest and the welfare of poor subsistence households. However, improved market opportunities require public investment in either transportation systems or in the dissemination of market information, or both. Reduced transportation costs translate into increases in the local values of all goods and services delivered to the market. This can cause further depletion of forest resources as the increased demand price induces additional collection for sale in markets outside the local community. However, improved market access generally increases other opportunity as well, including opportunity for agricultural products and also for labor in employment alternatives that are more rewarding than employment in local subsistence forest collection. In fact, improved market opportunity generally attracts labor away from forest collection. For those households whose labor opportunity has changed, their demand for low-value, labor intensive, forest products generally declines, and the depletion of the local natural forest declines with it. The forest may begin to recover. In sum, better and broader market opportunity generally induces greater diversification in the local economy, and economic diversification probably reduces (but does not eliminate) the extremes in economic cycles and, therefore, some of the safety net demands for forest resources. As it does, the natural forest begins to recover.[19]

The Local Impacts of Commercial Forest Operations

Turn now from subsistence communities to commercial forest operations. There can be no doubt that forests are a source for regional economic development in some cases. Nevertheless, there also can be no doubt that the availability of a standing forest resource is an insufficient condition for regional development—as discussed in chapter 6 and also earlier in this chapter.

The notions persist, however, that public policy encouraging commercial forest activity can induce regional development, and also significantly dampen wide swings in local economic activity and, thereby, assure a more stable local economy. Both are important policy perspectives because the peoples who inhabit forested regions are often poor and good development policy might make a difference in their lives, and because boom and bust cycles in economic activity often characterize their resource-based local economies. Public policy encourag-

---

18 See Sikor and Nguyen (2007) for an example from Vietnam's central highlands or more general discussions in Hobley (2007) and Angelsen and Wunder (2003).
19 See discussions in Illukpitiya and Yanagida (2008), Scheer et al. (2004), Dewi et al. (2005), Belcher and Schreckenberg (2007), Wilkie and Godoy (1996), Gunatilake and Chakravorty (2001), and Byron and Arnold (1999) for both a range of examples and general discussion.

ing forest activity may have beneficial effects on local activity in these situations, but it is important to understand the conditions under which this is possible, as well as the limits to its effectiveness, and also the misinformation that deludes us in some cases.

Take first the perception that harvests of public timber induce local or regional development. The decision to harvest public timber, or private timber for that matter, does have a beneficial effect on the overall regional economy. It supports employment for loggers and millworkers. These loggers and millworkers, and millowners too, spend their wages and profits locally, thereby increasing local demand in the retail sector as well as local demand for the firms that supply materials for the logging and wood processing industries.

The merit of the decision to harvest depends, however, on both the net value of the standing timber and the question of who pays for its harvest. Of course, private landowners and independent loggers must cover their own costs. They will not participate in these activities unless they obtain net gain from their action. They gain when prices in the external markets for their products increase or when the costs of their externally purchased inputs decrease. When there are net gains, then landowners receive stumpage fees, loggers and millworkers receive wages, millowners profit, and the entire community gains as all three—landowners, loggers and millworkers, and millowners—increase their expenditures in other sectors of the local community.

The case for publicly owned forests can be different. The public agencies can, and often do, transfer funds from other public sources to support timber sales and harvests that, otherwise, would be financially unrewarding. Many public forestry agencies rely on an uneconomic accounting system such as the allowable cut system discussed in an appendix to chapter 3. Others may simply use regional development as a justification for transferring funds to local timber operations.[20] In either case, the result is the same, an exogenous source of public funds subsidizes the local timber management and harvest activities—and local harvesting can continue only as long as the greater public in the rest of the country is willing to allow the transfer of its general public funds to support activity in the particular forested region in question. If or when the transfer ever comes to an end, then public timber activity in the local region in question can no longer exceed the level that is justifiable on its own financial merit. At this time the publicly subsidized inducement for regional development will cease.

The argument for public support for regional development is often made to sound better by citing a multiplier effect. The multiplier measures the effect of additional rounds of spending in the local economy due to an initial exogenous injection. It includes the initial payment for local goods and services associated with the public activity, loggers' wages for example, plus the secondary effect on the local economy of local purchases of retail goods by the loggers' families,

---

20 See Hultkranz (1991) for the example of Swedish subsidies for silviculture and forest road construction designed to effect regional employment and seasonal unemployment.

and then the tertiary effect of additional wage payments to retail employees who work additional hours to satisfy the increased demands from logger's families, and so forth. The positive effect on the local economy diminishes with each subsequent round of expenditure by local employees and businesses because, in each round, some wages are removed as savings and some expenditures leave the local economy for goods and services produced outside the region. Multipliers depend on the size of the region and the extent of its self-sufficiency, with larger and more self-sufficient regions having larger multipliers. Nevertheless, regional multipliers in the neighborhood of 1.6 are common for the first year of any exogenous injection. It should be clear that multiplying any exogenous expenditure by 1.6 and counting the result as a project benefit makes any project of a public agency appear more attractive.

The problem with the multiplier effect does not lie in the accounting. There is no doubt that exogenous financial injections into local economies do result in secondary, tertiary, and further expenditures and, thereby, cause the local economy to expand. The problem is that any expenditure of the same public funds anywhere and in any economic activity will have its comparable multiplier effect. Therefore, the multiplier provides no comparative advantage for one particular local public activity in preference to any other, and it would be less confusing to eliminate it from the consideration of alternative public projects, forestry or otherwise.

Assessing the impacts of public policy on community welfare: Income and employment are large components of the measure of any community's welfare. Therefore, a reliable assessment of the local economic effect of any activity, public or private, forestry or otherwise, requires an understanding of the sources of labor and capital applied to the activity in question, as well as their alternatives if they had not been employed in this activity. If, for example, an increment in timber harvests means additional employment for local workers who were previously unemployed and logging equipment that was idle, then the additional harvests induce one increment to regional development. On the other hand, if some of the new loggers had been actively employed in other local activities and others are attracted from activities outside the region, and additional logging equipment has to be purchased from manufacturers outside the region, then the new harvests induce lesser additions to the local economy.

Because the range of different substitution possibilities for the labor, capital and resources employed in forestry varies widely from region to region, it is impossible to draw general conclusions about the effect of any particular forest activity on regional development. We can, however, in a mathematical appendix, describe a fundamental organization for relating any forest activity in question to its labor and capital inputs, and these inputs to their opportunities for substitution. Data are available to estimate the model parameters for most regions and industries in most developed countries.

Daniels, Hyde, and Wear (1991) provide an example. They examined the potential for U.S. National Forest timber harvest decisions to stabilize cyclic economic activity in western Montana. Western Montana is mountainous country and its mountains impose high transportation costs for the products of forestry and ranching, the region's main industries. The U.S. Forest Service is the dominant timber supplier, providing more than 40 percent of all timber harvested throughout Montana and probably a larger share of harvests in the mountainous western part of the state at the time of the Daniels et al. analysis (Spoelma et al. 2008).[21] It is more likely that Forest Service harvest policy could have an effect on an economy like western Montana's than on that of some other region that is more diversified and that experiences the cheaper transportation that would allow greater integration into the broader external economy.

Of course, the Forest Service can control the volume of timber it offers for sale but it has less control over the volume harvested in any period of timber because those who win the auctions for public timber have some discretion on when they exercise their harvest rights. Nevertheless, Daniels et al. considered the impact if the Forest Service had the policy effect it desired, maintaining a constant flow of timber harvests in the presence of the greatest swing in wood product prices observed in recent years, an 18 percent decrease in the price of sawn lumber. In a less administered market, a lumber price decrease of this magnitude would be met with a decline in mill production and, therefore, declines in both mill revenues and mill demands for both logs and millworkers. The local mills would accept a greater, even or constant, flow of timber only if they could pay a lesser price for stumpage or delivered logs. In this event, the mills would maintain operation and, thereby, maintain a level of employment for loggers and millworkers but the collection of stumpage revenues by the Forest Service would decline.

Daniels et al. calculated that the effect of a stable harvest flow on the western Montana economy would be small. Logger and millworker employment would increase by 6–18 percent but total wage income for the entire community would be essentially unchanged because, even in this relatively undiversified economy, workers have other employment opportunities. The economic contribution of the wood processing sector would increase by 7–9 percent, but the net effect on the entire western Montana economy would be less than 0.1 percent. These small gains would be offset seven-to-ten-fold by a $6–11 million decrease in annual stumpage revenues. The U.S. Forest Service and the U.S. Treasury would absorb most of the loss, but the decline in stumpage prices induced by the expansion in Forest Service timber harvest would cause other, non-Forest Service, landowners to absorb some losses as well.[22]

---

21 Subsequently, with new restrictions on public timber harvest, US Forest Service timber harvests declined to 20–25 percent of all timber harvest in Montana by the mid-1990s. They declined even further to approximately 15 percent of all Montana harvests by 2004 (Spoelma et al. 2008).
22 In a related assessment, Lewis, Hunt, and Plantinga (2001) found that decreases in U.S. National Forest timber sales in 92 northern counties in the U.S. Northeast and Lake States had no effect on local employment. An increase in forest conservation, however, was related to a small increase in local employment.

Berck, Burton, and their colleagues took a different approach to examining the instability question and the effect on communities of policy intended to offset economic instability. They examined instability itself and its sources with examples from northern California and Oregon, two additional regions with dominant timber industries and a large public forestry presence.

Humboldt County in California, experiences two-and-one-half times as much variability in employment as other California counties, largely because there is little economic diversity in the county. Employment is concentrated in forestry. Berck et al. (1992) inquired whether greater economic diversity would make a difference by imposing the sectoral diversity of the state of California on a model of Humboldt County while maintaining the county's other economic characteristics. This difference decreased the variation in the county's employment by only 16 percent—causing Berck et al. to look for a different source of variability. They point to structure in general, or location and transportation costs in particular, as the obvious source. Furthermore, they conclude that since diversification is unlikely—the county just does not possess the characteristics that would attract other economic sectors—and since Humboldt County cannot change its location, then any policy that specifically targets forestry alone cannot have much effect and modifications in macroeconomic policy have greater potential.

In fact, subsequent observations from Oregon confirm this conclusion. Burton and Berck (1996) determine that the causal economic links run from macroeconomic variables to forest sector employment and then to stumpage cut or timber sold. Continuing these observations from Oregon, Burton (1997) compared sector specific policies with macroeconomic impacts on employment. In various hypothesis tests, she observes that macroeconomic forces, measured as either real gross domestic product, or as housing starts, have important effects on employment while public forest policy, measured as timber cut or sold, does not.

In sum, public forest policy itself does not significantly affect the probability of change in forestry employment. If it does not affect forestry employment, then it cannot have a significant impact on broader measures of total employment or income for the full community, no matter how important the forest sector is to the community or how diversified the community's economy may be.

Conclusions: The community effects of commercial forestry: Make no mistake: There are many components to the measure of local community welfare. Indications of education, health, and whatever is the opposite of social unrest are all important. The introduction to this chapter commented on the second dividend that some receive from living in local forested communities, a dividend in addition to their money income. Nevertheless, local income and employment remain crucial components of community welfare. They are the foci of our discussion.

Where commercial forest operations are financially rewarding, forestry can be a dominant sector, even so dominant as to support entire undiversified towns. Of course, in other cases, the forest sector is one of several contributing employment and income to the local economy. However, since forest activity must occur where trees grow and can be harvested inexpensively, especially during the first two stages of forest development, commercial forest activity tends to occur in more remote regions typified by undiversified local economies. These economies tend to experience boom and bust cycles in response to the vicissitudes of the external world's demand for their few products. The local community has no control over these external demands and, because the local economy is not diversified, its households have few local employment and income alternatives during the bust part of these cycles.

This suggests two topics of interest subordinate to the question of community impacts of commercial forestry, one having to do with the opportunity for commercial forestry to lead local development and the other having to do with diminishing the problems the community experiences during the most difficult parts of the economic cycle.

The first is easy. Where commercial activity is financially rewarding, then it will exist. It may have to overcome significant transportation costs but, where it does, it pays wages that offer new opportunity to the generally small local population, and it attracts new employees from outside the region. We have seen this during the first and second forests of the U.S. South, and in the towns of Tailandia and Breves in Brazil. We have anticipated the same for some parts of Laos as that country improves its roads and its transportation costs decline.

The employment and income instability that occur in undiversified forest-based communities create a more difficult problem. The natural tendency is to confront it directly, as some governments and public forestry agencies attempt with subsidies and guaranteed timber harvest levels. Such policies do insure a flow of raw material to local wood processors who can then maintain a higher level of employment and production than they would otherwise until prices recover in the markets for their products.

However, the resulting contribution to the local economy is small because many forest sector employees are mobile and readily find employment in other sectors, even in an undiversified economy. Further along in the development process, in the third stage of forest development, local economies are more diversified—as we have seen in the South's third and fourth forests, in the town of Paragominas in Brazil, and in Longyou County in China. With greater diversification, the full local community is less subject to the economic fluctuations of any single sector like forestry, and those forest workers who are affected have even greater alternative employment opportunity.

The problem for communities with undiversified local economies remains, and there are many communities around the world with such forest-based economies. If direct attempts to influence their employment and income in difficult times are largely unsuccessful, and structural adjustments; particularly improvements in the transportation system and, therefore, decreases in transportation costs; are not immediately likely, then the remaining policy alternative is look outside the community itself to those broader external markets that have strong effects on the local community. Policies that generate improvements in aggregate demand or more selective policies that induce expansion in the housing market and its demand for wood products are examples. The effects of such macroeconomic policies, however, take awhile to pass down through other sectors to the forest sector and its dependent communities and, even then, their effect on forest sector employment and income are spread across all communities producing forest products, some of which experience greater economic hardship and are more needful of the policy intervention than others.

So—what is the solution for those undiversified communities that are both reliant on forest sector employment and subject to the vicissitudes of external markets for forest products? Aside from economic development that will see some of these communities decline and even disappear as both production and transportation costs decrease for commercial forest activity in other regions in the third stage of development; regions like the Piedmont with its fourth forest in the U.S. South, Paragominas in Brazil, or the more-populated regions where Vietnam's forest development is occurring most rapidly; there isn't one. Economic cycles will continue to be part of their experience for the numerous more remote and less diversified forest-based communities.

Conclusions

In a sense, little has changed from the three-stage pattern of forest development first discussed in chapter 2. Whether discussing markets or policy, one or another class of landowners or, as in this chapter, the effects of forests on entire local human communities, the particular stage of development sets much of the discussion.

For communities in regions in the first stage of development, the natural forest provides support for subsistence households and supplementary income for local commercial agricultural households and workers. Subsistence households rely on the forest for a range of wood and non-wood products that form a part, sometimes a large part, of their regular household consumption. Some of these households also collect selected forest products for exchange in the local markets where they receive a share of their very limited cash income. Often it is the poorest households in any subsistence community for whom the collection of these products is most important, and during the worst of times the forest and its products can serve as a social safety net for these poorest local households.

Members of other households in other communities participate in rudimentary logging and sawmill operations—usually when the local demand for sawnwood is strong and when their labor is not as needed in their household's agricultural activities. For these households, the forestry activity provides a welcome supplement to the household's agricultural earnings.

Communities in the second stage of forest development experience a range of dependence on their forests. Some households in these communities continue to rely on logging and rudimentary sawmills for supplementary employment and income. For other communities, forestry may be one of a small number of important sectors in an undiversified economy. Communities in western Montana or northern California in the United States are examples. In these communities, forestry offers full-time employment for some households. In still other communities, forestry may be the only source of employment and income for all households in the community. Company towns are an example. Communities in both cases tend to experience the boom or bust cycles common to many remote resource-based communities. Fisheries, mining, even some agriculture and some tourism-based communities share this experience with forest-based communities. It is due to their relative remoteness and their focused economic activity and, despite various attempts, policies designed to limit the negative effects of economic cycles on these undiversified communities have only most limited success.

The third stage of forest development is characterized by managed forests, often at some distance from any region's remaining natural forests. The economies of communities in these regions are more diversified. They tend to include multiple economic sectors and forestry is only one generally smaller sector. The forest sector itself is often specialized, with a variety of sawmill, plywood, and other wood processing facilities and, because the demands for labor by these facilities are different, those who provide labor to this sector often specialize themselves. Economic cycles are not so damaging for these communities as for those in the second stage of forest development because the income and employment effects of a downturn in one of the community's sectors are offset by labor opportunity in other sectors which are not all at the same stage in their economic cycles. Nevertheless, even for communities and regions in the third stage of forest development, a sustainable forest sector does not follow automatically. The quality of local and national institutions, especially those that provide for the security of long-term land tenure and the stability of financial markets is also crucial.

Literature Cited

Amacher, G., W. Hyde, and B. Joshee. 1993. Joint production and consumption in traditional households: Fuelwood and agricultural residues in two districts of Nepal. *Journal of Development Studies* 30(1): 206–225.

Amacher, G., W. Hyde, and K. Kanel. 1999. Nepali fuelwood consumption and production: Regional and household distinctions, substitution, and successful intervention. *Journal of Development Studies* 35(4): 138–163.

Angelsen, A., and S. Wunder. 2003. *Exploring the forest-poverty link: key concepts, issues, and research implications.* Occasional paper no. 40. Bogor, Indonesia: Center for International Forestry Research.

Arnold, M, G. Kohlin, R. Persson, and G. Shepherd. 2003. *Fuelwood revisited: What has changed in the last decade?* Occasional paper no. 39. Bogor, Indonesia: Center for International Forestry Research.

Arnold, M., G. Kohlin, and R. Persson. 2006. Woodfuels, livelihoods, and policy interventions: changing perspectives. *World Development* 34(3): 596–611.

Belcher, B., and K. Schreckenberg. 2007. Commercialization of non-timber forest products: a reality check. *Development Policy Review* 25(3):355-377

Belcher, B., M. Ruiz-Perez, and R. Achdiawan. 2005. Global patterns and trends in the use and management of commercial NTFPs: implications for livelihoods and conservation. *World Development* 33(9): 1435–1452.

Berck, P., D. Burton, G. Goldman, and J. Geohagan. 1992. Instability in forest and forestry communities. In P. Nemetz, ed., *Emerging issues in forest policy*, Vancouver: University of British Columbia Press, pp. 315–338.

Bluffstone, R. 1995. The effect of labor markets on deforestation in developing countries under open access: An example from rural Nepal. *Journal of Environmental Economics and Management* 29(1): 42–63.

Boyd, R., and W. Hyde. 1989. *Forestry Sector intervention: The impacts of public regulation on social welfare.* Ames: Iowa State University Press, pp. 48–89.

Bulte, E., R., Damania, and R. Deacon. (2005). Resource intensity, institutions, and development. *World Development* 33(7): 1029–1044.

Burton, D. 1997. An astructural analysis of national forest policy and employment. *American Journal of Agricultural Economics* 79(3): 964–974.

Burton, D., and P. Berck. 1992. Statistical causation and national forest policy in Oregon. *Forest Science* 42(1): 86–92.

Byron, N., and M. Arnold. 1999. What futures for the people of the tropical forests? *World Development* 27(5): 789–805.

CFIC (China's Forest Information Center). 1998. *Development of China's wood industry.* Beijing: Chinese Academy of Forestry.

Chomitz, K. 2007. *At loggerheads? Agricultural expansion, poverty reduction and environment in the tropical forests.* Washington, DC: World Bank.

Dangi, R., and W. Hyde. 2001. When does community forestry improve forest management? *Nepal Journal of Forestry* 12(1): 1–19.

Daniels, S., W. Hyde, and D. Wear. 1991. Distributive effects of Forest Service attempts to maintain community stability. *Forest Science* 37(1): 245–260.

Dasgupta, S. U. Deichmann, C. Messner, and D. Wheeler. 2005. Where is the poverty-environment nexus? Evidence from Cambodia, Lao PDR, and Vietnam. *World Development* 33(4): 617–638.

Dewi, S., B. Belcher, and A. Puntodewo. 2005. Village economic opportunity, forest dependence, and rural livelihoods in East Kalimantan, Indonesia. *World Development* 33(9): 1419–1434.

Ekbom, A., and J. Bojo. 1999. *Poverty and environment: Evidence of links and integration in the country assistance strategy process.* Africa Region discussion paper no. 4. Washington, DC: World Bank.

Escobal, J., and U. Aldana. 2003. Are nontimber forest products the antidotes to rainforest degradation in Madre de Dios, Peru. *World Development* 31(11): 1873–1877.

FAO/UN (Food and Agriculture Organization of the United Nations). 2001. Global forest resources assessment 2000. Forestry paper 140. Rome: FAO.

Fisher, M., and G. Shively. 2005. Can income programs reduce tropical forest pressure? Income shocks and forest use in Malawi. *World Development* 33(7): 1115–1128.

Foster, A., M. Rosenzweig, and J. Behrman. 1997. *Population and deforestation: Management of village common land in India.* Draft manuscript, Department of Economics, University of Pennsylvania, Philadelphia.

Godoy, R., and K. Bawa. 1993. The economic value and sustainable harvest of plants and animals from the tropical forests: Assumptions, hypotheses and methods. *Economic Botany* 47: 215–219.

Godoy, R. D. Wilkie, H. Overman, G. Cubas, and J. Demmer. 2000. Valuation of consumption and sale of forest goods from a Central American rainforest. *Nature* 406: 62–63.

Gopalakrishnan, C., W. Wickramasinghe, H. Gunatilake, and P. Illukpitiya. 2005. Estimating the demand for non-timber forest products among rural communities: A case study from the Sinharaja rain forest region, Sri Lanka. *Agroforestry Systems* 65: 13–22.

Gunatilake, H. 1998. The role of rural development in protecting tropical rainforests: Evidence from Sri Lanka. *Journal of Environmental Management* 53: 273–292.

Gunatilake, H., and U. Chakravorty. 2001. *Forest harvesting by local communities: A comparative dynamic analysis with an empirical application.* Unpublished research paper, Agricultural Economics Department, University of Peradeniya, Sri Lanka.

Gutierrez, R., Ruiz Perez, X. Yang, M. Fu., Geriletu, and D. Wu. 2009. Changing contribution of forests to livelihoods: Evidence from Daxi village, Zhejiang Province, China. *International Forestry Review* 11(3): 319–330.

Hegan, L., G. Hauer, and M. Luckert. 2003. Is the tragedy of the commons likely?: Investigating factors preventing the dissipation of fuelwood rents in Zimbabwe. *Land Economics* 79(2): 181–197.

Hobley, M. 2007. Where in the world is there pro-poor forest policy and tenure reform? Washington, DC: Rights and Resources Initiative.

Hultkranz, L. 1991. Effects on employment and seasonal unemployment of subsidies to forestry in northern Sweden. *Scandanavian Journal of Forestry Research* 6: 243–291.

Hyde, W., J. Xu, and B. Belcher. 2003. Introduction. In W. Hyde, B. Belcher, and J. Xu, eds., *China's forests: Global lessons from market reforms*. Washington, DC: Resources for the Future, pp. 1–21.

Hyde, W. 2006. *Forest development and its impact on rural poverty: Global experience and projections for the countries of the Mekong River Basin*. Unpublished report submitted to Asian Development Bank REAP 6515.

Hyde, W., and G. Kohlin. 2000. Social forestry reconsidered. *Silva Fennica* 34(3): 285–315.

Illukpitaya, P., and J. Yanagida. 2008. Role of income diversification in protecting natural forests: Evidence from rural households in forest margins of Sri Lanka. *Agroforestry Systems* 74: 51–62.

ITTO Tropical Forest Update. 2007. Tapping the potential of communities. 17(4): 2.

Jodha, N. 2000. Common property resources and the dynamics of rural poverty: field evidence from the dry regions of India. In W. Hyde, G. Amacher and colleagues, *Economics of forestry and rural development: An empirical introduction from Asia*. Ann Arbor: University of Michigan Press, pp. 203–222.

Kusters, K., R. Achdiawan. B. Belcher, and M. Ruiz-Perez. 2006. Balancing development and conservation? An assessment of livelihood and environmental outcomes of nontimber forest product trade in Asia, Africa, and Latin America. *Ecology and Society* 11(2): article 20.

Lewis, D., G. Hunt, and A. Plantinga. 2001. *Public conservation land and employment growth in the northern forest region*. Unpublished working paper. Department of Resource Economics and Policy, University of Maine, Orono.

Lin, J. 1992. Rural reforms and agricultural growth in China. *American Economic Review* 81: 34–51.

Linde-Rahr, M. 2005. Extractive non-timber forestry and agriculture in rural Vietnam. *Economic and Development Economics* 10: 363–379.

Magrath, W. 2004. EAP forest strategy: Data and thoughts for discussion. Unpublished PowerPoint presentation. Washington, DC: World Bank.

MOF (Ministry of Forestry). 1995. *China forestry action plan*. Beijing: China Forestry Press.

Mekonnen, A. 1998. *Rural household fuel production and consumption in Ethiopia: A case study in rural energy and afforestation*. Doctoral dissertation, Economics Department, Gothenburg University, Sweden.

Monela, G., G. Kajembe, A. Kaoneka, and G. Kowero. 1998. *Household livelihood strategies in Miombo woodlands, emerging trends*. Unpublished research report, Forest Economics Dept., Sokoine University of Agriculture, Morogoro, Tanzania.

Nasi, R., D. Brown, D. Wilkie, E. Bennett, C. Tutin, G. van Tol, and T. Christophersen. 2007. *Conservation and the use of wildlife-based resources: The bushmeat crisis*. Technical series no. 33. Bogor, Indonesia: Center for International Forestry Research.

Neumann, R., and E. Hirsch. 2000. *Commercialisation of non-timber forest products: Review and analysis of research*. Bogor, Indonesia: Center for International Forestry Research.

Peters, C., A. Gentry, and R. Mendelsohn. 1989. Valuation of an Amazon rainforest. *Nature* 339: 655–656.

Prebble, C. 1999. Fruits of the forest. *Tropical Forest Update* 9(1): 1.

Prebisch, R. 1950. *The economic development of Latin America and its principal problems*. New York: United Nations.

Power, T. 1980. *The economic value of the quality of life*. Boulder, Co: Westview Press.

Reddy, S., and S. Chakravarty. 1999, Forest dependence and income distribution in a subsistence economy: Evidence from India. *World Development* 27 (7): 1141–1149.

Robinson, E., and G. Kajembe. 2009. *Changing access to forest resources in Tanzania*. EfD discussion paper 09-p10. Washington, DC: Resources for the Future.

Ruiz-Pérez, M., M. Fu, J. Xie, B. Belcher, and M. Zhong. 1996. Policy change in China: The effects on the bamboo sector in Anji County. *Journal of Forest Economics* 2(2): 149–176.

Ruiz-Pérez, M., M. Zhong, B. Belcher, C. Xie, and M. Fu. 1999. The role of bamboo plantations in rural development: The case of Anji County, Zhejiang, China. *World Development* 27(1): 101–114.

Ruiz-Pérez, M., and B. Belcher. 2001. Comparison of Bamboo production systems in six counties in China. In F. Maoyi, M. Ruiz Pérez, and Y. Xiausheng, eds., *Proceedings of the workshop on*

China social Economics: Marketing and policy of the Bamboo sector. Beijing, China: China Forestry Publishing House, pp. 18–54.

Ruiz-Pérez, M., M. Fu, X. Yang, and B. Belcher. 2001. Toward a more environmentally friendly bamboo forestry in China. *Journal of Forestry* 99 (7):14–20.

Scherr, S. 1995. Economic factors in farmer adoption of agroforestry: Patterns observed in western Kenya. *World Development* 23(5): 787–804.

Scherr, S., A. White, and D. Kaimowitz. 2004. *A new agenda for forest conservation and poverty reduction: Making markets work for low-income producers.* Washington, DC: Forest Trends.

Schreckenberg, K., E. Marshall, A. Newton, J. Rushton, and D. te Velde. 2005. *Commercialization of non-timber forest products: Factors influencing success.* Unpublished report for the Forestry Research Programme of the UK Department for International Development.

Sheil, D., and S. Wunder. 2002. The value of tropical forest to local communities: complications, caveats, and cautions. *Conservation Ecology* 6(2): 9.

Sikor, T., and T. Nguyen. 2007. Why may forest devolution not benefit the rural poor? Forest entitlements in Vietnam's central highlands. *World Development* 35(11): 2010–2025.

Singh, I., I. Squire, and J. Strauss. 1986. The basic model: Theory, empirical results, and policy conclusions. In I. Singh, L. Squire, and H. Strauss, eds., *Agricultural household models.* Baltimore, MD: Johns Hopkins University Press. pp. 39–69.

Spoelma, T., T. Morgan, T. Dillon, A. Chase, C. Keegan, III, and L. DeBlander. 2008. *Montana's forest products industry and timber harvest, 2004.* USDA Forest Service Resource Bulletin RMRS-RB-8.

Stevens, J. 1978. *The Oregon woods products labor force.* Unpublished manuscript, Agricultural Economics Department, Oregon State University, Corvallis.

Stoian, D. 2005. Making the best of two worlds: Rural and peri-urban livelihood options sustained by nontimber forest products from the Bolivian Amazon. *World Development* 33(9): 1473–1490.

Sunderlin, W. A. Angelsen, and S. Wunder. 2004. *Forests and poverty alleviation.* Unpublished discussion paper. Bogor, Indonesia: Center for International Forestry Research.

Tachibana, T., T. Nguyen, and K. Otsuka. 2001. Agricultural intensification versus extensification: A case study of deforestation in the northern-hill region of Vietnam. *Journal of Environmental Economics and Management* 41(1): 44–69.

van der Ploeg, F. 2011. Natural resources: Curse or blessing. *Journal of Economic Literature* 49(2): 366–420.

Vedeld, P., A. Angelsen, E. Sjaastad, and G. Kobugabe Berg. 2004. *Counting on the environment: forest incomes and the rural poor.* Environmental Economics Series Number 98. Washington, DC: World Bank.

Viana, V. M., R. Mello, L. deMorais, and N. Mendes. 1996. *Ecology and management of Brazil nut plantations in extractive reserves in Xapuri, Acre.* Unpublished research paper.

Whitelaw, E., E. Niemi, and C. Batten. 1990. *Transition strategies for timber-dependent communities.* Washington, DC: The Wilderness Society.

Wilkie, D., and R. Godoy. 1996. Trade, indigenous rain forest economies and biological diversity. Model predictions and directions for research. In M Ruiz-Perez and J. Arnold, eds., *Current issues in non-timber forest products research.* Bogor, Indonesia: Center for International Forest Research, pp. 83–102.

World Bank. 2001. *World development report 2000/2001: Attacking poverty.* Oxford, UK: Oxford University Press.

World Bank. 2004. *Sustaining forests—A development strategy.* Washington, DC: World Bank.

Yin, R., and D. Newman. 1997. Impacts of rural reforms: The case of the Chinese forest sector. *Environment and Development Economics* 2(3): 289–303.

Zhang, L., C. Saint-Pierre, and H. Liu. 2004. *Poverty and environmental dynamics: Challenges and opportunities for China.* Beijing: Chinese Academy of Sciences. Summary report of an international workshop organized by the National Development and Reform Commission and China's Center for Agricultural Policy, Institute of Geographical Sciences and Natural Resources Research, Chinese Academy of Sciences.

Appendix 11A:
Nonseparable Household Production and Consumption

When workers or households produce for their own consumption and also buy or sell in the market, then their activities become more complex than those described by the standard supply and demand functions which are related only by a common price term. That is, their production and consumption are no longer "separable." This is the case for the subsistence farmer and the subsistence collector of forest products whose own household consumes some of the family's production, but whose family may also buy or sell some of the same agricultural or forest products in the local market. The relationships necessary for empirical estimation in these situations are not widely recognized in forestry but they are necessary for understanding the behavior of subsistence collectors of fuelwood, forage, fodder, fruits and nuts, forest herbs, and other products of the natural forest, and for conducting inquiry into the scarcity of these products.

Consider a representative household that maximizes a continuous, monotonic, quasi-concave utility function subject to budget, time, and non-negativity constraints. Household utility U is a function of goods and services that use forest products as inputs (e.g., heating and cooking with fuelwood, cooking with fruits and herbs, etc.), the household endowment of labor L, household consumption of other goods X, and various local demographic characteristics Ω that may be important to household preferences.

$$max_{Q,L,X} \; U(Q,T\text{-}L,X;\Omega)$$
$$s.t. \; p_X X + p_F F_p = p_F F_s + I$$
$$Q \geq 0 \quad (11a.1)$$
$$F_c \geq 0, \; F_p \geq 0, \; F_s \geq 0$$
$$L \geq 0, \; T\text{-}L \geq 0$$

where F_c, F_p, and F_s are forest products collected, purchased, and sold, respectively. T is total time the household can allocate to all activities and L refers specifically to labor used for the collection of forest products. The p_i are prices and I is exogenous income from all sources other than forest products, both on- and off-farm.

Forest resources are intermediate inputs in household utility

$$Q = Q(F_h, \Theta) \quad (11a.2)$$

where F_h is the total consumption of forest products in the household and Θ is a vector of technologies that affect the household consumption of forest products (e.g., stove quality). Total household consumption of these products is

$$F_h = F_c + F_p - F_s \quad (11a.3)$$

The production function F_c for forest products is continuous and quasi-concave

$$F_c = F_c(L,A;\Omega) \quad (11a.4)$$

where L and A are vectors of the household's variable and fixed factors of production, respectively. Forest production, as it is essentially collection, is a labor intensive activity. (We could complicate production only slightly, without altering the generality of the model, by adding variable capital—such as baskets, knives, and spades.)

Substituting eqs. (11a.3) and (11a.4) into the budget constraint yields the production augmented budget constraint.

$$p_X X + p_F F_p = p_F F_s(L,A;\Omega) + I \qquad (11a.5)$$

The household maximizes utility subject to time, non-negativity, and the new budget constraint. The first order conditions for utility maximization are

$$\partial U/\partial X = \lambda\, p_X$$
$$\partial U/\partial L = \lambda\, p_X\, \partial F_s/\partial L + \mu$$
$$(\partial U/\partial Q)\,(\partial Q/\partial F_p) = \lambda\, p_F \qquad (11a.6)$$
$$(\partial U/\partial Q)\,(\partial Q/\partial F_s) = -\lambda\, p_F$$
$$\lambda\, p_F\, \partial F_c/\partial L = 0$$

and the budget constraint. λ is the Lagrangian multiplier associated with the marginal utility of income and μ is the Lagrangian associated with the constraint on household time. The second condition in eq. (11a.6) shows that the price of household labor is an endogenous value equal to the value of the household's marginal product of labor used in the collection of forest products. The third and fourth conditions show that those households that purchase and sell forest products do so at a market price. The second-fourth conditions also show that those households that collect for their own consumption face more complex decisions involving labor opportunities that depend on household preferences, technologies and marginal utilities. Goetz (1992) and Amacher, Hyde, and Kanel (1996) show that the opportunity costs for household collection of forest products can be significantly different from the market price for the same goods. This means that conclusions about the scarcity of forest products that rely only on market information may diverge from conclusions that also incorporate evidence on the household's collection for its own domestic consumption.

The first order conditions and the budget constraint provide most of the information necessary to derive the household's labor supply for forest production, its market demand and supply of these products, and the household's consumption of other goods.

The final condition in eq. (11a.6) implies that the household maximizes its net income conditional to its chosen supply of labor. The household's net income, in cash terms, can be defined as

$$N(p_F, A; L) = G(p_F, A; L) - p_A A \qquad (11a.7)$$

where $\qquad G(\,.\,) = max\,[p_F F_s(L,A;\Omega)]$

$G(\,.\,)$ has the properties of a variable profit function. The revised household budget constraint now becomes

$$p_x X + p_F F_p = N(p_F, A; L) + I \qquad (11a.8)$$

A set of output supply and input demand functions can be created by applying Hotelling's Lemma to eq. (11a.7)

$$HF_h = \partial N / \partial p_F = HF_h(p_F, A; L) \qquad (11a.9a)$$

$$HL = \partial N / \partial w = HL(p_F, A; L) \qquad (11a.9b)$$

where HF_h and HL are the conditional net income maximizing supply and demand choices, and w is the price of variable inputs (e.g., labor).

Consumption

The consumption equations are derived in the same manner. The revised budget constraint, eq. (11a.8), leads to necessary conditions identical to eq. (11a.6), except that the second necessary condition takes the form

$$\partial U / \partial L = \lambda \, \partial N / \partial L$$

where $\omega = \lambda \, \partial N / \partial L$ is the unobserved "virtual" or shadow price of labor used in collecting forest products. Therefore, when the household maximizes its utility, $\partial N / \partial L$ is the value of its marginal product.

It should now be clear that utility maximization creates a set of household choices of the form

$$X = X(p_X, \omega, p_F, I) \qquad (11a.10a)$$

$$L = L(p_X, \omega, p_F, I) \qquad (11a.10b)$$

$$F_p = F_p(p_X, \omega, p_F, I) \qquad (11a.10c)$$

$$F_s = F_s(p_X, \omega, p_F, I) \qquad (11a.10d)$$

where the terms in w contain λ.

The production and consumption sides of the model can be integrated by substituting the second condition of eq. (11a.10) into the conditional net income maximizing supply and demand choices, eqs. (11a.9).

$$HF_h = HF_h[p_F, A, L(p_X, \omega, p_F, I)] \qquad (11a.11a)$$

$$HL = HL[p_F, A, L(p_X, \omega, p_F, I)] \qquad (11a.11b)$$

The last two are the structural equations of the model. From these two equations it is clear that the household production and consumption decisions are interdependent—or "nonseparable". That is, changes in the exogenous factors affecting consumption induce changes in labor supply and, consequently, in the household's choices of inputs and outputs. Similarly, production shocks induce changes in the virtual price of labor, which affects the household's supply of labor and its choice of consumer goods.[23]

---

23 These last two sentences refer to the mathematical demonstration of the forest as a social safety net.

Empirical Estimation

There are a variety of possibilities with regard to functional form for the empirical specifications. As one approach we might assume that the production function is of Cobb-Douglas form. This means that the conditional profit function is log linear, as are the empirical specifications of eqs. (11a.11). Following Thornton (1991) and maximizing the Stone-Geary form of the utility function, subject to the non-linear budget constraint, eq. (11a.8), creates a non-linear estimation problem that can be solved by assuming budget linearity at the point of utility maximization. The predicted value of the household's marginal product of labor that emerges becomes a proxy for the unobserved virtual wage.

The empirical forms of the household forest product choices corresponding to eqs. (11a.10) become

$$p_F F_p = F_p(\omega, p_F, A, I, \Omega; \varepsilon_p) \qquad (11a.12a)$$
$$F_s = F_s(\omega, p_F, A, I, \Omega; \varepsilon_s) \qquad (11a.12b)$$

where the ε_i are mean zero errors. Making p_x the numeraire allows us to remove it from the system. The virtual wage term in each condition of eq. (11a.12) can be calculated as $\omega = \lambda\, F_c/L$, where λ is the estimated coefficient on the labor term in the production function and F_c is the predicted measure of household production.[24]

The market equations for forest products are (11a.12a), the market expenditure for forest products, and (11a.12b) the market supply for forest products. The first is linear (as discussed above) and the second is log-log since it was derived from a Cobb-Douglas production function.

Finally, some households collect for their own consumption but neither sell nor purchase in the market. This creates a censored data dependent variable that can be accommodated with a Tobit model for the purchase and sale equations, 11a.12a and 11a.12b.

Literature Cited

Amacher, G., W. Hyde, and K. Kanel. 1996. Household fuelwood demand and supply in Nepal: Choice between cash outlays and labor opportunity. *World Development* 24(11): 1725–1736.

Goetz. S. 1992. A selectivity model for household food marketing in sub-Saharan Africa. *American Journal of Agricultural Economics* 74(2): 444–452.

Thornton, J. 1994. Estimating the choice behavior of self-employed business proprietors: An application to dairy farming. *Southern Economic Journal* 87(4): 579–595.

---

24 We could specify labor supply similarly—but that is unnecessary where the focus of the analysis remains on forest product collection, sales, and purchase.

Appendix 11B:
The Generalized Economy of a Forest-Based Community[25]

The elements of any individual community's general economy can be characterized in its equations of change. These are the equations that explain changes in the demand for the community's final products relative to changes in their prices and, on the supply side, changes in the supply of the community's final products relative to changes in their input factor shares and changes in the use of these factors relative to changes in their prices.

More formally, households in a community with a large forest products sector consume two goods, wood products X and generalized all-other-goods Y. Prices of each, p_X and p_X, are set in the broader external market and the local forest-based community is a price taker with respect to both goods.

$$dp_Y/p_Y = 0 \text{ and } dp_X/p_X = -c_1 \tag{11b.1}$$

We are interested in instability in the wood products market. The market for all-other-goods is stable. The wood products market receives an external shock equal to some constant c_1. That is, prices in the external market either spike or (with a negative sign) plummet. It is the negative effect on the local economy of such an external shock that policymakers desire to avoid.

The community also manufactures the same two goods, wood products and a summary all-other-goods. It uses three inputs in its manufacture of wood products: labor L, capital K, and stumpage S[26]. The manufacture of all-other-goods uses only the first two. Both production activities are competitive and both operate with constant returns to scale. These common and reasonable assumptions allow us to write the conditions for changes in product supply relative to changes in input shares as

$$dX/X = \theta_{KX}(dK_X/K_X) + \theta_{LX}(dL_X/L_X) + \theta_{SX}(dS_X/S_X) \tag{11b.2}$$

and $$dY/Y = \theta_{KX}(K_Y/K_Y) + \theta_{LY}(dL_Y/L_Y) \tag{11b.3}$$

where θ_{ij} is the initial share of input i in the total cost of producing output j (e.g., $\theta_{KX} = r_Y K_Y/p_X X$).

The conditions of competition and constant returns mean that the changes in the factors of production relative to their prices are:

$$d(K_Y/L_Y)/(K_Y/L_Y) = \sigma_{KL\cdot Y}[d(r_Y/w)/(r_Y/w)] \tag{11b.4}$$

---

25 The formulation in this appendix follows Daniels et al. (1991). It is general for the economic short run where most factors of production are fixed within the forest-based community but labor can move between sectors. Later, the discussion of eq. (11b.16) reconsiders the possibility of new capital facilities.

26 A simple modification would distinguish between two different sources of stumpage, S_1 and S_2 for communities with two landowner classes (e.g., public and private) who behave very differently. In this case it would also be necessary to identify the elasticity of substitution between their two similar products.

where σ_{KL-Y} is the substitution elasticity between capital and labor in the production of Y, and r and w are the unit costs of capital and labor (the return on capital and the wage rate), respectively. Eq. (11b.4) can be rewritten as

$$dK_Y/K_Y - dL_Y/L_Y = -\sigma_{KL-Y}(dr_Y/r_Y - dw/w) \quad (11b.5)$$

Similarly,

$$dL_X/L_X - dK_X/K_X = (1-\theta_{SX})\sigma_{LK-X}(dr_X/r_X - dw/w) + \theta_{SX}\sigma_{LS-X}(ds/s - dw/w)$$
$$- \sigma_{SX}\sigma_{KS-X}(ds/s - dr_X/x) \quad (11b.6)$$

$$dL_X/L_X - dS_X/S_X = (1-\theta_{KX})\sigma_{LS-X}(ds/s - dw/w) + \theta_{KX}\sigma_{LK-X}(dr_X/r_X - dw/w)$$
$$- \theta_{SX}\sigma_{SK-X}(dr_X/r_X - ds/s) \quad (11b.7)$$

$$dK_X/K_X - dS_X/S_X = (1-\theta_{LX})\sigma_{KS-X}(ds/s - dr_X/r_X) + \theta_{LX}\sigma_{KL-X}(dw/w - dr_X/r_X)$$
$$- \theta_{LX}\sigma_{SL-X}(dw/w - ds/s) \quad (11b.8)$$

where s is the unit stumpage price and the σ_{ij-x} are partial substitution elasticities. Eq. (11b.8) is the difference between eqs. (11b.6) and (11b.7). Therefore, these equations are linearly related and only two of the three are necessary for further calculation.

Furthermore, under the same conditions of competition and constant returns, factor payments in each sector just exhaust the sector's total receipts. Therefore,

$$P_Y dY + Y dp_Y = w dL_Y + L_Y dw + r_Y dK_Y + K_Y dr_Y \quad (11b.9)$$

In addition, the factor marginal products equal their marginal costs divided by the output prices (e.g., $MP_{L-Y} = w/p_Y$)

$$dp_Y = w dL_Y + r_Y dK_Y \quad (11b.10)$$

Subtracting eq. (11b.10) from eq. (11b.9) and dividing by Y yields the expression

$$dp_Y = (L_Y/Y)dw + (K_Y/Y)dr_Y \quad (11b.11)$$

or
$$dp_Y/p_Y = \theta_{LY}(dw/w) + \theta_{KY}(dr_Y/r_Y) \quad (11b.12)$$

Similarly,

$$dp_X/p_X = \theta_{LX}(dw/w) + \theta_{KX}(dr_X/r_X) + \theta_{SX}(ds/s) \quad (11b.13)$$

Finally, the expression for stumpage supply is

$$(ds/s)\, e_s = dS_X/S_X \quad (11b.14)$$

where e_s is the stumpage supply elasticity.

To this point, there are eight equations (11b.2, 11b.3, 11b.5-7, and 11b.12-14) and eleven unknowns (d_x, d_y, dK_X, dL_X, dS_X, dK_Y, dL_Y, dr_X, dr_Y, dw, and ds). Underlying assumptions about the forest-based community specify three of the latter.

$$dr_Y = 0 \quad (11b.15)$$
$$dK_X = 0 \quad (11b.16)$$

$$dL_X = dL_Y \qquad (11b.17)$$

That is, the greater exogenous market determines the return to capital in the production of all-other-goods, r_y, and the community is a price taker with respect to it. Capital facilities such as sawmills in the wood products sector are immobile during the short run of the business cycle. Therefore, K_x is fixed, but the return on capital employed by the local wood industry can vary. Specifically, some of it can be unemployed at any time. Labor supply in the community is stable, but it can move between sectors as the wage adjusts. Labor neither emigrates from the community during the period of inquiry, nor does new labor immigrate into the community from outside.

The labor supply assumption, eq. (11b.17), is restrictive because, in fact, workers are mobile. Some will emigrate in search of better opportunity when local wages are low, and others will immigrate into the region from outside when local wages are more attractive. However, when the policy intent is to assist the local economy, then it is the local population and local workers who are important. Restricting their opportunity to emigrate reduces their opportunity, but biases upward the local impact of any externally determined adjustment in timber supply.

The system now involves eight equations and eight unknowns. It is a solvable system. By attaching estimates for any seven unknowns, the value of the eighth can be determined. For example, we can assess the effect of a government policy or a corporate decision to maintain a level of stumpage production—as Daniels *et al.* (1991) did for western Montana, and as we might for an even less diversified logging and sawmill town in, perhaps, Cameroon, Indonesia's East Kalimantan, or Brazil's Para. We could also assess the effect of closing an existing mill or of opening a new mill. The latter only requires setting eq. (11b.16) at a new positive level sufficient to include the addition of the new capital facility. The results for any assessment distinguish between effects on labor (changes in the numbers employed and their wages in both sectors), on changes in the returns obtained by the owners of stumpage and of capital, and on aggregate community welfare (the total of adjustments in all wages and all returns to capital in both productive activities).

Literature Cited

Daniels, S., W. Hyde, and D. Wear. 1991. Distributive effects of Forest Service attempts to maintain community stability. *Forest Science* 37(1): 245–260.

12
SUMMARY, CONCLUSIONS, POLICY IMPLICATIONS

What can we conclude about the characteristics of forestry that make it different from other natural resources and from other sectors of the economy, about the participation of forests and the forest sector over the course of economic development, and about the market activities and policy decisions that are effective in improving the condition of forests and the various forest-based resources and ecosystem services? What are the important roles for public agencies in the management of forests, and what tasks lie ahead for interested citizens?

By focusing on forest land and marginal shifts in land use, we can conclude, first and foremost, that the economic characteristics of forests are not uniform. As forests are not all biologically similar, they also are not similar in their economic characteristics. Forests differ in important measure with the stage of local economic development and their differences create unusual contrast between forestry and the economic features of other primary natural resources. Moreover, the characteristic forests for each the three different stages of their economic development respond in distinctive, and often contrasting, manner to the same market and policy variables. This creates unusual complexity for forest policy. One preferred policy does not generally fit all forests and all forest lands, and this crucial observation is lost in many discussions of forest policy. It severely complicates broad attempts at institution-wide policy for global development agencies and global non-governmental organizations (NGOs), and similarly complicates national forest policy for numerous larger countries with different regions in two or even all three stages of forest development.

Stages of Development and the Margins of Economic Activity

Chapter 2 traced the changing characteristics of forests through three distinct stages of regional economic development. For forests, development in any region and at any moment in historic time begins with the exploitation of previously unclaimed and essentially free material at the frontier of a biologically mature natural forest. At this moment in time, forestry is a land clearing activity, whether for the economic value of the forest's resources or for the value of the land itself once it is cleared of trees and ready for subsequent agricultural use.

As local development proceeds, the demand for the economic resources of the forest persists, and further exploitation pushes the frontier of economic activity beyond, sometimes well beyond, the extensive margin of land use for agriculture and other non-forest land uses. The developing region now exhibits a segment of open access land between the margin of extensive agricultural use and the frontier of the mature natural forest. This open access land has been previously exploited for its valuable forest resources. What degraded resources remain within it are of insufficient value for current exploitation and they are also of insufficient value to overcome the cost of establishing and maintaining claims for their secure and permanent use.

The total area in this segment of open access land may be large or small and either modestly or almost entirely depleted of its forest resources depending on the ease of establishing secure property rights and on the competing opportunities for local employment. The property rights to the degraded segment of land, and to the natural forest beyond it, are not secure because its forest resources tend to be low-valued and geographically dispersed. The total area of land within this region of open access is larger and the volume of standing forest per hectare within it is smaller where the local employment alternatives to extractive exploitation are scarcer and, therefore, small scale extraction of forest resources remains rewarding for many. Conversely, the land area in open access is smaller and its residual forest resource is larger where alternative employment opportunities away from the forest are greater and local households have less incentive for its further exploitation.

Of course, better employment opportunity and more effective formal institutions are generally associated with regions and countries of greater overall development. Therefore, the segments of open access forest tend to be smaller in regions and countries of greater overall development. Nevertheless, open access exploitation does exist even in developed countries—as, for example, millions of dollars of illegal timber harvests in the United States and Canada attest. However, the largest and most depleted segments of open access forest and wasteland occur in some of the poorest parts of the world. In India, for example, the land area officially designated as wasteland exceeds the area officially designated as forest. As the government of India can readily remind the rest of us, the struggle to overcome forest degradation is very much related to the struggle to overcome poverty.

Beyond this segment of open access, a mature forest still exists. In fact, this mature forest may be great in area, extending many kilometers and many hectares into the totally undeveloped interior of sparsely populated regions like Siberia, the Amazon, northern Canada, Alaska, and the Congo. Exploitation of the narrow economic frontier at the edge of this mature forest that is closest to any center of economic activity continues, often for many years. It continues until the costs of its ever more distant exploitation (collection, harvest and shipment costs) exceed the costs of growing managed forests on lands with secure property rights that are more easily accessible to those centers of regional human development. At this time, forestry enters its final, third, stage of development with commercial management on some lands; at least a modicum of open access exploitation on other less accessible, previously harvested, lands; and some continuing removal of mature forest resources at the frontier when and where the harvest of those resources can compete with the more expensive management but lesser shipment costs of the more uniform products from the more accessible managed forests. As in the first two stages of forest development, some amount, often a large and growing amount, of mature natural forest still remains beyond the frontier of commercial extraction.

All regions in the third stage progressed from the first and second stages at some earlier time in their history, and regions in each of the three stages exist in various places around the world today. In fact, while some regions of many countries are in the third stage with increasing areas in managed forest, some of the same countries contain other regions still in the first two stages of lesser forest development. Western Oregon and Washington in the United States are in the third stage with large tracts of managed forest plantation, while eastern Oregon and Washington on the other side of the Cascade Mountains and less accessible to the larger coastal population centers are in the less developed second stage where almost all timber harvests occur in the natural forest. For a second example, the third stage of forest development with active and even aggressive forest management and some forest plantations in the coast and Piedmont of the U.S. South contrasts with lesser development and fewer harvests from natural forests in the mountains farther inland in the same region. Similarly, for yet another example, the vital third stage forest sector in the south of Finland and the Nordic countries contrasts with less active harvesting from the northern forests of those countries and from the forests of adjacent Russian Korelia, forests that grow without significant management input and which are less accessible to the populated commercial and industrial centers in the south.

All three stages occurring in one broad geographic region or another at the same moment in time, as they do in various countries in the modern world, implies multiple margins of economic activity. Regions in the initial, first, stage of forest development have a single margin of activity at the forest frontier, and agricultural expansion is the primary source of forest depletion during this stage. Removal of the native forest is the only economic activity at this margin.

This same forest frontier remains the primary economic margin for regions in the second stage of development but commercially valuable forest resources rather than agricultural land conversion become the primary justification for additional forest removal. Some extraction may also continue from the previously harvested open access lands of the second stage as harvest technologies improve or as the depleted residual resources in these lands recover to a minimum economic size.

Regions in the third stage share the two margins of the second stage but their managed forests add their own intensive and extensive margins. The intensive margin of managed forestry can include forest plantations, and even advanced thinning, fertilizing, and pruning activities for a few high-valued commercial operations. Natural stands regenerated as part of a planned silvicultural process (but not from any variety of nursery-grown seedlings) characterize the extensive margin of managed forestry.

These four active economic margins are not unique to forest resources, but the four taken together do make forest resources different from most other economic activities. The first marginal activity at the frontier is comparable to much of the economic activity in minerals or ocean fisheries. The second, recovery of residual material from degraded open access land, is somewhat comparable to those hardrock mineral operations that have returned to the waste piles of prior extraction to reclaim previously uneconomic material. The third and fourth, intensive and extensive, margins of active management are comparable to the Ricardian margins of agricultural activity. The unusual characteristic of forest resources is that, unlike most other economic activity, they experience all four margins in some regions of some countries, and certainly some regions of six continents.

The four margins are general for commercial timber and for most other forest resources as well. That is, timber, fruits and nuts, and even forest recreation and some managed wildlife, all display the four economic margins. The specific physical location and even the geographic breadth of these margins vary, however, from one forest resource to another. For example, different timber species vary in their local economic valuation and, therefore, in the extent of their managed stands and also in the specific location of their frontier harvest activity. The economic frontier for one species may be at the edge of the natural forest as even an untrained observer recognizes it. The frontier for a less common and higher valued species may be deeper into the interior of the natural forest as more accessible clumps of that species were removed at some prior time and what remains of the species are more isolated stands deeper within the natural forest interior. The economic frontier for native fruits very likely occurs at yet a different interior location. Of course, forest recreation occurs wherever users find the appealing natural characteristics that attract visiting sightseers, picnickers, hikers, campers or whatever. However, even for forest recreation, more accessible sites of any particular set of characteristics attract greater use.

Taken altogether, this means that the same market factors and policy decisions can affect forest use at four different margins. Not all parts of the world contain all four margins for any one forest resource at the same moment in time. But where all four are active, they are unlikely to be of the same physical extent or regional economic importance. This means that the anticipated effects of market factors and policy decisions on the common measures of forests in one part of the world can be nonexistent, insignificant, or even contrary, to those on the forests of another part of the world. Failure to recognize the four margins, the different regions in which each exist, and the related differences in market and policy effects on them explains the failed expectations of many well-intended forest policies.

Market and Policy Effects on the Forest

The three-stage discussion of forest development shows the importance of separating the global forest into two, natural and managed, components. It also shows how poverty and relatively undeveloped institutions, specifically those associated with establishing and maintaining property rights, are essential ingredients of the inland and upland extension of the natural forest frontier, and of both the expanding limits of the area of open access and the magnitude of its degradation. It should be clear that the opposite, economic development, coincides with the emergence of managed forestry and, eventually, forest plantations and their substitution for the commercial products of the natural forest. Furthermore, economic development also coincides with improvements in the effectiveness of the institutions that assure property rights as well as with a human population that places a relatively greater value on environmental services and, therefore, has an increasing incentive to protect the natural forest. In sum, general economic development and policies supporting economic development have a crucial favorable affect limiting further degradation of the natural forest environment and, eventually, supporting its recovery.

Nevertheless, a caution is in order. The three-stage sequence of forest development is not a statement that rising relative prices for forest products will solve the problems of forest degradation and deforestation. Rising relative prices are an incentive to increase timber harvests during the first two stages. They are an incentive to increase degradation in the open access region and deforestation at the frontier. It is only in the third stage of forest development, when the cost of managing more accessible forestland becomes competitive with the cost of extraction and shipment from the less accessible natural forest frontier, that rising forest product prices and rising environmental values induce substitution away from some harvesting on the natural forest. It is only in this third stage that the natural forest may begin to recover.

This is not just theory. Recent empirical assessments of China's island province of Hainan over the course of the twentieth century and, then, the broader experience of China's thirty-one diverse provinces (including municipalities and autonomous regions) since market reforms and rapid economic development beginning in 1978 confirm the three-stage pattern of forest development, including first the decline of the natural forest, and then, later, the increasing importance of managed forests and, finally, the decline in forest degradation and improvement in the natural forest environment that eventually coincides with economic development. More casual observations from Western Europe, North America, and India's Punjab provide further evidence of the importance of economic development in achieving these desirable forest outcomes. The policy conclusion to be taken from these observations is that economic development eventually creates beneficial effects on the natural forest environment (see chapter 6).

Of course, policies that are more focused and selective than those supporting general economic development also have their effects on the forest. These include regional, agricultural, industrial, and macroeconomic policies not designed for their effects on forests but which have spillover effects on the forest, as well as those other policies directly intended to alter forest use and the forest itself.

Regional Development and Local Forest Administration

Two varieties of regional policy have important effects: roads and the design of local community institutions. New roads, including railroads, can open vast areas for economic development and forest exploitation—as they did for inland western United States and Canada in the late nineteenth century and for Brazil's Amazon and Thailand's northeast more recently. Of course, careful road design can also direct use away from selective areas that merit protection and roads do improve opportunity away from the forest for otherwise isolated rural populations. Therefore, roads (indeed, full transportation networks) can have both positive and negative effects on the forest, and their appropriate design is crucial to anticipating their net effects on the forest environment.

Increased local (rather than national) control of the natural forest is a theme of current policy interest in many countries. It is the perspective of many global development agencies and NGOs that local community control of the forest is more effective than national forest administration. In fact, local control is effective where the greater forest values at risk are local and where the community holds a unified perspective of the importance of those values. On the other hand, local community management is not as successful where the values at risk are national or global, as for unique recreational opportunity or biodiversity (e.g., Yellowstone or the Serengeti), or where different local values compete. An example of the latter occurs where some members of a local community value the forest as a control on water runoff and downstream erosion, while others in the community view its trees as a source of fuel. The former wish to preserve the forest; the latter wish to harvest it.

Greater local than national values suggest that local management will be better informed, less expensive, and more effective than national administration of the resource. On the other hand, greater national or global values suggest that local management will provide less than an optimal level of protection for the resource. Competing local values, together with the ease of access that characterizes natural forests, suggest either difficult community negotiation or open access depletion. Therefore, as with most policy of any design for any objective whatsoever, the effectiveness of transfers from national administration to local forest management depends on the characteristics of the communities selected for the transfers and of the local resources in question. Local community management does improve national forest administration in some cases, but even successful local management where the values at risk are entirely local depends on the capacity of the local institutions. Furthermore, examples from an array of private and community administration of forest resources seem to argue that, where it is successful, local management depends on local choice rather than the imposition of preferred institutions from outside. In sum, local community management has its effective examples, and they are plentiful, but local management certainly is not a panacea for the multitude of challenges facing natural forests today (see chapter 4).

Of course, national administration of the forest is not a universal solution either. Indeed national administration often introduces its own institutional timber management problem, one that is unique to forestry. The "allowable cut" policy followed by most public forestry agencies is a specialized biological timber harvest calculation that intends to maintain timber harvests at a level commensurate with the forest's natural rate of biological growth. In fact, allowable cut harvests are calculated regardless of the economic merit of timber activity at any particular location. The result is often a limitation of harvests in prime economic locations but an extension of the harvest activity into uneconomic and unsustainable regions. Economic rents build up in the regions administered by national forestry agencies while, at the same time, publicly funded timber roads and "deficit" timber sales permit harvesting in other regions that private commercial timber operations would never harvest on their own. The debate surrounding the allowable cut calculation is well known among public agency foresters who, because of their biological training, may be less informed on economic principles. The debate and the calculation itself seems to be entirely unfamiliar, however, to other policy analysts, environmentalists, and the general public, all of whom might find millions of dollars of potential savings for their public treasuries and millions of hectares of environmentally favorable adjustments from revision of this misguided public agency practice, even while finding increased commercial timber volumes in some other locations. It would be good if some of them took the time to understand the calculation (see appendix 4A).

Direct Forest Policies

When we discuss forest policies we usually think of taxes, subsidies, and regulations that directly target the forest or, particularly, the owners of private forest lands. We often begin by thinking of the effect of these direct forest policies on production from a given forest base while overlooking the affected shifts in marginal land use. In fact, both effects are important. However, previous paragraphs, and some that follow, argue that policies with other, non-forest, targets but which spillover to also affect forest land use may have even more pervasive effects.

Nevertheless, direct forest policies do have their own effects. Chapter 4 reviewed a wide range of forest policies, but it is clear that policies designed to ensure reforestation following timber harvests and, in some cases, to expand the area in forestry, are most common. These include subsidies (known as forest incentive payments) and reforestation incentives such as free or discounted seedlings. Requirements on forest management such as those that restrict harvesting adjacent to watercourses or in certain biologically diverse areas, or those that limit the application of certain herbicides are also common in many developed countries.

The former, subsidies and reforestation incentives, have a small positive effect on the margins of managed forests but much of the policy expense is wasted because these managed forests are already commercially viable. The additional incentive is unnecessary for managers who already obtain a good financial return on their forest investment. On the other hand, these policies have nothing beyond an occasional cursory and temporary effect on the natural forest frontier where reforestation is not financially viable. Therefore, they have no effect on forests in regions in the first two stages of development.

Where public policy provides free or discounted seedlings, for example, and the private forests are commercially viable, we observe that landowners gratefully accept the public largesse—and they continue their forest operations much as they would have without the public assistance. Where private forests are not commercially viable, landowners may accept the public's seedlings, but we have numerous observations where the survival rates of these seedlings are low or nonexistent—because the longer-term economic incentive is absent. It is only for a few private lands at the margin, lands that are at the cusp of becoming viable for commercial forestry, that free or discounted seedlings can make a difference.

The latter category of forest policies, restrictions on certain management and harvest activities, have an environmentally-preserving effect on managed forests. They can also affect harvest practices on the natural forests during all three stages of forest development—although the greatest land area in natural forest is administered by public forestry agencies and these agencies generally operate under their own rules and regulations.

Market and Policy Spillovers from Adjacent Sectors

Some of the strongest effects on forests spillover from policies whose primary targets are the agriculture and wood processing sectors, or from market adjustments in either of those two sectors. Many of these adjacent sector effects on the forest are not well understood and some go largely unrecognized.

Agriculture policies and market adjustments affect frontier forests during the first stage of forest development. Cheap land policies during nineteenth- century western expansion in the United States and Canada are an example, as was Brazil's policy inducement for agricultural expansion into the Amazon in the 1990s.

Agricultural policies and markets have little effect, however, during the second stage of forest development. Their effect during the third stage is complex, but it tends to concentrate at the intensive margin of managed forestry where greater agricultural profits and agricultural support policies provide a competitive advantage for agriculture over land use for managed forestry. This advantage constrains the natural progression into third stage managed forests and, as it does, extends the period of dependence on the products of the natural forest.

Globally, the greatest agricultural policy effect on forests may originate with the immense system of agricultural support policies in North America, Western Europe, and Japan. These policies provide an agricultural advantage in competition with managed forests in the countries of their origin. In addition, as they encourage production in the developed countries, they depress global agricultural prices and diminish the incentives for commercial agricultural activity in the developing countries. It is likely that many of those who would obtain employment in the commercial agriculture sector of developing countries, instead, in the absence of a larger domestic commercial agricultural sector, must rely on an existence that is crucially dependent on subsistence agricultural opportunity within the natural forest. Therefore, it is quite possible that developed country agricultural support policy is a significant source of tropical and developing country forest degradation and deforestation. Of course, this is only an hypothesis. It is an hypothesis that does receive limited attention—but it has never been fully examined and surely it deserves greater attention (see chapter 4).

Some incentives for the wood processing sector are different. They potentially have an opposite, forest conserving, effect. The wood processing industries are the primary source of deforestation during the second stage of forest development, but this is likely to change as regions move into the third stage.

Some industrial establishments require specialized wood inputs which are more conveniently recovered from the more uniform stands of managed forests. For others, notably in the capital-intensive pulp and paper industry, the mobility of their fixed investment in plant and equipment (the mill) is limited and the wood resource is only a small share of total operating costs. It is more effective for these establishments to rely on the development of nearby managed forests than on an increasingly distant natural forest. For virtually all wood processing industries, newer processing technologies permit greater recovery of useful material from any log and any hectare of forest. Altogether, this may mean that technological advance in the wood processing sector moderates any increasing demand it may place on the natural forest frontier. Therefore, technological change in this sector, and policies encouraging it, probably have a forest-conserving effect. This too is a largely untested hypothesis—but certainly another that is worthy of greater inquiry (see chapters 3, 4, and 7).

The Macroeconomic Environment and Economic Growth

The overall macroeconomic environment and economic growth are crucial determinants of both the short-run and long-run well-being of forests and forestry—just as they are determinants of the performance of the entire national economy. Forest policy analysts, however, tend to focus more narrowly on forestry markets and policies and they give these broader characteristics of the economy too little attention. Similarly, macroeconomic policy analysts have broader regional, national, and global responsibilities and, for them, the specific effects of macroeconomic policies on the forest are of less interest. Nevertheless, the broader measures of economic performance and more specific monetary and fiscal policies as well, while they may not target the forest directly, can have strong effects on the forest, effects that are often greater than those of forestry markets and those more selective policies that target the forest directly. Our understanding of all that happens to forests would be greater if these broader macroeconomic market and policy impacts on the forest received greater attention.

Economic and political stability: General economic and political stability is a primary example. Stability is a central issue in any discussion of national economic growth and development. Macroeconomists understand its importance, but it is generally beyond the realm of forest policy analysts, most of whom have been trained in the developed countries (or with the educational material of the developed countries) where economic and political stability are the norm. However, much of the global interest in forestry today focuses on developing countries, particularly tropical countries, where economic and political stability is not such a common experience and where, not coincidentally, the rates of forest degradation and deforestation are greatest.

Where instability exists, it increases preferences for unsustainable short-term activities. Loggers and millowners with access to the forest harvest more rapidly and farther into the natural forest, taking all that is of value while the opportunity still exists—because the local instability suggests that the opportunity to harvest that exists today may not exist tomorrow. Furthermore, these loggers and millowners do not invest in more efficient harvesting and processing equipment, and they certainly do not invest in uncertain long-term activities like forest management when the likelihood of future return on their investment is most uncertain. They harvest rapidly and deplete whatever local forest is financially accessible to them. Consider any country during its period of political and military volatility, or during a period of widely fluctuating national finance. Declining rates of investment and unusual rates of deforestation accompany these periods.

Or consider the converse, stability and economic growth. In fact, a measure of stability is a pre-condition for economic growth. Along with economic growth, the local institutions, including those that assure property rights, also develop—and improve in their effectiveness. The region of open access forest declines, investment in more efficient logging and milling technologies increases, and the opportunity for eventual third stage managed forestry improves. The overall forest environment improves. Eventually, as aggregate economic well-being improves, the demands for non-market forest resource services grow relative to those for wood products and the forest environment improves further yet. Consider the contrasts in the economic growth and forest environment between countries like Thailand and Myanmar, or between North and South Korea, or consider the impact on forest industrial investment and forest growth of recent stability and economic growth in a rapidly developing country like Vietnam (see chapter 6).

Finally, with stability and growth, and with time, new technologies become effective, and the topic of technological change deserves yet further inquiry. New technologies affect forestry and the sectors that compete with it at the four land use margins in a complex variety of ways. Just those new technologies strictly related to forestry; new harvesting and wood processing technologies and, in the third stage, new biological forest management technologies; effectively substitute for a portion of society's demand for the forest's raw materials. The substitution is a long slow process, but it can be very important. In stable and growing economies, the rate of adoption of new harvesting and wood processing technologies over the course of the twentieth century was greater than the biological growth rate of most forests. This is a crucial point. Since economic growth and the adoption of new technologies go hand-in-hand, one likely net effect of the more rapid rate of improvement in harvesting and wood processing technologies is a further round of improvement in the forest environment (see chapter 3).

Macroeconomic policy: Within a stable political and economic environment, central governments can begin implementing selective fiscal and monetary policies with reasonable expectations for their effectiveness. Some governments, usually with encouragement from multilateral financial institutions, have also introduced more general economy-wide structural reforms. Consider each of these in turn; first fiscal and monetary policies, and then structural reform.

Fiscal policies relate to public expenditures. They are a tool for improving the aggregate economy during times when general private consumption and private investment lag. They can have an effect on the forest sector when the increased public expenditure triggers expansion in sectors of the economy that consume forest products. The construction industry is often a primary target of fiscal policy and its use of wood products can, after certain time lags, lead to a short-term increase in timber harvests.

Monetary policy refers to the central bank's means for easing credit in times of lagging private investment. Its effect on the forest, like that of fiscal policy, is also indirect and limited. It depends on the demand, generated by the renewed private investment, for the products of the forest industry. Where domestic housing shares in the renewed investment, as it very often does, then this particular market's demand for lumber and other wood products can have an eventual effect on the forest.

Structural adjustments are wide-ranging programs of reforms in overall domestic policy, often imposed by multi-lateral financial institutions as a precondition for their assistance in improving a weakened national economy. Their introduction is an attempt to strengthen a country's economic fundamentals and, thereby, improve the conditions for growth, but the effects of their component reforms on the forest can be mixed.

Brazil's example from the last quarter of the twentieth century is instructive. In Brazil's case, reductions in subsidized agricultural credit and in expenditures for road construction reduced further deforestation and improved the conditions of the forest at the natural frontier. However, export incentives designed to improve Brazil's balance of payments led to appreciating prices for more capital-intensive commercial agricultural products like soybeans. This increased the land value of those small farms that could be converted to larger commercial agricultural operations and decreased the real agricultural wage. Land substituted for labor. Both agricultural workers and small farmers lost. They migrated to the frontier where they converted Amazonian forest to smallholder agriculture. In sum, the package of reforms in Brazil's structural adjustment program included specific components with opposing effects on the natural forest.

Brazil's program is only illustrative and not indicative of a general pattern for structural adjustment programs in general. It does show that, since structural adjustments are designed for their effects on the aggregate economy, their impacts on the forest depend on the selective components that make up the overall package of reforms. Therefore, the specific forest effects must vary with each particular country and its specific set of structural reforms (see chapter 6).

What is clear is that those interested in either the forest or in macroeconomic policy should not overlook the links between the two. The relationship between them adds an important factor for any consideration of appropriate policy.

Market and Policy Effects on Local Forest-dependent Communities

Altogether, this suggests that, within a stable political and economic environment, the various fiscal and monetary policies may have only a small impact on a nation's total forests and on all of its local forest dependent communities. However, when we consider the characteristics of those communities, it becomes clear that, for many of them, their typical dependence on only one or two sectors, forestry and often agriculture or mining, and their remoteness from larger communities that support a greater diversity of economic activity, can cause them to suffer significant losses in employment and income in response to even small adjustments in the broader regional and national markets for their products.

In fact, boom and bust cycles are not an uncommon experience for natural resource based communities in general and forest policy analysts are aware of this problem as it affects forest-based communities in particular. They have recommended regional development programs, often subsidized by the central government, and controls on the physical flows of raw logs to the mills as a means of assuring steady employment for these communities. The evidence, however, is that the first generally succeeds only as long as the central government is willing to continue the subsidy. Meanwhile, many development programs fail because they cannot address the fundamental problems, the community's remote location and its lack of economic diversity, the characteristics that cause it to be so strongly affected, even by modest adjustments in the broader external economy's demands for the products of the specific community's one or two primary sectors.

The second recommendation, controls that attempt to assure constant and regular physical flows of raw materials like logs, while helping maintain mill production and employment, do little for community-wide employment because the labor force in these communities moves easily between its two or three primary sectors. Meanwhile, maintaining a flow of logs that is not responsive to market prices only exaggerates the losses in stumpage revenues that local landowners must absorb.

In fact, the actual empirical evidence argues that macroeconomic forces, measured as real gross national product or housing starts, seem to have the greater favorable impact on employment and, by extension, on overall community well-being in these more remote forest-based communities (see chapter 11).

Policy Design to Protect and Enhance Forests and Forest Resource Services

To this point, the discussion has focused on the array of policy (tax, subsidy, and regulation) and market effects that impact forest resources and forest-based communities. These are of unquestioned importance for an understanding of the condition of the world's forests. All of them occur many times over in both developed and developing countries around the world. However, a large part of the modern interest in forests turns the question around. Rather than asking the effect of each class of policies, it inquires, instead, of the most effective policy design for creating some desired effect, and the most discussed effects today probably have to do with one or another of five high profile topics: carbon sequestration and global climate change, biodiversity and critical habitat, environmental tourism and forest-based recreation, erosion control and watershed protection, and sustainable forestry and the control of deforestation.

Carbon Sequestration and Global Climate Change

There is no scientific doubt that our global climate is warming. Some doubt the role of mankind, but there is no doubt that carbon emission is one source of the warming. There is also no doubt that forests play a role in the changing carbon balance, adding to atmospheric carbon when wood fiber burns but withdrawing and storing atmospheric carbon as young trees and forests grow. Since mature trees and forests add little growth and, therefore, do not significantly alter the carbon balance, this means that using forests as a control on climate change requires a focus on the forest-land use margins—where either harvests of mature trees and some burning emit carbon or expanding forests and their young growth sequester carbon.

Although all four forest margins can shift with time and with development, the usual policy focus is on the margin identified with the natural forest frontier. This margin is most active during the first two stages of forest development and, in today's world, the developing regions experiencing change at this margin are generally in tropical countries. The global policy intent is to decrease forest removal and land conversion in these regions and countries, and even to replace it with new forest growth reclaiming some land and storing some carbon. (This is a short-term solution that can sequester carbon only so long as land is available for conversion back to forest. A longer term solution must be dependent on controlling, other, greater sources of carbon emission, sources such as the consumption of fossil fuels. And this, of course, is beyond the role of forestry or forest policy.)

A policy of restoring forests at the natural frontier might be implemented by having the more concerned developed countries pay the tropical developing countries to protect and expand their natural forests. In fact, this is a common suggestion and some developed country institutions have arranged financial transfers of this sort—although there seems to be some confusion between protecting the forest margin and protecting the full natural forest. Of course, the latter is not threatened and payments to protect it are unnecessary. The problem lies in enforcement at the natural forest frontier. Even this frontier is remote, the protected resources are individually of low value to the local population, and no country has the financial or human resources to protect them from all encroachment.

Therefore, carbon and climate policy focusing on the forest may not be especially effective. Taxing the greater sources of carbon emissions, very largely the developed country consumption of fossil fuels, would probably be more effective.

Biodiversity and Critical Habitat

The protection of biodiversity is based on the idea that currently undeveloped, and even unknown, species possess characteristics of potential future value. Preserving their habitats helps preserve the species and the option to obtain future advantage as we learn more about them.

Most agree with some level of protection, particularly for the remaining undeveloped habitat of both known and yet-to-be identified species, and there are various examples of successful habitat protection in the developed countries. The currently threatened habitat is largely in the undeveloped natural forest, and a large share of the international policy discussion focuses on that in the more diverse tropical moist forest. The tropical moist forest contains more than one-half of all species of flora and fauna, including many that are still unknown. This particular forest is all the more important because it tends to occur in regions within the first two stages of forest development. Therefore, it and its endangered habitats are more threatened than those of either the tropical dry forest or the temperate and boreal forests.

The first task is to identify threatened habitats at the margins of current land use and to restrict those uses that are inconsistent with the protection of these habitats. The second task is to conduct the resource inventories and biological research necessary to identify currently unknown habitat beyond the frontier and to design future development away from these areas.

The problem for the design of effective policy, once more, is a mismatch between those most willing to pay for the identification and protection of these habitats and those most directly and immediately affected. Those most willing to pay are the better-off populations of the developed countries. Those most directly affected are the poor rural populations living in or near the threatened frontier who often find personal short-term advantage from extractive activities in these regions. Once more, the cost of protecting against local encroachment, extraction, and damage to the endangered habitat, is generally larger than most developing country governments and their resource management agencies can afford.

On the other hand, there is a small contrasting example. There are a very few large international drug companies that compensate developing countries to protect their tropical forests in exchange for a share of any eventual gain from developing new uses from the protected species habitat and species. The success of these arrangements in protecting endangered marginal forest remains an empirical question.

Environmental Tourism and Forest-based Recreation

A broad range of the human population participates in environmental tourism and forest recreation at a vast array of physical sites—from wealthier globe-trotting tourists to local picnickers, hikers, birdwatchers, fishermen, etc. at sites ranging from Yellowstone and the Serengeti to local village parks. Once more, the economic problem is either one of protecting islands of specialized forested sites within lands that are valuable for other, extractive uses, or identifying forestlands that are inaccessible for extractive land uses (beyond the commercial frontier of the natural forest) but uniquely attractive for nature tourism, and then protecting these sites before they become accessible for whatever extractive uses that may be incompatible.

For the most unusual sites, places such as Yellowstone and the Serengeti, fees can be charged at points of limited access and the collected revenues can be used to establish boundaries and systems for monitoring and enforcing the exclusion of undesirable uses. Two problems remain, however, even in the case of these unique resources. First, the fee is general. It makes no distinction for particular resources or for specialized services within the site covered by the fee. Sometimes this problem can be addressed by placing the general management of all resources under one integrated operation, a national park service, for example, and competitively allocating concessions for specialized services like hotels and guides within each park boundary. Each concession then discriminates by charging for its own specialized service.

The second problem exists even when this first problem is solved. While tourists who arrive from long distances can be charged or excluded at the national border or the park boundary, the exclusion of local users is more difficult. Therefore, monitoring and enforcing restrictions against competing local uses of the unique resource is more difficult. Poaching within East Africa's game parks and illegal timber harvesting within the boundaries of Southeast Asia's natural reserves are examples. A partial solution to this problem can often be obtained by establishing an interest in the park's tourist services within the local population, or even some sharing of collected revenues for the park with local interests. However, the determination of effective revenue sharing mechanisms is not an easy task and, even with them, some amount of local trespass is still likely.

For those forest resources that are not unique and do not attract global tourists the most common non-consumptive users are members of the local community. Local institutions are generally better-suited to manage these resources and we observe many successfully protected village parks and forest sanctuaries around the world.

Erosion Control and Watershed Protection

Erosion control and watershed protection incorporate all the services of trees and watersheds in managing wind, water, and soil movement. Like carbon sequestration, watershed protection can be divided into two broad classes of activities—those that require new conservation interventions such as tree planting to deter wind and water erosion, and those that maintain the services of existing forested watersheds and limit their deterioration. The difference in economic effect is that watershed values tend to be local or regional, and those who benefit are individually or as a group easier to identify, while carbon sequestration is a global public value. Depending on the particular watershed management activity, it can be of greater benefit to an individual landowner who has the ability to make the conservation investment and improve the productivity of his or her own land, or it can yield greater benefit to a range of downstream or other off-site land users within the same watershed.

Many of the first class of watershed management activities, those requiring new conservation investments, are responses to human development. They are a means of improving the productivity of existing (often agricultural) land uses. These investments typically occur on private lands. Therefore, the resulting increases in the manager's own private long-run productivity are often sufficient to induce the necessary conservation investments.

The second class of watershed management activities protects the upland watersheds or the coastal wetlands for the benefit of off-site residents of the same general region. For example, logging, grazing livestock, and the upland collection of fuelwood are common sources of upland erosion and downstream sedimentation. They tend to occur within either the open access degraded forest or the neighborhood of the natural forest frontier. Therefore, since costs are greater than the rewards of management at these locations, unfettered private management will be unsuccessful and regulation or public ownership along with a degree of monitoring and enforcement (M&E) is the usual means for insuring the common watershed benefits for the local community. Where most or all members of the local community share in the common watershed benefits, local residents also share a common incentive to protect the watershed, and monitoring and enforcement may be a relatively simple matter. In other cases, where the local incentives are dissimilar and members of the community compete for different uses of the land, M&E consumes more resources and protecting the public benefits is a more difficult task.

Finally, some cases require broader regional or national oversight and neither private nor local collective action will be sufficient in these cases. These include an increasing number of examples of communities that pay for the protection of upstream watershed services. Some of these (e.g., New York City's payments for its Catskill Mountains watershed or the bonds sold to insure the condition of the Panama Canal for those large corporations that must ship their products through the Canal) are spectacular.

Sustainable Forestry and the Control of Deforestation

The concern for sustainable forestry grew out of a much older concern that the world was running out of natural resources. Of course, it has not run out of resources in general nor of timber in particular. In fact, the timber stocks in many parts of the world are greater today than they were 100, or even 50, years ago. However, residual doubts exist in the minds of many regarding the potential for timber shortages and others, while not so concerned about the depletion of market-based forest products, are concerned that we may be depleting our stock of global means to provide the non-market ecosystem services of forests. These arguments are the basis for modern discussions of forest sustainability and attempts to control deforestation and forest degradation.

Perhaps the most useful discussion for this purpose would focus on the idea of "sustainable options". With this focus, the objective would be to "maintain, in perpetuity, options for all the various uses of forest resources, market and non-market, consumptive and non-consumptive, known and unknown". This would mean controlling environmental destruction. It would mean maintaining for the future the potential for all the different uses of the land and other forest resources. It would also mean using the forest to help maintain other future options by using it to help control erosion, to protect critical habitat and important aesthetic resources, and to limit global climate change. Such a statement of sustainability would allow some shifts of forestland to agriculture, others from agriculture back to forest, and still others from natural forest to managed forest so long as both the land's productive base and the genetic material of the forest biota remain undamaged. Relative values will change with time and preferred patterns of land use will change with them. We can do little about this, but we can insure that changes in relative values and changes in land use do not destroy opportunities for new and different use of the land and the forest resource in the future.

In the context of the three-stage model of forest development, this perspective of sustainability is consistent with minimizing the area of degraded open access forest while locally regulating specific eroding watersheds, critical habitats, and important aesthetic resources, both within and outside the degraded area. The fundamental means for minimizing the degraded area involve reducing the cost of establishing and maintaining property rights and attracting extractive human activity away from the forest. The first requires finding the least cost bundle of property rights and the institutions that can provide it. The second requires careful design of roads and other infrastructure in order to (a) reduce the human impact on critical areas and (b) provide alternative external labor opportunity for local households who otherwise would be more dependent for their own livelihood on the natural forest and its various extractive resources.

Stating the argument a different way: Poverty is a crucial source of forest degradation and forest depletion. Economic development induces improvements in the forest environment as it shifts land into sustainable (third stage) activities. In fact, economic development is likely to have a second round of beneficial effects as well. Improved wages and better labor opportunities create the first round. Then, along with improvements in overall welfare, the local institutions also tend to become more effective over the course of general economic development. Their budgets probably increase, and they improve in their ability to insure property rights and in their ability to manage economic transitions and provide for economic stability. Both improved institutions and a more stable economy lower the transactions costs for establishing and maintaining property rights and cause a second round of reductions in the degraded area.

In fact, we know that an area of open access forest exists, even in the most economically developed countries, and forest trespass and theft occurs, even in developed countries. However, the amount of illegal activity is probably greater in less developed regions and countries. Countries undergoing rapid change in economic welfare provide good examples, and the most notable of these in recent years have been the countries of the former Soviet Union. Many of these countries suffered serious economic decline and instability as they as they adjusted to new arrangements independent of the former Soviet Union. The effectiveness of their formal institutions declined as well. Illegal logging increased dramatically in many of these countries, and it increased simultaneously with the decline in overall economic welfare.

The broader historic evidence is also consistent with these arguments. Countries draw down their stocks of natural resources like trees and forests as they enter periods of initial forest development. However, they also build back their stocks of forest after some point in the development process. For example, in the last 60 years the natural forest cover has increased dramatically in countries and regions as different as Switzerland, France, Denmark, the northeast of the United States, and India's Punjab, and similarly in the last 30 years for China. In each of these regions and countries, agricultural land use has remained relatively constant or even declined over the period of economic growth and the increase in forest cover has been greater than any decrease in agricultural land area. The only explanation must be that, in a period of economic development, forest cover has expanded into areas of previously degraded land.

In sum, sustainability in the sense of maintaining options for land and resource use over time is important everywhere, but it may be most difficult to insure in the less-developed regions of the world. These tend to be characterized by the first two stages of forest development where depletion of some non-market value from the natural forest is inevitable. The institutions responsible for managing the natural forests of these regions may be neither well-established nor well-funded. Therefore, establishing priority locations and resources for protection is especially important for less-developed regions and countries. However, rural economic development and the alleviation of rural poverty must be central to long-run improvements in forest sustainability, and also to any attempt to decrease the rate of global deforestation. Rural development is everyone's objective, not just the objective of forest policy. Accomplishing it is not an easy task, but it is certainly no more difficult than trying to accomplish sustainable forestry and decrease the rate of deforestation through the imposition of government regulations on the use of relatively low valued and widely dispersed forest resources by a scattered and poor rural population.

In fact, combining these observations with those of the impact of macroeconomic performance on forests and forest-based communities leads to the broad and general hypothesis that economic stability and growth in the aggregate economy is the best path to more sustainable forestry in general and to a quality natural forest environment in particular (see appendix 4A).

Characteristic Categories of Land Ownership

The latter section of the book reviewed the four characteristic distinctions in forest ownerships: industrial, institutional, non-industrial private, and public. The patterns of land ownership in each of these four categories accompany the three-stage pattern of forest development in predictable ways—although with nuances that are important for understanding the behavior of these different ownerships and for administering and evaluating policy.

The Forest Industry

The forest industry is composed of the various wood processing establishments—sawmills and plywood operations, pulp mills, papermaking facilities, etc.—that also manage forests primarily to supply woody raw material for their processing facilities. These industrial establishments may obtain harvest concessions on the lands of other owners, but they only begin to own and manage their own private forests during the third stage of forest development when the price incentives are sufficient to justify private ownership and, even then, only in regions where the competition for wood supply is sufficient to generate the establishment's long-term interest in protecting the flow of timber to its mills at a competitive price.

For those industrial establishments with concessions to timber harvests in the first and second stage of forest development, and for the (generally) public agencies with permanent responsibility for the lands that support them, the problem is to arrange an acceptable contractual arrangement between the two parties. The public agency is interested in the maximum recovery of revenues from use of the forest that is consistent with protecting against the negative off-site effects of timber harvesting (such as the erosion and downstream sedimentation that can accompany timber harvesting) and guaranteeing the long-run environmental viability of the land and of the fauna and flora native to it. In contrast, the industrial establishment's interest is in the least cost flow of timber over the financial lifetime of its mill. The differences in these two opposing interests, as well as the specialized environmental differences of each harvest site, create complexity for the contract for any particular logging site or longer term logging concession.

Beyond this, anyone with a broader public interest who is inclined to inquire into the efficiency of public agency management should also inquire into the public's best interest for the final allocation of the economic rents recovered during the harvest activity. Which party, industrial concessionaire or public agency, disposes of these revenues in a manner keeping with the public's best interest? Where agency efficiency or the corruption of public officials are serious considerations, as they are in many forested regions, then public harvest policy deserves greater attention, and it is not inconceivable that the industrial concessionaire's distribution of its profits is more in keeping with the public interest than the public agency's distribution of timber revenues (see chapter 5).

For industrial establishments in regions in the third stage of forest development, the second condition, the importance of local competition for their supply of wood inputs, is crucial. Smaller operations with lesser financial investment are not as concerned with the long-term flow of their primary wood resource. Some sawmills, for example, are small and mobile. Their capital facilities easily transport to a location with more or better timber when the local supply dwindles.

Larger operations, with greater capital investment, especially those with smaller ratios of primary resource costs to total production costs, have greater interest in protecting the continuous local operation of their greatest capital expenditure, their mill. The pulp and paper mills of the Nordic countries and North America are among the most capital intensive operations of any industry anywhere in the world. Their capital investments are multiple orders of magnitude greater than that of the average sawmill in the same countries, and their woody raw material is a much smaller share of total production costs. (The cost of virgin wood fiber is less than 6 percent of the total capital invested in an average pulp and paper facility in the United States but more than 45 percent of the total cost of capital for an average sawmill—and an even larger share of the total cost of invested capital for many smaller sawmills.) Therefore, these pulp and paper operations are more likely than lumber companies in the same regions to own their own forestlands. Furthermore, for the larger pulp and paper operations, land ownership becomes relatively even more important where the competition for wood fiber is greater. Therefore, the extent of industrial land ownership has been greater, for example, in coastal Georgia in the United States than in Canada's Quebec. The industry is large and technologically advanced in both regions, but the competition for virgin wood fiber is greater in Georgia.

The skill sets of millworkers reinforce these differences between the less intensive manufactured capital, more wood resource intensive, more mobile sawmill industry, and the pulp and paper industry with its opposite characteristics. Sawmill workers tend to be more variable in their skill sets and many of them move easily from sawmill employment to agricultural or other employment as market conditions change. Employment in the pulp and papermills, however, tends to be more specialized and these mills cannot easily replace their much smaller numbers of higher-skilled workers.

In sum, smaller and more mobile industrial establishments in the wood products industries tend to respond to progress through the first two stages of forest development and the associated decreases in diminishing locally available wood supply by changing the location of their operations, by moving with the changing economic opportunity and the shifting forest frontier. More capital intensive operations, in the presence of increasing scarcity, tend to remain in the same geographic locations and begin either investing in their own timber management or contracting with others to grow wood fiber for them (see chapter 7).

Institutional Forest Landowners

Institutional landowners are establishments such as pension funds, insurance companies, banks, and foundations that hold forestland, along with a great variety of other investments, in large financial portfolios. They are relatively newer participants in forestry and their participation is greatest in those regions in the third stage of forest development where the forest industry is technologically advanced and the markets for timber and forestland are competitive.

Two factors explain their existence. Industrial landowners began to recognize that the financial returns on their forestlands were less than the returns on their greater investments in wood processing facilities. For many industrial firms, land ownership was not a good investment. On the other hand, institutions with diversified financial portfolios began to recognize that the returns on forestland are somewhat counter-cyclical. Therefore, these investments provide balance in a diversified portfolio with other more regularly cyclical investments. Furthermore, investments in forestland have a good risk-return profile. That is, they provide an acceptable return for less than a market level of risk.

These factors, taken together, led many industrial landowners in the United States to divest themselves of forestland in the later years of the twentieth century just as the large financial institutions began purchasing large blocks of forestland. More than 12 million hectares transferred ownership over a period of 25 years. The same lands continue to provide a flow of timber to the industry today—as the financial institutions sell their mature timber to the industry. The only difference is that large financial institutions find greater advantage in forest ownership than the former industrial landowners did.

After an initial period and numerous transfers of land ownership, the best opportunities for further transfers are becoming more limited. The financial institutions, initially most active in the U.S. South and Southeast, have begun looking elsewhere. They have now expanded their holdings of forestland into Canada, South America, Australia, and New Zealand. Some evidence suggests that China's forests share similar countercyclical and risk-return characteristics. They too may attract investment from the large financial institutions.

The land transfers to financial institutions simplified the task of the foresters who manage these lands. While under industrial ownership, the industry's foresters had to justify management to their senior industrial supervisors as a necessity for the mills despite a financial return that failed to meet industry standards, and the senior financial managers of the industrial firms could always point to other firms in the same industry that operated with less land. The forest manager's task is more straightforward under institutional ownership. The objective is cleaner. Simply maximize the return from the land and its forest resources, primarily timber, without particular concern for supply to any one firm's mills or without a requirement to provide timber revenues to offset corporate losses during that one firm's financially less successful years (see chapter 8).

Non-industrial Private Forest Ownerships

NIPF landowners manage small parcels of forest but without any associated wood processing facilities. They are a diverse group, including some who manage their forests for commercial timber value or household subsistence use but others who manage for a range of recreational, amenity, and other non-timber uses. Perhaps a majority, while emphasizing either one or another single timber or non-timber objective, still obtain some personal gain from both the timber and the non-timber products and services of their small forestlands.

Small landowners exist within all three stages in the pattern of forest development. Subsistence farmers in regions in the first stage do not exercise property rights to forests but some may manage trees in their homegardens and all of the pioneer settlers in the first stage have access to the resources of the open access forest. Farm landowners in the second and third stages of forest development may also manage trees in homegardens or in shelterbelts along the periphery of their agricultural properties and some possess small parcels of commercially sub-marginal forestland contained within the boundaries of their agricultural lands. Some landowners in the third stage possess full rights to small commercial forest properties. And, of course, others possess rights to recreational properties wherever the natural landscape attracts recreational use within any of the three stages of development.

Farm landowners, in both developed and developing countries, manage their forests for commercial gain where it is profitable, but their greater comparative advantage usually lies in their agricultural activities. Forest management is generally a secondary activity for them. Conveniently, forest management is an activity they can delay till seasons of lesser agricultural demand for their labor. For those developing country farm landowners with off-farm labor opportunities, forestry can be a means to obtain some return from their (previously) agricultural lands while they spend the majority of their time and effort in the more financially rewarding off-farm activity.

For farm landowners, the timing of the timber harvest decision can also be variable. They often delay harvests until times of unusual household financial need. In this case, the forest provides an insurance against unusual circumstances for the NIPF landowner much like the insurance that the forests of industrial and institutional landowners provide in times of higher than usual log prices or adverse cyclical portfolio returns, respectively.

When household need does not determine the harvest timing decision, farm landowners do respond to timber market prices. However, they tend to be relatively less aware of local timber markets than of their primary agricultural markets. Therefore, they may delay harvesting until the timber markets surpass some reservation price that triggers their personal interest, a price that each farmer implicitly sets as a minimum value for his or her commercial timber. This reservation price must also be sufficient to cover the costs of those isolated parcels of normally sub-marginal commercial forest contained within the boundaries of their agricultural lands.

Non-farm owners of NIP forests, unlike farm landowners, tend to be absentee landowners who do not live on their forest properties. For many of them, the timber value of their forests tends to be less important than the real estate value plus the entirely personal value associated with their own experience on the land. They may harvest timber, but less frequently than farm landowners, and timber market prices are often no more motivating than, for example, the desire to remove a few trees in order to open a scenic vista for a recreational home or to improve wildlife habitat.

The total of NIPF lands (along with the public forests) are a significant focus of modern forest policy. NIP forests may account for more than half of the global forest, but they are individually small and varied, too varied for their owners to lobby successfully against the special interests of the forest industry, whose strong interest is in obtaining a cheap and reliable wood supply from the NIP lands, or the environmentalists or professional foresters, who each have their own perspectives of what comprises "good" forestry. Since none of these, the industry, environmentalists, or professional foresters, share the objectives of many NIPF landowners, but all three desire more and "better" forests, they often support public policy that imposes management practices and harvest and reforestation decisions that are not entirely consistent with NIPF landowner objectives. For example, policy may offer free seedlings, public technical assistance, and financial incentives for forest management and NIP landowners may willingly accept all three, but these landowners may harvest less volume and less frequently than policymakers anticipate.

Policy directed toward the NIPF lands would improve with better understanding of the distinctions between NIP landowners and their ownership objectives. Separate and independent assessments of farm and non-farm landowners or assessments distinguishing between landowners with stumpage supply objectives and those with primarily recreation and amenity objectives for their lands are crucial to improved understanding of this class of forest landowners. Moreover, the recognition of these distinctions is increasingly important as the groups of farm landowners and small forest owners with commercial timber objectives are declining while the group of owners with recreation and amenity objectives is expanding almost everywhere in the world. This means that it is increasingly necessary to modify our policy expectations to reflect the declining timber price responsiveness of NIPF landowners in general, and especially the lesser price responsiveness of non-farm landowners. Successful policy for the NIPF lands and landowners must be increasingly measured in terms of something other than a general willingness to add increments to NIPF timber supply from the undifferentiated combination of both groups of NIPF landowners.

The Public Forestlands

The public forestry agencies generally have three responsibilities; forestry research, technical assistance for private forest landowners, and the management of large tracts of public land. In the first two, the agencies' obligations are to those producers and consumers of forest-based goods and services who are not well-served by the market. This generally means small landowners and consumers of forest products who may be at an information disadvantage in their transactions with large industrial consumers of wood. Increasingly, today, it also means producing research and sharing information about the environmental services of forests, services whose value is not fully reflected in the market but whose merit is increasingly recognized and important in the modern world.

The third responsibility, management of the public's forestlands, covers the largest total global land area of any of the four ownership categories. The central government forest management agencies tend to be larger landholders than any other single private or corporate forest landholder in most countries. The lands they manage are more diverse than most private forestlands because, while the public lands often include some commercial forest, their greatest total area tends to be in the lands beyond where most potential investors would find reward from private ownership. As many public forest lands are less accessible and less commercially desirable, they generally include a greater variety of terrain and, therefore, greater biotic variety as well.

Their diversity would be a disadvantage for industrial landowners with their more focused and specialized objectives, and probably a disadvantage for institutional landowners for the same reason. The greater numbers and variety of non-industrial private landowners would individually find advantage in some, but not all, of the diverse features of the public lands.

This diversity can create difficulties for public land managers. For example, it is most certainly a source of greater environmental risk than that experienced by managers of private forestlands. However, the diversity of the public lands also provides the means for satisfying the very widest range of public agency responsibilities; including the protection of endangered species and the production of a great variety in outdoor recreational activities, the widest possible range of habitat for native fish and game, etc. In this respect, diversity is also an advantage for the public forestlands and the social responsibilities of their managers.

From the perspective of economic value, there are three broad categories of public forest: (1) commercial forests, (2) forests with positive but non-market value, and (3) forests with economic value that is less than the cost of their maintenance. The public forestry agencies might wisely manage the first and second according to the principles of good economics, recognizing both market and non-market values and costs, and also recognizing that multiple outputs is the economically efficient "highest and best" use on some of these lands. The public agencies have a stewardship responsibility for the third category. Stewardship means protecting these lower-net-value lands for their potential future options and also ensuring that activity on these lands, whether natural or caused by mankind, does not create long-term detrimental effects, either on- or off-site. This means, for example, controlling wildfires from spreading to more valuable adjacent properties and limiting the potential for erosion on the public lands to be a source of downstream sedimentation and loss of greater value.

Many public agencies would claim that this is precisely how they do operate. However, few do follow market criteria for the timber resource (see appendix 3b.) and, in the U.S. case, established public law limits their ability to follow market signals for the grazing of domestic livestock.

Furthermore, many public forestry agencies tend to appeal to additional criteria like income distribution, community stability, regional development, and multiple use as if these too are independent management objectives irrespective of their economic merit; and then the agencies seek yet further public input from a wide range of clientele groups. The intention is certainly noble, but the result must be confusion and uncertainty. How does one measure these additional criteria and additional inputs and when does one objective or one interest group take precedence over the others? Indeed, when is one of these criteria comparable to another and how would the agency demonstrate comparable treatment? Without formal definition as to how these additional criteria and the inputs of the various different clientele groups are to be included in the decision calculus, managers are left without clear instruction and the public is left with reason to object regardless of any final agency decision. We have seen the result in the United States: almost continuous litigation, until either the courts or partisan politics often take the decision away from the agency and its professional managers.

This experience should lead everyone—land managers, analysts, and the interested public—to prefer a simpler and clearer set of decision criteria. The effects of many attempts to use forestry as a means for income distribution, community stability, etc., have been weak at best, and sometimes contrary. Therefore, it seems that a return to economic efficiency—broadly defined to include both market and non-market values and ecosystem services—as the guiding management criterion might be well-advised. Of course, within the guidance of economic criteria for land use decisions, it will still be important to trace the potential impacts of those macroeconomic and adjacent sector market and policy effects that are often overlooked but which can have such strong impacts on forests in general, and on the more remote public forests and their neighboring human communities in particular (see chapter 10).

Community Impacts

We followed the pattern of forest development through the U.S. South over the course of the twentieth century, and the pattern of similar but more recent and rapid development for communities in Brazil's eastern Amazon, and then in China. The China example was for bamboo, a product that uses similar land and labor inputs to timber and also substitutes for timber in many of their mutual products. Bamboo markets are largely unadministered in China and, therefore, they are probably a good illustration of what timber markets would be if the government played a less active role in their management. In all three examples, we traced the pattern of industrial growth, and also the patterns of forestry's use of land and labor inputs, as well as the patterns of forestry's association with its supporting local community.

In a sense, little has changed from the beginning of our discussion of the three-stage pattern of forest development in chapter 2. Whether discussing markets or policy, one or another class of landowners, or the effects of forests on entire local human communities, the particular stage of development sets much of the discussion.

For communities in regions in the first stage of development, the natural forest provides support for subsistence households and supplementary income for local commercial agricultural households and workers. Subsistence households rely on the forest for a range of wood and non-wood products that form a part, sometimes a large part, of their regular household consumption. Some of these households also collect selected forest products for exchange in the local markets where they receive a share of their very limited cash income. Often it is the poorest households in any subsistence community for whom the collection of these products is most important, and during the worst of times (e.g., crop failures, loss of cash income) the forest and its products can serve as a social safety net for these poorest local households. Women and men (and children) from these households tend to reflect the expected economic behavior in their participation in all household activities, including the collection of the forest's products. That is, neither women nor men seem to have a unique role with regard to activities associated with the forest. Rather both women and men allocate their personal labor to activities and at times when it brings greatest advantage to the household.

Members of other households in other communities participate in rudimentary logging and sawmill operations—usually when the local demand for sawnwood is strong and when their labor is not as necessary for their household's agricultural activities. For these households, the forestry activity provides a welcome supplement to the household's agricultural earnings.

Communities in the second stage of forest development experience a range of dependence on their forests. Some households in these communities continue to rely on logging and rudimentary sawmills for supplementary employment and income. For other communities, forestry may be one of a small number of important sectors in an undiversified economy. Communities in western Montana or northern California in the United States are examples. In these communities, forestry offers full-time employment for some households. In still other communities, forestry may be the only source of employment and income for all households in the community. Company towns are an example. Communities in both cases tend to experience the boom or bust cycles common to many remote resource-based communities. Fisheries, mining, even some agriculture and some tourism-based communities share this experience with forest-based communities. It is due to their relative remoteness and their focused economic activity and, despite various attempts, policies designed to limit the negative effects of these cycles have only most limited success.

The simple existence of a standing natural forest is not sufficient to insure development, but the existence of a natural resource can be one source of development where the necessary markets and transportation systems also exist. In these cases, forestry has a role in improving the lot of some of the poorest rural inhabitants. However, it is clear from our U.S., Brazil, and China examples that, while the poor do gain from an active forest sector, those who gain the most are not the poorest and, furthermore, the local poor obtain their greatest gains when they can avail themselves of new and better employment opportunities. Some of these new employment opportunities may be in processing activities for forest products. Others may be in a developing commercial agricultural sector, or in more distant employment of various sorts.

The third stage of forest development is characterized by managed forests, often at some distance from any region's remaining natural forests. The economies of communities in these regions are more diversified. Forestry is only one generally smaller sector among many, and the others are generally larger. As a result of the greater regional economic diversification, households in these communities have a greater variety of employment alternatives. The forest sector itself is often specialized, with a variety of sawmill, plywood, and other wood processing facilities and, because the demands for labor by these facilities are different, those who provide labor to this sector often specialize. Economic cycles are not so damaging for these communities as for those in the second stage of forest development because the income and employment effects of a downturn in one of the community's sectors are often offset by labor opportunity in other sectors which are not all at the same stage in their economic cycles (see chapter 11).

One Final Caution: Data

The forestry data remain a problem for any economic or economics-related assessment. The official forestry data for all countries are collected to physical standards for the timber resource, standards that are unique to each country. Some countries are beginning to add physical measures of other forest products and services as well. Nevertheless, the available data focus very largely on the timber resource. Furthermore, the fundamental physical measures for inclusion in the timber data are not always similar from country to country and, where they are similar, the minimum standards vary by orders of magnitude. The Food and Agriculture Organization of the UN (FAO) integrates these various data into comparable national and international measures of forest area and volume. Nevertheless, the FAO data are still physical data. Physical data are a common limitation for assessments of most natural resources: fisheries and wildlife, minerals, oil and gas, even air and water; and market and policy analysts active in forestry, as those inquiring into performance in these other natural resource sectors, must be aware of the limitations of the available data, as well as of the modifications in the data or their analyses that are necessary to create reliable assessments.

The terms used with the official physical data can also be misleading. Forests are often reported as "productive," "commercial," or "managed" forest area or volume. These seemingly market-oriented terms have different meaning for economic analysis than for the inventory specialists and biological forest managers who collect and assemble the data. In general, "productive" and "commercial" forests include large areas that satisfy some physical measure but are not necessarily productive or commercial by economic standards. "Managed" forests include lands described within a broad planning document, but not necessarily within the region of any active on-the-ground silvicultural management.

While some of the lands and trees in these categories are not economic, other lands and trees that are not included in the official national forest inventories are economic, at least for some purposes. These latter include trees in fencerows, homegardens and agroforestry plots, along roadsides, in many town and city parks, etc. These trees seldom, if ever, become part of the commercial timber supply, but they include the great majority of trees in many places. This is no small point. Fly over New Jersey in the United States or the island of Java in Indonesia, for example. The tree cover appears almost continuous, yet the official forest in both regions is minimal. The trees that do not appear in official forest statistics provide services like shade and protect the soil from wind and water erosion. They are important for carbon sequestration and concerns about global change—as all trees sequester carbon, not just those in the official government measure of forest. Bangladesh, for the opposite sort of example, has very little forest cover that is so easily apparent. Yet the source of woodfuel, a crucial commodity for Bangladesh's many poor households, is almost entirely from these unaccounted trees. In summary, reliable forest policy must match the resource service which is the policy objective with the measure of forest and trees.

Furthermore, these physical forestry data make no distinctions for forests at the different land use margins discussed in this book. Yet, we have seen that the same market factors and policy decisions can have different, and even, opposite, effects at the different margins. Moreover, not all parts of the world contain forests at all four economic margins and, where all four do exist, they may not be of the same physical extent or local economic importance. Therefore, the anticipated market and policy affects on the common physical measures for one part of the world can be nonexistent, or even contrary, to those on the forests of another part of the world. Failure to adjust the common forestry data for the four margins and, therefore, failure to recognize the differences that markets and policies have on these margins causes many well-intended and, otherwise, carefully considered policy and management decisions to have unexpected or even perverse outcomes. (See appendix 2a.)

Summary Policy Implications

The distinctions between natural and managed forests, between the first two and the third stages of forest development, and between natural forests at the economic frontier and those well into the hinterland beyond that frontier are crucial for policy. One size does not fit all. Disregard for these distinctions is the source of the ineffectiveness of numerous public forestry activities.

Price is only the most obvious example. Increasing prices are an incentive to increase the level of harvests from the natural forest. They are an incentive to increase harvests from managed forests as well, but they are also an incentive to increase the level of silvicultural management and expand the land area in managed forestry as well. They have no such effect on the natural forest. That is, increasing prices are neither an effective incentive to increase management in the natural forest nor to add land to the forest base in regions in the first two stages of forest development.

Similarly, the standard public forest policies—forest incentive payments, technical assistance or extension forestry, and free or discounted seedlings and publicly supported nurseries—have an effect only on the managed forests in regions in the third stage of development or regions at the cusp of entry into the managed forestry of the third stage. Where public support assists with the planting of new seedlings in regions in the first two stages, it is not surprising that we often observe that those seedlings wither unattended in the ground.

Consider those other forest or forest-related policies that attract more current attention. Certification, the concept of incentives, even requirements, for lumber from guaranteed sustainable forest operations, may seem favorable for the forest environment. The case for certification is an easy one for northern and western Europe where most of the forests are in the managed third stage of development. It is not a measure of success that many of these forests are certified. The case is similar for existing managed forests in other parts of the world. Forest certification is less likely, and even dubious where it does occur, for those parts of the world where most timber harvests originate from natural forests in the first two stages of development. Since the latter is descriptive of many tropical forests, required certification could even be an imposition and a deterrent to further development in some regions. Therefore, the effect of required certification could easily be an unintended bias of the developed world for its own products and against the products and the welfare of some parts of the developing world.

Protection for biodiversity, watersheds, or whatever forest ecosystem services must be measured in marginal forest lands protected—not in total area under protection. That is, the majority of lands under actual forest management as well as the majority of the natural forest, that beyond the accessible economic frontier, is neither threatened nor modified in any favorable way by most policy. These lands provide ecosystem services regardless of formal rules. Therefore, the effectiveness of programs to protect forest ecosystem services must be measured by additions to and removals from the four threatened economic margins—with particular emphasis for many ecosystem services on the threatened natural forest frontier.

The actual measurement of the forest, itself, is important. First, officially "managed" forest is often only an indication of land contained within the boundaries of a forest planning document, not an indication of on-the-ground applications of silvicultural treatment. Some of the land within these planning boundaries may be well beyond any intended physical management now or in the foreseeable future. Second, official government measures of "forest" generally exclude trees along roadsides, fences, in smaller municipal parks, backyards, and homegardens, even most orchards, and they often exclude trees in remote arid regions. Yet these trees too provide some varieties of forest ecosystem services. The previous section of this final chapter reviewed these points in only slightly more detail. The crucial summary point is that reliable forest policy must match the resource service which is the policy objective with the measure of forest and trees.

There are other examples, but the five discussed above make the point. Those who encourage their own preferred forest policies without regard for the distinctions between managed and natural forests, between the first two and the third stages of forest development, and between those resources at the economic frontier of natural forests and those in the hinterland well beyond the economic frontier are almost certain to be shortchanging their own effort and the limited funds of those they represent.

INDEX

Page locators in *italics* indicate figures and tables.

accessibility: and habitat protection, 158–160; and logging costs, 211–212
advice, technical assistance and forest incentives, *117*, 119
afforestation, 118–119
Africa: central African non-commercial natural forests, 33; and managed forests, 32
aggregate economic growth effects, 233–244, *236*, *240*, 246
agricultural extension, 119
agricultural households, 20
agricultural land value: and forest resource assessments (FRAs), 57*n*31; new forest frontier, stage I, 15–18, *16*; and property rights, 45–46, *46*; and property taxes, 114–115
agricultural policy: effects on forest and forestry, 11; incentives and carbon sequestration, 157; macroeconomics and forest impacts, 225–227; spillover effects, 131–135, *133*
agricultural technologies: biological growth and long-term forest effects, 73–78; and developing forest frontiers, *75*; and infrastructure improvements, 79–80, *79*; land-saving technologies on mature forest frontiers, 76–77, *76*; and logging technology improvements, *79*, 80; neutral and land-using technologies on mature forest frontiers, 77–78, *77*; and new forest frontiers, *74*
Alaska, non-commercial natural forests, 33
allowable cut: allowable cut effect (ACE), 103; allowable cut limits, 122–124; forest management and financial returns, 103–106; and frontier forest timber rents, 178; model for determining, 100–101; model qualifications and variations, 101–103; and public forestland management challenges, 381–382
Alternatives to Slash and Burn (ASB), 18*n*5
Amazon region: eastern Amazon industrial forestry, 264–272, *268*; and multiple market values, 41–43, *43*; and non-commercial natural forests, 33; roads and access to frontier forests, 177–178; and subsistence communities, 406–407. *See also* Brazil
Anji County bamboo production, 390–392, *391*, *392*
Asia, East Asian financial crisis, 202, 203, 217–218, 401
Australia, community management and forest development, 144*n*28
"avoided deforestation", 154

bamboo development in China, 390–398, *393*, *394*
bequests, NIPF landowner motivations, 332–333
bidding: and competitive market prices, 184–185; and sawmills' market power, 276*n*6
biodiversity: biodiversity protection and national income account errors, 254; and community management, 147; and physical forest stock estimates, 60, 60–61; policy objectives, 157–160, 445; stand age and non-market environmental values, 11
Brazil: and agricultural settlement policies, 132; eastern Amazon industrial forestry, 264–272, *268*; and forest certification, 128;

Brazil (*continued*): and forest degradation, 27n14; and reforestation requirements, 124–125; structural adjustments and domestic economy shocks, 222–223

Brazil nuts and multiple market values, 42–43, *42*

Breves, eastern Amazon industrial forestry, 266–267, *268*

bribes, forest concession monitoring and enforcement, 181n3

British Columbia: and costs of illegal logging, 26; logging bans and export restrictions, 216; and shifts in production locations, 126–127

bulk, factor costs and industry location, 304–306

Cambodia: anticipation of development in, 401–403, *402*; logging bans and harvest restrictions, 124; and planned economies, 220–221; and social and institutional change, 219–220

Canada: and agricultural settlement policies, 132; forest concessions, 189–190; and forest incentives, 115; global export value of forest products, 202; and non-commercial natural forests, 33; softwood exports, 213, 213n9

capital asset pricing model (CAPM), 311–312, *312*, 313, 316–317

capital depreciation: and environmental accounting, 257–258; Philippine environmental accounting, 258–262, *260–261*

capital gains: and income taxes, 111–113, 112n1; and institutional investors, 308

capital-labor ratio (K/L), sawmills, 279, 279n9

capital-saving technologies and rates of technological change, 88–89

carbon sequestration: and community management, 147; and forest management policies, 40, 444–445; and national income account errors, 253; and physical forest stock estimates, 60–61; policy objectives, 153–157

Catskill Mountains, watershed protection, 361–362

cell phones and long-run forest development, 79n13

certification, forest certification, 127–130

change, dynamic forest development patterns, 13–14

charges. *See* standards and charges

Chile: and forest incentives, 115, 116, 117–118; and reforestation requirements, 124–125

China: aggregate economic growth effects on Hainan Island, 234–238, *236*, 244; and bamboo development, 390–398, *393*, *394*; community management and incomplete property rights, 146–147; and environmental Kuznets' curve (EKC) for forestry, 242–243; environmental loss and regional trade, 207n7, 209–210; and forest degradation, 27, 27n14; and forest incentives, 116; logging bans and export restrictions, 215; mature forest frontier, stage III, 29–30; and NIPF landowner motivations, 318, 331–332; and paper industry, 291, 292; planned economies and domestic economy shocks, 220–221; and recreational forest use, 367; regional development programs and macroeconomic forest impacts, 227; and removal of lands from agricultural use, 134; scale constraints on NIPF forest management, 329–330; watershed management and protection, 164

clearcuts, 120, 125

climate change, 153–157, 373

coastal wetland protection, 163

collusion, 185n4

commercial forestry, local community impacts, 414–420

commercial gain, NIPF landowner motivations, 322–330

community impacts. *See* local human communities

community management: limitations on, 145–147; and property rights, 45–46, *46*, 141–142, 141n24, *142*; successful management, 142–144, 142n25

competition for labor, 131n19

competition for land, 131–135, *133*

concessionaires. *See* loggers

conservation: capital gains and income taxes, 113; consumptive use and subsistence communities, 413–414; investments and watershed protection, 162–163; NIPF landowner motivations, 330–332; and reversion of cropland to forest, 134

construction and fiscal policy, 223–224

consumer welfare, 207–209, *209*

consumption of forest products. *See* forest development stages

continuous forest inventories (CFIs), 58

contracts, rents, and royalties: overview, 194–195, 200; contractual instrument effectiveness, 10; lump sum contracts, 195–196; mature timber concessions, 174–176, *175*, 183–184; uniform fixed royalties, 196–200

Costa Rica, logging bans and export restrictions, 215

costs: eastern Amazon industrial forestry, 265–266, 267, *268*, 269–270; and forest certification, 127–130; forest concession monitoring and enforcement, 180–181; illegal logging, 26;

independent logging operations (NAICS 113310), 293–294, *296*; license fees and public forest recreation, 363–365, *363*; logging costs and potential environmental damage, 211–212; marginal harvest and delivery costs, mature forests, 30, 30*n*18; mature forest regional trade effects, 205–207, *206*; silviculture prescriptions, 125–127; and timber concessions, 174–176, *175*, 178–180, *179*; timber harvest permits, 123; timber production and management of public forestlands, 357; time bias and property taxes, 114–115; transaction costs and forest development patterns, 13, 17, *19*, *21*, *24*, *31*, 41. *See also* opportunity costs; property rights

critical habitat policy objectives, 157–160

currency devaluation, 225

cut and run harvests, 114

cutover lands, 69, 114

data. *See* forestry data

dauerwald (continuous forest), 104–105

death taxes, 113

deforestation: deforestation control policy objectives, 164–170, *166*, *168*, 447–449; developing forest frontier, stage II, 22; environmental Kuznets' curve (EKC) for forestry, 203–204, 239–243, *240*, 246; and forest degradation, 22–25, *24*; and forest inventories, 58; and illegal logging, 25–28; and infrastructure development, 140–141; and multiple market values, 43; structural adjustments and domestic economy shocks, 223; sustainable forestry and deforestation control, 164–170, *166*, *168*; and voluntary carbon markets, 155–156

degraded forest: and deforestation, 22–25, *24*; global forest inventory data, 34–36, *35*; local institutions and rural labor opportunity, 36–39, *38*; and multiple market values, 43

delivered prices: eastern Amazon industrial forestry, *268*, 271; and harvest restrictions, 123–124; and sawmill costs, 276; and timber concessions, 174–176, *175*; and transportation costs, 55–56

Denver metropolitan area, managed forests, 31–32

desired behaviors and forest regulation design, 121–122

developing countries, agricultural policies, 134, 134*n*22

developing forest frontier, stage II: described, 20–28, *21*; land use impacts summary, 85, *86*; and technological change, 80–82, *81*. *See also* forest development stages

development. *See* economic growth and development; local human communities

direct financial assistance, 116–118, *117*

direct policy instruments. *See* policy

distribution of timber revenues, 186–187

diversification, forest sector as national economy component, 233

doi moi reforms, 220

domestic economy shocks: overview, 218–219; Dutch disease, 221; social and institutional change, 219–221; structural adjustments, 221–223

domestic programs, macroeconomics and forest impacts, 227–228

dominant resource value, 355–356

Dutch disease, domestic economy shocks, 218, 221

East Asian financial crisis, 202, 203, 217–218, 401

economic development. *See* economic growth and development; macroeconomics; trade effects

economic forest stocks, physical stock data vs. economic stock assessments, 58–61

economic geography: overview, 2–3, 13–15, 32–34, 47, 48–49; developing forest frontier, stage II, 20–28, *21*; and environmental landscape variety, 46–48, *48*; forest degradation, 22–25, *24*, 36–39, *38*; and forest policy, 39–40; global forest data, 34–36, *35*; illegal logging, 25–28; market-valued opportunity costs, 43–44, *44*; mature forest frontier, stage III, 28–32, *31*; and multiple market values, 41–43, *42*; new forest frontier, stage I, 15–20, *16*; and non-market values, 43–45, *44*; and property rights, 45–46; and shifting cultivation, 18–20, *19*. *See also* forest development stages

economic growth and development: overview, 228; aggregate economic growth effects overview, 233–234; environmental Kuznets' curve (EKC) for forestry, 239–243, *240*, 246; and forest degradation and depletion, 166–170, *166*, *168*; Hainan Island, 234–238, *236*, 244; relative size of forest sector, 229–233, *230–231*, *232*. *See also* national income accounts

El Salvador, managed forests, 32

Employee Retirement Income Security Act of 1974 (ERISA), 308

enforcement. *See* monitoring and enforcement

engineered wood members (NAICS 321213), 281–283

England, managed forests, 30–31

environmental accounting, 256–262, *260–261*

environmental charges, 121–122

environmental damage and loss, regional trade, 205–207, *206*, 207*n*7, 209–212, 212*n*8
environmental Kuznets' curve (EKC) for forestry, 203–204, 239–243, *240*, 246
environmental landscape and forest development patterns, 46–48, *48*
environmental performance bonds and awards, 184
environmental protection: and early harvests, 126; endangered habitat and forest management policies, 40; endangered species and forest regulations, 121; and forest certification, 127–130; and "green space" taxes, 115; landowner concerns, 176–178; and lump sum contracts, 195–196; and management of public forestlands, 357–358, 360–361; regional transfer of regulatory effects, 126–127; and uniform fixed royalties, 198–200. *See also* monitoring and enforcement
environmental tourism and forest-based recreation, 160–162, 446
erosion control: and environmental Kuznets' curve (EKC) for forestry, 241; and forest regulation design, 121; and national income account errors, 254; and policy objectives, 162–164, 446–447
"estate crops" in Indonesia, 30
estate taxes, 113
Estonia, forest degradation and depletion, 167–169
European Union, agricultural policy and macroeconomic forest impacts, 226–227, 226*n*24
even-aged stands, 98
ex situ preservation, 160
exchange rate policy, 224–225
exogenous shocks, 216–218
export restrictions, 215–216
exporting region, mature forest regional trade effects, 205–209, *206*, *208*, *209*
extraction activities and costs: and forest degradation, 22–25, *24*; global forest inventory data, 34–36, *35*; and long run forest development, 90–91; new forest frontier, stage I, 47

factor costs and industry location, 304–306
family forests, 322–323, 323*n*3, 326–327
Faustmann model: appropriate context for, 98–99; described, 93–95; effect of labor and capital inputs on managed forests, 114*n*3; and monetary policy, 224; optimality conditions, 95–96; special cases, 97–98; and timber harvest permit costs, 123*n*9; variations on, 96–97

fiber costs and rates of technological change, 88–89
fiber sources and forest management practices, 72–73
field pine, 69
financial returns, forest management and allowable cut, 103–106
Finland: forestry and GDP, 201; scale constraints on NIPF forest management, 330*n*14; and shifting cultivation, 19*n*6; and shifts in production locations, 126–127; and silviculture prescriptions, 125–126
"first growth" forests, 64–65
fiscal policy, macroeconomics and forest impacts, 223–224
Food and Agriculture Organization (FAO), 57–58
forest-based community, generalized economy, 429–431
forest-based environmental services, 253
forest-based recreation, 160–162
forest certification, 127–130
forest concessions: overview, 173–174, 192–193; contract time period or duration, 183–184; environmental performance bonds and awards, 184; in Indonesia, 190–192, 190–191*n*9; landowner objectives and logger objectives, 176–180, *179*; limited numbers of bidders, 184–185; lump sum contracts, 178–179, 181–182, 195–196; monitoring and enforcement, 180–183; in nineteenth-century Germany, 187; in Ontario, Canada, 189–190; and private landowners in the U.S. South, 188–189; specialized contract considerations, 183–187; stumpage price contracts, 179–180; timber concessions and forest development model, 174–176, *175*; timber revenue distribution, 186–187; uniform fixed royalties, *179*, 182–183, 196–200; and the U.S. Forest Service, 188
forest cover: and aggregate economic growth effects, 238; environmental Kuznets' curve (EKC) for forestry, 203–204, 239–243, *240*, 246; and forest resource assessments (FRAs), 57, *57*; in U.S. South, 68
forest degradation: consumptive use and subsistence communities, 413–414; and forest development patterns, 22–28, *24*, 36–39, *38*; and multiple market values, 42–43, *42*. *See also* degraded forest
forest depletion, 10
forest development stages: in Africa, 32; and agricultural policies, 131–135, *133*; in China, 29–30; developing forest frontier, stage II, 20–28, *21*; economic geography overview, 2–4, 13–15, 32–34, 47, 48–49; in El Salvador,

32; in England, 30–31; environmental Kuznets' curve (EKC) for forestry, 239–241; and environmental landscape variety, 46–48, *48*; and forest degradation, 22–25, *24*, 36–39, *38*, 42–43, *42*; and forest policy, 39–40; and global forest data, 34–36, *35*; and illegal logging, 25–28; in Indonesia, 30; and industrial forestry, 297–302; land use impacts summary, 85, *86*; and local human communities, 420–421; local market and policy implications, 4; macroeconomic market and policy implications, 4–5; and margins of economic activity, 432–435; market and policy effects on forests, 436–443; market-valued opportunity costs, 43–44, *44*; mature forest frontier, stage III, 28–32, *31*; mature forest harvest contracts, 4–5; and multiple market values, 41–43, *42*; new forest frontier, stage I, 15–20, *16*; NIPF lands and landowner distinctions, 319–322, *321*; and non-market values, 43–45, *44*; and permanent agricultural households, 20; and property rights, 45–46; and shifting cultivation, 18–20, *19*; in South and Southeast Asia, 32; in the U.S., 31–32
forest extension programs, 119
forest extraction and logging technology improvements, 79, 80
forest growth, 34n20, 58
forest incentive payments (FIP), 116–118, *117*, 345, 345n22
forest land value, 45–46, *46*
forest products: global export value, 202; mature forest frontier, stage III, 28–29, 28–29n16; natural and managed forests, 2. *See also* manufactured wood products (NAICS 321)
forest recreation, 60, 330–338, *334*, 362–366, *363*, *366*
forest regeneration, 96–97
forest resource assessments (FRAs), 56–61, *57*
forest sector as national economy component, 229–233, *230–231*, *232*
forest stocks: and continuous forest inventories (CFIs), 58, 59–60, 62; forest resource assessments (FRAs), 56–58, *57*; physical forest stock revisions, 61–62; physical stock data vs. economic stock assessments, 58–61
forest value gradients, 41, 41n25
forestland owners. *See* industrial forestry; non-industrial private forest (NIPF) landowners; public landowners
forestry consultants, 325
forestry data: analysis challenges, 12, 14–15, 458–459; data envelopment analysis (DEA), 378–379; global forest data and forest development patterns, 34–36, *35*; physical stock measures, problem, and solution, 56–62, *57*; price and quantity data, problem, and solutions, 53–56
fossil fuel consumption, 155
free seedlings, 118–119, 134
frontier forest: overview, 2–3; and allowable cut model, 105–106; economic welfare and land use impacts, 204–216, *206*, *208*, *209*; and environmental concerns, 176–177; and habitat protection, 159–160; and industrial forestry, 298–300. *See also* forest development stages
fuelwood: factor costs and industry location, 304–306; and forest stock inventories in South and Southeast Asia, 59; and national income account errors, 253; and NIPF landowners, 322, 322n2, 333; and Philippine environmental accounting, 258–259, 258–259n42; and physical forest stock estimates, 60–61; and subsistence communities, 410–412
furniture and related product manufacturing (NAICS 337), 68, 283–286, *284*, *285*

gender, fuelwood and subsistence communities, 410–412
General Agreement on Trade and Tariffs (GATT), 213–214
Generalized System of Preferences, 214
Germany, nineteenth-century Germany forest concessions, 187
Ghana, reforestation requirements, 124–125
ginseng harvest, 25
global climate change, 153–157, 373
global interests and community management, 147, 147n31
global positioning systems (GPS), 79n13
government activity valuation, 253, 253n35, 255–256
grassland and livestock grazing, 45–46, *46*, 358–359
"green banks", landowner needs and harvest scheduling, 326–327
Green Revolution, 65
"green space" taxes, 115
gross domestic product (GDP): described, 201; forestry and global GDP, 201–202; and national income accounts, 250–251, *251*
gross national product (GNP), United States, 250–251, *251*, *252*
gross output value (GOV), 235, 237
Guatemala, forest certification, 129

harvest: and delivered prices for logs, 55–56; global harvest overview, 32–34, 34n20; restrictions, 122–124; scheduling and landowner needs, 326–327, 335–338;

harvest (*continued*): scheduling and market opportunities, 323–326; and shipment restrictions, 122–124; and silviculture prescriptions, 125–127; volumes and sustainability, 64–65
herbicides, 120, 125
high grading, 179
Honduras, community management and forest development, 145–146
Household Responsibility System (HRS), 235, *236*
household services, national income accounting, 253–255
Humboldt County, California, 418

illegal logging: in Cambodia, 220; and contract structure, 180, 183; environmental performance bonds and performance awards, 184; forest concession monitoring and enforcement, 180–183, 180–181n2, 181n3; in Indonesia, 190, 192; and national income account errors, 253, 254; open access and deforestation, 25–28; and stumpage price contracts, 180; and uniform fixed royalties, 199–200
importing region, mature forest regional trade effects, 205–209, *206*, *208*, *209*
imports and tariffs, 213–214
in situ value: frontier timber vs. managed forests, 177; public forestland, 350–351; and timber concessions, 175–176, *175*
inaccessible mature natural forest, 34–36, *35*
incentives: overview, 115–116, *117*; agricultural incentives and carbon sequestration, 157; and agricultural policies, 132–135, *133*; financial assistance, 116–118; forest policy, 115–119; free seedlings, 118–119; technical assistance, 119
income, environmental Kuznets' curve (EKC) for forestry, 239–243, *240*
income redistribution, 118
income taxes and forest policy, 111–113, *112*
independent logging operations (NAICS 113310), 293–297, *296*
India: community management and forest development, 142n25, 144n28, 145n29; and forest degradation and depletion, 27, 27n14, 169; and harvest restrictions, 123
Indonesia: and East Asian financial crisis, 217–218; and forest certification, 128–129; forest concessions, 190–192, 190–191n9; and forest degradation, 28; and harvest restrictions, 123; log export bans, 136; logging bans and export restrictions, 215–216; mature forest frontier, stage III, 30; and non-commercial natural forests, 33; subsistence settlements, developing forest frontier, stage II, 27, 27n13; transmigration policy, 132
industrial forestry: overview, 9–10, 12, 263–264, 272–273, 297–302, 450–451; and allowable cut model, 105; eastern Amazon, 264–272, *268*; engineered wood members (NAICS 321213), 281–283; factor costs and industry location, 304–306; factor-output ratios, 306; and forest development stages, 297–302; furniture and related product manufacturing (NAICS 337), 283–286, *284*, *285*; independent logging operations (NAICS 113310), 293–297, *296*; input proportions, 304–306; manufactured wood products (NAICS 321) overview, 273, *275*, *277*, 282; paper (NAICS 322121) and newsprint (NAICS 322122), 291–293; pulp and paper manufacturing (NAICS 322) overview, 286–287, *288*, *289*; pulp (NAICS 322110), 287–291; reconstituted wood products (NAICS 321219), 281–283; sawmill industry (NAICS 321113), 273–279; timber plantations and Faustmann model, 96; trusses (NAICS 321214), 281–283; veneer and plywood (NAICS 321211 and 321212), 279–281
infrastructure improvements: defined, 138; and developing forest frontiers, 80–82, *82*; and forest development, 138–141, *139*; and habitat protection, 159; on Hainan Island, 237; and long-run forest development, 79–80, *79*, 79n13; and mature forest frontiers, 82–84, *83*
inheritance taxes, 113
instability, macroeconomic forest impacts, 11, 245
institutional investors: overview, 9–10, 307–308, 451–452; asset performance, risk vs. average returns, *311*; capital asset pricing model (CAPM), 316–317; forest sector as national economy component, 232–233; investment implications and future prospects, 313–316; and pension funds, 5–6; revenues and taxes, 302–313, *311*, *312*; timberland portfolios, *312*
institutions: forest development and forest policy, 141–147, *142*; public forestland organization and administration, 378–379
international lending, 221–223
International Monetary Fund (IMF), 190–191, 222
International Statistical Industrial Classification (ISIC), 272
international trade. *See* trade effects
investment. *See* institutional investors
Ireland, removal of lands from agricultural use, 134

Japan, agricultural policy and macroeconomic forest impacts, 226–227
Java, timber production overview, 33
joint production, public forestland management, 368
jululing trees, harvest restrictions, 123

Kenya, environmental tourism, 161
Korea, trade effects on forest environment, 212
kraft paper mills, 69

labor: and capital investment policies, 227–228; capital-labor ratio (K/L), sawmills, 279, 279n9; competition for, 131n19; and deforestation, 140–141, 166, *166*; global forestry employment, 201; and independent logging operations (NAICS 113310), 295–297, *296*; paper and newsprint industry, 293, 293n17; productivity and mechanization of pulp and paper industry, 71
labor-saving technologies, 88–89
laminated beams, 281
land conversion: "avoided deforestation", 154; in Brazilian Amazon, 74–75, *74*, 75n10
land-saving technologies on mature forest frontiers, 76–77, *76*
land use: land use impacts and trade effects, 204–216, *206*, *208*, *209*; land use margins and long run forest development, 90–91; marginal shifts and economic geography forestry framework, described, 2–3, *4*; property rights and watershed protection, 162–163
land-using technologies on mature forest frontiers, 77–78, *77*
land value: and agricultural policies, 132–133, *133*; developing forest frontier, stage II, 20–22, *21*; and forest degradation, 22–25, *24*; and forest incentive programs, 116–119, *117*; and income taxes, 111–113, *112*; logging bans and wood products industry spillover policies, 135–137, *135*; mature forest frontier, stage III, 30–32, *31*; and mature forest regional trade effects, 205–209, *206*, *208*, *209*; new forest frontier, stage I, 15–18, *16*; NIPF lands and landowner distinctions, 319–322, *321*; and non-market values, 43–45, *44*; and property rights, 45–46, *46*, 113–115, 141–142, *142*; and reforestation requirements, 125; and road improvement effects, 138–141, *139*; and shifting cultivation, 18–20, *19*; and silviculture prescriptions, 125–127; sustainable forestry and deforestation control, 164–170, *166*, *168*
landowners. *See* industrial forestry; non-industrial private forest (NIPF) landowners; public landowners

Laos: anticipation of development in, 403–405, *404–405*; planned economies and domestic economy shocks, 220–221
least cost plus loss models, 385
Liberia, institutional change and domestic economy shocks, 220
license fees and public forest recreation, 363–365, *363*
limited numbers of bidders, forest concessions, 184–185
local human communities: overview, 7, 386–390, 420–421, 456–458; and bamboo development in China, 390–398, *393*, *394*; and environmental tourism policy objectives, 161–162; forest development benefits, 395–398; and future development in Cambodia, 401–403, *402*; and future development in Laos, 403–405, *404–405*; and future development in Vietnam, 398–401, *399–400*; generalized economy of forest-based community, 429–431; local impacts of commercial forest operations, 414–420; market and policy effects on, 443; nonseparable household production and consumption, 425–428; subsistence community impacts, 406–414, *409*; and watershed protection, 162–164
location. *See* economic geography; forest development stages
log storage, sawmill industry, 276–278, 278n8
loggers: forest concession objectives, 178–180, *179*; and landowner objectives, 176
logging: independent logging operations (NAICS 113310), 293–297, *296*; quotas, 122–124; road construction and use, 125. *See also* industrial forestry
logging bans: and bamboo development, 390; in Cambodia, 124; described, 122; and export restrictions, 215–216; and wood products industry spillover policies, 135–137, *135*
logging technology improvements: and developing forest frontiers, 80–82, *82*; and economic stock assessments, 59–60; and forest extraction, *79*, 80; and mature forest frontiers, 82–84, *83*
logs: and delivered prices, 55–56, 174–176, *175*; factor costs and industry location, 304–306; and market exchange comparison points, 54; price and quantity data, 53–54
long run forest development: overview, 12, 64–67, 73; infrastructure and management developments, 78–87; rates of technological change, 87–89; technological change and biological growth, 73–78; technological change and the U.S. South, 67–73;

long run forest development (*continued*): and unique characteristics of forestry, 89–91
long-term and permanent trade effects, 202, 202*n*3
longleaf pine, 68–69
low-value timber, 41, *42*
lumber prices and transportation costs, 55–56
lump sum contracts, 178–179, *179*, 181–182, 195–196

macroeconomics: overview, 201, 203–204, 216, 244–246, 440–443; and agricultural policies, 225–227; commercial forestry effects on local communities, 418–420; domestic economy shocks, 218–223; domestic fiscal policy, 223–224; domestic monetary policy, 224; and domestic programs, 227–228; Dutch disease, 218, 221; exchange rate policy, 224–225; exogenous shocks, 216–218; policy hypotheses, 245–246; political and economic stability, 440–442; social and institutional change, 219–221; structural adjustments, 221–223. *See also* national income accounts
mahogany, multiple market values, 41–42, *42*
Malawi: and forest degradation, 27, 27*n*14; NIPF landowners and forest management discretion, 328
managed forest: and allowable cut model, 105; and environmental landscape variety, 46–48, *48*; global forest inventory data, 34–36, *35*; mature forest frontier, stage III, 28–32, *31*; natural stands and Faustmann model, 96–97; policy and forest development patterns, 39–40; timber plantations and Faustmann model, 93–96
management: overview, 89–90; allowable cut model, 103–106; community management and property rights, 45–46, *46*; costs and silviculture prescriptions, 125–127; and developing forest frontiers, 80–82, *81*; economic incentives for, 71–73; economic welfare and land use impacts, 204–216, *206*, *208*, *209*; and environmental tourism policy objectives, 160–162; and forest certification, 127–130; forest concession monitoring and enforcement, 182–183; and forest regulations, 120; fundamental questions, 10; global forest inventory data, 34–36, *35*; in Indonesia, 190–192; land use impacts summary, 85, 86–87; limitations on, 13; and macroeconomic policy, 244–246; mature forest frontier, stage III, 29–32, *31*, 82–84, *83*; nineteenth-century Germany forest estates, 187; and NIPF landowners, 327–330; in Ontario, Canada, 189–190; private landowners in U.S. South, 188–189; property rights and forest development, 141–142, 141*n*24, *142*; public land management challenges, 351, 369–373, *372*, 379–382; public land stewardship, 353–354; public lands timber management, 357–358; and rates of technological change, 88; and technological change, 65–66; time bias and property taxes, 114–115; U.S. Forest Service, 188. *See also* forest concessions; public landowners
"mancha" (mahogany), multiple market values, 41, *42*
manufactured wood products (NAICS 321): overview, 67, 273, *275*, *277*, *282*; engineered wood members (NAICS 321213), 281–283; industry and firm size measures, *275*; major inputs and operating characteristics, *277*; raw material consumption and product shipment growth, *282*; reconstituted wood products (NAICS 321219), 281–283; sawmill industry (NAICS 321113), 273–279; trusses (NAICS 321214), 281–283; veneer and plywood (NAICS 321211 and 321212), 279–281
marginal harvest and delivery costs, mature forests, 30, 30*n*18
marginal land use costs and public landowners, 41*n*24
marginal land use shifts and Faustmann model, 93
market access: developing forest frontier, stage II, 20–28, *21*; and forest incentive programs, 116–119, *117*; and income taxes, 111–113, *112*; mature forest frontier, stage III, 30–32, *31*; new forest frontier, stage I, 15–18, *16*; NIPF landowners, 323–326; NIPF lands and landowner distinctions, 319–322, *321*; and non-market values, 43–45, *44*; and property rights, 45–46, *46*; property rights and forest development, 141–142, *142*; roads and forest development, 140; and shifting cultivation, 19–20, *19*; sustainable forestry and deforestation control, 164–170, *166*, *168*; and timber concessions, 174–176, *175*, 178–180, *179*
market exchange points, 54
market prices: pulp price and quantity data, 54; technology transfer and public land management, 374. *See also* delivered prices
market values: and forest development patterns, 41–43, *42*; market-valued opportunity costs, 43–44, *44*. *See also* management
mature forest frontier, stage III: described, 28–32, *31*; land use impacts summary, 85, 86–87, *86*; and technological change, 82–84, *83*. *See also* forest development stages

mature natural forest: and environmental landscape variety, 46–48, *48*; global forest inventory data, 34–36, *35*; new forest frontier, stage I, 15–20, *16*; regional trade effects, 205–207, *206*

Mekong River Basin, anticipation of development in, 398–406, *399–400, 402, 404–405*

Mexico: community management and forest development, 141*n*24; and forest certification, 129; trade liberalization effects, 205*n*6

migrant and subsistence settlements, 27, 27*n*13

minerals management, public lands, 359–361

Ministry of Forestry, Indonesia, 190–192, 190–191*n*9

mobility, sawmill industry (NAICS 321113), 273–274

monetary policy, macroeconomics and forest impacts, 224

monitoring and enforcement: environmental protection and landowner objectives, 178; forest concessions, 180–183, 180–181*n*2, 181*n*3; and lump sum contracts, 195–196; and uniform fixed royalties, 198–200

motives, NIPF landowners: overview, 319, 322; bequests, 332–333; forest management discretion, 327–328; forest management scale constraints, 328–330; landowner needs, 326–327; market opportunity, 323–326; multiple motives, 333–338, *334*; recreation/leisure and conservation, 330–332

multiple market values, 41–43, *42*

multiple uses, 7, 152–153, 153*n*33, 366–368

mushroom collection, 25*n*10

Myanmar, trade effects on forest environment, 212

national guidelines, variation in forest cover definitions, 57, *57*

national income accounts: overview, 250–251, *251, 252*; account errors, 251–256; account omissions, 253–255; environmental accounting, 256–262, *260–261*; modified Philippine income and product accounts for 1988, *260–261*; U.S. GNP and gross national expenditure for 2002, *252*; U.S. summary measures of national income for 2002, *251*; value of government activity, 255–256

national income (NI), 250, *251*

natural forest frontier. *See* frontier forest

natural hazards, 369–373, *372*

natural regeneration, Faustmann model, 96–97

Nepal: community management and forest development, 142*n*25, 145, 145*n*29; and environmental tourism, 161; and forest degradation, 27, 27*n*14; national income

account errors, 253*n*35; NIPF landowners and forest management discretion, 328; and watershed protection, 163–164

net national product (NNP), 250, *251*

neutral and land-using technologies on mature forest frontiers, 77

neutral taxes, 111

new forest frontier, stage I: and allowable cut model, 105–106; described, 15–20, *16*; land use impacts summary, 85, *86*. *See also* forest development stages; frontier forest

"new public domain" lands, 114

New York City, watershed protection, 164, 361–362

Nicaragua, community management and forest development, 144*n*28

non-commercial natural forests, 33

non-consumable forest products, 239–241

non-industrial private forest (NIPF) landowners: overview, 5–7, 318–319, 338–341, 452–454; bequest motives, 332–333; farm vs. non-farm owners, 339–341, 339*n*21; forest management discretion, 327–328; forest management scale constraints, 328–330; landowner needs, 326–327; landowner objectives overview, 322–323; management fallacies, 344–348; market opportunity, 323–326; multiple motives, 333–338, *334*; NIPF lands vs. landowners, 319–322, *321*; recreation/leisure and conservation motives, 330–332

non-market environmental service, 251–253

non-market forest products and forest degradation, 24–25, 25*n*10

non-market values: and forest development patterns, 43–45, *44*; and public forestland management, 351, 365–366, *366*

non-neutral taxes, 111

non-point pollution, 122

non-tariff barriers (NTBs), 214–215, 214*n*13

non-timber forest products and values: and Faustmann model, 98; local communities and forest dependence, 406–414, *409*; and multiple market values, 41–43, 41*n*25, *42*; and physical forest stock estimates, 60

nonseparable household production and consumption, 425–428

Nordic countries: and forest certification, 128; and forest incentives, 115; and long run forest development, 73; reforestation requirements, 124–125

North American Industrial Classification System (NAICS), 272, 272*n*3

Norway: and environmental accounting, 256–257; environmental loss and regional trade, 207*n*7; sawmills' market power, 276*n*6

472 Index

nurseries, forest management practices, 71–72
nutrient depletion and shifting cultivation, 19–20

"old growth" forests, 64
oligopsony, 185n4
Ontario, Canada, forest concessions, 189–190
Ontario Ministry of Natural Resources (OMNR), 189–190
open access forest: and environmental landscape variety, 46–48, *48*; and forest degradation, 20–28, *21*, *24*, 36–39, *38*; illegal logging and deforestation, 25–28; and regional trade, 204
opportunity costs: and forest degradation, 23–24, *24*; labor and infrastructure development, 140–141; and non-market values, 43–45, *44*
optimal harvest: and Faustmann model, 97, 97n23; optimal age or timber rotation, 9; and property taxes, 114–115
oriented strand board (OSB), 71, 273

palm trees, 124
Panama Canal, watershed protection, 362
paper. *See* pulp and paper manufacturing (NAICS 322)
Papua New Guinea: and forest certification, 129n17; and non-commercial natural forests, 33
Paragominas, eastern Amazon industrial forestry, *268*, 269–270
People's Republic of (North) Korea (PRK), trade effects on forest environment, 212
performance awards, 184
permanent agricultural households, 20
permit costs, timber harvest permits, 123
pest populations and shifting cultivation, 19–20
pesticides, 120, 125
the Philippines: environmental accounting, 258–262, *260–261*; land transfers and incomplete property rights, 146, 147n30; subsistence settlements, developing forest frontier, stage II, 27, 27n13
physical forest data. *See* forestry data
physical standards on imported goods, 214–215
Pinchot, Gifford, 382
planned economies, 220–221
plantations: and Faustmann model, 93; U.S. South's Third and Fourth Forests, 70–73
plywood industry: Indonesia and East Asian financial crisis, 217–218; Indonesia and log export bans, 136; research support, 136–137; responses to resource availability, 12; and U.S. South forests, 68; veneer and plywood (NAICS 321211 and 321212), 279–281
Poland state forestry agency, stochastic frontier analysis, 379

policy: overview, 7–8, 108–110, 130–131, 147–148, 436–437, 459–461; agricultural policy spillover effects, 131–135, *133*, 439–440; community welfare impacts, 416–418; consumptive uses and non-forest employment, 10–11; direct financial assistance, 116–118, *117*; direct policy forest effects, 438–439; effects on local human communities, 443; forest certification, 127–130; and forest development patterns, 39–40, 39–40; free seedlings, 118–119; fundamental questions, 11; harvest and shipment restrictions, 122–124; incentives, 115–119; income taxes, 111–113, *112*; infrastructure and forest development, 138–141, *139*; institutions and forest development, 141–147, *142*; and macroeconomics, 440–443; NIPF landowner myths, 344–348; policy objectives overview, 109–110; property taxes, 113–115; reforestation requirements, 124–125; regional development and local forest administration, 437–438; regulation design, 121–122; regulations overview, 119–121, *120*; silvicultural prescriptions, 125–127; standards and charges, 110–111; taxes, 111–115; technical assistance, 119; trade policy instruments, 212–216; and wildfire control, 370–373, *372*; wood products industry policy spillover effects, 135–138, *135*, 439–440. *See also* forest concessions; policy objectives
policy objectives: overview, 109–110, 152–153; biodiversity and critical habitat, 157–160, 445; carbon sequestration and global climate change, 153–157, 444–445; environmental tourism and forest-based recreation, 160–162, 446; erosion control and watershed protection, 162–164, 446–447; sustainable forestry and deforestation control, 164–170, *166*, *168*, 447–449
pollutants: and beneficial effects of trade on forest environment, 212n8; pollution control policy and wood products industry, 137
Portland, Oregon, watershed protection, 163
poverty: and forest degradation and depletion, 166–170, *166*, *168*. *See also* local human communities
price and quantity data: and continuous forest inventories (CFIs), 58, 59–60, 62; described, 53–54; forest resource assessments (FRAs), 56–58, *57*; market and delivery prices for logs, 55–56; physical forest stock revisions, 61–62; physical stock data vs. economic stock assessments, 58–61
price supports, *133*, 134, 226–227
private forests, resource supply concerns, 70

Index **473**

process-oriented technological change, 65n1
processing technology, economic stock assessments, 59–60
production costs and rates of technological change, 65–66, 65n1, 87
productivity measures. *See* national income accounts
property rights: and carbon sequestration, 156; community management and forest development, 141–142, *142*; cost of maintaining in new forest frontiers, 17–18; developing forest frontier, stage II, 20–28, *21*; differing bundles of, 45–46, *46*; and forest incentive programs, 116–119, *117*; and income taxes, 111–113, *112*; mature forest frontier, stage III, 30–32, *31*; and multiple market values, 41–43, *42*; NIPF lands and landowner distinctions, 319–322, *321*; and non-market values, 43–45, *44*; and silviculture prescriptions, 126; sustainable forestry and deforestation control, 164–170, *166*, *168*. *See also* forest concessions
property taxes, 113–115
public agencies. *See* public landowners
public assistance, NIPF landowner myths, 347–348
public goods: and carbon sequestration, 155; and community management, 147; and forestry research, 376–378
public landowners: overview, 5–7, 7n1, 350–352, 379–382, 454–456; forest recreation, 362–366, *363*, *366*; grassland and livestock grazing, 358–359; least cost plus loss models, 385; management challenges, 351, 369–373, *372*, 379–382; and marginal land use costs, 41n24; minerals management, 359–361; and multiple uses, 366–368; natural hazards, 369–373, *372*; and "new public domain" lands, 114; public forestry research, 376–378; public organization and administration, 378–379; single resource value, 355–356; state-owned forests and timber revenue distribution, 186–187; stewardship, 353–355; technology transfer, 373–375; timber management, 357–358; watershed management, 361–362
public market interventions. *See* policy
public recreation, valuation of government activity, 255–256
pulp and paper manufacturing (NAICS 322): and forest management practices, 71–73; mechanization of, 71; in Ontario, Canada, 189–190; overview, 286–287, *288*, *289*; paper (NAICS 322121) and newsprint (NAICS 322122), 291–293; price and quantity data, 53–54; pulp (NAICS 322110), 287–291; pulpmills and Faustmann model, 96; pulpwood market exchange comparison points, 54; pulpwood prices and transportation costs, 55–56; and resource availability, 12; and technological change, 65; and U.S. South's Second Forest, 68–69; and U.S. South's Third and Fourth Forests, 70–71

quotas, logging quotas, 122–124

railroads, 68–69, 140
real estate investment trusts (REITs), 5, 314, 315
real estate limited partnerships (RELPs), 308
reconstituted wood products (NAICS 321219), 281–283
recreation: forest recreation, 362–366, *363*, *366*; NIPF landowner motivations, 330–332, 333–338, *334*; and valuation of government activity, 255–256, *366*
red-cockaded woodpecker, 159
Reducing Emissions for Deforestation and forest Degradation (REDD), 154, 155
Redwood National Park, 161, 319, 367
reforestation: free seedlings, 118–119; policy requirements, 124–125
regional development: macroeconomics and forest impacts, 227; roads and forest development, 140
regional trade associations, 213–214
regulation: carbon sequestration and global climate change, 155–156; and government activity valuation, 255–256; public forestland management challenges, 380–382
regulations, private forestlands: overview, 119–121, *120*; forest certification, 127–130; harvest and shipment restrictions, 122–124; and public objectives, *120*; reforestation requirements, 124–125; regulation design, 121–122; silvicultural prescriptions, 125–127. *See also* management; spillover effects
renewability and long range forest development, 91
rents: and frontier forest timber values, 177–178; and lump sum contracts, 178–179, *179*, 195–196; mature timber concessions, 174–176, *175*; and stumpage price contracts, 179–180; and uniform fixed royalties, 196–200
Republic of (South) Korea (PRK), trade effects on forest environment, 212
research: and forest management practices, 71–72, 72n6; public forestry research, 376–378
resource availability, 12
"resource curse", 389n4
revenues and institutional investors, 308–309
risk: and illegal logging, 26; and institutional investors, 310–313, *311*

river transport, U.S. South's First Forest, 68
roads: and access to frontier forests, 177–178; infrastructure and forest development, 138–141, *139*
Roanoake Rapids mill, 69
roundwood, price and quantity data, 53–54
royalties: severance taxes, 111; uniform fixed royalties, *179*, 182–183, 196–200
rubber tapping, 145–146

sandalwood harvest restrictions, 123
Sarawak, forest sector as national economy component, 229
sawmills: eastern Amazon industrial forestry, 266, 267, *268*, 270; and independent logging operations (NAICS 113310), 294–295; mobility of, 70; and resource availability, 12; sawmill industry (NAICS 321113), 273–279; and U.S. South's Second Forest, 69
sawnwood: Indonesia and East Asian financial crisis, 218; mature forest regional trade effects, 205–207, *206*; price and quantity data, 54
scale economies and sawmills, 274
"second growth" forests, 64–65
selective harvest and forest degradation, 22–25, *24*
severance taxes, 111
shifting cultivation, 18–20, *19*
shipment restrictions, 124
short-term trade effects, 202, 202*n*3
Siberia, non-commercial natural forests, 33
silviculture practices: , and industrial investors, 9–10; and Faustmann model, 94; and forest policy, 125–127; NIPF landowner myths, 344–347; and private forestland regulations, 125–127; technology transfer and public land management, 374–375
single resource market value. *See* management
size and diversity of public lands, management challenges, 379–380
social and institutional change, domestic economy shocks, 219–221
"social safety net", landowner needs and harvest scheduling, 326–327
Solomon Islands, forestry and GDP, 201
South Africa, forest certification, 128
South America: and degraded forests, *38*, 39; and subsistence settlements, 27, 27*n*13. *See also* Amazon region
South and Southeast Asia: and degraded forests, 37–39, *38*; and managed forests, 32
southeastern India, forest degradation, 23–24, 25
southwestern Virginia, forest degradation, 24–25
specialized contract considerations, forest concessions, 183–187
spillover effects: overview, 131, 137–138, 439–440; agricultural policy, 131–135, *133*; wood products industry policy, 135–138, *135*
spruce budworm, 369–370
Sri Lanka, harvest restrictions, 123
stability and macroeconomic forest impacts, 245
stand age and non-market environmental values, 11
standards and charges: direct financial assistance, 116–118, *117*; forest certification, 127–130; and forest policy, described, 110–111; free seedlings, 118–119; harvest and shipment restrictions, 122–124; incentives, 115–119; income taxes, 111–113, *112*; property taxes, 113–115; reforestation requirements, 124–125; silvicultural prescriptions, 125–127; taxes, 111–115; technical assistance, 119. *See also* regulations, private forestlands
Statistical Industrial Classification (SIC), 272, 272*n*3
stewardship, public land management, 353–355
stochastic frontier analysis, 378–379
streamside management, 125
structural adjustments, domestic economy shocks, 221–223
structural particleboard (SPB), 282
stumpage: eastern Amazon industrial forestry stumpage prices, *268*, 271, 271*n*2; and market exchange comparison points, 54; and market prices for logs, 55–56; price and quantity data, 53; stumpage payments, 10; stumpage price contracts, 179–180, *179*; stumpage value and U.S. Forest Service timber sales, 188
sub-Saharan Africa: and degraded forests, *38*, 39; property rights and forest development, 141
subsidies: agricultural input subsidies, 132–133, *133*; agricultural policy and macroeconomic forest impacts, 226–227, 226*n*24
subsistence economies: and commercial forest operations, 414–420; developing forest frontier, stage II, 27, 27*n*13; and forest development patterns, 15, 15*n*4; local community impacts, 406–414, *409*; and national income account omissions, 254–255; NIPF landowner motivations, 322–330
sulfate pulping, 69
Sumatra, subsistence settlements, 27, 27*n*13
supply: regulatory effects on, 122–124; and U.S. South's Second Forest, 70
sustainability: aggregate economic growth effects on Hainan Island, 238; agricultural land value, new forest frontier, stage I, 17; and deforestation control, 164–170, *166*, *168*; and "first growth" forests, 64–65; sustainable forestry and deforestation control, *68*, 164–170, *166*, 447–449. *See also* local human communities

Swan Valley campground management, 365–366

Tailandia, eastern Amazon industrial forestry, 265–266, *268*
Tanzania: global interests and community management, 147*n*31; and subsistence communities, 407
tariffs, 213–214
Tax Reform Act of 1976, 308
taxes: bequests and NIPF landowner motivations, 332–333; carbon sequestration and global climate change, 155–156; forest policy, 111–115; income taxes and forest policy, 111–113, *112*; and institutional investors, 308–309; labor and capital investment policies, 227–228; property taxes and forest policy, 113–115
technical assistance and forest incentives, *117*, 119
technological change: agricultural technologies on developing forest frontiers, *75*; agricultural technologies on new forest frontiers, *74*; biological growth and long-term forest effects, 73–78; and developing forest frontiers, 80–82, *82*; and forest research, 376; land use impacts summary, 84–86, *85*; long run forest development in the U.S. South, 67–73; and long run forest development overview, 65–66, 65*n*1; long-term effects on timber production, 11–12; and mature forest frontiers, 82–84, *83*; mechanization of pulp and paper industry, 71; rates of, 87–89; and sawmills, 274, 276. *See also* agricultural technologies
technology transfer and public land management, 352, 373–375
Thailand: community management and forest development, 142*n*25; effects of trade on forest environment, 212; logging bans and export restrictions, 215; property rights and forest development, 141; roads and access to frontier forests, 177–178
three-stage forest development model. *See* forest development stages
timber concessions. *See* forest concessions
timber harvest. *See* harvest
Timber Mart South, 325
timber prices: overview, 14; aggregate economic growth effects on Hainan Island, 235–236, *236*, 237–238; competitive market prices and limited number of bidders, 184–185; timber supply price elasticity, 125–126
timber production: and NIPF landowner motivations, 333–338, *334*; and public forestland management, 357–358; and U.S. South forests, 68
timber revenue distribution, 186–187
timber sales, 173

timber shortages, 164–170, *166*, *168*
timberland investment management organizations (TIMOs), 309, 314, 315
time bias, property taxes and forest management, 114–115
time, forest concessions contract duration, 183–184
"tipping" of trees, 24–25
tourism: and community management, 147; environmental tourism policy objectives, 160–162; tourist concessions, 173
trade effects: overview, 11, 202–203, *204*, 244–246; aggregate economic growth effects, 233–243; agricultural policy, 225–227; domestic economy shocks, 218–223; domestic fiscal and monetary policy, 223–225; economic growth and development, 228–243, *230–231*, *232*, *236*, *240*; economic welfare and land use impacts, 204–216, *206*, *208*, *209*; environmental Kuznets' curve (EKC) for forestry, 239–243, *240*; exogenous shocks, 216–218; Hainan Island aggregate economic growth effects, 234–238, *236*, 244; other domestic programs, 227–228; policy instruments, 212–216; relative size of forest sector, 229–233, *230–231*, *232*. *See also* national income accounts
transaction costs: and forest development patterns, 17, *19*, *21*, *24*, *31*, 41; improved agricultural technology and increased agricultural productivity, 74–78, *74–77*; and infrastructure improvements, 79–80, *79*, 79*n*13; and logging technology improvements, *79*, 80; and multiple market values, 41–43, *43*; property rights and transaction costs, 13
transportation costs: and delivered prices for logs, 55–56; factor costs and industry location, 304–306; and market exchange comparison points for different forest products, 54
trespass, and scale constraints on NIPF forest management, 328–329
tropical forests and forest degradation, 23, *24*
truck transport, U.S. South's Second Forest, 69
trusses (NAICS 321214), 281–283

uncertainty: and harvest restrictions, 123; regional trade effects, 209–211; social and institutional change and domestic economy shocks, 219–221
unearned increments, revised harvest policies and timber harvest revenues, 186
uneven-aged stands, 98
uniform fixed royalties, *179*, 182–183, 196–200
United Kingdom, forest incentives, 115, 118
United States: agricultural policy and

macroeconomic forest impacts, 226–227, 226n24; and agricultural settlement policies, 132; capital gains and income taxes, 111–113, 112n1; and costs of illegal logging, 26; Denver metropolitan area and managed forests, 31–32; and environmental accounting, 256–257; environmental loss and regional trade, 207n7; and forest incentives, 16, 115, 117, 118; global export value of forest products, 202; and inaccessible forest, 36, 36n22; national economy forest sector component, *231*; national income accounts, 250–251, *251*, *252*; and NIPF landowners, 318; plywood industry research support, 136–137; reforestation requirements, 124–125; regional development programs and macroeconomic forest impacts, 227; regional economy forest sectors, *232*; and removal of lands from agricultural use, 134; tariffs on Canadian softwood imports, 213, 213n9; timber production overview, 33. *See also* institutional investors; manufactured wood products (NAICS 321); U.S. South

unprotected forest, global forest inventory data, 34–36, *35*

upland watershed protection, 163

urban poor, housing effects of harvest restrictions, 124

U.S. Forest Service: and forest concessions, 188; and public forestland management, 365–368, *366*; public forestland management challenges, 381–382; roads and access to frontier forests, 177–178; and timber harvests in western Montana, 416–417; and valuation of government activity, 255–256; wildfire and management of public forestlands, 357, 369–373, *372*

U.S. National Park System, 351, 363–364, 379

U.S. South: agricultural mechanization and decreased farm labor demand, 77; forest concessions, 188–189; and institutional investors, 315; and plywood industry research support, 136–137; regional forest industry overview, 67–68; and shifts in production locations, 126–127; and silviculture prescriptions, 125–126; the South's First Forest, 68–69; the South's four forests, 68–73; the South's Second Forest, 69–70; the South's Third and Fourth Forest plantations, 70–73; the South's Third and Fourth Forests, 70–73; technological change and long run forest development, 67–73

utilization improvements: and developing forest frontiers, 80–82, *82*; land use impacts summary, *85*, *86*; and mature forest frontiers, 82–84, *83*

Vanuatu, forestry and GDP, 201

veneer and plywood (NAICS 321211 and 321212), 279–281

Vietnam: anticipation of development in, 398–401, *399–400*; community management and forest development, 145n29; and environmental Kuznets' curve (EKC) for forestry, 243; income and environmental Kuznets' curve (EKC) for forestry, 243; non-market forest products and forest degradation, 25n10; planned economies and domestic economy shocks, 220–221

virgin wood, and paper industry, 291–292

Virginia: and forest degradation, 24–25; state forest reforestation requirements, 125

virola timber, 266

voluntary carbon markets, 155–156

"Volvo effect", landowner needs and harvest scheduling, 326–327

wages, labor and capital investment policies, 227–228, 228n26, 245

water quality and forest regulations, 121

watershed protection: and policy objectives, 162–164; and public forestland management, 163–164, 361–362

watershed values and physical forest stock estimates, 60

Western European countries: and forest certification, 127–128; and reforestation requirements, 124–125

western pine beetle, 369–370, 370n13

Weyerhauser Company, 5

wildfire and management of public forestlands, 357, 369–373, *372*

wildland-urban interface, 371–372, *371*

wildlife and fish, physical forest stock estimates, 60

women and the poor, forest depletion, 10, 410–412

wood processing technologies and technological change, 88–89

wood products industry, policy spillover effects, 135–138, *135*

World Trade Organization (WTO), 214nn11–12

yield taxes, 115

Zambia: community management and forest development, 145n29; subsistence settlements, developing forest frontier, stage II, 27, 27n13

Zhejiang province bamboo production, 390–395, *391*, *392*